Science and Application of Nanotubes

FUNDAMENTAL MATERIALS RESEARCH

Series Editor: M. F. Thorpe, *Michigan State University*
East Lansing, Michigan

ACCESS IN NANOPOROUS MATERIALS
Edited by Thomas J. Pinnavaia and M. F. Thorpe

DYNAMICS OF CRYSTAL SURFACES AND INTERFACES
Edited by P. M. Duxbury and T. J. Pence

ELECTRONIC PROPERTIES OF SOLIDS USING CLUSTER METHODS
Edited by T. A. Kaplan and S. D. Mahanti

LOCAL STRUCTURE FROM DIFFRACTION
Edited by S. J. L. Billings and M. F. Thorpe

PHYSICS OF MANGANITES
Edited by T. A. Kaplan and S. D. Mahanti

RIGIDITY THEORY AND APPLICATIONS
Edited by M. F. Thorpe and P. M. Duxbury

SCIENCE AND APPLICATION OF NANOTUBES
Edited by D. Tománek and R. J. Enbody

A Continuation Order Plan is available for this series. A continuation order will bring delivery of each new volume immediately upon publication. Volumes are billed only upon actual shipment. For further information please contact the publisher.

Science and Application of Nanotubes

Edited by

D. Tománek and R. J. Enbody

Michigan State University
East Lansing, Michigan

Kluwer Academic / Plenum Publishers
New York, Boston, Dordrecht, London, Moscow

Proceedings of Nanotube '99, an International Conference, held July 24–27, 1999, in East Lansing, Michigan

ISBN 0-306-46372-5

233 Spring Street, New York, New York 10013

http://www.wkap.nl

10 9 8 7 6 5 4 3 2 1

A C.I.P. record for this book is available from the Library of Congress

Printed in the United States of America

SERIES PREFACE

This series of books, which is published at the rate of about one per year, addresses fundamental problems in materials science. The contents cover a broad range of topics from small clusters of atoms to engineering materials and involve *chemistry*, *physics*, *materials science,* and *engineering*, with length scales ranging from Ångstroms up to millimeters. The emphasis is on basic science rather than on applications. Each book focuses on a single area of current interest and brings together leading experts to give an up-to-date discussion of their work and the work of others. Each article contains enough references that the interested reader can access the relevant literature. Thanks are given to the Center for Fundamental Materials Research at Michigan State University for supporting this series.

M.F. Thorpe, Series Editor
E-mail: thorpe@pa.msu.edu
East Lansing, Michigan

PREFACE

It is hard to believe that not quite ten years ago, namely in 1991, nanotubes of carbon were discovered by Sumio Iijima in deposits on the electrodes of the same carbon arc apparatus that was used to produce fullerenes such as the C_{60} "buckyball". Nanotubes of carbon or other materials, consisting of hollow cylinders that are only a few nanometers in diameter, yet up to millimeters long, are amazing structures that self-assemble under extreme conditions. Their quasi-one-dimensional character and virtual absence of atomic defects give rise to a plethora of unusual phenomena. The present Proceedings volume represents state-of-the-art nanotube research, witnessing the self-assembly process of these nanostructures that is as amazing as their morphology and mechanical, chemical, and electronic properties.

The idea behind the Nanotube'99 (NT'99) workshop was to provide a forum for a free exchange of ideas in the rapidly developing nanotube field. The four-day NT'99 workshop, which was held at the East Lansing Marriott hotel from July 24-27, 1999, could be viewed as an intense brain-storming session of Physicists, Chemists and Engineers, Theorists and Experimentalists, who met to readjust the road map of nanotube research. The unusually frank and open-minded atmosphere of the workshop was a fruitful ground for discerning between promising and less promising approaches, and reevaluating the interpretations of experimental and theoretical data. The common enthusiasm for nanotubes helped the diverse community to find a common language, while enriching the field by shared expertise.

To facilitate an informal exchange of ideas in a research field that depends in a fundamental way on a collaboration between nations and cultures, including those of Academia and Industry, we often went new ways. The internet, in particular the world-wide web, was used as the primary medium to announce the NT'99 workshop, thus reaching out beyond the scientific community to everybody interested and excited about new discoveries and technologies. The web turned out to be a very efficient means of communication, in particular with participants from the former Soviet block. Web-based registration, submission, and posting of abstracts (including graphics), and continuously updated workshop information were appreciated by many.

A web-based pre-workshop discussion forum, called the "Virtual Workshop" and linked to the workshop site at the URL *http://www.pa.msu.edu/cmp/csc/vw99/*, attracted significant attention by raising the most important problems in the research field and addressing existing controversies. Observed interest in particular topics, inferred from the number of "hits" on the web, was very useful for determining the most suitable format. At the same time, this publicly accessible, pre-workshop discussion greatly enhanced discussions at the workshop.

The response to our announcement surpassed our expectations, as close to two hundred persons from twenty-seven countries expressed their interest in attending NT'99. The attendees included leading experts as well as novices in the field, a sculptor and a poet who have found inspiration in nanotubes. Believing firmly that innovative research leads to commercial applications, we tried our best to strengthen connections between Academia and Industry. We were very happy to host participants from leading universities next to those from high-tech companies such as Zyvex, Samsung, IBM, Allied Signal, General Motors, Toyota, and others.

We resisted the temptation to accommodate the large number of contributions in parallel sessions, but rather devoted the first two days to thematically ordered "poster plus" presentations. Most participants liked the unusual format of these sessions that started with a "two viewgraph/two minute" presentation of each contributor, summarizing the main message of their research. This established a personal basis for a more detailed discussion of the results at the posters. Each session was concluded by a plenary discussion with the panel of contributors, moderated by the session chair. This format allowed the same general question to be addressed by several panelists who were able to offer insight based on their particular techniques and results.

The rather informal first part of the meeting was followed by two days devoted to invited talks. We found that the informal atmosphere established in the first two days allowed free-wheeling discussions to continue over the ideas in the invited talks. We believe that the openness and communicative effort of the participants made this workshop a real success, and it is reflected in the wide range and quality of contributions in this Proceedings volume.

NT'99 would not have been possible without financial support. We would like to thank Michigan State University for generously co-sponsoring this workshop, and the Center for Fundamental Materials Research at MSU for contributing to the cost of producing these proceedings. We also appreciate the contribution of sponsors towards the cost of refreshments. The efforts of Lorie Neuman and Janet King, who organized the workshop and proceedings, are greatly appreciated. We would also like to acknowledge the invaluable help of Savas Berber and Young-Kyun Kwon in the organization of the workshop and formatting of the proceedings contributions.

David Tománek
Richard J. Enbody
East Lansing, Michigan

CONTENTS

Part I: Morphology, Characterization, and Formation of Nanotubes

Material Ethereal

by Peter Butzloff
University of North Texas
Department of Materials Science
P.O. Box 5308, Denton, TX 76203

We speculate
but underrate
what mystery we wrangle,
a Nanotube
from carbon crude,
that nature did entangle.

Two kinds or type
of nano-pipe
can possibly appeal:
A Multi-Wall
or Single-Wall
of carbon might congeal.

At nano volts
with little jolts
our Nanotube we measure.
It bends to shape
with twist and plait
a circuit sold for treasure.

In Multi-Wall
electrons crawl
resisting with displeasure,
but Single-Walls
like metal hauls
electrons at our pleasure.

At carbons bound
and chiral wound
electrons shot in action
streak and shine
a structure fine
and pattern with diffraction.

In Multi-Walls
like twisted halls
electron spins can couple.
Now concentrate
such tubes by weight
to soften signals supple.

A Single-Wall
type tubule tall
finds rare the indication
of spinning slave;
electrons crave
magnetic dissipation.

At length alight
from angle right
vibration shifts so sweetly
a signal bright
of Raman light
and frequency discretely.

A stick or stone
may break a bone.
but pull a tube this stately,
then tensile trait
at length would rate
of forces large and greatly.

Oh Nanotube
of carbon crude,
so cumbersome we trundle
through pass of phase,
by time decays,
and character to bundle!

FILLING CARBON NANOTUBES USING AN ARC DISCHARGE

A. Loiseau[1], N. Demoncy[1,2], O. Stéphan[3], C. Colliex[3], and H. Pascard[2]

[1]LEM, UMR n° 104 ONERA-CNRS, Onera, B.P. 72, 92322 Châtillon Cedex, France

[2]LSI, CEA-CNRS, Ecole Polytechnique, 91128 Palaiseau Cedex, France

[3]LPS, CNRS URA 002, Université de Paris Sud, Bât 510, 91405 Orsay Cedex, France

ABSTRACT

Nanowires have been obtained by arcing a graphite cathode and a graphite anode doped with powders of various elements among transition metals (Cr, Ni, Co, Fe, Pd), rare earth metals (Sm, Gd, Dy, Yb) and covalent elements (S, Ge, Se, Sb). As far as metals are concerned, it is shown, using combined structural and chemical analysis modes of transmission electron microscopy, that fillings are due to the addition of trace amounts of sulfur to the doping elements. A general growth mechanism based on a catalytic process involving carbon, a metal and sulfur is proposed and accounts for the microstructure and the chemical composition of the different kinds of fillings.

INTRODUCTION

Filling carbon nanotubes is highly attractive for the development of nanomaterials with new electronic or magnetic properties and has focused many efforts since the theoretical prediction that nanotubes should display a unique capillary behaviour [1] . This prediction has been confirmed by the experiment achieved by Ajayan et al. [2] who succeeded in opening and filling carbon nanotubes by heating them in air in presence of molten Pb . Since this pioneering work , different methods of filling have been explored.

The first route to produce filled carbon nanotubes is to use pre-existing nanotubes to open them by oxydation and to fill them by physical or chemical techniques. In the physical method, capillarity forces induce the filling of a molten material having a low surface tension [3]. Elements such as S, Cs, Rb, Se or the oxides of V, Pb and Bi are known to have this property. Fillings have been experimentally obtained with oxides like PbO, Bi_2O_3 [4], V_2O_5 [5], MoO_3 [6], and with salts of metals like Ag, Co, Pb [7, 8]. The chemical method uses wet chemistry [9] . In that case, the metal oxides or metals that are encapsulated are discrete crystals unlike the long continuous fillings obtained by molten media or arc-discharge as described below. Finally, it is also interesting to mention that carbon nanotubes have been

Science and Application of Nanotubes, edited by Tománek and Enbody
Kluwer Academic / Plenum Publishers, New York, 2000

converted to carbide nanorods, which are not encapsulated, by reaction with oxide or iodide species [10].

The second route consists to produce the nanotubes and their fillings simultaneously. The main method of synthesis is a modified arc-discharge method in which the anode is doped with a chosen element. The arc-discharge method, which consists in establishing an electric arc between two graphite electrodes in a He atmosphere, was first used by Krätschmer et al. [11] to produce fullerenes in large quantities and was later optimized for the synthesis of carbon nanotubes [12,13]. The modified arc-discharge method led to the formation of endohedral fullerenes [14] and then of nanoparticles encapsulated in graphitic shells [15,16]. The first filled nanotubes produced by this method contained yttrium carbide [17,18,19]. Other carbides (Gd, La, Mn, Nb, Ta, Mo) were later inserted into nanotubes by the same method [20 - 23]. As far as nanotubes filled with pure elements are concerned, partial fillings of pure Ni inside bamboo-shape carbon nanotubes have first been reported [24]. In 1994, Ajayan et al. showed that long pure Mn fillings can be formed by doping the anode with pure Mn metal [25] and recently, pure Cu and Ge nanowires have been synthesized using a hydrogen arc [26,27].

Alternative methods of direct synthesis of filled nanotubes have recently been developped and have shown their efficiency. Sn containing nanotubes have been formed by electrolysis of graphite electrodes immersed in molten $LiCl/SnCl_2$ mixtures [23,28] and similar results have been obtained by adding trace amounts ot Bi and Pb powders to LiCl [29]. Futhermore, needle like nanotubes containing encapsulated Ni have been produced by pyrolysis at 950°C of sandwich structures consisting of alternating thin films of C60 and Ni [30].

In this work, we have used the modified arc-discharge technique to synthetize various filled nanotubes by successively doping the anode with several elements listed here: Ti, V, Cr, Mn, Fe, Co, Ni, Cu, Zn, Y, Nb, Mo, Ru, Rh, Pd, Ag, Cd, Hf, Ta, W, Re, Os, Ir, Pt, Au, Sm, Gd, Dy, Er, Yb, B, Al, In, Ge, Sn, Pb, Sb, Bi, S, Se, Te. Long continuous fillings, often exceeding one micron in length and therefore being true encapsulated nanowires, have been obtained for twelve elements belonging to different groups : transition metals (Cr, Ni, Re, Au), rare earths (Sm, Gd, Dy, Yb) and covalent elements (S, Ge, Se, Sb). Partial fillings were also obtained with elements like Mn, Co, Fe, Pd, Nb, Hf, Os, B, Te, Bi.

Both the structure and the chemical composition of these encapsulated nanowires have been investigated in detail by Transmission Electron Microscopy (TEM): diffraction patterns (SAED), high resolution imaging (HRTEM) and Electron Energy Loss Spectroscopy (EELS) using the line-spectrum mode implanted on a dedicated scanning transmission electron microscope which provides elemental concentration profiles [34]. This technique was proved to be essential and particularly adapted for characterizing filled carbon nanotubes and understanding their formation. We have shown that filling with metals is due to the presence of a small amount of sulfur, which was initially present as an impurity in the graphite electrodes and we propose a coherent and general scheme for the formation and the growth of the nanotubes filled with metals emphasizing the role played by sulfur in the different steps of the process.

The paper is organized as follows. Experimental procedures are described in section 2. Section 3 presents the structural and chemical analysis of different kinds of nanowires. Section 4 is focused on the special case of Cr to clearly demonstrate that the presence of sulfur is essential to fill the nanotubes. Finally the section 5 is devoted to the discussion of the results and of the growth mechanism.

EXPERIMENTAL PROCEDURES

Synthesis of the filled nanotubes

The arc discharge method and the experimental conditions used for the synthesis have been described in [31]. In brief, we used two graphite electrodes (purity 99.4 %) 9 mm in diameter. The anode was drilled with a 6-mm hole and packed with a mixture of graphite and chosen element powders. The arcing conditions were 100-110 A dc, 20-30 V, under a 0.6 bar He atmosphere, during 30-60 minutes. The deposit formed on the cathode and containing filled nanotubes and nanoparticles encapsulated in graphitic shells was ground, ultrasonically dispersed in ethanol and dropped on holey carbon grids for electron microscopy observations.

Observation and analysis

HRTEM and standard EELS characterisations were performed using a JEOL 4000FX working at 400 kV, equipped with a Gatan 666 parallel collection electron energy-loss spectrometer. High spatial resolution EELS analyses were performed using a dedicated scanning transmission electron microscope (STEM) VG HB501, working at 100 kV, equipped with a field-emission source and a parallel collection electron energy-loss spectrometer. This instrument provides EELS spectra with a typical 0.7 eV resolution recorded on subnanometer areas (see details in [35]). We made intensive use of the « line-spectrum » mode which consists to acquire systematic series of EELS spectra at successive positions typically 0.3 to 1.0 nm apart in one dimension. This technique has proved its efficiency to characterize inhomogeneous nanotubes [25, 34, 36] since it provides with elemental profiles along or across the nanotube. The acquisition time per spectrum is reduced to 1s to prevent beam irradiation effects. Each spectrum is processed from this line-scan to extract the local chemical composition and quantitative elemental profiles as follows: the K- or L-edges of the different elements present in the nanotube are integrated over a fixed energy window after substraction of a power-law background, all the edges are then normalized by inelastic cross sections.

STRUCTURAL AND CHEMICAL ANALYSIS

Figure 1 presents a selection of TEM images of filled nanotubes that have been successfully produced when the anode was doped with a transition metal like Cr, Ni (fig. 1a, 1b), a rare earth like Sm, Dy (fig. 1c, 1d) and a covalent element like S, Sb (fig. 1e, 1f). In these examples, the nanotubes are completely filled from their tips and their length ranges from a few hundred nanometers to a few microns.

Figure 2 displays examples of partially filled nanotubes found for Co, Pd and Se: in these cases, the filling is discontinuous and consists of a sequence of particles located at different places along the nanotube. Each particle never exceeds a few hundred nanometers. These nanotubes are hollow at one tip and often capped by a particle at the other tip. The structural characteristics of both kinds of nanotubes, i.e. the number of graphitic layers, the degree of graphitization, the crystallinity of the filling material, depend on the chosen element as described in [31, 32]. The variety of the microstructures is illustrated by the examples of Fig.3. The filling material can be a single crystal (fig. 3a,c) or microcrystalline (fig.3b). Note that both situations can be found for the same element. This is the case for Yb (fig. 3b,c) and for Ge (fig3a, fig7). The material wrapping the filling can be well graphitized concentric walls (fig. 3d) or more or less turbostratic carbon (fig.3e, f). Layers are in general well

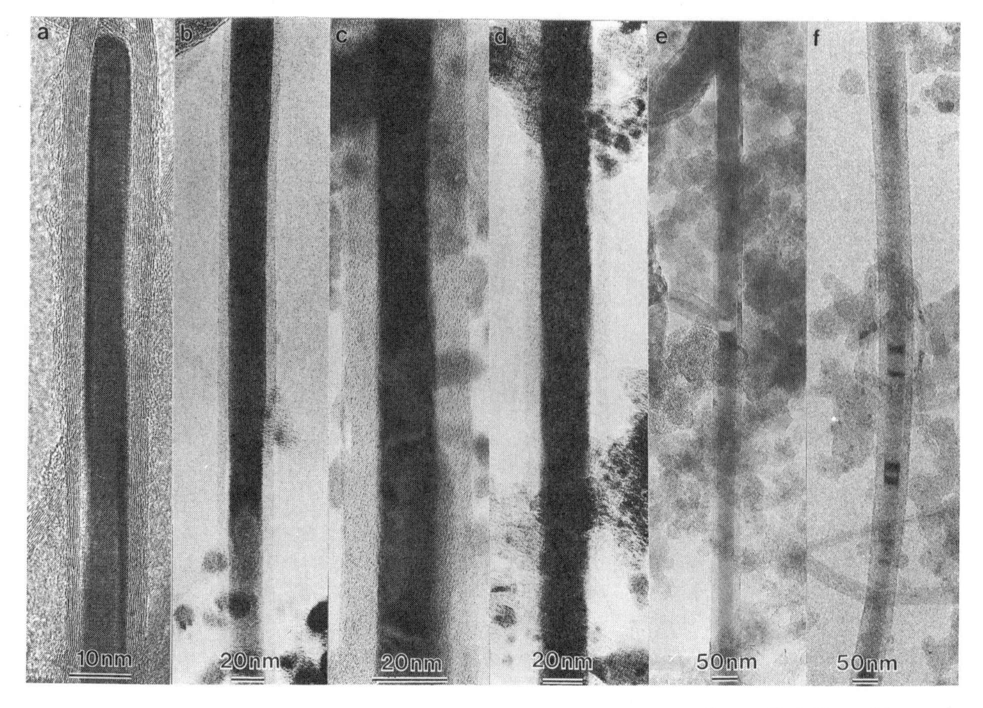

Figure 1 : TEM images of filled carbon nanotubes produced by arc-discharge when a 99.4 % graphite anode is doped with : (a) Cr, (b) Ni, (c) Sm, (d) Dy, (e) S, (f) Sb.

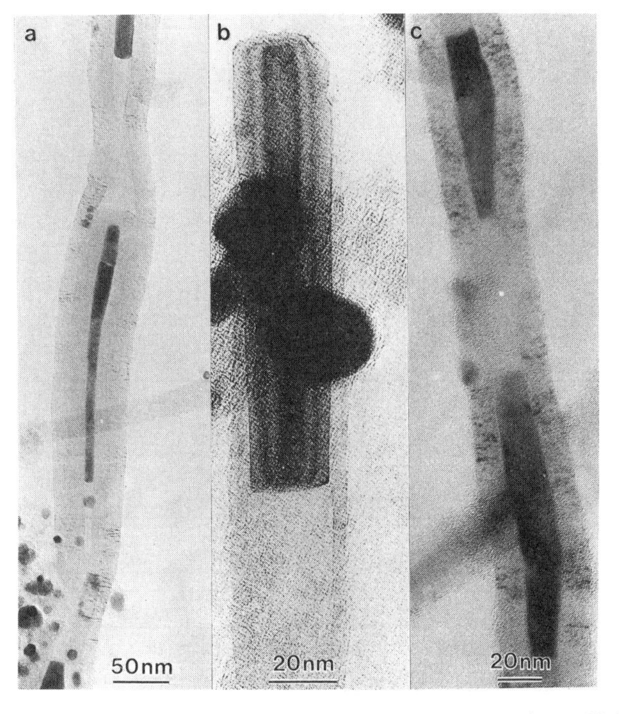

Figure 2 : TEM images of filled carbon nanotubes produced by arc-discharge when a 99.4 % graphite anode is doped with : (a) Pd, (b) Co, (c) Se.

aligned along the tube axis, except in a few cases found for Sb (fig. 3f), Bi and Te, for which the carbon envelope consists of a stacking of graphitic cones. The number of the layers can varies from 2-3 to a few tens.

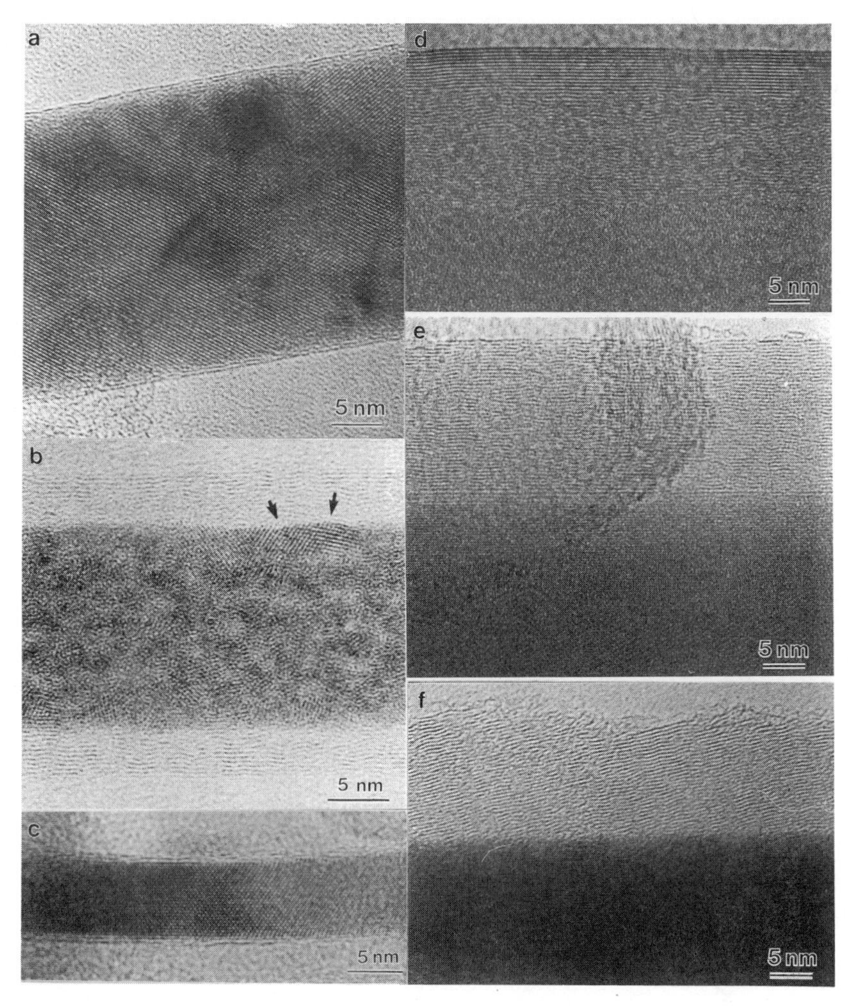

Figure 3 : HRTEM images of filled carbon nanotubes. The anode dopant is Ge in (a), Yb in (b) and (c) B in (d) and S in (e) and Sb in (f). The graphitic walls of the nanotubes are imaged by fringes separated by approximately 0.34 nm. The filling is a single crystal with definite orientation relationships with the graphitic layers in (a), (c) and e) whereas it is polycrystalline in the other cases. In (c), arrows indicate crystallites in Bragg orientation and imaged by a set of lattice. The degree of graphitization of the carbon layers and their number highly depend on the element doping the anode.

The encapsulated crystals were initially thought to be carbides since in most cases the corresponding selected area electron diffraction (SAED) patterns were not consistent with the structure of the pure elements [31, 32]. The chemical composition of the fillings has been then investigated in a rather systematic way using high spatial resolution. Surprisingly, these nanoanalyses have revealed that the nanowires obtained with metals are not carbon-rich as initially assumed but contain sulfur. The source of sulfur was found to be the graphite electrodes (99.4 %) : a refined analysis of this graphite has revealed that the major impurities are Fe (0.3 %) and S (≈ 0.25%). The analysis of different cases is presented in detail in [37] and shows that important amounts of sulfur were found in numerous filling materials along with the inserted element. We focus here on two representative examples: Cr and Ni.

The case of Cr is emphasized in Figure 4. As shown in Fig.1a, Cr-based nanowires are very long and straight. A majority of them was found to contain sulfur in a ratio close to 1:1± 0.2 (the quantification of the sulfur amount cannot be very precise as explained in [37]), excluding any other element. Figure 4 presents an example of concentration profiles deduced from a line-spectrum across the filled nanotube shown in inset. The C profile is characteristic of a hollow carbon nanotube [38] and is perfectly anticorrelated to both the S and Cr intensity profiles.

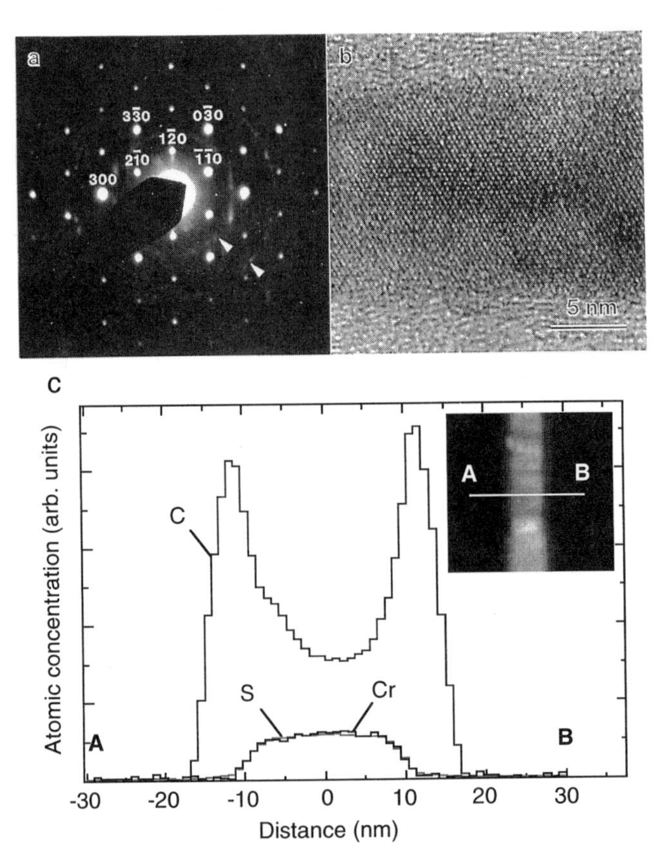

Figure 4 : Chemical and structural analysis of a filled nanotube obtained with a 99.4% graphite anode doped with Cr. (a) Selected area electron diffraction pattern of a conveniently oriented nanotube. It is consistent with a <001> projection of the trigonal chromium sulfides Cr5S6 or Cr2S3 (which have the same space group and very close lattice parameters). (b) Corresponding HRTEM image showing the 0.3 nm periodicities due to <110> type reflections. (c) Concentration profiles of C, Cr and S deduced from the intensity of the C-K, Cr-L2,3 edges and the S-L2,3 edges of the EELS spectrum as the beam is scanned across the nanotube. Inset : Annular dark field image of the nanowire. The white bar represents the line of scan.

This strong phase separation proves that the tubular layers wrapping the nanowire are free of sulfur and of metal and that the filling material does not contain carbon. Furthermore, S and Cr profiles are very correlated, suggesting that the filling is an homogeneous chromium sulfide. Longitudinal line-spectra have confirmed the chemical homogeneity over long parts of the nanowires. The structure of these sulfides has been identified from SAED and HRTEM analyses to be the trigonal compounds Cr_5S_6 or Cr_2S_3 which have a composition nicely consistent with that deduced from EELS spectra [37]. The trigonal symmetry is clearly seen in Figure 4 which shows both the SAED pattern and the corresponding HRTEM image of a filling oriented with its trigonal axis parallel to the electron beam. One remarkable feature of the fillings is that they are always a single crystal epitaxed on the graphite layers : the tube axis is systematically found to be parallel to a <100> direction of the sulfide structure indicating that strong orientational relationships exist between the nanotube and the inner crystal during its formation.

The second typical example concerns experiments with Ni and is presented in Figures 5 and 6. Ni-based nanowires are also very long and contain sulfur. As for Cr, transverse EELS line-spectra show that sulfur is homogeneously mixed with nickel inside the nanotubes.

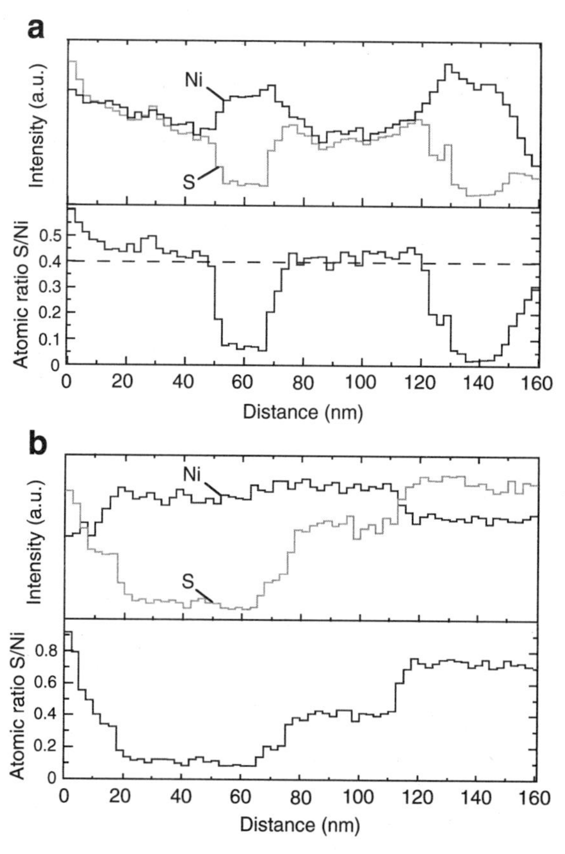

Figure 5: Chemical analysis of a filled nanotube obtained with a 99.4% graphite anode doped with Ni. (a) and (b) Longitudinal electron energy-loss line-spectra along the axis of two different nanotubes. The Ni and S concentration profiles were deduced from the intensities of the Ni-$L_{2,3}$ and S-$L_{2,3}$ edges. In (b), the left part corresponds to a sulfur rich region close to the tip of the nanotube.

However, in constrast with Cr, the concentration in sulfur is not homogeneous along the tube axis as shown by the longitudinal line-spectra of Fig. 5. The filling consists of a succession of crystallites alternatively pure Ni and nickel sulfides with a S/Ni ratio ≈ 0.4. More important amounts of sulfur are often found, in particular at the tip of nanowires (see the left part of fig. 5b). The compositional changes are associated with structural and and morphological changes of the filling material as shown in Fig.6a. In this example, the filling displays three crystallites having different structures and imaged by different arrays of fringes; the grain boundaries are indicated with arrows on the image. In the areas where S/Ni ratio ≈ 0.4, it is likely that a metastable sulfide has been formed, since no stable nickel sulfide

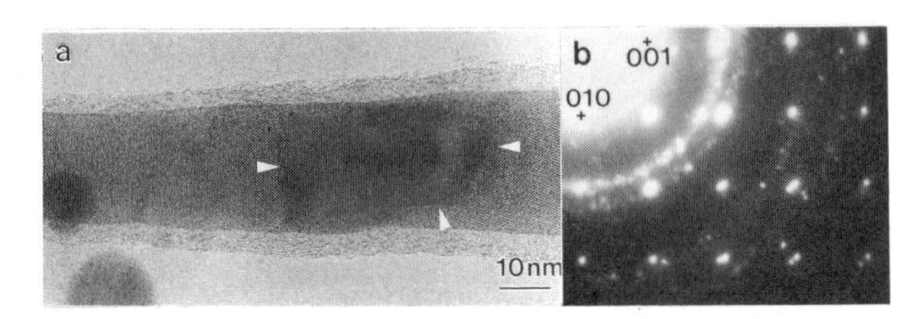

Figure 6: Structural analysis of a filled nanotube obtained with a 99.4% graphite anode doped with Ni. (a) HRTEM image. Arrows indicate a grain which has a different structure than the surrounding crystallites. (b) SAED pattern consistent with the (100) zone axis of the pseudo cubic Ni3S2 phase.

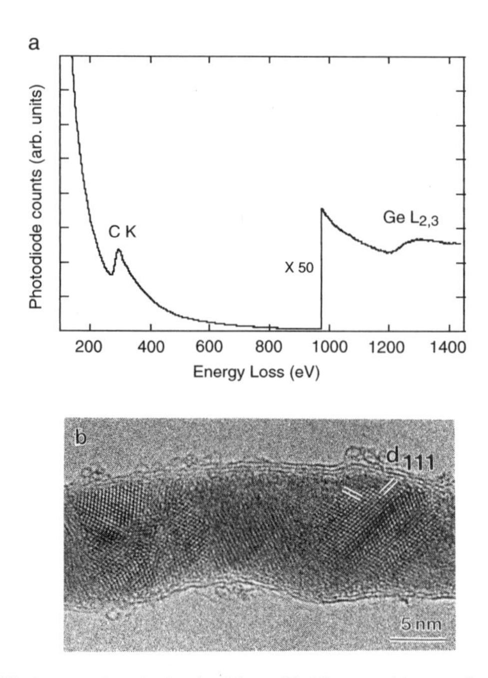

Figure 7: Analysis of a filled nanotube obtained with a 99.4% graphite anode doped with Ge. (a) EELS spectrum obtained on a part of a nanowire. The C-K, Ge-L2,3 edges are clearly seen and no other edge is detected. (b) HRTEM image of the filling material which is polycrystalline. Pure Ge microcrystals in a <110> projection can be seen with typical microtwins and stacking faults of the <111> dense planes.

corresponding to such a S/Ni ratio is known. In the more S-enriched areas, the pseudo cubic compound Ni_3S_2 has been identified from SAED analyses [37]. Figure 6b presents the SAED pattern corresponding to the (100) zone axis of this sulfide and exhibiting a pseudo four fold symmetry. Its chemical composition corresponds to a S/Ni ratio close to 0.7, consistent with the EELS analysis of the S-rich areas seen in Fig. 5b.

The only exception to this spectacular concentration phenomenon of sulfur inside the nanotubes was observed for Ge as shown in Fig.7a. Fillings are free of sulfur since only the C-K edge (285 eV) and the Ge-$L_{2,3}$ edge (1217 eV) are present on the EELS spectra.SAED and HRTEM analyses lead to conclude that the nanowires have the structure of pure Ge. In that case, nanowires are very thin and encapsulated in only 2 or 3 graphitic layers . They can be single crystalline (Fig.2a) or polycrystalline (Fig.7b) and in that case, shown in Fig.7b, the crystallites contain typical microtwins or stacking faults located in the dense <111> atomic planes. It is worth noticing that the nanowires recently produced using an hydrogen arc display very similar features [27].

To summarize, the results presented in this section show a spectacular concentration phenomenon of sulfur inside the nanotubes and suggest that the growth of nanowires results from the presence in the arc of three elements : carbon, a metal and sulfur.

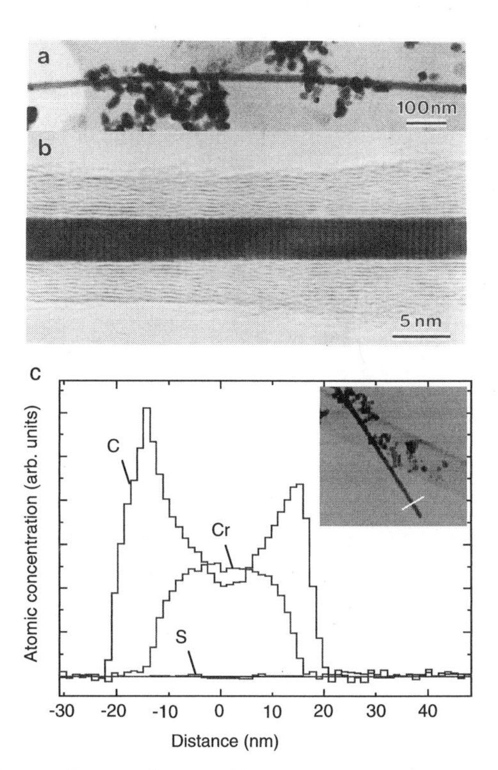

Figure 8: (a) HRTEM image of a nanowire coated by carbon obtained with a high purity graphite anode doped with Cr and S in a S/Cr=0.5% atomic ratio. (b) Magnification showing the lattice image of the carbon layers and of the filling crystal which is single crystalline and in epitaxial relationship with the graphitic walls. (c) Concentration profiles of C, Cr and S across a carbon nanotube filled with pure Cr.

ANALYSIS OF THE ROLE OF SULFUR

In order to understand the roles played by sulfur and the metal, we performed two kinds of experiments.

First we used high purity carbon rods (99.997%) and we successively doped the anode with Co (99.99%), Ni (99.9%), Cr (99.95%), Dy (99.9%) and S (sublimed). The other experimental conditions [31] were unchanged. It is striking that no filled nanotube was found in the cathode deposit. Co and Ni yielded to the formation of single-walled nanotubes as usually observed [39] whereas only empty multi-walled nanotubes were produced with Cr, Dy and S.

In the second kind of experiments, we have focused on the case of Cr which produced the longest nanowires and we have added controlled trace amounts of sulfur to Cr in a S/Cr=0.5% atomic ratio (i.e. S/C=0.1% in weight). This doping resulted in the abundant formation on the cathode of true nanowires encapsulated in carbon nanotubes (fig.8). However nanotubes are slightly less abundant than when using the 99.4 % graphite. The images of Fig.8 attest that the structural characteristics of these nanowires are very similar to those obtained with the 99.4 graphite (see fig. 1a). They are completely filled from their tips, their length also often exceeds one micron as shown in fig.8 and fillings are single-crystals as previously. EELS nanoanalysis reveals two kinds of fillings depending if they contain sulfur or not. The first kind corresponds to a small fraction of nanowires and contain sulfur in the same ratio than those obtained with the 99.4 graphite. SAED patterns have confirmed that in that case filling was a trigonal sulfide Cr_5S_6 or Cr_2S_3 . However for a majority of nanowires, which defines the second kind, the fraction of S is below the limit of detection of the spectrometer. Most of these nanotubes exhibit the concentration profile of fig. 8c which reveals that they are filled with pure Cr. In a few cases presented in [37], concentration profiles indicate that the filling is a chromium carbide such the orthorhombic Cr_3C_2 phase.

We can conclude from these experiments, that the addition of a very small amount of sulfur to Cr provokes the formation of long encapsulated nanowires having various chemical compositions, including pure Cr. These results definitely prove that S is crucial for filling carbon nanotubes with a metal, what we discuss in the next section.

DISCUSSION: GROWTH MECHANISM

Three remarks have to be made for introducing the discussion on the growth mechanism we propose in this section. First, in contrast with other synthesis methods, it is likely that, according to the micronic length of numerous encapsulated nanowires, the nanotube and its filling form almost simultaneously : the filling by capillarity of empty nanotubes after their formation on the cathode is difficult to envisage on such a length and the variety of structural characteristics depending on the chosen element confirms this assumption.

The second remark concerns the temperature of formation of the nanowires. In our experiments, one of the key factors is the important temperature gradient on the cathode which is created by the efficient water-cooling of the electrodes. The temperature between the electrodes can exceed 4000°C whereas on the rear part of the cathode, the temperature is lower than 1000°C. Filled nanotubes are found in localized regions of the soot around the cathode which depend on the chosen metal. We can therefore assume that they grow on the cathode in a definite temperature range, between 1000°C and 2000°C. Such temperatures are higher than those used in the synthesis of catalytically grown carbon fibers which are usually below 1000°C [see e.g. 42]. Furthermore they are close or above the melting point of most of the metals used in our experiments.

In the third remark, we would like to point out on the morphological similarities that the filled nanotubes, in particular those partially filled, share morphological similarities with carbon nanofibres catalytically grown by chemical vapor deposition (CVD) [40 - 42]. The conical shape of elongated particles capping the partially filled nanotubes as well as the widened shape of long nanowires shown in fig.9 are representative features reminding us of catalytically grown fibers. Therefore the growth mechanism of carbon fibers can give clues for the growth of filled nanotubes as already proposed in [25,33] although temperatures of formation are significantly different according to the second remark.

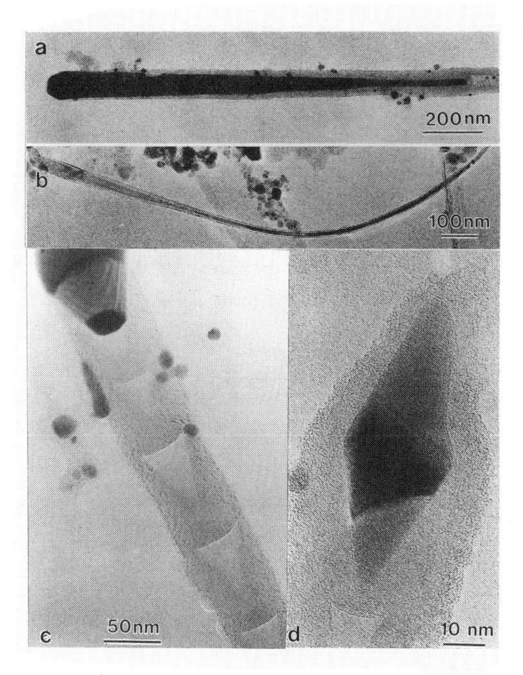

Figure 9: (a) TEM image of a Ni-containing nanotube obtained with a 99.4 % graphite anode, having a typical widened shape and a faceted tip. (b) TEM image of a long filled nanotube, obtained with a high purity carbon anode doped with Cr and S in a S/Cr = 0.5 % atomic ratio, also having a slightly widened shape. (c) Bamboo-shape nanotube, obtained with a 99.4 % graphite anode doped with Pd, with a conical particle at the tip. (d) Conical tube capped by a biconical particle obtained with a 99.4 % graphite anode doped with Co.

Catalytic growth of carbon fibers has been discussed by different authors and its mechanism is qualitatively well understood [see 40 - 44]. Based on it, we propose for the growth of the filled nanotubes the mechanism which is sketched in fig.10. The growth starts from a carbon-rich metal particle deposited on the cathode. Carbon diffuses through the metallic particle and precipitates at the rear face of the particle resulting in the graphitic layers of the multiwall nanotube. However this process arises at much higher temperatures than for carbon fibers, between 1000° and 2000°C, such as the metallic particle is almost in a liquid state. In such a way the catalyst particle can flow inside the tube as it grows. This is the second step of the growth process. The solid or molten state of the catalyst makes the difference between filled nanotubes and tubes or nanofibers capped by a particle. This mechanism is supported by the feature, shown in fig. 11, of a filled nanotube having an empty and conical tip. It is likely that the nanotube was first capped by a conical metallic particle, from which the graphitic layers have emerged. In a second step, the metallic particle, being liquid, has flowed inside the hollow channel of the nanotube leaving the tip open.

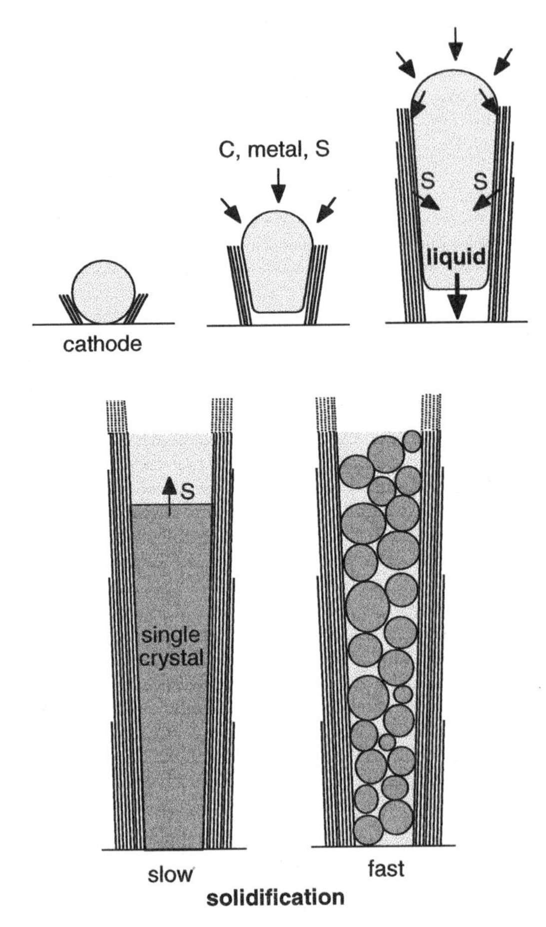

Figure 10: Schematic growth mechanism proposed for the formation of a carbon nanotube filled with a metal on the cathode of an arc-discharge experiment in presence of sulfur (see text for details).

The specific role of sulfur in this process is the following. Below 2000°C, sulfur is known to promote the graphitization of carbon materials such as heavy petroleum products by acting as a cross-linker [45 -47]. Once the graphitization is achieved it is released from the graphitc structure. We suggest that the same phenomenon occurs here. It is likely that C and S easily combine in the vapor phase, maybe forming S-rich carbon clusters. Sulfur enhances the catalytic activity of the metal for the graphitization of the carbon at the surface of the metallic particle. This effect has been experimentally proved for metal like Co [48]. After what, it is released from the carbon layers and is trapped in the metallic particle because of its strong affinity with metals as sketched in Fig.10. Since it reduces melting points of metals, it helps the metallic material to remain in the liquid state and to flow inside the growing nanotube. Furthermore, the spreading and wetting properties of the metallic phase over the graphitic walls are certainly enhanced by the presence of sulfur. The growth continues as long as the carbon, metal and sulfur species supply and temperature conditions are maintained. This process results in a progressive increase of the S concentration in the nanowire. The total amount of S is therefore directly related to the amount of carbon involved in the nanotube and is determined by the length and the number of carbon layers. This explains the difference in concentration of sulfur observed between the nanowires and the nanoparticles encapsulated in a few graphitic layers which are always observed free of sulfur [37]. The same explanation accounts for the absence of sulfur in the Ge nanowires. This is due the reduced number of graphitic layers wrapping the nanowire.

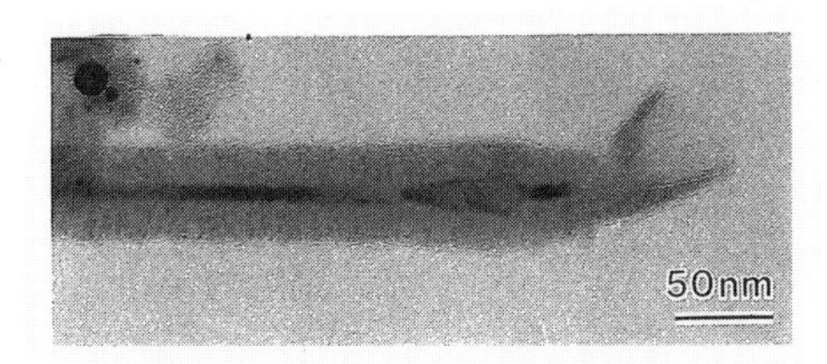

Figure 11: TEM images of filled nanotubes with a conical open tip. The doping element of the anode was Nb.

The final step of the process is the solidification of the filling material. The final microstructure and the chemical composition of the nanowires are determined by the solidification process and the cooling conditions as shown in Fig.10. A rapid quench will lead to a microcrystalline filling. On the contrary, if the solidification is slow enough, a solidification front can be established so that the formation of long single-crystals will be possible. In that case, the process is apparented to a directional solidification as in the Bridgman technique which is a well known technique used for the growth of single crystals of metallic systems. The nanotube plays the role of the furnace used in the Bridgman technique. Since sulfur is in general not soluble in metals, segregation phenomena occur at the solid-liquid interface and determine the structure and the composition of the crystallites. These features can be simply understood by considering equilibrium phase diagrams S-metal. We discuss in the following two kinds of situations.

We first consider the schematic phase diagram drawn in Fig. 12a. It is a simplified representation of the phase diagrams of S-M systems where M is a metal such as Ni, Co, Fe, Pd [49]. It is characterized by two features. First, S is completely soluble in the liquid metallic phase whatever its composition, and almost insoluble in the solid metallic phase, leading to the existence of different sulfides, S1, S2..., depending on the concentration in sulfur. Second, the melting point of the metal decreases as the concentration of sulfur of the liquid phase increases up to the concentration of the eutectic point E which is close to the melting point of the first sulfide S1. In such a case, as the solidification starts upon cooling (see the arrow), the sulfur is rejected in the liquid phase through the solidification front and a pure metallic crystallite first grows. In the same time, the composition in sulfur of the liquid phase increases up to the composition of the eutectic point. Then the growth of the definite sulfide S1 replaces that of the pure metallic phase. If the composition of the liquid phase is higher than that of S1, sulfides like S2 are formed. In both cases, the precipitation of the sulfide provokes a rapid decrease of the concentration in sulfur of the liquid phase and the growth of the pure metal takes place again. As a consequence of this recurrent process, the nanowire is a succession of pure metal crystallites alternating with metallic sulfides. This is typically what we have observed for Ni.

We now consider the schematic phase diagram drawn in Fig. 12b. The difference with the previous case lies in the existence of a miscibility gap in the liquid state. This means that two liquid phases exist depending on the concentration in sulfur, an almost pure metallic liquid phase L1 and a S-rich liquid phase L2. In the solid state, the stable phases are the pure metal and definite sulfides such as S1. This situation occurs for elements such as Cr or Ge

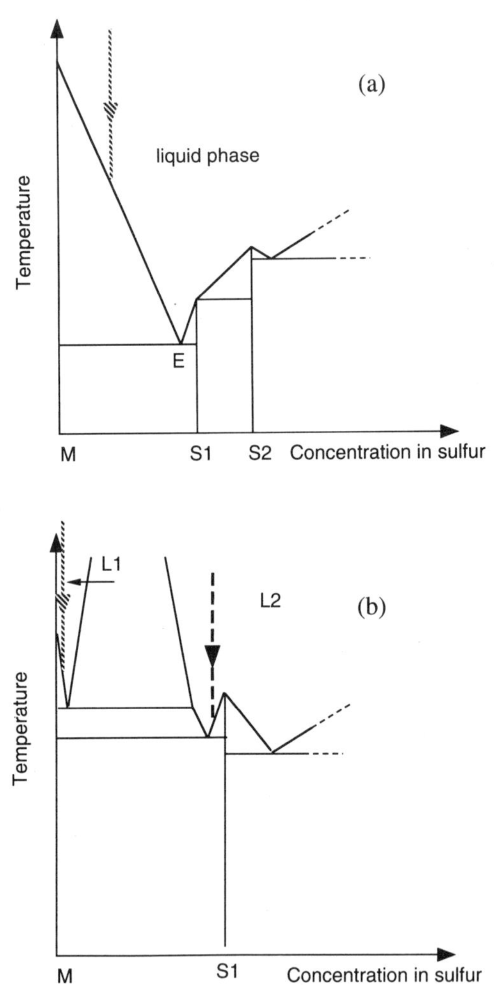

Figure 12: (a) Simplified representation of the phase diagrams of S-M systems where M is a metal like Ni, Co, Fe. In these cases, S and M form in the molten state a liquid solution and the addition of sulfur to the metal results in an important decrease of the melting point down to the eutectic point E. The solid phases are the almost pure metal or definite compounds of composition S1 and S2. The arrow indicates the typical quench which is considered for a metallic particle having a given sulfur concentration. (b) Simplified phase diagrams of S-M systems where M is a metal like Cr or Ge. In these cases, there is a miscibility gap in the molten state resulting in the existence of two liquid phases, L1 and L2. The vertical arrows indicate the quenches which are considered for both phases. The solidification process leads to the formation of the pure metal in the first case and to the sulfide S1 in the second case.

[49].Two liquid phases can therefore be formed for the filling material. Their solidification, outlined by the dashed and the dotted arrows in fig.12, leads to the crystallisation of either the pure metal phase or the sulfide S1. This process nicely accounts for the two kinds of fillings observed in the case of Cr in the experiment where the anode has been doped by trace amounts of sulfur. In the experiment with the 99.4% graphite, the amount of sulfur is higher so that the L2 liquid phase and consequently the sulfide CrS are always formed. In the case of Ge, GeS compounds are not formed because the number of the encapsulating carbon layers and their length are not sufficient to obtain sulfur rich liquid particles of Ge.

CONCLUSION

In conclusion we have shown that the presence of sulfur in catalytic quantity is crucial for the production using an electric arc of abundant metal based nanowires which can be free of sulfur. We propose that sulfur enhances the catalytic activity of the metal as far as graphitization is concerned and helps the filling materials to remain in the liquid state and to flow inside the growing nanotubes. The role of sulfur is certainly not unique and it is likely that elements like selenium, hydrogen, oxygen probably can act in a similar way on filling processes. It is striking that , using an hydrogen arc, very similar Ge nanowires have recently been produced [29]. Furthermore the discussion of the growth mechanism suggest that the structure and the chemistry of the nanowires can be understood by considering the thermodynamical data of the corresponding metal-S systems. It is worth mentioning that similar considerations based on the BN-C equilibrium phase diagram also accounts for the formation and the self-organisation of mixed BN-C nanotubes [50]. In that case also, the observed microstructures of the nanotubes strongly support a growth from a liquid phase towards thermodynamically stable structures. One can therefore consider to use equilibrium phase diagrams and thermodynamic data to develop a nanometallurgy within the electric arc chamber and that it would be possible to control the formation of definite nanowires in the future.

REFERENCES

1. M.R. Pederson and J.Q. Broughton, *Phys. Rev. Lett.* 69: 2689 (1992)
2. P.M. Ajayan and S. Iijima, *Nature* 361: 333 (1993)
3. E. Dujardin, T.W. Ebbesen, H. Hiura and K. Tanigaki, *Science* 265:1850 (1994)
4. P.M. Ajayan, T.W. Ebbesen, T. Ichihashi, S. Iijima, K. Tanigaki and H. Hiura, *Nature* 362:522 (1993)
5. P.M. Ajayan, O. Stéphan, Ph. Redlich and C. Colliex, *Nature* 375:564 (1995)
6. Y.K. Chen, M.L.H. Green and S.C. Tsang, *Chem. Commun.*, 2489 (1996)
7. D. Ugarte, A. Châtelain and W.A. de Heer, *Science* 274:1897 (1996);
8. D. Ugarte, T. Stöckli, J.-M. Bonard, A. Châtelain, W. A. de Heer, *Appl. Phys.* A 67:101 (1998)
9. S.C. Tsang, Y.K. Chen, P.J.F. Harris and M.L.H. Green, *Nature* 372:159 (1994); Y.K. Chen, A. Chu, J. Cook, M.L.H. Green, P.J.F. Harris, R. Heesom, M. Humphries, J. Sloan, S.C. Tsang and J.F.C. Turner, *J. Mater. Chem.* 7: 545 (1997)
10. H. Dai, E.W. Wong, Y.Z. Lu, S. Fan and C.M. Lieber, *Nature* 375:769 (1995)
11. W. Krätschmer, L.D. Lamb, K. Foristopoulos and D.R. Huffman, *Nature* 347:354 (1990)
12. S. Iijima, *Nature* 354:56 (1991)
13. T.W. Ebbesen and P.M. Ajayan, *Nature* 358:220 (1992)
14. R.D. Johnson, M.S. de Vries, J. Salem, D.S. Bethune and C.S. Yannoni, *Nature* 355:239 (1992)
15. R.S. Ruoff, D.C. Lorents, B. Chan, R. Malhotra and S. Subramoney, *Science* 259:346 (1993)
16. M. Tomita, Y. Saito and T. Hayashi, *Jap. J. Appl. Phys.* 32:L280 (1993)
17. S. Seraphin, D. Zhou, J.Jiao, J.C. Withers and R. Loufty, *Nature* 362:503 (1993)
18. S. Seraphin, D. Zhou, J.Jiao, J.C. Withers and R. Loufty, *Appl. Phys. Lett.* 63:2073 (1993)
19. J.M. Cowley and M. Liu, *Micron* 25:53 (1994)

20. S. Subramoney, R.S. Ruoff, D.C. Lorents, B. Chan, R. Malhotra, M.J. Dyer and K. Parvin, *Carbon* 32:507 (1994)
21. J.M. Cowley and M. Liu, *Carbon* 33: 225 (1995)
22. J.M. Cowley and M. Liu, *Carbon* 33:749 (1995)
23. M. Terrones, W.K. Hsu, A. Schidler, H. Terrones, N. Grobert, J.P. Hare, Y.Q. Zhu, M. Schwoerer, K. Prassides, H.W. Kroto, D.R.M. Walton, *Appl. Phys. A* 66:1 (1998)
24. Y. Saito and T. Yoshikawa, *J. of Crystal Growth* 134:154 (1993)
25. P.M. Ajayan, C. Colliex, J.M. Lambert, P. Bernier, L. Barbedette, M. Tencé and O. Stéphan, *Phys. Rev. Lett.* 72:1722 (1994)
26. A.A. Setlur, J.M. Lauerhaas, J.Y. Dai and R.P.H. Chang, *Appl. Phys. Lett.* 69: 345 (1996)
27. J.Y. Dai, J.M. Lauerhaas, A.A. Setlur and R.P.H. Chang, *Chem. Phys. Lett.* 258: 547 (1996)
28. W.K. Hsu, M. Terrones, J.P. Hare, N. Grobert, H.W. Kroto and D.R.M. Walton, *Proceedings of the International Winterschool on Electronic Properties of Novel Materials, Molecular Nanostructures*, 381 (Kirchberg 1997)
29. W.K. Hsu, J. Li, M. Terrones, H. Terrones, N. Grobert, Y.Q. Zhu, S. Trasobares, J.P. Hare, C.J. Pickett, H.W. Kroto, D.R.M. Walton, submitted to *Chem. Phys. Lett.*
30. N. Grobert, M. Terrones, A.J. Osborne, H. Terrones, W. K. Hsu, S. Trasobares, Y.Q. Zhu, J.P. Hare, H.W. Kroto, D.R.M. Walton, *Appl. Phys. A* in press
31. C. Guerret-Piécourt, Y. Le Bouar, A. Loiseau and H. Pascard, *Nature* 372:761 (1994)
32. A. Loiseau and H. Pascard, *Chem. Phys. Lett.* 256:246 (1996)
33. A. Loiseau, *Full. Science and Tech.* 4:1263 (1996)
34. M. Tencé, M. Quartuccio and C. Colliex, *Ultramicroscopy* 58:42 (1995)
35. C. Colliex, *J. Electron Micros.* 45:44 (1996)
36. K. Suenaga, C. Colliex, N. Demoncy, A. Loiseau, H. Pascard, F. Willaime, *Science* 278:653 (1997)
37. N. Demoncy, O. Stéphan, N. Brun, C. Colliex, A. Loiseau, H. Pascard, *Eur. Phys. J. B* 4:147 (1998)
38. O. Stéphan, *Ph.D. Thesis, Université d'Orsay, France* (1996)
39. D.S. Bethune, C.H. Kiang, M.S. de Vries, G. Gorman, R. Savoy, J. Vasquez and R. Beyers, *Nature* 363:605 (1993)
40. N.M. Rodriguez, *J. Mater. Res.* 8: 3233 (1993)
41. M.S. Dresselhaus, G. Dresselhaus, K. Sugihara, I.L. Spain and H.A. Goldberg in *Graphite Fibers and Filaments* edited by M. Cardona (Springer, 1988)
42. M. Audier, A. Oberlin and M. Coulon, *J. Cryst. Growth* 55: 549 (1981)
43. R.T.K. Baker, M.A. Barber, P.S. Harris, F.S. Feates and R.J. Waite, *J. Catal.* 26:51 (1972)
44. G.G. Tibbets, *J. Cryst. Growth* 66:632 (1984)
45. A. Oberlin, *Carbon* 22:521 (1984)
46. E. Fitzer and S. Weisenburger, *Carbon* 14:195 (1976)
47. X. Bourrat, A. Oberlin and J.C. Escalier, *Fuel* 542:521 (1987)
48. M.S. Kim, N.M. Rodriguez and R.T.K. Baker, *J. Catal.* 143:449 (1993)
49. *Binary Alloy Phase Diagrams* edited by T.B. Massalski (ASM International, 1990)
50. K. Suenaga, F. Willaime, A. Loiseau, C. Colliex, *Appl. Phys. A* 68:301 (1999) .

SIMULATION OF STM IMAGES AND STS SPECTRA OF CARBON NANOTUBES

Ph. Lambin,[1] V. Meunier,[1] and A. Rubio[2]

[1]*Physics Department, Facultés Universitaires N-D Paix*
61 Rue de Bruxelles, B-5000 Namur, Belgium.
[2]*Departamento Fisica Teorica, Universidad de Valladolid*
E-47011 Valladolid, Spain.

INTRODUCTION

During the last two years, a number of exciting results have been obtained from scanning tunneling microscopy (STM) and spectroscopy (STS) on single-wall carbon nanotubes [1,2,3]. In important issue in these studies, started in 1993 [4,5,6,7], is the characterization of the atomic and electronic structures of the nanotubes. STS provided a beautiful confirmation of the theoretical prediction that the nanotubes can be either metallic or semiconductor, depending on the two indices n and m of their wrapping vector. The precise determination of these two indices with STM remains a challenge, however, which demands the achievement of atomic resolution. Still, geometric distortions of the lattice exist that need to be taken into account for a proper determination of the helicity of chiral nanotubes [8]. Even the measurement of the tube diameter with STM is not easy due to a tip shape convolution effect [9,10].

As it is already true with conventional electron microscopy, more especially with a transmission microscope, STM image simulation may help the interpretation of the observations. In some cases, best fit to experimental STM data might provide reliable structural parameters of the imaged system, the n and m indices of a single-wall nanotube for instance. In other cases, the simulation of the STM image of a defect or the prediction of its STS characteristics may help to identify that defect. In this respect, the role of the pentagons that should be present in the closing cap of a nanotube has already been explored by comparing STS data to computed densities of states [11]. Electronic structure calculations [12] have also been carried out recently in support to the remarkable observation by STS of the electron standing wave pattern that develops in finite-length armchair nanotubes [13]. Topographic STM images of a pentagon–heptagon pair [8] and Stones-Wales defects [14] have already been simulated, giving a clear picture of what they could look like. All these examples testify to the importance of a theoretical investigation of the spectroscopic properties and STM images of carbon nanotubes.

In the present work, a theoretical modeling of scanning tunneling microscopy of the nanotubes is presented. This theory is based on the standard perturbation formulation of elastic tunneling within a tight-binding description of the π-electrons of the nanotubes. The tip is treated as a single atom with an s wavefunction. The model has

Science and Application of Nanotubes, edited by Tománek and Enbody
Kluwer Academic / Plenum Publishers, New York, 2000

been tested on planar graphite and transposed to the cylindrical geometry. The simplicity of the method makes it possible to investigate the effects of network curvature and the signature of topological defects in a systematic way. Several illustrations of topographic STM images and current-voltage characteristics are given, and compared to available experimental data and other results from *ab-initio* calculations.

STM THEORY

Consider an occupied electronic state on the tip, α, and an unoccupied state, β, of the sample and introduce at time $t = 0$ a stationary, weak coupling interaction v between the two systems. Time-dependent perturbation theory then shows that for large t, the probability for an electron to go from α to β vanishes unless the two states lie exactly at the same energy:

$$P_{\alpha\to\beta}(t) \sim \frac{2\pi}{\hbar} t \, |\langle\alpha|v|\beta\rangle|^2 \delta(E_\beta - E_\alpha) \ . \tag{1}$$

Provided there is an external wire enabling the circulation of the electrons, a stationary current set us across the gap between tip (t) and sample (s). This current is given by the time derivative of the transfer probability multiplied by the electron charge. The total current is the sum over all states, labeled by their energy E, multiplied by the difference of occupation numbers $f(E)$ on both sides of the junction:

$$\mathcal{I} = \frac{2\pi e}{\hbar} \int_{-\infty}^{+\infty} dE \, [f_t(E) - f_s(E)] \sum_{\alpha,\beta} |\langle\alpha|v|\beta\rangle|^2 \delta(E - E_\alpha)\delta(E - E_\beta) \ . \tag{2}$$

Eq. 2 is the well-known starting expression of almost all theories of elastic tunnel processes.

In tight-binding (TB), assuming one orbital per site to simplify, the electronic states on the isolated tip and sample are linear combinations of the corresponding atomic orbitals:

$$|\alpha\rangle = \sum_{I\in t} \chi_I^\alpha |\eta_I\rangle \ , \ |\beta\rangle = \sum_{J\in s} \psi_J^\beta |\theta_J\rangle \ . \tag{3}$$

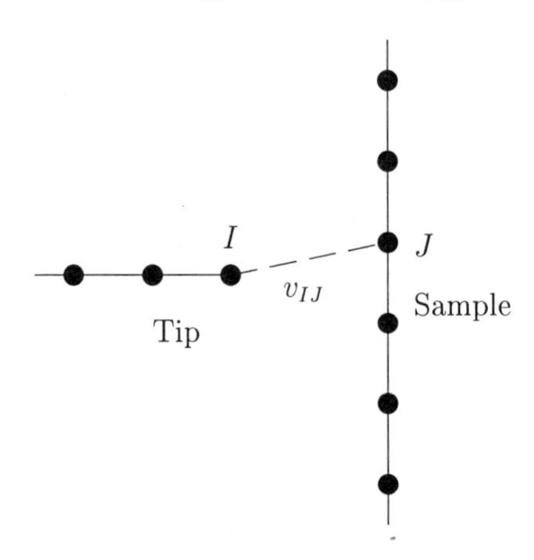

Figure 1: Schematic representation of the tip-sample interface in STM.

Inserting these expressions into the square matrix element $|\langle\alpha|v|\beta\rangle|^2$ in eq. 2 leads to quantities of the following kind

$$n^s_{JJ'}(E) = \sum_{\beta\in s} \psi^{\beta *}_J \delta(E - E_\beta)\psi^\beta_{J'} = \frac{-1}{\pi} \text{ Im } R^s_{JJ'}(E + i0) \tag{4}$$

that under time reversal symmetry are the imaginary parts of the Green function (or resolvent operator) elements between pair of sites[15]. With these notations at hand, the current at 0 K is

$$\mathcal{I} = (2\pi)^2 \frac{e}{h} \int_{-eV}^{0} dE \sum_{I,I'\in t} \sum_{J,J'\in s} v_{IJ} v^*_{I'J'} n^t_{II'}(E^t_F + eV + E) n^s_{JJ'}(E^s_F + E) \tag{5}$$

where V is the tip–sample bias potential ($e > 0$), the E_F's are the Fermi levels of the unperturbed systems, and $v_{IJ} = \langle\chi_I|v|\theta_J\rangle$ is a tip–sample coupling element, see fig. 1. The energy levels of the tip sites have been shifted to accommodate the bias and contact potential of the junction.

For the sake of illustration, considering just one coupling element between the ending atom of the tip and a single atom of the sample reduces the double summations in eq. 5 to one term with diagonal elements n^t_{II} and n^s_{JJ}, namely the unperturbed local densities of states on the two coupled atoms. The current is then a convolution product of the two densities of states over the bias window. When this bias is small, the low-voltage conductance becomes

$$\frac{d\mathcal{I}}{dV} \approx (2\pi)^2 \frac{e^2}{h} |v_{IJ}|^2 n^t_{II}(E^t_F + eV) n^s_{JJ}(E^s_F) \; . \tag{6}$$

This resembles the Tersoff-Hamann result [16], except that the densities of states are computed on their own atomic sites. Here the quantity that decreases exponentially with the distance is the tip–sample interaction v_{IJ}.

For the applications that follow, the STM current was calculated with eq. 5 by considering just one tip atom I at the apex, with an s-wave orbital like in Tersoff-Hamann theory. The corresponding tip density of states n^t_{II} was taken to be a Gaussian function of 6 eV full width at half maximum. The sample is a carbon nanotube with one π orbital per atom. The tip–sample coupling interactions were taken as sp Slater-Koster hopping terms having the following expression:

$$v_{IJ} = v_0 \, w_{IJ} \, e^{-d_{IJ}/\lambda} \cos\theta_{IJ} \tag{7}$$

$$w_{IJ} = e^{-ad^2_{IJ}} / \sum_{J'} e^{-ad^2_{IJ'}} \tag{8}$$

where d_{IJ} is the distance between the tip atom I and the sample atom J, θ_{IJ} is the angle between the orientation of the π orbital on site J and the IJ direction. The gaussian weight factor w_{IJ} was introduced for convergence reasons. The parameters used are $\lambda = 0.85$ Å and $a = 0.6$ Å^{-2}. The parameter v_0 is of the order of 1 eV, its actual value does not matter as long as absolute values of the current are not required.

When the sample is a perfect nanotube with not too many atoms per unit cells (a few hundreds at the most), the quantities $n^s_{JJ'}(E)$ can directly be computed from their definition, eq. 4, the β states then denoting Bloch states in the reciprocal unit cell over which the summation is performed. When the sample is not crystalline, the last equality in eq. 4 can be used by computing the Green function elements with the recursion technique. This technique, originally designed for the calculation of diagonal elements of the Green function [17], can be adapted to non-diagonal elements [18]. In case of a real, symmetric Hamiltonian with one orbital per atom like here, off-diagonal elements can be obtained from the following identity involving three diagonal elements of the Green function

$$R_{JJ'}(z) = R_{J'J}(z) = \langle\frac{\theta_J + \theta_{J'}}{\sqrt{2}}|(z-H)^{-1}|\frac{\theta_J + \theta_{J'}}{\sqrt{2}}\rangle - \frac{1}{2}[R_{JJ}(z) + R_{J'J'}(z)] \tag{9}$$

where the first term is related to a fictitious orbital defined as the symmetric combination of the orbitals on the J and J' sites. The calculations were performed with 200 (STM images) or 300 (STS spectra) levels of continued fraction.

When tractable, *ab-initio* calculations of the STM topographic image were also performed within the simplest model as proposed by Tersoff and Hamann [16]. The tunneling current is computed directly from first-order perturbation theory, as above, where the essential quantity to be computed is the transfer matrix element between tip states and sample states. In this simplified model, similar to the one described in eq. 5 above, the STM current for an external applied bias voltage V is directly proportional to the spatial-local-density of states (LDOS) integrated between the Fermi level of the tip and sample [19], that is:

$$\mathcal{I}(\mathbf{r}, V) \propto \int_{E_F^s - eV}^{E_F^s} dE\ \rho_{LDOS}(\mathbf{r}, E) , \tag{10}$$

with

$$\rho_{LDOS}(\mathbf{r}, E) = \sum_{\beta} \mid \psi_\beta(\mathbf{r}) \mid^2 \delta(E_\beta - E) , \tag{11}$$

where ψ_β and E_β are the electron wavefunction and eigenvalue of state β, respectively. We then approximate the constant current images as iso-surfaces of $\mathcal{I}(\mathbf{r}, V)$. In writing the intensity in this way we have made further assumptions, namely that both the tunneling matrix elements and the tip density of states in the energy window are constant. Again, the differential conductivity, $d\mathcal{I}/dV$ as measured in STS experiments, gives a direct access to the electronic level structure of the nanotube (the LDOS).

GRAPHITE

As a first application of the TB formalism, the STM current was computed for the case of a single sheet of graphene for three positions of the tip: (a) on top of an atom, (b) above the center of a C-C bond, and (c) above the center of an hexagon. As shown in fig. 2, the I–V curves obtained on the first two sites are close to each other, whereas the current is much smaller with the third geometry. The small value of the current above the hexagon center is due to the contribution of the non-diagonal elements of the Green function that couple the pairs of atoms located on the six-membeed cycle. The asymmetry of the I–V curves reflects the asymmetry of the tip density of states with respect to the Fermi level (see eq. 6) and is also conditioned by the off-diagonal elements of the sample Green function.

When coupling two or more graphene layers with the Bernal graphite stacking, the two atoms per unit cell become unequivalent: atom A has a neighbor directly beneath whereas atom B does not. In a small interval around the Fermi level, the LDOS on site A is much smaller than that on B [20]. As a result, the STM current at small bias is larger when the tip is above an atom B, which therefore appears as a protrusion in the constant-current image. This interpretation is considered as the ad-hoc explanation of the fact that only every other two atom is seen in the STM images of graphite [21].

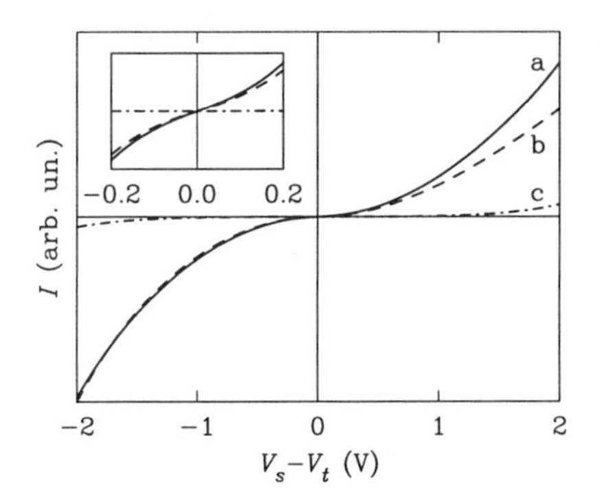

Figure 2: STM current versus sample voltage for a single sheet of graphene when the tip is located 0.5 nm above (a) an atom, (b) the center of a C-C bond, (c) the center of an hexagon. Details of the curves between -0.2 and 0.2 V are shown in the inset.

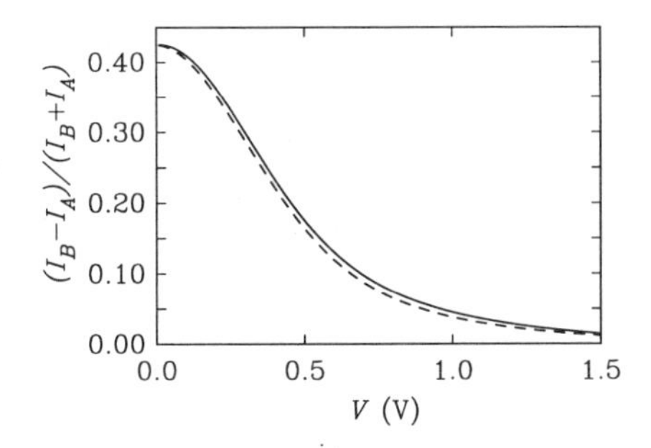

Figure 3: Site asymmetry of the tunneling current above site B and site A in Bernal graphite when the tip is at a distance of 0.5 nm. V is the bias potential, the full curve corresponds to a positive sample potential and the dashed curve to a positive tip potential.

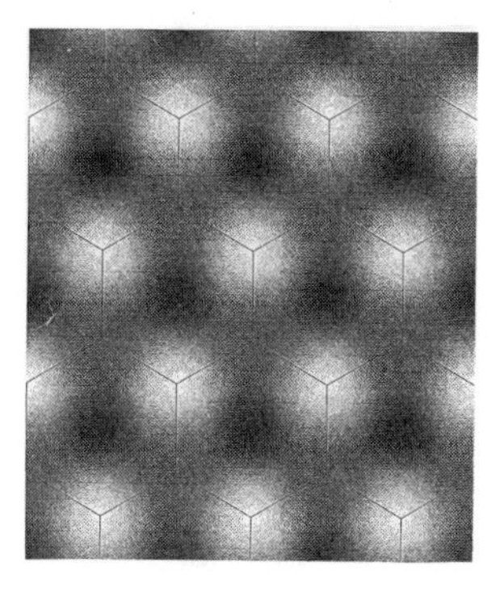

Figure 4: STM current map at constant height (0.5 nm) above a multilayer graphite surface. The tip potential is 0.25 V, which corresponds to a current ratio $I_B/I_A = 2$. The B atoms are clearly resolved, whereas the A atoms do not come out.

The site asymmetry computed by eq. 5 is plotted against the bias potential in fig. 3 for both signs of polarity. There is a good agreement with the calculations of Tománek and Louie[21] who obtained $(I_B - I_A)/(I_B + I_A) = 0.3$ at 0.2 V. The current distribution at 0.5 nm above graphite is shown in fig. 4 for a tip potential equal to 0.25 V. There are no marked local maxima of the current at the locations of the A atoms, only the B atoms are seen (as white features). By increasing the bias, the differences between the densities of states on A and B sites become less important and the asymmetry washes out gradually.

PERFECT NANOTUBES

STS spectra

Recent STS measurements on single-wall carbon nanotubes (SWNT) [1,2,22] have confirmed the early prediction that these nanotubes can be metallic or semiconductor, in proportion 1:3, depending on the indices n and m of their wrapping vector [23,24,25]. At this end, $d\mathcal{I}/dV$ (or $d\log\mathcal{I}/d\log V$) as determined from STS was used as a representation of the nanotube density of states. These current derivatives plotted against V exhibit either a metallic plateau or a semiconductor band gap, which scale like the reciprocal of the diameter, and a series of peaks corresponding to Van Hove singularities. Computed $d\mathcal{I}/dV$ curves are shown in fig. 5 for a series of SWNT's, which directly compare with published experimental data [1]. If some fine details of the curves vary with the actual pair of indices, the gross features of the densities of states depend only on whether $n-m$ is a multiple of 3 (metal) or not (semiconductor) [14,26,27]. The metallic plateau of a conducting nanotube is three times as large as the bang-gap of a semiconducting nanotube of equivalent diameter. The width of the gap and metallic plateau is given by $E_g = 2\gamma_0 d_{CC}/d$ and $W = 6\gamma_0 d_{CC}/d$, respectively, with d the nanotube diameter, d_{CC} the carbon-carbon bond length and γ_0 the π hopping interaction between C 2p orbitals. The exact value of this parameter is in the range 2.5 eV (determined by DFT) – 3.0 eV (experimental value for graphite). In all our TB calculations, we used $\gamma_0 = 2.75$ eV.

The electron wave function in a nanotube can be imaged by STS, as shown recently on a short piece ($\sim$ 30 nm) of an armchair nanotube[13]. Due to the discretization of the allowed wave vectors in a finite-length nanotube, standing-wave electronic patterns are obtained. In the vicinity of the HOMO and LUMO states of a finite-length armchair nanotube, the local density of states is modulated along the axis by $\sin^2(2\pi x/\lambda_F)$ or $\cos^2(2\pi x/\lambda_F)$ envelopes, with $\lambda_F = 3a_0 = 7.5$ Å the Fermi wavelength[12]. The resulting oscillation of the local DOS has been observed experimentally[13]. Modulation of the local DOS produced by the reflection of Bloch waves at the end cap of an infinite nanotube has been studied theoretically by Kane and Mele[28]. In an armchair nanotube, they also found λ_F-osillations for the states close to the Fermi level, in agreement with earlier tight-binding calculations [29,30].

In relation with the experimental observations reported here above[13], tight-binding calculations were performed on a 33.4 nm long (10,10) nanotube capped on both sides. STS spectra were computed while scanning the tube at a constant height of 0.5 nm along a path parallel to the axis (dotted line in fig. 6). In the central panel of fig. 6, curve (b) shows the λ_F period oscillation of $d\mathcal{I}/dV$ superposed to a function having the period a_0 of the lattice. This explains the peak sub-structure observed in the interval λ_F, which reproduces the experimental data fairly well[13]. The sample potential (0.2V) of curve (a) shows an unoccupied state localized on the cap [30]. The corresponding

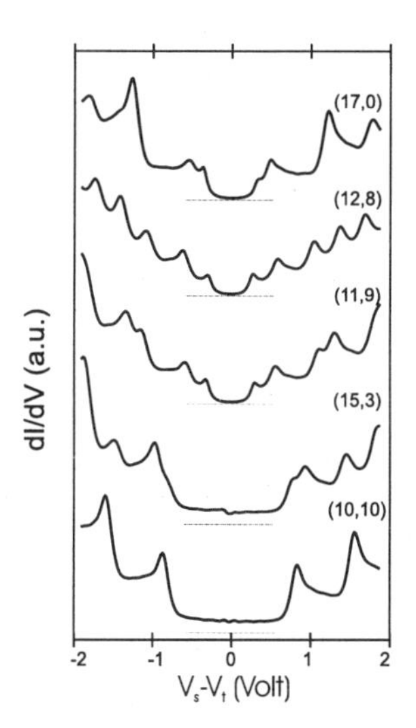

Figure 5: Computed differential conductance dI/dV of five SWNT's with equivalent diameter. The (10,10) and (15,3) nanotubes are metals, the others are semiconductors.

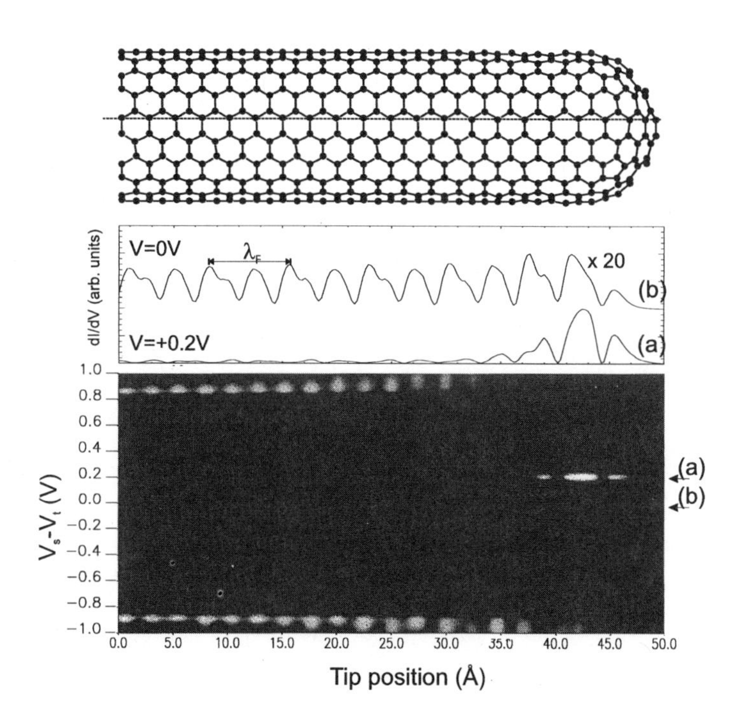

Figure 6: STS spectra of a piece of (10,10) nanotube capped by two C_{240} halves. Top: portion of tubule on which the STS computation was performed, along a scan line visualized by the dashed line. Bottom: map representing dI/dV in gray scale against the tip x coordinate and the bias potential. Middle: line-cuts labeled (a) and (b), representing dI/dV *versus* x for $V_s - V_t = 0.2$ V and $= 0$ V respectively.

local density of states rapidly decreases as one proceeds into the tube. In the bottom panel of fig. 6, the states at $\sim \pm 0.9$V correspond to Van Hove singularities located on both sides of the metallic plateau in the density of states. These states, together with the tip-localized state, are by far the most intense features. With the contrast and resolution achieved in these calculations, the discretized structure of the spectrum is not revealed by the gray-scale map[?]. The bulk properties of the (10,10) nanotube are recovered 4 nm away from the tip, considering the spatial dispersion of the Van Hove peaks. This is in agreement with DFT calculations for finite-length tubes[14] that showed a smooth transition from a level structure characteristic of a molecular-wire to that of a delocalized one-dimensional system that seems to be completed for tube-lengths of the order of 5 nm. The STS spectra of the finite-size tubes is in very good agreement with the more sophisticated *ab-initio* calculations.

For these tight-binding calculations and the other illustrated below, there were no substrate to support the nanotube. When the nanotube is deposited on gold, there is a charge transfer which lowers the Fermi level of the nanotube by ~0.25 eV [1]. Our *ab-initio* calculations [14] show that the interaction with the substrate or with other tubes does not alter the STM images observed or computed for isolated tubes, even if the tubes are bound to the substrate by charge transfer. However, tube/substrate or tube/tube interactions modify the electronic spectrum in several ways: They open a small "pseudogap" in the tube states at the Fermi level (pseudogap of the order of 0.1 eV, see below); they make more accentuated the electron-hole asymmetry in the DOS and smooth out the van Hove singularities. However, even if the electronic level structure is sensitive to external perturbations, the whole set of STM images of armchair carbon nanotubes can be understood in terms of the isolated SWNT wave-functions.

STM images

When the STM tip follows the curvature of a nanotube, the tunneling current tends to follow the shortest path, which is normal to the tube, and the corresponding "off-z" component of the current increases as the tip moves aside to the tube[31]. Due to this, the imaged lattice appears stretched in the direction normal to the tube axis. As shown in fig. 7, an atom of the nanotube that projects at location (x', y') on the horizontal plane is imaged when the tip has horizontal coordinates $x = x'$ and $y = y'(r + \Delta)/r$ when it is assumed that the tunneling current is radial. Here x is the coordinate parallel to the axis (perpendicular to the drawing), y is normal to the axis, r is the tube radius, and Δ is the tip-nanotube distance. The projected lattice is therefore stretched by the factor $(r + \Delta)/r$ in the y direction[8]. Due to this distortion, the angles between the three zig-zag chains of C atoms measured on an STM image deviate from $\pi/6$. The distortion can be corrected for by squeezing the y coordinates in such a way as to restore the correct angles[32].

When a point-like tip scans a nanotube at constant current and moves aside to the tube, its z coordinate decreases rapidly with increasing y (see fig. 7) because the tip follows a cylinder of radius $r + \Delta$. By comparison with this geometric corrugation of the nanotube, of the order of the tube diameter, the atomic corrugation is much smaller, of the order of 0.1 nm. In the image simulation, it is easily possible to get rid of the geometric corrugation of the nanotube by representing the variations of the radial distance $\rho = r + \Delta$ of the tip versus its x and y coordinates. In this representation, the topographic contrast does not deteriorate as the tip moves aside to the topmost part of the nanotube as it should with the conventional $z(x, y)$ image. On the topmost part of the nanotube, the $\rho(x, y)$ map is close to the experimental data.

Fig. 8 shows computed STM images of an armchair (10,10) nanotube and a zig-zag

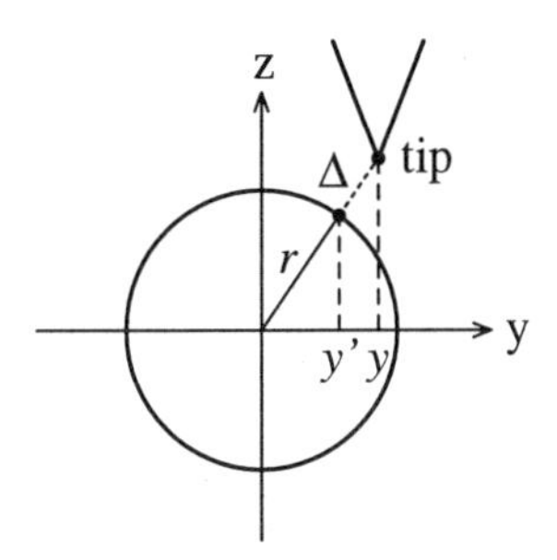

Figure 7: Relation between the position of the STM tip (y) and that of an imaged atom (y') of the nanotube when the tunneling current follows the shortest path between tip and nanotube (dashed line).

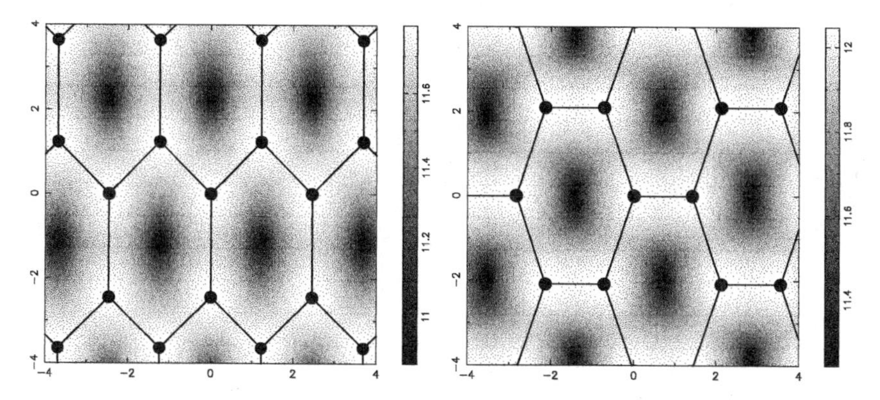

Figure 8: Gray-scale representation of the radial distance $\rho(x, y)$ at constant current of the STM tip apex above the topmost part of (10,10) (left) and (18,0) (right) non-chiral nanotubes. According to the simple rule $n - m = \mathcal{M}(3)$, both nanotubes are metallic. The tip potential is 0.3 V. The axis of the nanotubes is parallel to the horizontal x direction. All coordinates are in Å.

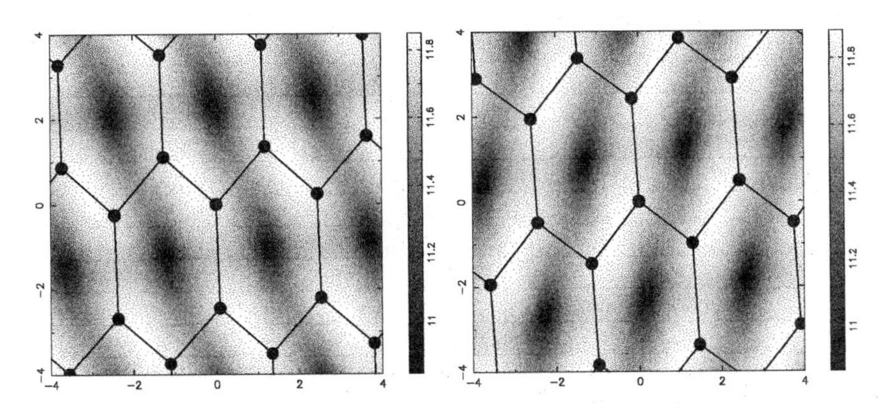

Figure 9: Same as in fig. 8 for the (11,9) (left) and (12,8) (right) chiral nanotubes. Both nanotubes are semiconductors, with a band gap of ~0.6 eV. The tip potential is 0.5 V.

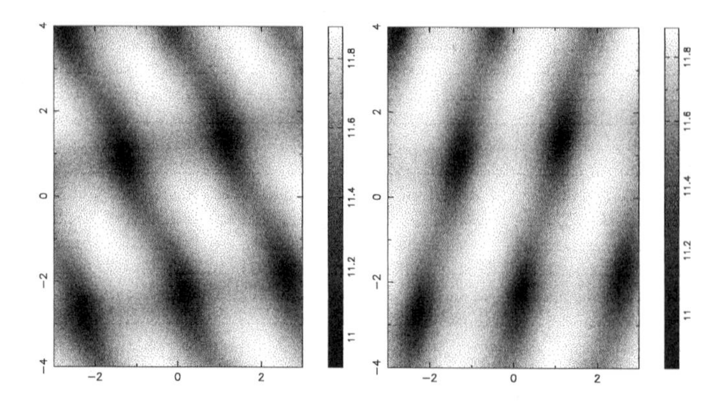

Figure 10: Computed STM images of the (12,8) semiconducting nanotube for two opposite bias potentials (left: $V_t - V_s = -0.35$ V, right: $V_t - V_s = +0.35$ V).

(18,0) one. In these two-dimensional plots of the radial distance of the tip, $\rho(x, y)$, the coordinate x parallel to the axis is along the horizontal direction. Like in graphite, the centers of the hexagons correspond to sharp dips in the corrugation. This is confirmed by DFT calculations based on the Tersoff-Hamann theory [12,14]. In the present calculations, the distance of the tip to the atom at the center of the images was set to 0.5 nm. The geometric distortion discussed above represents as much as 70% (inflation of the ordinates by a factor of 1.7). Experimentally, distortion of 20 - 60 % has been observed on single-wall nanotubes with diameter in the range 1.2 - 1.5 nm [32]. The effects of the distortion are clearly visible on the honeycomb lattice that was superimposed on the image. The distortion is also responsible for the elongate shape of the corrugation deeps at the center of the hexagons. In the armchair nanotube, the largest protrusion are realized on the atoms, all the bonds look the same. In the zig-zag nanotube, the largest protrusions are realized on the bonds parallel to the axis. There is a saddle point at the center of the inclined bonds.

In the case of chiral nanotubes, one kind of bonds protrudes more than the others, like with the zig-zag geometry, but not always the ones closest to the axial direction (see fig. 9). These bonds form spiral patterns around the nanotube. The elongate holes at the center of the hexagons are no longer aligned with the circumference, as often observed in the experimental images [1,2,3]. The helicities (pitch angle) measured in fig. 9 are 6 and 12° for (11,9) and (12,8), respectively, the real values being 3.3 and 6.6°, respectively. The 70% distortion of the image is responsible for this deviation.

Recently, Kane and Mele [28] showed from tight-binding DOS calculations that the band-edge states of a semiconducting nanotube have complementary structures when imaged at positive and negative bias. This effect, confirmed by our STM image calculations, is illustrated in fig. 10 for the (12,8) nanotube. By switching the tip potential from -0.35 to +0.35 V (slightly away from $\pm E_g/2$), the protruding bonds change from one zig-zag chain of atoms to another. This anisotropy is related to the differences between bonding and anti-bonding wavefunctions. The handedness of the spiral pattern in the STM image of a semiconducting nanotube not only depend on the sign of the bias but also on whether $\text{mod}(n - m, 3) = 1$ or 2. This effect is clearly shown in fig. 9 where the two kinds of helicities are illustrated.

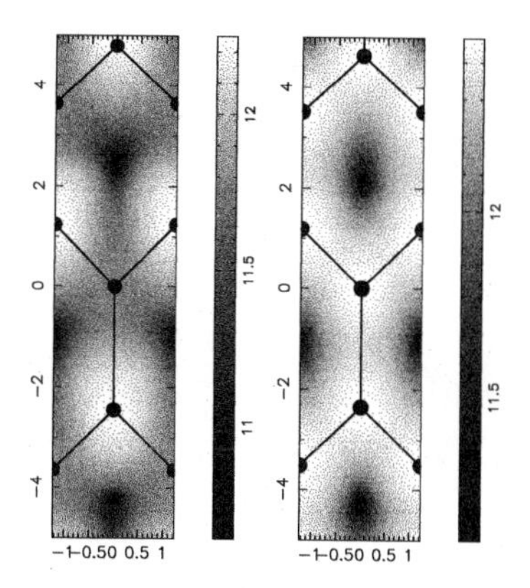

Figure 11: Computed STM images of two-layer nanotubes using the same representation as in fig. 8. Left : (5,5)@(10,10), right: (6,6)@(11,11). The tip potential is 0.1 V.

MULTI-WALL NANOTUBES AND ROPES

In a perfect, infinitely long SWNT, all the atoms are equivalent. There is therefore no reason why a site asymmetry could develop on these nanotubes so as to explain a triangular lattice that is often observed [33]. The situation may change in the presence of defects or end caps [28]. In multi-wall nanotubes, the interaction between the π orbitals over the last two layers affects the STM image, sometimes like in graphite where only every other two atoms of the external layer are clearly imaged [34]. In other cases, there is a modulation of the image intensity because a perfect lattice coherence cannot be realized between two cylindrical graphitic sheets. When the last two layers have different helicities, a Moiré super-pattern can form at the surface of the tube along the axis [4].

Calculations were carried out on bi-layer using the tight-binding formalism already developed for multilayer nanotubes [35]. As a general rule, the STM images computed were found similar to those of single-wall nanotubes: no modulation of intensities could be detected and all the atoms looked essentially the same in most of the cases. However, a site asymmetry similar to that of graphite was revealed in the particular case of (5,5)@(10,10). An image of this system is shown on the left-hand side of fig. 11 where one atom protrubes 0.4 Å above the other. This atomic contrast is similar to the one realized in graphite at low voltage (see fig. 4). With a less symmetrical bi-layer like (6,6)@(11,11) illustrated in the right-hand side of fig. 11, there is no such asymmetry.

The explanation of the results described above is the following. In multiwall armchair nanotubes, the interlayer coupling perturbs the density of states of the individual layers in a small interval around the Fermi level [35,36,37]. At the Fermi energy, the perturbed wave function is a mixing of the states that cross the Fermi level, and these states belong to the A_1 or A_2 irreducible representations of the full symmetry group of the individual layers [38]. The important fact is that by mixing these states, a bi-partition of the honeycomb network is realized on each layer, with increased LCAO coefficients on one lattice and reduced ones on the other [39]. In other words, there are two kinds of unequivalent atoms in the multilayer, but the difference between them is only significant in a small interval around the Fermi energy, just like in graphite. The half width of this interval is of the order of the interlayer coupling, 0.1 eV. The difference between

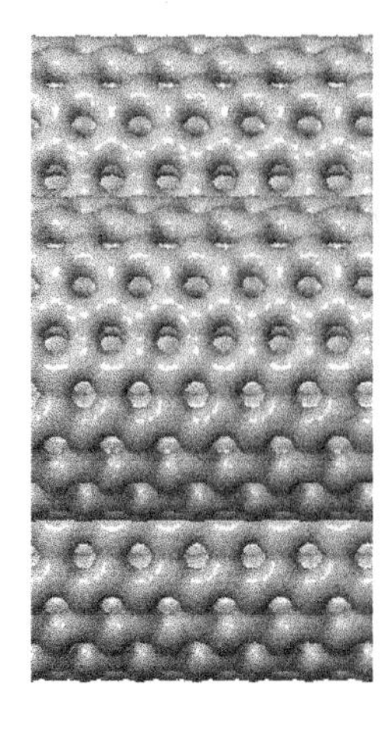

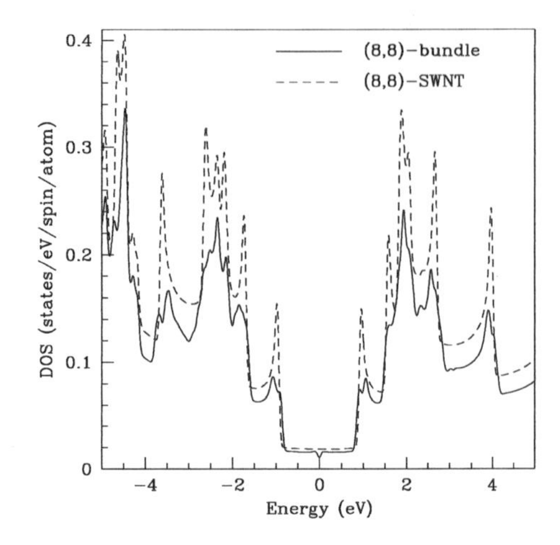

Figure 12: STS and STM topographic 3-d image computed *ab initio* for a tube bundle formed by three (8,8) armchair nanotubes (of ~ 1.1 nm diameter). The applied voltage is 0.5 V. The STM image is an iso-surfave of the tunneling current (eq. 10).

the two atoms is also only significant when the two layers have a common rotational symmetry group, otherwise the matrix elements of the perturbation brought about by the interlayer coupling becomes too small, due to cancellation of terms all around a circumference. For instance, the (5,5)@(10,10) system preserves the five-fold symmetry, whereas (6,6)@(11,11) has no rotational symmetry. As a consequence, there are two unequivalent atoms in the first system, and virtually no such bi-partition in the second case. This makes all the difference between the two STM images shown in fig. 11. Let us finally remark that a special arrangement of the two layers in (5,5)@(10,10) exists that conserves the symmetry D_{5d} of the internal tubule [40]. In this case, the mixing between the A_1 and A_2 states is forbidden for symmetry reason and all the atoms remain equivalent. DFT calculations of the STS and STM image of a three layer nanotube (5,5)@(10,10)@(15,15) confirm this result [14]. We checked that changing the polarity of the applied voltage does not introduce appreciable changes in the STM-topographic image.

In fig. 12, we present the calculated 3D-STM image obtained *ab initio* for a bundle made of three (8,8) nanotubes for an external voltage of +0.5 V. The fact that the position in energy of the peaks in the density of states is not strongly modified explains the success of comparing the isolated SWNT spectra with experiment. However the shape of the spectra (relative intensities) is strongly affected by tube-tube interactions.

DEFECTS IN NANOTUBES

Twisted nanotube

It is clear that a torsional twist of a nanotube affects its apparent helicity. In particular, single-wall nanotubes in a rope can be twisted, as revealed recently by STM [41]. This is an important distorsion, because twisting a nanotube affects its electronic properties and may drive an armchair nanotube in a semiconducting state [42]. It turns out that a twisted achiral nanotube can be distinguished from a perfect, chiral nanotube. This was demonstrated experimentally on a twisted armchair nanotube, presumably (10,10),

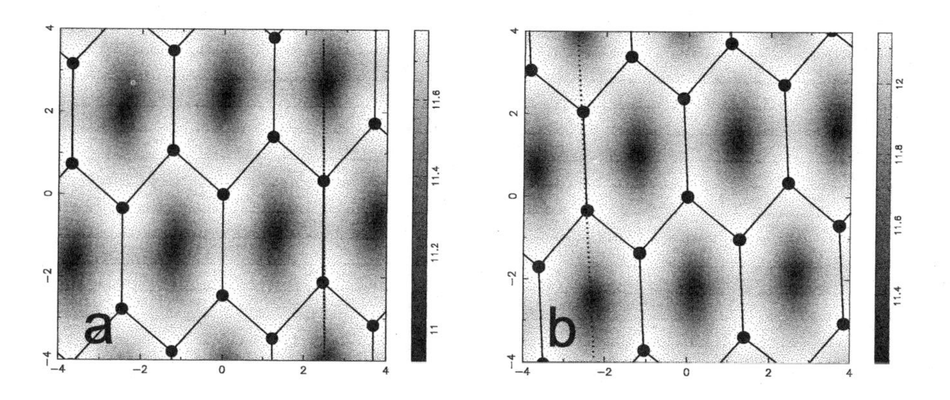

Figure 13: Topographic STM images $\rho(x, y)$ at constant current of (a) a (10,10) nanotube uniformly twisted at 4.7° pitch angle and (b) a perfect (12,9) nanotube having the same helicity. The tip potential is 0.5 V.

located on the top of a rope [41]. In the experimental STM image, the apparent helicity of the nanotube was 4°, which corresponds to a uniform twist by 360° over a distance of 63 nm. One third of the bonds of the twisted tube remained perpendicular to the axis, and this would not be the case with a chiral nanotube.

The STM images shown in fig. 13 were computed for a (10,10) nanotube, twisted with a pitch angle of 4.7°, and a chiral (12,9) nanotube that presents the same helicity. Due to the geometric distorsion discussed above, the apparent helicity of both nanotubes is 7.5°. For the two nanotubes in fig. 13, the atomic corrugation is the same and the images look the same, although the elliptical holes at the center of the hexagons of the twisted (10,10) tubule are slightly distorted as compared to the shapes found in (12,9). The main difference between the two images is that a line joining the centers of next-nearest-neighbor hexagons along the circumference (dotted line in fig. 13) clearly remains perpendicular to the axis in the twisted (10,10) nanotube, in agreement with experiment, whereas they do not in (12,9).

Pentagon–heptagon pair

Pentagon–heptagon pairs provide a natural way of connecting two different nanotubes [43] or changing the diameter of a nanotube [44]. There are many observations of nanotubes by high-resolution electron microscopy [44,45] and also STM [46] where the presence of pentagon–heptagon pairs is highly suspected. In principle, STM and STS could provide a proof for the existence of these odd-membered rings if by chance they were located on the topmost part of the nanotube. Computer simulation of the STS spectrum [11] or the STM image [8] of pentagon and heptagon can help to identify the signature of these defects.

Fig. 14 shows a computed STM image of a (12,0)/(9,0) junction between two zigzag nanotubes based on a pentagon–heptagon pair parallel to the axis [47]. The two defects are separated by two hexagons in a conical section that joins the two nanotubes with diameters 0.94 and 0.70 nm. The pentagon is responsible for the large protrusion near the center of the STM image, whereas the heptagon is barely visible. Because the odd-membered rings affect more strongly the anti-bonding states then the bonding ones [29,48], they induce electron–hole asymmetry. By changing the polarity of the tip, occupied ($V_t - V_s > 0$) or unoccupied ($V_t - V_s < 0$) states of the nanotubes are probed, and these come out differently. Physically, the patterns in the images can be interpreted as the result of interferences between the waves scattered by the defect.

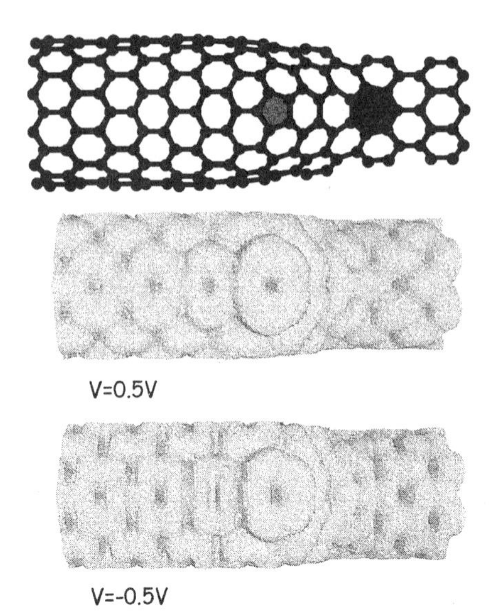

Figure 14: Top: (12,0)/(9,0) junction between two zig-zag nanotubes. Bottom: 3-d representation of the STM tip height at constant current above the junction for two opposite polarizations of the tip ($V = V_t - V_s$).

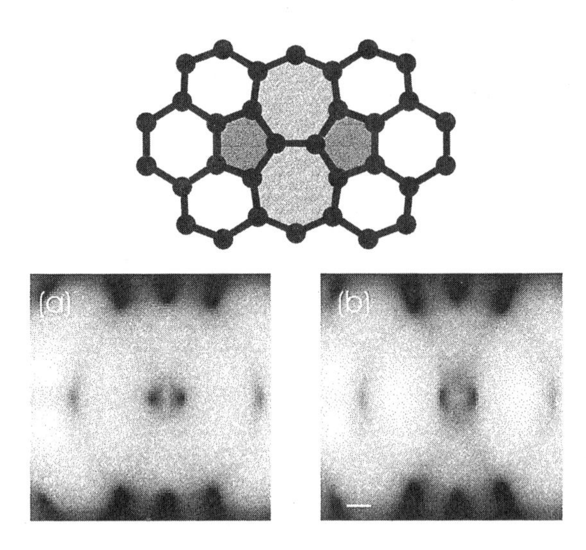

Figure 15: Top: Atomic configuration of the Stone-Wales defect on a (10,10) nanotube whose axis is along the horizontal direction. Bottom: 3-d representations of the constant-current height of the STM tip for two polarizations (a) $V_s - V_t$ = -0.5 V and (b) +0.5 V. The small horizontal bar in the image (b) represents 1 Å.

Stone-Wales defect

A Stone-Wales defect can be viewed as the transformation of four hexagons in two pentagons and two heptagons after $\pi/2$ rotation of a C-C bond [49]. Transformations of this sort may be responsible for the spherical shape of the carbon onions [50]. In nanotubes, molecular-dynamics calculations indicate that Stones-Wales defects could form under high tensile strain [51]. STM calculations have been performed for a Stone-Wales defect in (10,10) nanotube. In the configuration considered here, the two pentagons are parallel to the nanotube axis (see fig. 15). Due to the mirror symmetry along a plane perpendicular to the tube axis of the Stone-Wales defect, the STM pattern keeps that symmetry. This would not be the case for other relative orientations of the defect.

The two holes at the center of the STM images in fig. 15 correspond to the two atoms shared by the seven-membered rings. The are two other holes in the heptagons, located near the atoms opposite to the common edge (at 6 and 12 o'clock). The protrusions on the left- and right-hand sides of the center are located on the five-membered rings, near the pentagonal bonds that are normal to the axis. These protrusions are especially important when the tip is negatively polarized (fig. 15b). Computed STS images of the Stone-Wales defect on a (6,6) have been published elsewhere [14]. These calculations show that the Stone-Wales defect can be, indeed, experimentally accessible by STS measurements and 3D-mapping.

CONCLUSIONS

The computation of the STS spectra of carbon nanotubes indicates that the dI/dV curves reproduce well the essential features of local density of states. The simulation of the STM topographic profiles show that the nanotube images with atomic resolution are affected by a geometric distortion induced by the curvature of the tube. This distortion stretches the honeycomb network in the direction normal to the axis, and influences the measured helicity. Like single-wall tubules, multi-wall nanotubes do not present a site asymmetry similar to that of multilayer graphite, except for special, symmetric configurations. Geometrical and topographic defects have typical signatures in STM that might identify them. In particular, a somewhat larger protrusion is predicted to occur on a pentagon, more especially with a negative bias of the tip.

ACKNOWLEDGMENT

This work has been partly funded by the Belgian Interuniversity Research Project on Reduced Dimensionality Systems (PAI-IUAP P4-10). V.M. acknowledges the FRIA organization for financial support. A.R. acknowledges financial support from DGES (Grants: PB95-0720 and PB95-0202) European Community TMR contract ERBFMRX-CT96-0067 (DG12-MIHT). Computer time for the *ab-initio* calculations was provided by the C^4 (Centre de Computació i Comunicacions de Catalunya). This work has benefited from helpful discussions with Amand Lucas, Jean-Pol Vigneron, Laszlo Biró, Géza Márk, Cees Dekker, Liesbeth Venema, and Philip Kim.

REFERENCES

1. J.W.G. Wildoer, L.C. Venema, A.G. Rinzler, R.E. Smalley, and C. Dekker, Nature **391**, 59 (1998).

2. T.W. Odom, J.L. Huang, Ph. Kim, and Ch.M. Lieber, Nature **391**, 62 (1998).
3. A. Hassanien, M. Tokumoto, Y. Kumazawa, H. Kataura, Y. Maniwa, S. Suzuki, and Y. Achiba, Appl. Phys. Lett. **73**, 3839 (1998).
4. M. Ge and K. Sattler, Science **260**, 515 (1993).
5. M.J. Gallagher, D. Chen, B.P. Jacobsen, D. Sarid, L.D. Lamb, F.A. Tinker, J. Jiao, D.R. Huffman, S. Seraphin, and D. Zhou, Surf. Sci. **281**, L335 (1993).
6. T.W. Ebbesen, H. Hiura, J. Fujita, Y. Ochiai, S. Matsui, and K. Tanigaki, Chem. Phys. Lett. **209**, 83 (1993).
7. Z. Zhang and C.M. Lieber, Appl. Phys. Lett. **62**, 2792 (1993).
8. V. Meunier and Ph. Lambin, Phys. Rev. Lett. **81**, 5888 (1998).
9. L.P. Biró, S. Lazarescu, Ph. Lambin, P.A. Thiry, A. Fonseca, J. B.Nagy, and A.A. Lucas, Phys. Rev. B **56**, 12490 (1997).
10. G.I. Márk, L.P. Biró, and J. Gyulai, Phys. Rev. B **58**, 12645 (1998).
11. D.L. Carroll, P. Redlich, P.M. Ajayan, J.C. Charlier, X. Blase, A. De Vita, and R. Car, Phys. Rev. Lett. **78**, 2811 (1997).
12. A. Rubio, D. Sanchez-Portal, E. Artacho, P. Ordejon, and J.M. Soler, Phys. Rev. Lett. **82**, 3520 (1999).
13. L.C. Venema, J.W.G. Wildoer, J.W. Janssen, S.J. Tans, H.L.J. Temminck Tuinstra, L.P. Kouwenhoven, and C. Dekker, Science **283**, 52 (1999).
14. A. Rubio, Appl. Phys. A **68**, 275 (1999).
15. K. Kobayashi and M. Tsukada, J. Vac. Sci. Technol. A **8**, 170 (1990).
16. J. Tersoff and D.R. Hamann, Phys. Rev. Lett. **50**, 1998 (1983).
17. R. Haydock, V. Heine, and M.J. Kelly, J. Phys. C Solid St. Phys. **5**, 2845 (1972).
18. J. Inoue, A. Okada, and Y. Ohta, J. Phys.: Condens. Matter **5**, L465 (1995).
19. The ground-sate electronic structure of the nanotubes and the local density of states were computed with the pseudopotential density functional technique (DFT) [52,53] using plane-wave expansions with a cut-off energy of 48 Ry. The convergence with respect to both the plane-wave cutoff and super-cell size has been carefully checked. Norm-conserving pseudopotentials as proposed by Troullier and Martins [54] were used. As usual, the exchange and correlation effects were described within the local density approximation using the Perdew-Zunger parametrization of the Ceperley-Alder data [55].
20. J.C. Charlier, J.P. Michenaud, and Ph. Lambin, Phys. Rev. B **46**, 4540 (1992).
21. D. Tománek and S.G. Louie, Phys. Rev. B **37**, 8327 (1988).
22. P. Kim, T. Odom, J.L. Huang, and C.M. Lieber, Phys. Rev. Lett. **82**, 1225 (1999).
23. J.W. Mintmire, B.I. Dunlap, and C.T. White, Phys. Rev. Lett. **68**, 631 (1992).
24. N. Hamada, S.I. Sawada and A. Oshiyama, Phys. Rev. Lett. **68**, 1579 (1992).
25. K. Tanaka, K. Okahara, M. Okada, and T. Yamabe, Chem. Phys. Lett. **191**, 469 (1992).
26. J.W. Mintmire and C.T. White, Phys. Rev. Lett. **81**, 2506 (1998),.
27. J.C. Charlier and Ph. Lambin, Phys. Rev. B **57**, R15037 (1998).
28. C.L. Kane and E.J. Mele, Phys. Rev. B **59**, R12759 (1999).
29. Ph. Lambin, A. Fonseca, J.P. Vigneron, J.B.Nagy, and A.A. Lucas, Chem. Phys. Lett. **245**, 85 (1995).
30. Ph. Lambin, A.A. Lucas, and J.C. Charlier, J. Phys. Chem. Soliods **58**, 1833 (1997).
31. The electronic structure is easily resolved in shortest pieces of armchair nanotubes. See V. Meunier and Ph. Lambin, Phys. Rev. B (1999, in press).
32. M. Ge and K. Sattler, Appl. Phys. Lett. **65**, 2284 (1994).

33. L.C. Venema, V. Meunier, Ph. Lambin, and C. Dekker, Phys. Rev. B, submitted (July 1999).
34. L.C. Venema, J.W.G. Wildoer, C. Dekker, A.G. Rinzler, and R.E. Smalley, Appl. Phys. A **66**, S153 (1998).
35. N. Lin, J. Ding, S. Yang, and N. Cue, Carbon **34**, 1295 (1996).
36. Ph. Lambin, J.C. Charlier, and J.P. Michenaud, in "Progress in fullerene research", H. Kuzmany, J. Fink, M. Mehring, and S. Roth (Edits.), World Scientific, Singapore, 130 (1994).
37. R. Saito, G. Dresselhaus, and M.S. Dresselhaus, J. Appl. Phys. **73**, 494 (1993).
38. Y.K. Kwon and D. Tománek, Phys. Rev. B 58, R16001 (1998).
39. R. Saito, M. Fujita, G. Dresselhaus, and M.S. Dresselhaus, Phys. Rev. B **46**, 1804 (1992).
40. V. Meunier, Ph D Thesis, University of Namur, Belgium (in preparation).
41. J.C. Charlier and J.P. Michenaud, Phys. Rev. Lett. **70**, 1858 (1993).
42. W. Clauss, D.J. Bergeron, and A.T. Johnson, Phys. Rev. B **58**, R4266 (1998).
43. C.L. Kane and E.J. Mele, Phys. Rev. Lett. **78**, 1932 (1997).
44. B.I. Dunlap, Phys. Rev. B **46**, 1933 (1992).
45. S. Iijima, T. Ichihashi, and Y. Ando, Nature **356**, 776 (1992).
46. M. Terrones, W.K. Hsu, J.P. Hare, H.W. Kroto, H. Terrones, and D.R.M. Walton, Phil. Trans. R. Soc. London A **354**, 2025 (1996).
47. C. Dekker, Physics Today **52**, 22 (May 1999).
48. R. Saito, G. Dresselhaus, and M.S. Dresselhaus, Phys. Rev. B **53**, 2044 (1996).
49. L. Chico, V.H. Crespi, L.X. Benedict, S.G. Louie, and M.L. Cohen, Phys. Rev. Lett. **76**, 971 (1996).
50. A.J. Stone and D.J. Wales, Chem. Phys. Lett. **128**, 501 (1986).
51. H. Terrones and M. Terrones, J. Phys. Chem. Solids **58**, 1789 (1997).
52. M. Buongiorno Nardelli, B.I. Yakobson, and J. Bernholc, Phys. Rev. B **57**, R4277 (1998) ; P. Zhang, P.E. Lammert, and V.H. Crespi, Phys. Rev. Lett. **81**, 5346 (1998).
53. M.L. Cohen, Solid State Commun. **92**, 45 (1994) ; Phys. Scri. 1, 5 (1982) ; J. Ihm, A. Zunger, M.L. Cohen, J. Phys. C **12**, 4409 (1979).
54. W.E. Pickett, Comput. Phys. Rep. **9**, 115 (1989) ; M.C. Payne, M.P. Teter, D.C. Allan, T.A. Arias, J.D. Joannopoulos, Rev. Mod. Phys. **64**, 1045 (1992).
55. N. Troullier, J.L. Martins, Solid State Commun. 74 (1990) 613 ; Phys. Rev. B **43**, 1993 (1991).
56. D.M. Ceperley, B.J. Alder, Phys. Rev. Lett. **45**, 1196 (1980) ; J.P. Perdew, A. Zunger, Phys. Rev. B **23**, 5048 (1981).

APPLICATIONS RESEARCH ON VAPOR-GROWN CARBON FIBERS

G. G. Tibbetts[1], J. C. Finegan[1], J. J. McHugh[1]: J. -M. Ting[2], D. G. Glasgow[2], and M. L. Lake[2]

[1]Materials and Processes Laboratory, General Motors R&D Center, Warren, MI, 48009
[2]Applied Sciences, Inc., Cedarville, OH, 45314

VAPOR-GROWN CARBON FIBERS

The most venerable tubular carbon structure is filamentous carbon, which has been widely researched since the invention of the electron microscope in the early 1950's, when it was found that carbon filaments were widely distributed in combustion and pyrolysis products (1). Several decades of work elucidated the growth mechanism of these filaments, attributing it to the solution of carbon from pyrolyzing gas mixtures in a macroscopic catalytic particle and the further precipitation of the carbon onto a lengthening precipitate. Figure 1 outlines the growth of carbon filaments from catalytic particles and also diagrams how chemical vapor deposition of further carbon layers can increase the filament size from its normal submicron diameter to macroscopic dimensions.

The 1889 patent of Hughes and Chambers [2] describing the growth of "hair-like carbon filaments" which could be produced from a feedstock of hydrogen and methane in a sealed iron crucible seems to be the earliest reference to such macroscopic vapor-grown carbon fibers (VGCF).

In the 1970's and 1980's, researchers such as Koyama and Endo in Japan [3] and Tibbetts and Devour [4] in the US demonstrated good yields of vapor-grown carbon fibers by thermal decomposition of benzene or methane at about 1200°C. These structures were deliberately designed to mimic conventional carbon fibers of about 10 μm in diameter. Procedures for this growth were published and useful research was done on these materials [5], but, ultimately, the expense of producing the fibers limited their application.

The fundamental barrier was that the yield of macroscopic fiber per catalytic particle was destined to remain small as long as the most of the shorter fibers were incorporated in the thick vapor deposited carbon layers that typically covered the substrates.

In the late 1980's Koyama and Endo [6], Hatano, Ohsaki, and Arakawa [7], and others realized that to take advantage of the high yield of the filament formation and utilize the advantages of a continuous process, the CVD layer must be of minimum thickness and the fiber diameter must be kept small. In order to increase the efficiency of the process, nanometer sized catalytic particles were created *in situ*, avoiding any particle agglomeration wastage. Fibers could then move through the reactor with the gas stream and be collected as they exit.

With this general method, fibers with different lengths and diameters may be produced. They may incorporate a layer of CVD carbon of varying thickness and graphitization, if wished. A schematic of apparatus designed to produce such fibers is shown in Figure 2. This concept has been used in the pilot plant of Applied Sciences, Inc, of Cedarville, OH (www.apsci.com). Figure 3 shows two scanning electron micrographs of typical fibers produced in this apparatus. The fibers are about 200 nm in diameter and several micrometers long; they tend to be intertwined and interconnected.

The TEM shown in Figure 4 shows the duplex structure of the filaments, with the interference striations in the inner filament inferring that the inner structure is more highly crystallized than the CVD layer. The selected area diffraction pattern shown in the insert shows a broad (002) spot, implying modest orientation of the graphene planes, but some portions of the fibers show more complete ordering, as evidenced by other reflections. The fibers are very graphitizable; figure 5 exhibits a TEM of a graphitized fiber showing more

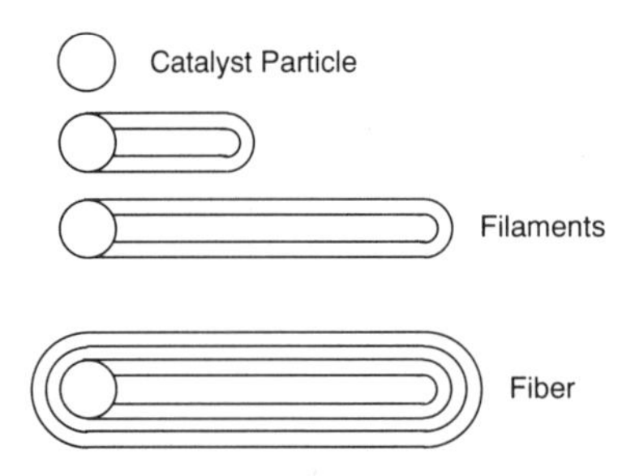

Figure 1. Vapor-grown carbon fibers are produced when a nanometer sized catalytic particle able to adsorb large amounts of carbon from the surrounding atmosphere and precipitate carbon filaments. Filaments thickened by chemical vapor deposition can become microscopic or even macroscopic fibers.

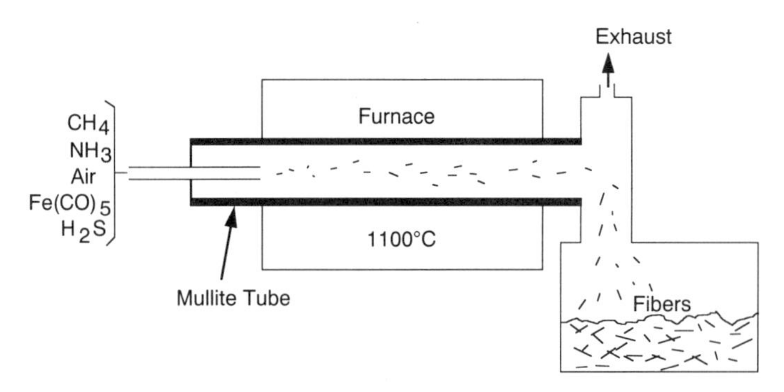

Figure 2. Apparatus for forming VCGF on a continuous basis. The feedstock is methane, with additional components of air to increase furnace temperature by combustion, ammonia to improve fiber quality, iron pentacarbonyl to produce the catalytic iron particles from which the filaments are grown, and hydrogen sulfide to assist filament nucleation

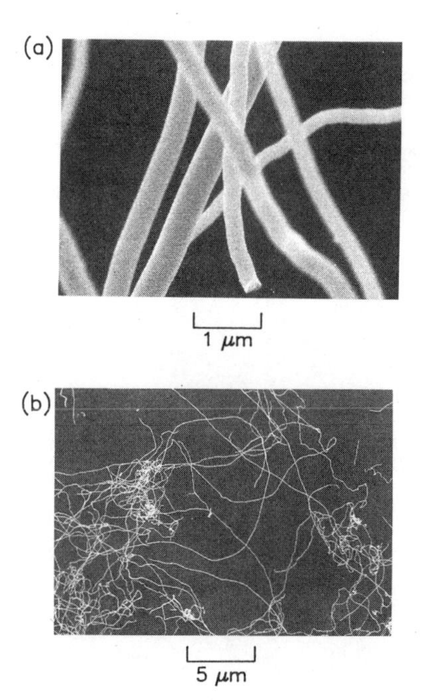

Figure 3. These SEM's show typical fibers produced in the ASI apparatus. Note that they are about 0.2 μm in diameter, many μm long, and entangled and interconnected

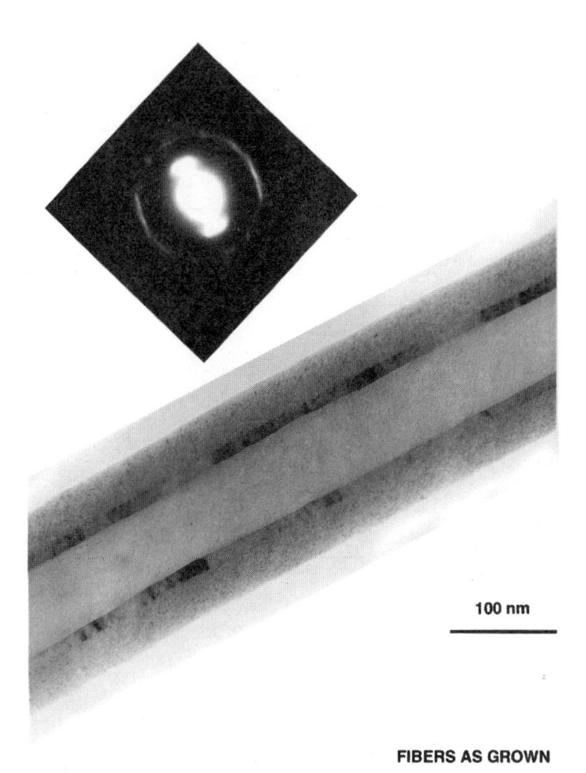

Figure 4. Transmission electron micrograph of a typical vapor-grown fiber showing the hollow core, graphitic filamentous interior, and vapor-deposited coating. The SAD pattern shows broad 002 spots and some higher order diffraction peaks

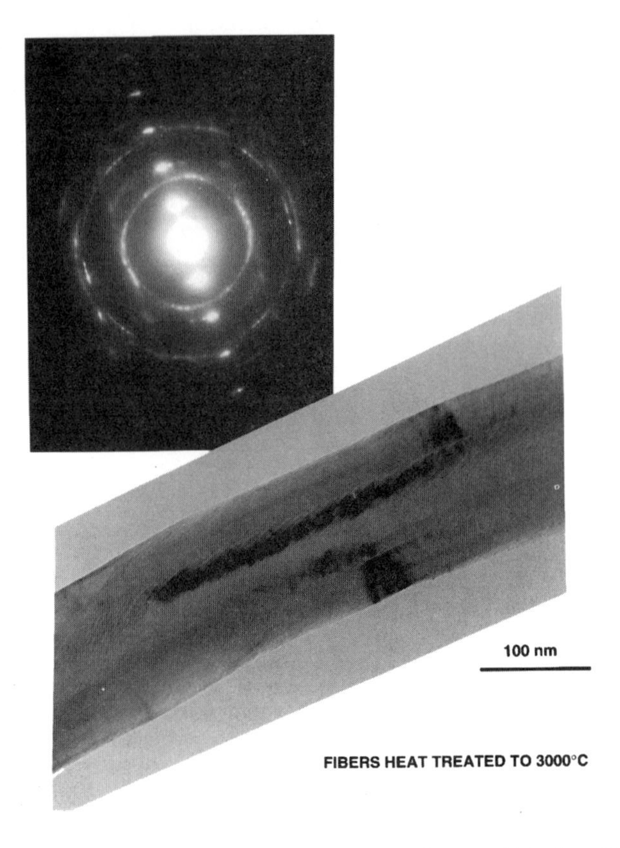

Figure 5. Transmission electron micrograph of a vapor-grown fiber graphitized at 3000°C, showing the hollow core, graphitic filamentous interior whose conical graphene planes are aligned at about 22° from the longitudinal axis, and vapor-deposited coating. The SAD pattern shows the material to be well graphitized

clearly that the filament's graphene planes are aligned about 22°from the longitudinal direction, while the CVD layers are longitudinal. These fibers have a diffraction pattern characteristic of more completely ordered graphite.

INFILTRATION

The first problem we found in fabricating composites of VCGF was that it was very difficult to infiltrate the polymer into the masses of low diameter fibers. Figure 6 shows photographs of Nylon 6,6 /VGCF composites; even though the individual fibers adhere well to the nylon, there are large areas of dry fiber which are not penetrated by the nylon. Why is it that infiltration of polymeric matrix into these fibers is such a problem? An examination of the basic equations of infiltration is instructive. The following discussion will show that the difficulties in forming void-free composites stem from the high viscous dynamic drag impeding the infiltration of polymer between the tiny fibers.

The work of Mortensen *et al* [8] has shown that the pressure P required to infiltrate a porous preform (or a loose clump of fibers) is the sum of the pressure required to overcome capillary effects, P_γ, and the pressure P_V required to overcome viscous drag,

$$P = P_\gamma + P_V \tag{1}$$

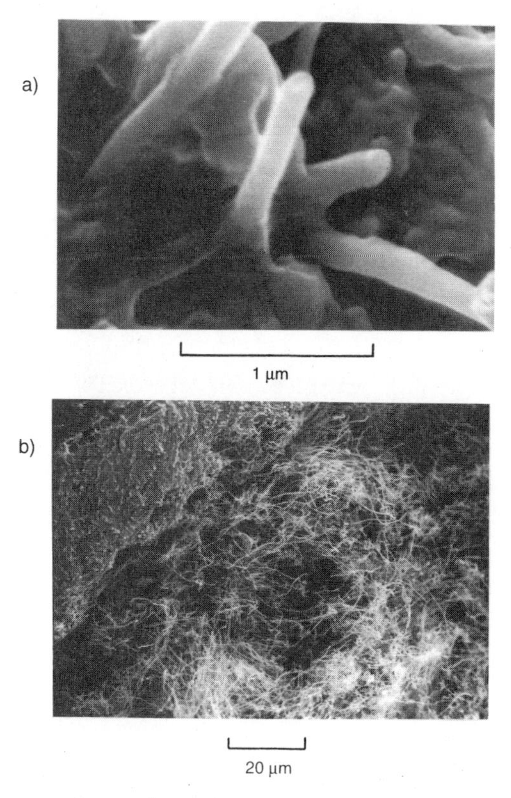

Figure 6. Scanning electron micrographs of the Nylon 6,6 composites showing (a) good adhesion of individual fibers to the polymer, (b) dry areas of fiber not completely infiltrated

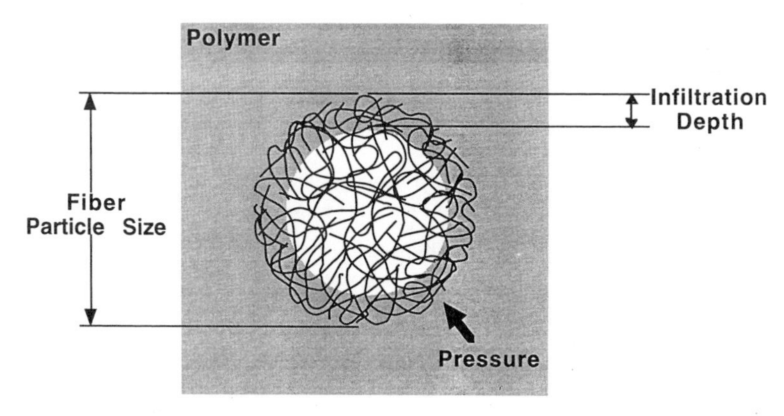

Figure 7. Diagram of a partially infiltrated fiber clump showing incomplete infiltration

The effective capillary pressure is given by

$$P_\gamma = S_f \sigma_{LA} \cos\theta \quad (2)$$

Here S_f is the surface area of the preform's interface per unit volume of composite, σ_{LA} is the polymer's surface energy (0.04 J/m^2 for liquid nylon), and θ is the contact angle. Once this rather small pressure has been supplied, the pressure gradient dP_V/dx generated across the infiltrating preform by polymer squeezed in at velocity v_0 will be given by Darcy's law, which may be written in one dimensional form as

$$\frac{dP_V}{dx} = -\frac{\mu v_0}{K}. \quad (3)$$

Where μ is the viscosity of the liquid polymer (≅52 Pa s for nylon) and K is the permeability of the preform. In the one-dimensional case, this equation may be integrated by using the "similarity solution", which assumes a fixed applied exterior pressure (10^5 Pa=1 atmosphere for a mixing duration of 10 minutes) on a clump of fibers (Figure 7). As the infiltration velocity decreases with time, the infiltration depth d becomes

$$d = \sqrt{\frac{2KP_0\tau}{\mu(1-v)}}. \quad (4)$$

after an infiltration duration τ at applied pressure P_0 and fiber volume fraction V. (We have suppressed the minus sign, which simply indicates that the pressure decreases as the infiltration front moves across the preform.)

For fibrous materials of radius r, the equation

$$K = \frac{2r^2\sqrt{2}}{9V}\left(1-\sqrt{\frac{4V}{\pi}}\right) \quad (5)$$

has been shown to represent the permeability fairly well [8]. The resulting dependence on r^2 means that K for 0.1 μm fibers is only 10^{-4} times the value for conventional 10 μm fibers. This extremely small value of K leads to very slow infiltration rates.

What is to be done about this? Dispersing and mixing the fibers before infiltration, infiltrating at high pressures, using less viscous polymers, and infiltrating for long periods will all help. Our most practical solution has been to decrease the average size of the clumps of fibers, making complete infiltration of the clumps easier. Significantly, equation 4 yields a 1 dimensional infiltration depth of 150 μm for 200 nm in diameter fibers, implying that the maximum diameter clump which may be infiltrated used these conditions is about 300 μm. This is in good agreement with the clump size measured with SEM for the As-grown fibers ball milled for 2 minutes.

For the thinner NT's, this problem will be more difficult than with VGCF, as the infiltration depth decreases with fiber diameter. For example, for SWNT's with a diameter of 1.4 nm, the infiltration depth under the above conditions will be 1 μm!

COMPARATIVE FIBER PROPERTIES

Although the modulus and tensile strength of 10 μm in diameter VGCF have been measured directly [10], the properties of submicron VGCF have not. As a first approximation, and probably a safe underestimate, we assume that modulus E and tensile strength T for both diameters are similar, giving the values listed in Table 1. The PAN fiber and fiberglass values are in the center of the range commonly produced and measured.

The values of tensile strength and modulus for VGCF were sufficiently good to inspire hopes for replacing fiberglass as reinforcement. For many applications, the modulus is the materials limiting property and is more important than the tensile strength; modulus is where VGCF fibers outclass fiberglass. The table also shows the exceptional properties recently measured for single walled nanotubes that would promise abundant applications in composite reinforcement. The 4000 Gpa measurement of the NT stiffness by Treacy *et al* [11] is somewhat indirect, but intriguingly above the value of 1000 Gpa usually measured for single crystal graphite. With such stiffness, one might expect NT's to rapidly replace glass fibers, whose stiffness is near 70 GPa.

Table 1. Comparison of mechanical properties of fibers.

	Fiberglass	Carbon (PAN)	VGCF	NT
E (Gpa)	70 [9]	350 [5]	240 [10]	4000 [11]
T (Gpa)	2.5	4	2.9	
ρ (g/cc)	2.5	1.9	2.1	

In the next section, we will review some elementary composite theory to relate these intrinsic fiber properties to measured and predicted composite properties.

COMPOSITE THEORY

In order to place the tensile strength and modulus measurements in proper perspective, model calculations were performed to predict the expected improvement in mechanical properties with fiber loading. Theoretical tensile strengths were obtained using a model developed by Baxter [12], which assembles three different fracture mechanisms; longitudinal fiber failure, matrix failure, and interfacial failure by using the Tsai-Hill criterion.

$$\left(\frac{\sigma_x}{\sigma_L}\right)^2+\left(\frac{\sigma_y}{\sigma_T}\right)^2-\left(\frac{\sigma_x\sigma_y}{\sigma_L^2}\right)+\left(\frac{\sigma_{xy}}{\tau}\right)^2=1. \qquad (6)$$

This equation relates the tensile stresses in a composite at fracture, with x being in the fiber direction and y transverse to the fiber, to the transverse tensile strength of the composite σ_T, the shear strength of the composite τ, and the rule of mixtures longitudinal tensile strength of the composite σ_L.[13]. In order to determine the latter, we must utilize a parameter l_c which describes how well the fiber couples to the matrix. Consider a fiber imbedded in a matrix which is under tensile stress (Figure 8). Force can be transmitted only through the fiber's surface, so that through any surface element of length dx located a distance x from the fiber's end, a force

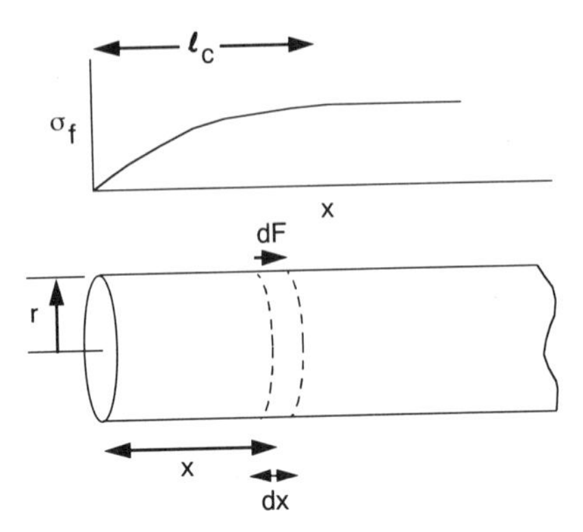

Figure 8. A fiber of radius r incorporated in a strained matrix is itself strained by shearing forces transmitted through the fiber exterior. The fiber strain is plotted above the fiber. It builds up to its maximum in a distance *lc*.

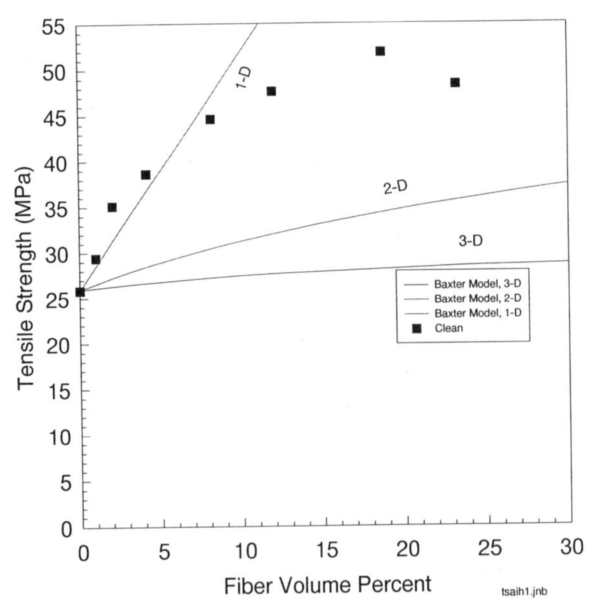

Figure 9. Tensile strength of polypropylene reinforced with various amounts of VGCF; the data are compared with calculations

$$dF = \pi r \tau_m dx \tag{7}$$

proportional to the element's surface area π r dx and the shear strength τ_m of the fiber-matrix interface is the maximum which may be transmitted to the fiber [13]. The maximum force which may be transmitted to the fiber is $F = \sigma_f \pi r^2$, where σ_f is the 2.92 GPa tensile strength of the VGCF [9]. In a simple linear approximation, the distance from the fiber end required to build up this stress is

$$l_c = \frac{\sigma_f r}{\tau_m}. \tag{8}$$

In our case, the preparation methods utilized to break up the fiber clumps shortens the average fiber length to 19 fiber diameters, while equation 8 gives an l_c of 98 diameters. Using these approximations, one may calculate the longitudinal tensile strength of a composite with fibers of average length shorter than the critical length l_c,

$$\sigma_L = V\sigma_f \frac{l}{2l_c} + (1-V)\sigma_m \tag{9}$$

Here σ_m is the tensile strength of the matrix and V is the volume fraction of fibers. To obtain tensile strength σ as a function of angle θ between the fiber direction and the applied stress, equation 6 may be solved to yield

$$\sigma(\theta) = \left(\frac{\cos^4\theta}{\sigma^2{}_L} + \left(\frac{1}{\tau^2} - \frac{1}{\sigma_L^2}\right)\sin^2\theta\cos^2\theta + \frac{\sin^4\theta}{\sigma_T^2} \right)^{-1/2} \tag{10}$$

In this expression, τ is taken as $1/\sqrt{3}$ times the matrix tensile strength (this sets σ equal to the matrix strength at 0 fiber volume). Calculations for 2 or 3 dimensionally random fibers then average $\sigma(\theta)$ over 2 or 3 dimensions.

Figure 9 shows that the models for 3, 2, and even 1-dimensionally random fibers in polypropylene underpredict the experimental results for tensile strength. The theoretical values are small partially because of the small values of l/l_c, and partially because of the angular averaging.

The Cox model [14] was used to calculate the Young's modulus of a composite with randomly oriented fibers. This model estimates stress transfer to fibers of length l, diameter d, modulus E_f, and volume fraction V in a matrix of modulus E_m and Poisson's ratio υ with the factor

$$\beta = \frac{l}{d}\sqrt{\frac{E_m}{(1+v)E_f \bullet \ln(\pi/4V)}}. \tag{11}$$

Combining this with the rule of mixtures yields a composite modulus E_c of

$$E_c = (1-v)E_m + \frac{1}{6} \bullet \left(\left[1 - \frac{\tanh\beta}{\beta}\right]\right)VE_f \tag{12}$$

The prefactor of 1/6 is for fibers randomly oriented in three dimensions, while 1/2 is appropriate for 2 dimensionally random fibers and 1 for oriented fibers.

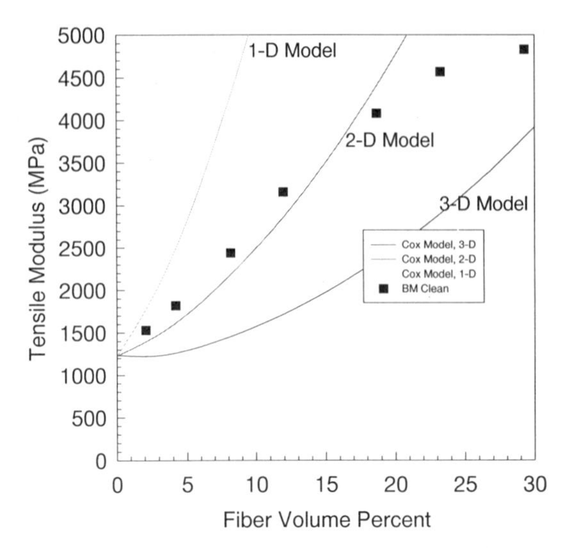

Figure 10. Young's modulus of polypropylene reinforced with various amounts of VGCF; the data are compared with calculations

Figure 10 shows that a Cox model calculation using a fiber modulus of 240 GPa [10], l/d=19, and ν=0.38 [15] grossly underestimates composite stiffness values obtained from experiments. A calculation for two dimensionally ordered fibers is in much better agreement with these data, though at higher volume fractions inadequate infiltration presumably keep the modulus from rising as rapidly as the theory. Therefore, both the tensile strength and the modulus model indicate a higher than expected degree of ordering of the fibers in these composites

SURFACE TREATMENTS

In our efforts to improve the properties of VGCF composites, we have tried numerous surface treatments for the VGCF to improve the mechanical properties. The following fiber designations refer to differing reactor gas mixtures used in production:

- Clean [15]: Relatively low gas space velocity through reactor. Cleanest fiber product with least aromatics.
- PR-18: Somewhat higher space velocity. Used primarily in rubber mixtures.
- PR-5: Still higher space velocity.
- Best shot [15]: Highest gas space velocity of methane-based fibers. Most aromatics on fiber surfaces. This mixture was designed for greatest feasible fiber production rate.

A high quality graphite may be produced by high temperature heat treatment:

- Graphitized fibers: Best Shot fibers were heated to 3000°C for 1 hour and gradually cooled in an inert gas atmosphere.

Four different wet chemical treatments were tested:

- Diamine salt of carboxylic acid, high (DH): Clean fibers were oxidized at 450°C for 60 minutes. 150 g of the oxidized fibers were mixed with a solution of 5 g of Armak 1192 (Akzo Nobel Chemicals) in 10 liters of propanol for 60 minutes. The fibers were dried in air overnight, followed by an oven drying at 38°C. Armak 1192 is a proprietary compound comprising two amine groups positioned near the end of a long hydrocarbon chain, each amine being fully neutralized by a large fatty acid. The 3 hydrocarbon chains are designed to bind well with hydrocarbons, while the amine groups bond well to oxygen.
- Diamine salt of carboxylic acid, low (DL): Prepared similarly to DH, except that only 2 g of Armak 1192 was used.
- Epoxy, high (EH): Clean (PR-1) fibers were oxidized at 450°C for 60 minutes. 150 g of the oxidized fibers were mixed with a solution of 2 g Epon 828 (Shell Chemicals) in 7 liters of propanol for 6 min and then dried overnight at 38°C.
- Epoxy, low (EL): Prepared similarly to EH, except only 0.5 g of the Epon 828 was used.

Finally, some samples of fibers were oxidized after production, using different procedures to increase the surface area, energy, and reactivity:

- Air-Etched fibers are Clean fibers oxidized in air at 450°C for 16 min and forced through a 0.16 cm sieve by a stainless steel wiper blade.
- CO_2 oxidized fibers are Best shot fibers oxidized with CO_2 in a tube furnace at 850°C for 15 minutes. Each of these materials was ball milled for 2 minutes using a Spex 8000 mixer mill as previously described [15].

Injection molding of mini-tensile specimens (ASTM Test Method D638 Type V) using VGCF/polypropylene was performed with a benchtop CSI MiniMAX Molder. The apparatus was equipped with a rotor, which may be submerged in a 12.7 mm diameter heated cup. Mixing was imparted by the rotary as well as by the vertical motion of the rotor, and extreme care was used to employ a similar mixing procedure for all samples. Samples with high fiber fractions required as much as 10 minutes mixing because the bulky fibers had to be gradually added to the melt in order not to overflow the cup. Samples with a low fiber volume fraction were subjected to the same thermal regimen. Further details on the injection procedure were presented in [15].

The specimens were mounted in the grips of an MTI tensile testing machine and stretched at 1 mm/min until failure occurred. The modulus was determined from the slope of the initial section of the stress-strain curve, while the tensile strength was determined from the ultimate load before separation of the two sections of the 2.54 cm dogbone.

Figure 11 shows the tensile strength and stiffness for 15 volume % PYROGRAF/polypropylene composites using reinforcing fibers produced by several methods and having the various surface treatments described in the previous section. These data show that it is possible to triple both the modulus and strength of the polypropylene resin (open circle) by adding only 15 volume % PYROGRAF. It also underscores the fact that some of the fiber production methods and surface treatments produce much better composites then others.

Let us first compare the reinforcing properties of the as-grown fibers produced by several different procedures. Figure 12 helps to rationalize these observations. It shows the tensile strength versus the graphitization index. Graphitization index g is a number

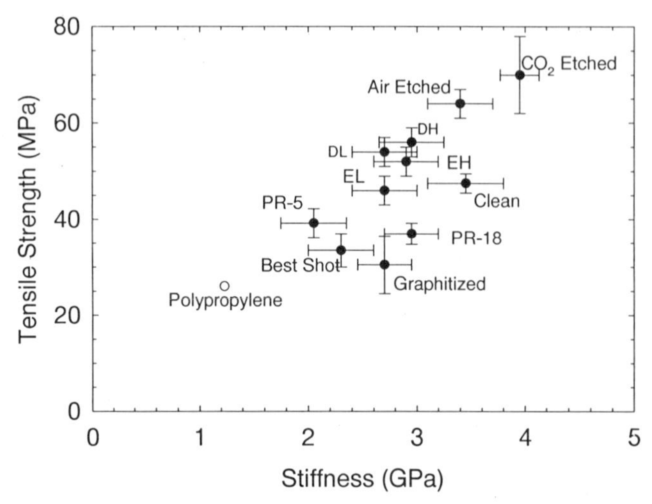

Figure 11. Tensile strength versus modulus for 15 vol% composites using different types of VGCF in polypropylene. The open circle shows the properties of polypropylene

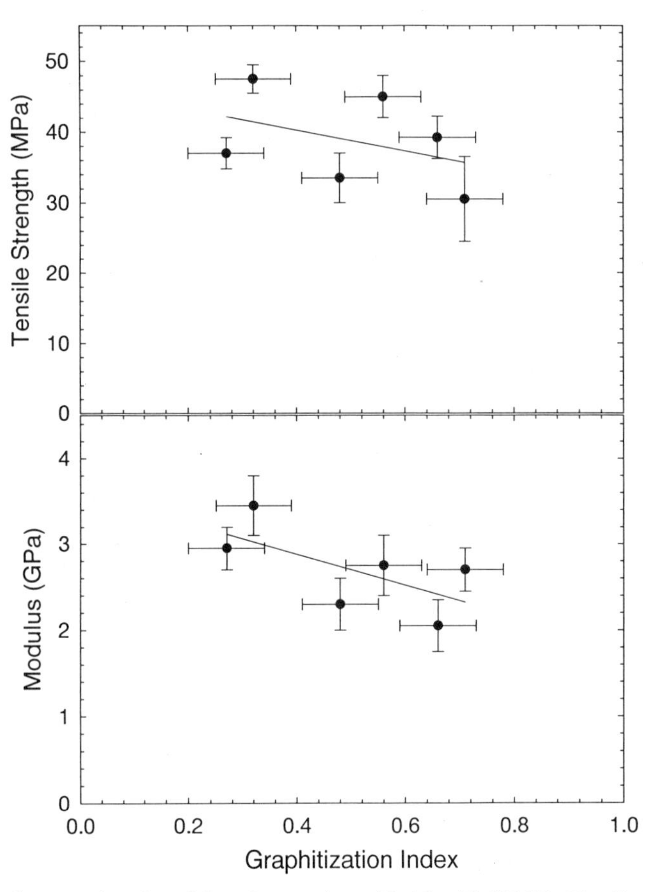

Figure 12. Tensile strength and modulus of composites with 15 vol% VGCF. The fibers were fabricated with several different graphitization indexes

calculated from the d_{002} lattice spacing (nm), generally determined from x-ray diffraction.

$$g = \frac{0.344 - d_{002}}{0.344 - 0.3354} \tag{13}$$

This number varies from 0 for completely disordered carbons to 1 for single crystal graphite. The plot shows that more graphitic fibers of higher graphitization index tend to make composites having both lower tensile strength and modulus. The data in Figure 12 exhibit large scatter because of the error in measuring graphitization in these low density fibers due to the difficulty in precisely defining the sample plane. Overall, fibers grown under conditions of lower feedstock residence time are more graphitic and adhere more poorly to polypropylene, an observation consistent with our earlier results.

Scanning electron micrographs show a significant difference in the fiber-matrix adhesion for the Air-Etched and Graphitized/polypropylene composites. Figures 13 and 14 show that Graphitized fibers project much further from the fracture surface than Air-Etched fibers do, implying that the interfacial shear strength of the Graphitized fibers is much less than that of the Air-Etched. It is evidently much easier for the graphitized fibers to pull out of the matrix as the composite fractures. This result is consistent with the higher strength of the Air-Etched/polypropylene versus the Graphitized/polypropylene composite. Note that the wetting angle of the polymer at the fiber interface appears approximately similar for the two cases, implying that the low interfacial strength is not merely a question of surface wetting. We can conclude from the micrographs presented Figures 13 and 14 that a higher graphitization index will decrease the fiber-matrix interphase adhesion and decrease composite strength and stiffness.

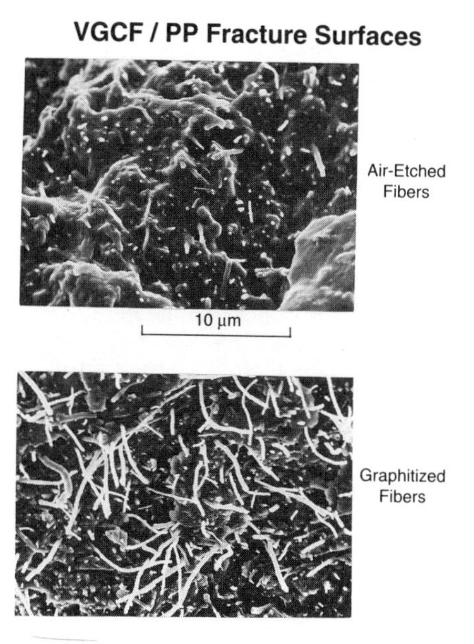

Figure 13. SEM's of the fracture surfaces of polypropylene composites fabricated from Air-Etched and Graphitized fibers

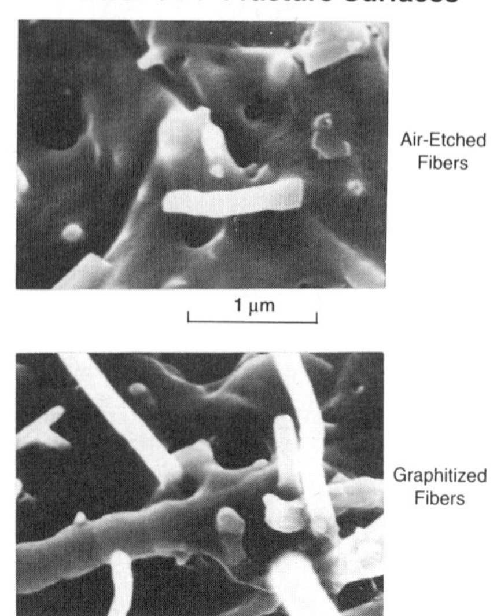

Figure 14. SEM's made at higher magnification of the fracture surfaces of polypropylene composites fabricated from Air-Etched and Graphitized fibers

Figure 11 also gives mixed encouragement to the notion that composite tensile properties may be somewhat improved by applying properly reactive chemical coatings to the raw fibers. Both the Epon 828 and the Armak 1192 seem capable of increasing the tensile strength of the composites in which they were used; moreover, high concentrations of both were more effective than low. However, all four treatments slightly decreased the measured composite modulus.

Perhaps the most salient feature of the data of Figure 11 is the higher strength (70 MPa) and stiffness (4 GPa) observed for the fibers which were CO_2 oxidized at 850°C compared to the graphitized PYROGRAF (strength 30 MPa and stiffness 2.6 Gpa). The air etched and sieved PYROGRAF (65 Mpa strength and 3.4 Gpa stiffness) does not have quite as good properties as the CO_2 etched, but is still far superior to the graphitized material.

In summary, the data obtained for the tensile strength and modulus of composites made from various varieties of fibers with various surface treatments on them may be rationalized by a few principles:

- More graphitic fibers adhere poorly to the polypropylene matrix compared to less graphitic fibers.
- Fiber matrix adhesion may be improved by moderately oxidizing the fibers either in air or CO_2 . This oxidation seems to become more effective as it increases the product of the external surface area and the surface energy of the fibers; however, excessive etching can be destructive. Chemically active coatings can modestly increase the tensile strength, while perhaps somewhat decreasing the modulus.

PROPERTIES IMPROVEMENTS

The table below compares the observed modulus improvements for glass fibers and VGCF, and it highlights several problems. Despite the substantially higher modulus of VGCF, the composite modulus improvement factor of 4 is not that much greater than the

2.8 value for chopped glass. That is because the glass fibers are so large compared to the composite thickness that they assume a 2 dimensional orientation rather than 3 dimensionally random orientation for the VGCF. Moreover, substantial improvement may be obtained with continuous glass fibers by utilizing a woven preform to boost the modulus improvement to 10.4!

Table 2. Modulus improvement factors for several types of composites.

	VGCF (random)	Glass (chopped)	Glass (weave)	NT (random)
Vol%,[Reference]	15 [16]	23 [17]	47 [17]	15
E/E_m	4	2.8	10.4	4.2
T/T_m	2.5	2.6	10.4	

The final column shows that if the modulus of NT's are assumed to be 4000 GPa, the theoretical Cox model composite improvement factor at 15 volume % will be only 4.2 (assuming 2-d orientation and l/d=19). So the stunningly large modulus which we expect for NT's may not be directly translated into spectacular composite properties unless their lengths can be perserved.

ELECTRICAL CONDUCTIVITY

Electrically conducting polymer-based composites are required in many applications, yet are difficult and expensive to make. Hitherto, in fabricating a component, the

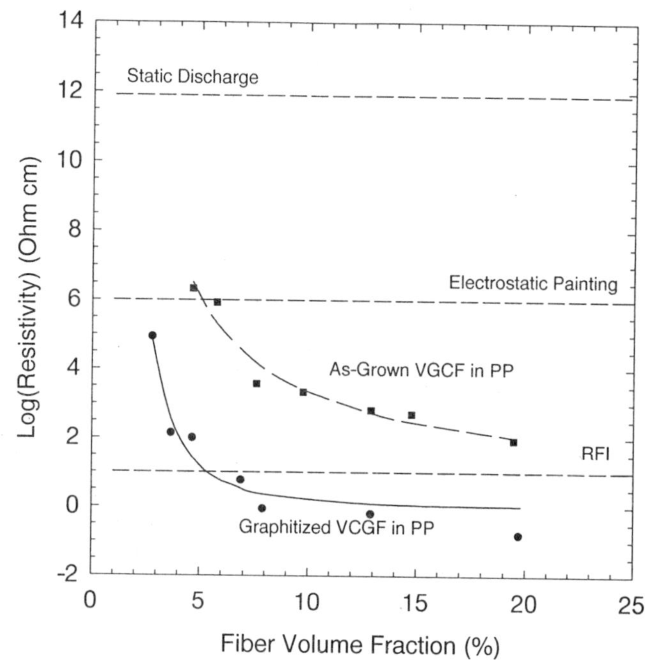

Figure15. Approximate resistivity upper bound required for several applications, compared with the electrical resistivity values for graphitized VGCF and as-grown VGCF composites in a polypropylene matrix

compounder had had a choice of adding, in order of decreasing cost: Hyperion Vapor-Grown Carbon Fibers [18], conventional carbon fibers [19], metallic particles [20], or carbon black [21].

Figure 15 shows the approximate resistivities required for the more common applications for which electrically conducting plastics are sought: static dissipation, electrostatic painting with no primer coat, and radio frequency interference (RFI) shielding [18]. The exact values required depend, of course, on the specific geometry of the pieces and the details of the application.

Figure 15 shows the resistivity Vs fiber volume fraction for graphitized VGCF and Clean (as-grown) VGCF/Polypropylene composites. Increasing the fiber volume fraction above 3% results in rapidly decreasing volume resistivity for the graphitized VGCF/polypropylene composites. The resistivity decreases from 7.9×10^{4} Ω cm at 3% fiber volume fraction to 1.5 Ω cm at 8% fiber volume fraction, far below that required for conducting primers and adhesion promoters. Moreover, the graphitized VGCF/polypropylene composites have a significantly lower resistivity than the clean VGCF/polypropylene composites (87 Ω cm) at 20% fiber volume fraction. The resistivity data points ρ are fitted to an expression derived from the percolation theory, [22]:

$$\rho = \rho_0 (V - V_c)^{-t} \tag{14}$$

where ρ_0 is a constant scaling factor, V is the fiber volume fraction, V_c is the critical volume fraction at percolation threshold, and t is the critical exponent at the percolation threshold.

Figure 15 shows that relatively small quantities of graphitized vapor-grown carbon fibers (VGCF) from Applied Sciences, Inc. [23] can allow the fabrication of extremely low resistivity composites of the thermoplastics polypropylene and nylon. The small fiber diameter of these VGCF makes these encouraging results possible.

ACKNOWLEDGMENTS

The authors would like to acknowledge Michael Balogh's skillful TEM work, Curt Wong's helpful SEM photographs, and the NIST ATP, who supported this work under Cooperative Agreement Award No. 70NANB5H1173, Vapor-Grown Carbon Fiber Composites for Automotive Applications.

REFERENCES

1. W. R. Davis, R. J. Slawson, and G. R. Rigby, Nature 171:756 (1953).
2. T. V. Hughes, and C. R. Chambers, US Patent No 405,480 (1889)
3. T. Koyama, Carbon 10:757 (1970)
4. G. G. Tibbetts, M. G. Devour, and E. J. Rodda, Carbon 25:367 (1987)
5. M. S. Dresselhaus, G. Dresselhaus, K. Sugihara, I. L. Spain, and H. A. Goldberg, *Graphite Fibers and Filaments*, Springer, Berlin (1988)
6. T. Koyama and M. Endo, Japanese patent number 1982-58,966 .
7. M. Hatano, T. Ohsaki, and K. Arakawa, Advancing Technology in Materials and Processes, National SAMPE Symposium 30:1467 (1985)
8. A. Mortensen, L. J. Masur, J. A. Cornie, and M. C. Flemings, Metall. Trans. A, 20A:2535 (1989)
9. C. Zweben, H. T. Hahn,T.-W. Chou, *Mechanical Properties of Composite Materials*, Technomic Pub., Lancaster, PA (1989)

10. G. G. Tibbetts, C. P. Beetz Jr., J. Phys. D:Appl. Ph. 20:292 (1987)
11. M. M. J. Treacy, T.W. Ebbesen, and J. M. Gibson, Nature,381:678 (1996)
12. W. E. Baxter, Metall. Trans. A, 23A:3045 (1992)
13. N. G. McCrum, C. P. Buckley, and C. B. Bucknall. Principles of Polymer Engineering, Oxford, NY, p. 250 (1988)
14. Cox, Brit. J. Appl. Phys. 3:72 (1982)
15. G. G. Tibbetts and J. J. McHugh, Jour Mat Res Soc. Accepted for publication (1999)
16. G. G. Tibbetts , J. C. Finegan, D. G. Glasgow, J.-M. Ting, and M. L. Lake, Extended Abstracts, 21st Biennial Carbon Conference, Charleston, July (1999)
17. Automotive Composites Consortium, Volume 1: Tests Conducted between 2/90 and 3/96, Materials Work Group, TR M96-01, August (1996)
18. B. Miller, Plastics World, September, p 73 (1996)
19. A. Dani and A.A. Ogale, Composites Science and Technology 56: 911 (1996)
20. T. Ota, M. Fukushima, Y. Ishigure, H. Unuma, T. Takashi, Y. Hikishi, and H. Suzuki, Jour. Mater. Sci Lett. 16:1182 (1997)
21. D. M. Bigg, Jour. Rheology 28:501 (1984)
22. P. Sheng, E. K.. Sichel, and J. L. Gittleman, Phys. Rev. Lett. 40:1197 (1978)
23. G. G. Tibbetts, C. A. Bernardo, D. W. Gorkiewicz, and R. L. Alig, Carbon 32:569 (1994)

THE GROWTH OF CARBON AND BORON NITRIDE NANOTUBES : A QUANTUM MOLECULAR DYNAMICS STUDY

Jean-Christophe Charlier[1], Xavier Blase[2], Alessandro De Vita[3,4], and Roberto Car[4,5]

[1] Unité de Physico-Chimie et de Physique des Matériaux,
Université Catholique de Louvain, Place Croix du Sud 1,
B-1348 Louvain-la-Neuve, Belgium.
[2] Département de Physique des Matériaux, U.M.R. n° 5586,
Université Claude Bernard, 43 bd. du 11 Novembre 1918,
F-69622 Villeurbanne Cedex, France.
[3] Istituto Nazionale di Fisica della Materia (INFM) and
Department of Material Engineering and Applied Chemistry,
University of Trieste, via Valerio 2, I-34149 Trieste, Italy.
[4] Institut Romand de Recherche Numérique en Physique des Matériaux,
PPH-Ecublens, CH-1015 Lausanne, Switzerland.
[5] Department of Chemistry and Princeton Materials Institute,
Princeton University, Princeton, New Jersey 08544, USA.

INTRODUCTION

A few years after their discovery in 1991 [1,2], carbon nanotubes are attracting much interest for their potential applications in highly performing nanoscale materials [3] and electronicdevices[4,5]. Synthesis techniques for carbon nanotubes have recently achieved high production yields as well as good control of the tubes multiplicity, shape and size [6]. Carbon nanotubes typically grow in an arc discharge at a temperature of ~3000 K. However, the mechanisms of nanotube formation and growth under such extreme conditions remain unclear [6]. The earliest models for growth of multi-walled nanotubes [7,8] were based on topological considerations and emphasised the role of pentagon and heptagon rings to curve inside or outside the straight hexagonal tubular network. The most debated issue, in later works, was whether these nanotubes are open- or close-ended during growth. In favour of the closed-end mechanism, it was proposed that tubes grow by addition of atoms onto the reactive pentagons present at the tip of the closed structure [9,10]. However, recent experimental studies suggest an open-end growth mechanism [11,12]. In this model the atoms located at the open end of the graphitic structure provide active sites for the capture of carbon ions, atoms or dimers from the plasma phase. Since any capped configuration is more stable than the open-end geometry, it was proposed that the latter could be stabilized by the high electric field present at the tip [13]. However, recent *ab initio* calculations [14,15] show that realistic electric fields cannot stabilize the growth of open-ended nanotubes. Moreover, there is controversy on whether in multi-

Science and Application of Nanotubes, edited by Tománek and Enbody
Kluwer Academic / Plenum Publishers, New York, 2000

walled nanotubes the inner or the outer tubes grow first [8,16] or if different tubes may grow together [17,18]. Finally, the growth of single-walled nanotubes requires the presence of metal catalysts, contrarily to the multi-walled case[12,19,20]. To date no single model seems to be capable of explaining all the experimental evidence.

1. MICROSCOPIC GROWTH MECHANISMS FOR CARBON NANOTUBES

In this work, we study the microscopic mechanisms underlying the growth of carbon (present section) and boron-nitride (next section) nanotubes by performing first-principles molecular dynamics simulations of single- and double-walled nanotubes [21,22]. In this approach [23], the forces acting on the atoms are derived from the instantaneous electronic ground state, which is accurately described within density functional theory in the local density approximation. In such a framework, the instantaneous electronic ground state is given within a formulation in which only valence electrons are explicitly taken into account. The interaction between valence electrons and nuclei plus frozen core electrons is described using norm-conserving pseudopotential for carbon, boron and nitrogen [24] and for hydrogen [25]. Periodic boundary conditions are adopted, using a supercell size for which the distance between repeated images is larger than 5 Å. This is sufficient to make negligible the interaction between images. The electronic wavefunctions at the Γ-point of the supercell Brillouin zone are expanded into plane waves with a kinetic energy cutoff of 40 Ry, which gives well converged ground-state properties of carbon systems [26].

In the present section, we perform calculations on finite tubular C nanotubes terminated on one side by an open end and on the other side by hydrogen atoms which passivate the dangling bonds. We consider two 120 carbon atoms systems representing a (10,0) « zigzag » (0.8 nm of diameter) (Fig.1.a-b) and a (5,5) « armchair » (0.7 nm of diameter) (Fig.2.a-b) single-walled nanotubes, and a 336 carbon atoms system representing a (10,0)@(18,0) « zigzag » (0.8 nm and 1.4 nm of diameter) double-walled nanotube (Fig.3.a-b) [27]. *Ab initio* calculations with such large systems and for a total simulation time of ~30 ps are made possible by massively parallel computing [28]. All nanotubes are initially relaxed to their open-end equilibrium geometries before being gradually heated up to 3500 K by performing constant temperature simulations [29].

1.1 SELF-CLOSURE OF ZIGZAG AND ARMCHAIR CARBON SWNTs

The choice of the (10,0) single-walled system is consistent with a dominant peak observed in the diameter frequency histogram of synthesised monolayer nanotubes [19]. In this case, the first observed structural rearrangement occurs at 300 K: it is a reconstruction of the tip edge characterised by the formation of triangles. This eliminates most of the dangling bonds and induces an initial inside-bending of the nanotube edge. At ~1500 K, the first pentagon is created from one of the top hexagonal rings, and leads to more substantial inside-bending. The extra carbon atom, dangling over the tube edge, moves onto the nearest trimer to form a square. The formation of two more pentagons with the same mechanism is observed at temperatures lower than 2500 K.

Finally at temperatures of ~3000 K, a global reconstruction of the tip develops, with the nanotube edge completely closing into a structure with no residual dangling bonds (Fig.1.c). This gives an energy gain of ~18 eV with respect to the initial open-end geometry. The closed structure obtained by molecular dynamics is still 4.6 eV higher in energy than the « ideal » C_{60}-hemisphere cap, containing only hexagons and isolated pentagons (as proposed in Fig.1.d). However, the timescale of our simulations prevents us from studying the further evolution of the tip geometry.

A similar calculation for a (5,5) armchair single-walled nanotube [27] also leads to tip closure (Fig.2). In this case the open nanotube edge consists of dimers (Fig.2.a-b), rather than single atoms as in Fig.1.a-b, and its chemical reactivity is lower. This is presumably the reason why we observe the first atomic rearrangement only at ~3000 K, with the formation of a pentagon plus a dangling atom (Fig.2.c). Due to the armchair symmetry no squares are formed, at variance with the zigzag case. On the other hand, a second pentagon is often created by the connection of the extra dangling carbon with a neighboring carbon dimer (see Fig.2.c). Successive processes of this kind lead to the self-closure of the

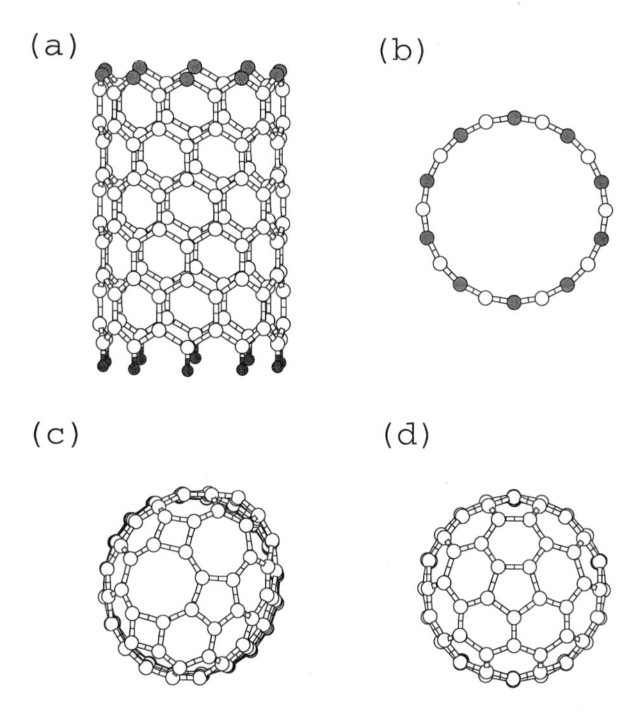

FIGURE 1 :Single-walled nanotube clusters. The tubes belong to the zigzag geometry and have a diameter of 0.8 *nm*. Each cluster contains 120 carbon atoms (*large white spheres* when the atomic coordination is 3: sp^2 bonding) and 10 hydrogen atoms (*small dark grey spheres*). Atoms which coordination is below 3 (sp^1 bonding and dangling atoms) or higher than 3 (sp^3 bonding) are represented with *large light grey* and *black spheres* respectively. The hydrogen atoms and the attached ring of carbons are kept fixed during the simulations. (a) Side view of the open-ended starting configuration at 0K, where 10 dangling bonds (*light grey spheres*) are present on the top edge of the zigzag geometry. (b) Top view of the same system illustrating the cylindrical geometry. (c) Top view of the close-ended configuration, after a thermalization period of 5 ps at 3500 K. The final cap topology is composed of 7 hexagons, 5 pentagons, 1 heptagon, 1 octagon, and 2 squares, verifying Euler's theorem. (d) Top view of a perfect capping topology (hemi-fullerenes) where only isolated pentagons and hexagons are used to close the structure.

nanotube (Fig.2.d) into a hemi- C_{60} (energy gain ~15 eV). We expect that similar self-closing processes should also occur for other nanotubes having a similar diameter, as it is the case for most single walled nanotubes synthesised so far [19]. We notice that the reactivity of closed nanotube tips is considerably reduced compared to that of open end nanotubes. It is therefore unlikely that single-walled nanotubes could grow by sustained incorporation of C atoms on the closed tip. This is in agreement with the finding that C atoms are not incorporated into C_{60} [30].

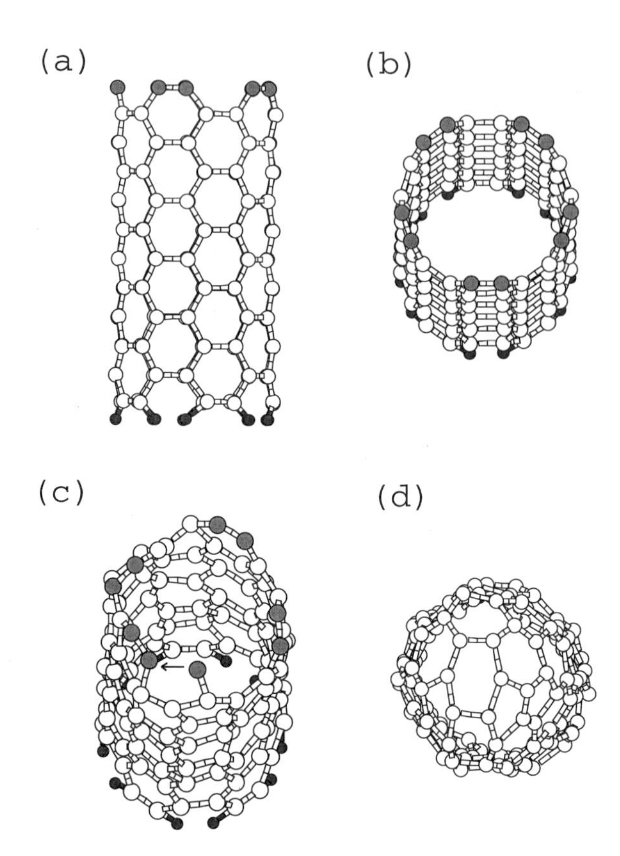

FIGURE 2 :The carbon nanotubes belong to the armchair geometry and have a diameter of 0.7 *n*m. Each cluster contains 120 carbon atoms and 10 hydrogen atoms. The black and white scale is defined in Fig.1. The hydrogen atoms and the attached ring of carbons are kept fixed during the simulations. (a) Side view of the open-ended starting configuration at 0K, where 10 dangling bonds coupled in dimers (*light grey spheres*) are present on the top edge of the armchair geometry. (b) Top view of the same system. (c) At 3000K, the first pentagon is created from one of the top hexagonal rings. Nine dangling bonds remain among which one extra dangling C atom. The latter is going to move aside, forming a second pentagon as indicated by the arrow. (d) Top view snapshot of the close-ended configuration at 3500 K. The final cap topology is composed of pentagons and hexagons only. Some pentagons are sharing the same edge, and are not isolated like in the topology described in Fig.1.d.

1.2 LIP-LIP INTERACTION IN CARBON MWNTs

In another set of calculations, we consider a (10,0)@(18,0) zigzag bilayer system as a prototypical example of multi-walled nanotubes. It consists of the (10,0) nanotube previously studied plus a concentric (18,0) tube (Fig.3.a-b). The interlayer distance is 3.3 Å, in good agreement with typical experimental values [2]. In order to study the effect of the second shell on edge stability we carry out simulations at temperatures ranging from 0 K to 3000 K. At 300 K, the topmost atoms of the inner and outer tube edges rapidly move towards each other forming several bonds to bridge the gap between adjacent edges. At the end, only two residual dangling bonds are left (see Fig.3.c). The resulting picture is one in which the open-ended bilayer structure is stabilized by « lip-lip » interactions as first proposed in [31], which inhibit the spontaneous dome closure of the inner tube observed in the previous simulations. Notice that the non-perfect reconstruction that we observe is favored by the incommensurability of the two graphene networks.

The lip-lip interactions stabilize the open-ended bilayer structure also at ~3000 K, that is a typical experimental growth temperature. However, the most striking feature of the simulation is that, at this of the edge structure makes it highly chemically reactive. During the short time span of this simulation (1.5 ps), we also observe isolated exchanges between atoms at the tube edge, suggesting low diffusional barriers. We report in Fig.3.d a snapshot of a typical configuration at ~3000 K.

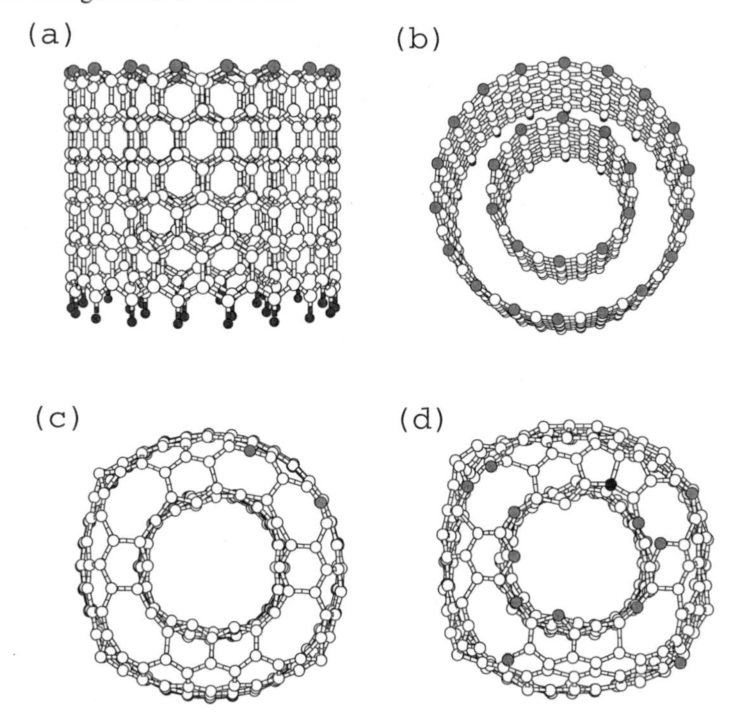

FIGURE 3 : Double-walled nanotube clusters. The tubes belong to the zigzag geometry and have diameters of 0.8 *n*m (inner shell) and 1.4 *n*m (outer shell), respectively. Each cluster contains 336 carbon atoms and 28 hydrogen atoms. The black and white scale is defined in Fig.1. (a) Side view of the open-ended starting bilayer system at 0K, where 28 dangling bonds are present on the two top edges. (b) Top view of the same carbon system. (c) Top view of « lip-lip » interactions between the two concentric shells, reducing the number of dangling bonds by the creation of 10 inter-shell and 3 intra-shell covalent bonds (triangles) at ~300 K. (d) Top view snapshot of the fluctuating edge of the bilayer system at ~3000 K, illustrating the high chemical reactivity of this region. This edge exhibits 12 dangling bonds, 2 pentagonal rings positioned side-by-side, and 1 four-fold coordinated atoms (*black sphere*), presenting the metastability of the nanotube top edge.

1.3 GROWTH MECHANISM FOR CARBON MWNTs

Since the residual dangling bonds and the continuous bond-breaking processes should provide active sites for absorption, we investigate the impact of carbon fragments on the fluctuating edge at ~3000 K. In one case, we consider two isolated C atoms approaching the edge with thermal velocity. In order to sample different impact scenarios, one of the atoms is projected towards a dangling bond site, while the other is projected towards the middle of a well formed hexagonal ring. We find that both atoms are incorporated in the fluctuating network of the nanotube edge within less than 1 ps. The above results suggest a possible growth mechanism in which incorporation of carbon atoms from the vapor phase is mediated by the fluctuating lip-lip bonds of the tube edge. This is reminiscent of the vapor-liquid-solid (VLS) model introduced in the sixties to explain the growth of silicon whiskers [32]. In this model, growth occurs by precipitation from a super-saturated liquid-metal-alloy droplet located at the top of the whisker, into which silicon atoms are preferentially absorbed from the vapor phase. The similarity between the growth of carbon nanotubes and the VLS model has also been pointed out by Saito *et al.* on the basis of their experimental findings for multi-walled nanotube growth in a purely carbon environment [17]. Solid carbon sublimates before it melts at ambient pressure, and therefore these authors suggest that some other disordered carbon form with high fluidity, possibly induced by ion irradiation, should replace the liquid droplet. In the microscopic model provided by our simulation, it is not so much the fluid nature of the edge but the fluctuating character of the lip-lip bonds that makes possible a rapid incorporation of carbon fragments.

2. FRUSTRATION EFFECTS AND GROWTH MECHANISMS FOR BORON-NITRIDE NANOTUBES

While carbon nanotubes [1] have attracted a lot of attention in the last few years, boron-nitride (BN) nanotubes [33-35] have been studied much less extensively. BN nanotubes are insulating with a ~5.5 eV band gap [36]. More generally, it has been shown that controlling the (x,y,z) stoichiometry of composite $B_xC_yN_z$ nanotubes can be used as a mean to tailor their electronic properties [37]. In addition, the lack of chemical reactivity displayed by BN nanotubes leads to the idea that BN nanotubes could be used as « protecting cages » or « molds » for any material encapsulated within [38]. These potential applications provide a strong motivation for pursuing the experimental and theoretical effort of better understanding BN tubular systems.

Recent experimental studies [34,35] have shown that the morphology of BN nanotubes differ significantly from the one of their carbon analogs. BN nanotubes tend to have amorphous-like tips or closed-flat caps, which are very rarely observed in carbon systems. In order to understand the origin of these differences, we have studied the growth mechanism of BN nanotubes by means of first-principles molecular dynamics simulations which was shown in the previous section to elucidate the microscopic processes of carbon nanotube growth. We find that the high energy cost of « frustrated » B-B and N-N bonds strongly affects the modalities of growth. In particular, pentagons and other odd-member rings are not stable at the growing tube edge at experimental temperatures. A metastable « open » tip structure with even-member rings only and no frustrated bonds is created when this is compatible with the tube network helicity (armchair nanotubes). The calculations indicate that single-wall tubes with this structure may grow uncatalyzed by chemisorption from the vapor phase. On the contrary, if the network helicity imposes the presence of frustrated or dangling bonds (zigzag nanotubes), these bonds are unstable, breaking and forming during the simulations. This leads to an amorphous tip structure, preventing growth. The results provide evidence in support of the topological models proposed in [34,35] for final BN tip geometries.

2.1 STABILITY OF BN NANOTUBES

Before studying the growth mechanism of BN nanotubes, and compare to the case of their carbon analogs, we first study the stability and geometry of infinite BN nanotubes on the basis of the present *ab-initio* approach. By minimizing both stress and Hellman-Feynman forces, we determine first the equilibrium geometry for *(n,0)* and *(n,n)* tubes with diameters ranging from 4 to 12 Å (index notations for the tubes refer to the convention of [27] as defined for graphitic nanotubes). The main relaxation effect is a buckling of the boron-nitrogen bond, together with a small contraction of the bond length (~1%). In the minimum energy structure, all the boron atoms are arranged in one cylinder and all the nitrogen atoms in a larger concentric one. Due to charge transfer from boron to nitrogen, the buckled tubular structure forms a dipolar shell. The distance between the inner « B-cylinder » and the outer « N-cylinder » is, at constant radius, mostly independent of tube helicity and decreases from 0.2 a.u. for the (4,4) tube to 0.1 a.u. for the (8,8) tube. As a result of this buckling, each boron atom is basically located on the plane formed by its three neighboring nitrogen atoms so that the sp^2 environment for the boron atom in the planar hexagonal structure is restored (at most, the N-B-N angles differ from 120° by 0.2% for the smallest tube). This tendency for three-fold coordinated column III atoms to seek 120° bond angle is extremely strong. For example, it explains the atomic relaxation of the (110) and (111)-2x2 surfaces of GaAs and other III-V compounds. On the other hand, the N-B-N angles approach the value of the bond angle of the s^2p^3 geometry. Buckling and bond length reduction induce a contraction of the tube along its axial direction by a maximum of 2% for the smallest tube studied.

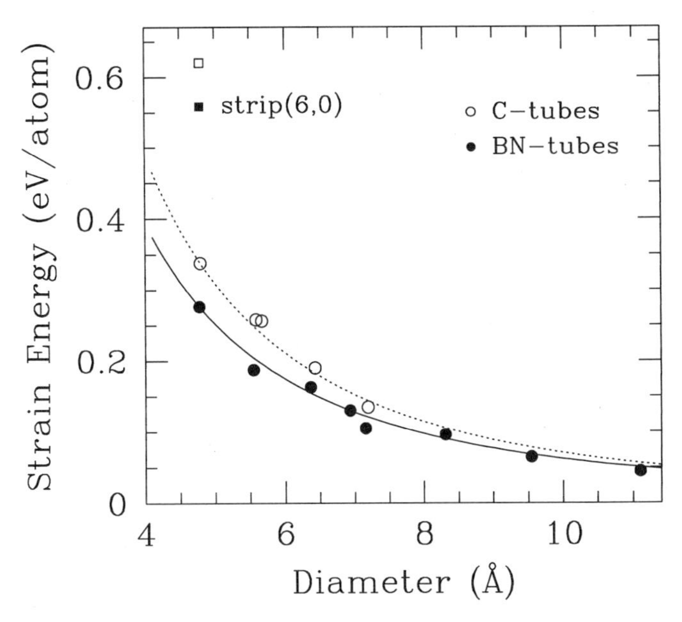

FIGURE 4 :Total energy of nanotubes in eV/atom as a function of tube diameter (in Å). The *black circles* represent the BN nanotube energies above the energy of an isolated hexagonal BN sheet. The *opened circles* represent the graphite nanotubes energies above the energy per atom of an isolated graphite sheet. The *solid* and *dashed* lines are guides to the eye. The energy of the strips corresponding to the (6,0) BN and carbon tubes are given respectively by the *filled* and *empty square*.

Energies per atom for the relaxed tubes are plotted in Fig.4. as a function of the tube diameter. The zero of energy is taken to be the energy per atom of an isolated hexagonal BN sheet. On the same graph, the energy per carbon atom above the graphite sheet energy is represented. As for graphitic tubes, BN tube energies follow the classical $1/R^2$ strain law, where R is the average radius of each tube. However, for the same radius, the calculated strain energy of BN nanotubes is smaller than the strain energy of graphitic tubes. This is related mostly to the buckling effect which reduces significantly the occupied band energy in the case of the BN compounds. Therefore, it is energetically more favorable to fold a hexagonal BN sheet onto a nanotube geometry than to form a carbon nanotube from a graphite sheet.

We also address the question of stability of a small tube versus opening into a strip of planar hexagonal BN. We performed total energy calculations for the strip corresponding to the small tube (6,0), allowing complete relaxation of the strip geometry. As shown on Fig.4 (filled square), the strip is less stable than the corresponding tube. As for carbon nanotubes, BN nanotubes with a radius larger than 4 Å are stable with respect to a strip.

2.2 FRUSTRATION ENERGIES

We now turn to the study of the growth mechanisms of BN nanotubes. Firstly, we evaluate the energy cost associated with a frustrated N-N or B-B bond as compared to a B-N bond. Starting from an isolated BN hexagonal sheet, we exchange two neighboring B and N atoms, creating 2 B-B bonds and 2 N-N bonds. The computed energy cost of this anti-site defect is 7.1 eV, after atomic relaxation. This corresponds to an average of 1.8 eV per frustrated bond. Secondly, we study various isomers of small C_{24} and $B_{12}N_{12}$ clusters. In particular, we compare the energies of two fullerene-like closed structures, one with pentagons and hexagons only (labeled {5,6}), the other with squares and hexagons (labeled {4,6}). In the case of carbon, we find that C_{24}{5,6} is more stable than C_{24}{4,6} by 1.5 eV, while in the case of BN compounds, the $B_{12}N_{12}${5,6} isomer is less stable than $B_{12}N_{12}${4,6} by 7.8 eV [39]. Assuming that elastic energies are equivalent for both C and BN compounds, we can estimate the cost of the 6 frustrated bonds present in the $B_{12}N_{12}${5,6} isomer to be 9.3 eV, that is ~1.6 eV per frustrated bond. This is consistent with the energy found above for a planar geometry.

2.3 THE ARMCHAIR BN SWNTs

We now turn to the study of the dynamics and growth mechanisms of BN nanotubes. We consider first the case of a (5,5) armchair nanotube. Starting from an « ideal » cleaved geometry (Fig.5.a), we gradually raise the temperature up to 3000 K which is typical of the experimental synthesis conditions.

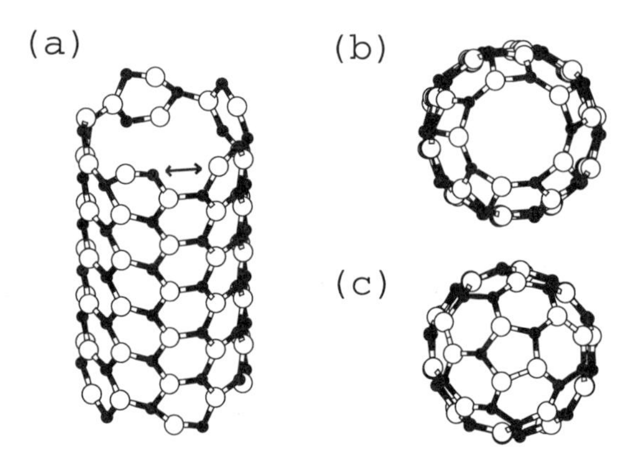

FIGURE 5 : Symbolic ball-and-stick representation of the (5,5) BN nanotube. *Large white* and *small black circles* are respectively B and N atoms. In (a), the initial configuration of the molecular dynamics simulation is shown (0~K). In (b), the final configuration at 3000~K exhibits the {4,6,10}-rings geometry discussed in the text. In (c), the hypothetical C_{60}-like capping is shown. The arrows indicate the motion of the atoms.

At relatively low temperature (~1500 K), the top-most BN hexagons bend inside connecting each other. This creates 5 squares and a decagon (Fig.5.b). In this configuration, all atoms are three-fold coordinated and no frustrated bonds are present. The energy of this BN {4,6,10} configuration at T=0 K is 5.7 eV lower than the energy of the cleaved starting configuration and 2.0 eV lower than the energy of an hypothetical BN-(5,5) nanotube capped with a perfect hemi-C_{60} structure (see Fig.5.c). Brought to 3000 K for more than 5 ps, the squares and the decagon never open and no further reorganization is observed. This suggests that tip structures containing several squares and a « large » even-member ring are quite stable in the case of BN based systems. An analogous molecular dynamics study has been performed for a BN-(4,4) nanotube. Similarly to what was found before, this leads to the creation of a {4,6,8} tip containing 4 squares and an octagon.

Further, we test the stability of the obtained caps against arrival of incoming B and N atoms from the plasma. We send a B atom at 3000 K thermal velocity along the axis of the BN-(4,4) nanotube onto the {4,6,8} tip structure.

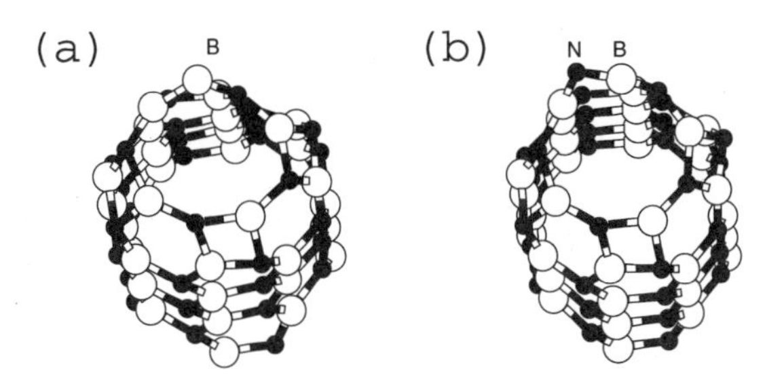

FIGURE 6 :Symbolic ball-and-stick representation of a (4,4) BN nanotube after incorporation of (a) a B atom and (b) a BN dimer. The incorporated atoms have been labeled by their chemical element symbol.

We find that the impinging atom is rapidly incorporated by re-opening a square (Fig.6.a) to create a pentagon. We repeat the same « experimen » with a BN dimer and, again, the dimer is rapidly incorporated by a square in order to form a perfect hexagon (Fig.6.b). The picture which emerges therefore from the present set of simulations is that armchair BN nanotubes can be stabilized into an intermediate « semi-opened » metastable cap, consisting of squares and a large even-member ring, which is able to incorporate incoming B or N atoms (or BN dimers) from the plasma phase. This contrasts significantly with the case of armchair carbon nanotubes where the formation of pentagons leads to closure onto a hemi-C_{60} cap which does not incorporate any incoming atoms, preventing further growth.

2.4 THE ZIGZAG BN SWNTs

We consider now the case of a (9,0) zigzag nanotube. In the chosen starting configuration, the final ring of atoms is made of B atoms only and important frustration effects can be expected in an attempt to close this configuration. For temperatures lower than 3000 K, the only observed process is a dimerization of some B atoms at the top (Fig.7.a). However, around 3000 K, a remarkable « diffusion » process at the tip of the nanotube brings N atoms above the terminal B-ring (Fig.7.b) in order to create a BN square (Fig.7.c). The formation of two more squares by the same mechanism is subsequently observed in the simulation. This first reorganization is followed by the formation of a « bridging » bond across the open end of the nanotube. However, contrarily to the armchair case, the system does not evolve towards a stable minimum energy configuration but samples several tip structures. We plot in Fig.7.d the configuration reached by the system after 3 ps of simulation at 3000 K.

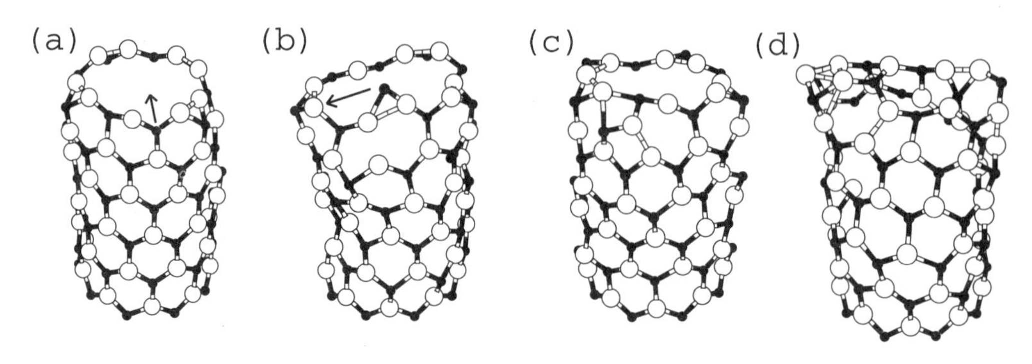

FIGURE 7 :Symbolic ball-and-stick representation of the (6,0) BN nanotube. In (a), the low-temperature dimerized configuration is shown. In (b) and (c), the flipping over of a N atom followed by the formation of a square is depicted. As a result, a frustrated pentagon has been « buried » inside the hexagonal network. In (d), we represent the final configuration reached by the system at the end of our simulation run. The arrows indicate the motion of the atoms.

2.5 GROWTH MECHANISMS FOR BN SWNTs

The zigzag BN nanotube open edge is therefore very unstable, evolving rapidly towards an amorphous-like tip. The creation of triangles in Fig.7.a indicates that the energy cost of two dangling bonds is larger than the cost associated with an homopolar bond. As a result, ~4-5 homopolar bonds are formed at the tip of our (9,0) system. At given « constant » frustration cost, we may expect that pentagons are preferred over triangles, driving the system towards tip closure. The rearrangements observed in Fig.7.b-c which lead to the replacement of triangles by pentagons and squares is indeed very similar to what was observed in the carbon case. However, rearrangements at the zigzag BN nanotube egde are not as stable as in the case of carbon nanotubes. This is because homopolar bonds are much weaker than B-N bonds. In the simulations, we observe these bonds to break and form frequently at temperatures ~3000 K. As a result, the zigzag BN edge does not reach any stable configuration as long as these frustrated bonds are present.

We can now compare the growth mechanisms which may be attributed to the two BN geometries investigated here. It is likely that an amorphous structure similar to the one found for the (9,0) geometry would naturally develop at some stage of the growth of any comparable zigzag tip. In turn, this seems to compromise the further growth of a well ordered nanotube structure. On the contrary, armchair nanotube edges can be stabilized into a well ordered « semi-opened » structure presenting no dangling bonds and no frustrated bonds. This structure is metastable and can revert to a growing hexagonal framework by incorporation of incoming atoms. This suggests that the growth of armchair nanotubes will be strongly favored as compared to the one of zigzag nanotubes. Further, the present simulations lead to the idea that armchair BN SWNTs may grow uncatalyzed, in great contrast with the carbon case. This conclusion is supported by the recent experimental observations of BN SWNTs without any trace of catalytic particle incorporated in (or protruding from) their tips [35].

We note that in the case of larger *(n,n)* nanotubes, the bending inside of the dangling dimers cannot lead to the formation of exactly n squares as in the previous cases. Indeed, the length of the BN bonds shared by the squares and the central large even-member ring will increase with increasing tube radius. However, local fluctuations of curvature at 3000 K can allow for the creation of one or several squares through the same mechanism described above. Since these squares do not survive the arrival of impinging atoms, the final nanotube closure will only occur under the unlikely event that such squares are incorporated in the hexagonal network. This is consistent with the topological models proposed in [34,35] which show a strong reduction of the local radius of curvature where the squares are present. As emphasized in these works, the presence of squares in the hexagonal network induces flat tips of the kind observed experimentally. On the basis of our simulations, we find that the number of squares in the final tip may be larger than three. This is an important result since it implies that large even-member rings such as octagons and decagons will be present at the tip of BN nanotubes. This possibility was pointed out in [34] on the basis of Euler's theorem for even-member ring systems. To explore the consequences of the presence of such large rings, we have performed static total energy calculations which show that Li atoms can diffuse without any barrier through the center of octagons and decagons. This suggests that BN nanotubes are potential candidates for a very pure metallic doping, in contrast with the case of carbon nanotubes for which re-opening of the cap by oxidation is necessary.

2.6 GROWTH MECHANISMS FOR BN MWNTs

In the case of carbon MWNTs, lip-lip interactions have been shown to stabilize the open edge of each nanotube by forming bridging bonds between concentric shells. These bridging bonds easily break to accomodate incoming atoms from the vapor phase, thus allowing for further growth. We expect that such a mechanism would apply also to BN concentric nanotubes, with the additional constraints introduced by frustration effects. As we saw above the open edge of individual armchair tubes is already metastable and lip-lip interactions are expected to further favour the growth. In the case of the zigzag geometry, lip-lip interactions may contribute to reduce the number of frustrated bonds (e.g. if a B-terminated edge is adjacent to a N-terminated edge). We recall however that previous simulations on carbon double-wall zigzag nanotubes have shown that, at 3000 K, triangles are dynamically created at the growing shell edge of each constituting nanotube. We expect that the « trimer-to-pentagon » switching process observed in Figs.7.a-c could also drive towards amorphization the zigzag « lip-lip edge », preventing the further growth of a well ordered MWNT. This suggests that most of the observed BN concentric nanotubes may be of the armchair type. Finally, we note that lip-lip interactions should favour the growth of multi-wall BN nanotubes in which all the concentric shells have the same helicity, since this is the best strategy to minimize the number of frustrated bonds during growth.

BN nanotubes with a large aspect ratio and an amorphous-like tip are frequently observed [34,35]. This indicates that, even in the case of BN nanotubes which can sustain growth (e.g. armchair nanotubes), a sudden non-stoichiometric fluctuation of plasma phase composition during growth would induce the formation of a large number of frustrated bonds at the growing edge, thus leading to amorphization, as shown above. We propose that arc discharge synthesis based on the used of stoichiometric BN electrodes, as compared to other techniques where B and N atoms are introduced separatly in the reaction chamber (see [35]), should enhance the homogeneity of the plasma. In this respect, an experimental study of the plasma composition for different experimental approaches [33-35] would be most helpful in optimizing the production of well formed BN nanotubes.

CONCLUSION

Quantum molecular dynamics is shown to be a valuable and very accurate tool used to elucidate microscopic growth mechanisms for carbon and BN nanotube systems. Our results for a double-walled carbon nanotube should be contrasted with those obtained for single-walled carbon nanotubes. In the carbon SWNTs case, dome closure rules out uncatalyzed growth. In contrast, for the double-walled carbon tube, the existence of a metastable edge structure can explain why MWNTs can grow by a non-catalytic process. More generally, the lip-lip interaction favors a growth in terms of tube pairs, which is consistent with the relative abundance of even-numbered walls observed in TEM images [1,11,40].

Our simulations do not allow us to address directly the issue of multi-walled nanotube closure. However, on the basis of our results for single-walled nanotubes, we can speculate that multiple dome closure may initiate when two pentagons form simultaneously at the growing edge of two adjacent walls. The rather low probability of such an event may explain why multi-walled carbon nanotubes tend to grow long.

In the case of BN nanotubes, different growth mechanisms can be expected for BN nanotubes with different chiralities. In particular, it appears that armchair BN SWNTs may grow uncatalyzed while zigzag tubes rapidly evolve into an amorphous-like structure. This difference originates in the ~1.6 eV average frustration energy associated with N-N or B-B bonds as compared to B-N bonds. BN nanotubes eventually close through the permanent insertion of squares in the bond network, inducing flat tips and the presence of large even-member rings such as octagons or decagons. The role played by frustration effects and that of square rings in determining disordered or ordered tube-edge structures are in agreement with the observed morphology of BN nanotube tips and with the helicity selection evidenced experimentally in the case of BN MWNTs.

ACKNOWLEDGEMENTS

J.C.C. acknowledges the *National Fund for Scientific Research* [FNRS] of Belgium for financial support. X.B. acknowledges support from the French *Centre National de la Recherche Scientifique* (CNRS). This work was partially supported through the Parallel Technology Program (PATP) between EPFL and Cray Research Inc., and by the Swiss NSF.

REFERENCES

1 S. Iijima, Nature **354**, 56 (1991).
2 T.W. Ebbesen and P.M. Ajayan, Nature **358**, 220 (1992).
3 M.M.J. Treacy, T.W. Ebbesen, J.M. Gibson, Nature **381**, 678 (1996).
4 W.A. de Heer, A. Châtelain, D. Ugarte, Science **270**, 1179 (1995).
5 S.J. Tans, M.H. Devoret, H. Dai, A. Thess, R.E. Smalley, L.J. Geerligs, and C. Dekker, Nature **386**, 474 (1997); S.J. Tans, A.R.M. Verschueren, C. Dekker, Nature **393**, 49 (1998).
6 T.W. Ebbesen, Physics Today **49**, 26 (1996).
7 S. Iijima, T. Ichihashi, Y. Ando, Nature **356**, 776 (1992).
8 S. Iijima, Mater. Sc. Eng. B **19**, 172 (1993).
9 M. Endo and H.W. Kroto, J. Phys. Chem. **96**, 6941 (1992).
10 R. Saito, G. Dresselhaus, M.S. Dresselhaus, Chem. Phys. Lett. **195**, 537 (1992).
11 S. Iijima, P.M. Ajayan, T. Ichihashi, Phys. Rev. Lett. **69**, 3100 (1992).

12 C.-H. Kiang and W.A. Goddard III, W.A. Phys. Rev. Lett. **76**, 2515 (1996).
13 R.E. Smalley, Mater. Sci. Eng. B **19**, 1 (1993).
14 A. Maiti, C.J. Brabec, C.M. Roland, J. Bernholc, Phys. Rev. Lett. **73**, 2468 (1994).
15 L. Lou, P. Nordlander, R.E. Smalley, Phys. Rev. B **52**, 1429 (1995).
16 X.F. Zhang, *et al.*, J. Crystal Growth **130**, 368 (1993).
17 Y. Saito, T. Yoshikawa, M. Inagaki, M. Tomida, X. Hayashi, Chem. Phys. Lett. **204**, 277 (1993).
18 E.G. Gamaly and T.W. Ebbesen, Phys. Rev. B **52**, 2083 (1995).
19 S. Iijima and T. Ichihashi, Nature **363**, 603 (1993).
20 D.S. Bethune, C.H. Klang, M.S. de Vries, G. Gorman, R. Savoy, J. Vazquez, R. Beyers, Nature **363**, 605 (1993).
21 J.-C. Charlier, A. De Vita, X. Blasc, R. Car, Science **275**, 646 (1997).
22 X. Blase, A. De Vita, J.-C. Charlier, R. Car, Phys. Rev. Lett. **80**, 1666 (1998).
23 R.Car, and M. Parrinello, Phys. Rev. Lett. **55**, 2471 (1985).
24 N. Troullier, and J.L. Martins, J.L. Phys. Rev. B **43**, 1993 (1991).
25 G.B. Bachelet, D.R. Hamann, M. Schlüter, M. Phys. Rev. B **26**, 4199 (1982).
26 The calculated intraplanar lattice constant of graphite, 2.46 Å, and the C-C bond lengths in C_{60}, 1.40 Å, and 1.47 Å, agree with experiment within 1%.
27 in the notation of: N. Hamada, S.-I. Sawada, A. Oshiyama, Phys. Rev. Lett. **68**, 1579 (1992). With standard notations, *(n,0)* and *(n,n)* nanotubes will be called respectively zigzag and armchair. Cutting these nanotubes perpendicularly to their axis will create two-coordinated « danglin » atoms in the first case and dangling dimers in the second one.
28 A. De Vita, A. Canning, R. Car, EPFL Supercomputing J. **6**, 22 (1994).
29 S. Nosé, Mol. Phys. **52**, 255 (1984); H.G. Hoover, Phys. Rev. A **31**, 1695 (1985).
30 B.R. Eggen, M.I. Heggie, G. Jungnickel, C.D. Latham, R. Jones, P.R. Briddon, Science **272**, 87 (1996).
31 T. Guo, P. Nikolaev, A.G. Rinzler, D. Tomànek, D.T. Colbert, R.E. Smalley, J. Phys. Chem. **99**, 10694 (1995).
32 R.S. Wagner and W.C. Ellis, Appl. Phys. Lett. **4**, 89 (1964).
33 N. G. Chopra, *et al.*, Science **269**, 966 (1995).
34 M. Terrones, *et al.*, Chem. Phys. Lett. **259**, 568 (1996).
35 A. Loiseau, *et al.*, Phys. Rev. Lett. **76**, 4737 (1996).
36 X. Blase, A. Rubio, S. G. Louie, and M. L. Cohen, Europhys. Lett. **28**, 335 (1994); *ibidem*, Phys. Rev. B. **51**, 6868 (1994).
37 X. Blase, J.-C. Charlier, A. De Vita and R. Car, Appl. Phys. Lett. **70**, 197 (1997), and references therein.
38 A. Rubio, Y. Myamoto, X. Blase, M.L. Cohen, and S.G. Louie, Phys. Rev. B **53**, 4023 (1996).
39 Similar results have been found at the level of MP2/DZP calculations. See: J. Jensen and H. Toftlund, Chem. Phys. Lett. **201**, 89 (1993). We refer the reader to this reference for figures of the studied clusters.
40 S. Seraphin, D. Zhou, J. Jiao, Acta Microscopica **3**, 45 (1994).

NANOSCOPIC HYBRID MATERIALS: THE SYNTHESIS, STRUCTURE AND PROPERTIES OF PEAPODS, CATS AND KIN

David E. Luzzi and Brian W. Smith

Department of Materials Science
University of Pennsylvania
Philadelphia, PA 19104-6272
USA

INTRODUCTION

Soon after the first production of multi-wall carbon nanotubes (MWNT) methods were sought to fill the tube cores with other materials. It can be presumed that the filling of nanotubes with a material possessing useful properties is one path to the production of functional nanotubes. The nanotube can then function as a container with good stability or also contribute to the properties of the final structure. The earlier work in this area has produced notable success in the opening and filling of MWNTs[1-9]. This prior work has involved the capillarity-based uptake of low surface tension liquids into MWNTs with large inner diameters produced using the carbon arc (CA) technique. The length of MWNT that has been continuously filled using these methods was short.

Given the ease and control with which they are now produced, it is of interest to determine if single-wall carbon nanotubes (SWNT) can be efficiently filled. Soon after the production of single wall carbon nanotubes (SWNTs) [10], it was noted that the diameter of the most abundant nanotube produced by the pulsed laser vaporization method was correctly-sized to encapsulate a C_{60} molecule [11]. Recently, we discovered that these unique hybrid structures (peapods) are a natural product in PLV-synthesized material that has been acid purified and annealed at 1100°C, albeit at low concentrations [12]. In this paper, we review these findings including the methods that can be used to synthesize these unique structures. We find that the filling of SWNTs with C_{60} (peapods) can be efficiently accomplished and that these materials are a precursor which can be further processed to form other types of filled SWNTs. The synthesis methods described herein provide a means to produce bulk quantities of these interesting materials for further study and point to the possibility of the general functionalizing of carbon nanotubes.

EXPERIMENTAL METHOD

The starting materials for this study were acid purified carbon nanotubes produced by either the pulsed laser vaporization (PLV) or carbon arc (CA) techniques. The PLV

Science and Application of Nanotubes, edited by Tománek and Enbody
Kluwer Academic / Plenum Publishers, New York, 2000

material had been synthesized by the laser ablation of a graphitic target impregnated with 1.2 at% each Ni/Co catalyst. This raw nanotube "felt" was refluxed in HNO_3 for 48 hours, rinsed and neutralized, suspended in surfactant, and filtered to form a thin paper [13]. The CA material was synthesized by carbon arc discharge using ~5 at% each Ni/Y catalyst [14] and then similarly purified [15]. Such wet chemical etching is known to open the ends of MWNTs [8] as well as attack the sidewalls of SWNTs [16]. One additional batch of PLV material had received an anneal of 1100 °C in vacuo prior to being received.

The as-purified materials were annealed under vacuums ranging from 20-40 μPa at temperatures between 100 and 1200 °C. These annealing treatments were carried out in a vacuum furnace or in-situ in a JEOL 2010F field-emission-gun transmission electron microscope (FEG-TEM). During in-situ anneals, temperature was monitored continuously via thermocouples. Only a few minutes were required to ramp between temperatures due to the small thermal mass of the heater. In addition, some PLV samples were annealed in a flowing Ar atmosphere at ambient pressure to a temperature of 600 °C.

The structure of both in-situ and ex-situ specimens was examined via TEM phase contrast imaging in either the FEG-TEM or in a JEOL 4000 high-resolution TEM (HRTEM). All microscopy was performed at an accelerating voltage of 100 kV to minimize the electron beam induced modification of the material. Microscopy specimens were prepared from nanotube paper by tearing away a small sliver and fixing it inside an oyster TEM grid, thereby forgoing additional chemical or thermal processing.

EXPERIMENTAL RESULTS AND DISCUSSION

In the phase imaging condition, all scattered and unscattered electrons that pass through the lens polepieces are used to produce the image. Due to its low atomic number, carbon scatters electrons weakly. Under phase imaging conditions, nanotubes can therefore be considered as weak phase objects. Images of weak phase objects will be two-dimensional projections along the electron beam direction of the three dimensional specimen potential convoluted with the point transfer function of the electron microscope. With the resolution of the microscope significantly better than the finest scale detail in an image, the image can be considered to be a direct magnification of the carbon shells. The intra-shell structure of the modified graphene sheet is below the resolution limit of the microscope and appears as a uniform contrast level (gray). Since the maximum scattering potential of the nanotube and C_{60} molecules exists where the structures lie tangent to the electron beam, the images will appear as a pair of parallel lines and a circle, respectively.

In Figure 1, images of sections of an empty SWNT and a SWNT containing a chain of C_{60} molecules, a peapod, are presented. These particular examples were found in the PLV material that had been annealed at 1100 °C. As expected, the empty SWNT in Figure 1a appears a parallel set of dark gray lines. The light contrast seen outside the SWNT is a Fresnel effect. In Figure 1b, a 1.4 nm diameter SWNT contains a self-assembled chain of collinear C_{60} molecules. The average spacing of C_{60} molecules within the chain is approximately 10 Å. However a tendency towards pairing of the molecules has been observed and Figure 1b is an example chosen to demonstrate this property. We hypothesize that this pairing can be the result of exposure of the C_{60} chains to UV or visible light, but this remains to be tested. It should be noted that, in contrast to the results presented below, this particular material was exposed to ambient light for a long time after annealing and prior to TEM characterization. Another extraordinary property of these hybrid structures apparent in Figure 1b is that the SWNT provides a scaffold that stabilizes individual C_{60} molecules in one location for times sufficiently long for experimental study.

The number of peapods in this material is quite small. The peapods tend to be segregated on the microscopic and macroscopic scales. On the microscopic scale,

Figure 1 High resolution electron micrographs of single wall carbon nanotubes (SWNT). The SWNTs appear as parallel dark lines separated by the diameter of 1.4 nm. The SWNTs are surrounded by vacuum. The light contrast outside and parallel to nanotube and the speckled contrast are due to Fresnel scattering and shot noise, respectively. a) an empty nanotube. b) a nanotube filled with a chain of C_{60} molecules seen in projection as circles. The arrow marks a closely spaced C_{60} pair.

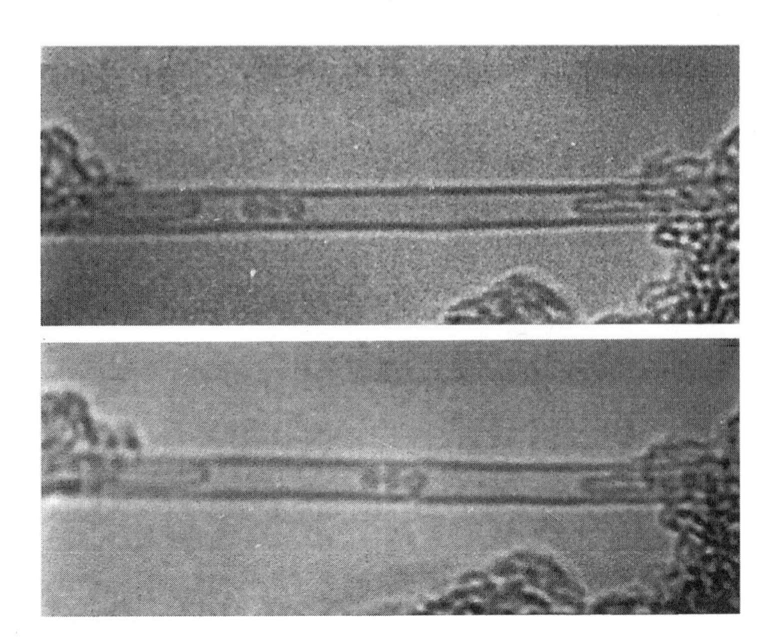

Figure 2 A nano-subway. The motion of a chain of three C_{60} molecules within a SWNT at room temperature is shown. The time between the images is approximately twenty seconds. Two long capsules are seen within the SWNT at either side of the chain. Three additional C_{60} molecules can be seen at the far right of the SWNT beyond the capsule. The focus condition is different between the two images accounting for the slight difference in contrast.

segregation is observed as regions in which few peapods can be found in the microscope adjacent to regions in which approximately 10% of observed SWNTs that are isolated or in small ropes contain C_{60}. Macroscopic segregation has been seen in UV-VIS experiments in which separate milligram weight samples were found to contain between 0.8% and 5% of SWNTs filled with C_{60} [17].

In Figure 2, a room temperature observation of the behavior of C_{60} encapsulated within a SWNT is shown. In the images, three C_{60} molecules are seen within a SWNT and separated from two long, capsule-like structures. As can be seen by comparing the two images, the three molecule chain shifts in position within the SWNT during the approximately twenty seconds between exposures demonstrating that the C_{60} molecules are mobile. The chain appears as nanoscopic subway train shuttling between two nano-stations. These and other capsule structures have been observed in this material at very low density.

What are these materials (cf. Figure 1b)? The encapsulated fullerenes, clearly governed by van der Waals interactions, can easily move within the surrounding SWNT[18]. Thus these materials cannot be considered as single molecules; nor are they fabricated composites since they will spontaneously form. Therefore, we have labeled this unique class of structures as nanoscopic hybrid materials.

Due to the large quantity of impurity present in as-synthesized PLV material, it has proven impossible to determine if peapods exist. As-synthesized CA material has been examined and no peapods have been found. In the as-received acid-purified PLV material, the SWNTs are coated with surfactant, which obscures the nanotubes and makes it difficult to observe peapods. Nevertheless, we have examined thousands of nanotubes in as-purified material and have never detected the telltale periodic contrast of peapods. An example of the material in this condition is shown in Figure 3.

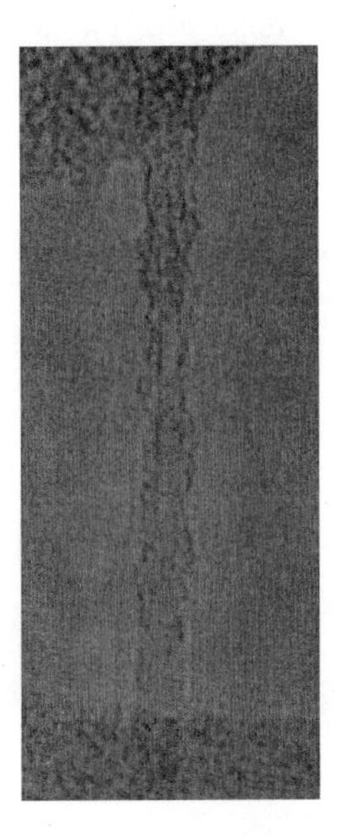

Figure 3 An HRTEM image of nanotubes in PLV material after annealing for 63 h at 225 °C. The damage due to the acid treatment is now visible. The arrow indicates a large defect in the sidewall of one SWNT.

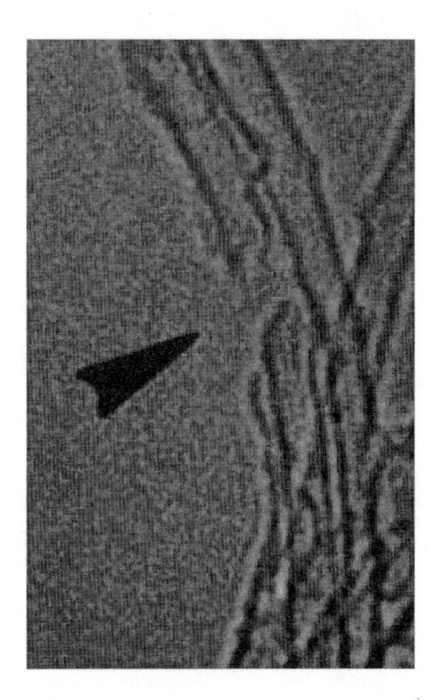

Figure 4 An HRTEM image of a SWNT in as-purified PLV material showing the coating of surrfactant.

In Figure 4, the same specimen as shown in Figure 3 is imaged after annealing in vacuum at 225°C for 63 hours. We have found that baking for at least 24 hours under these conditions removes the surfactant without modifying the nanotube material. In Figure 4, the nanotubes are seen to be free of most impurities. The walls of the nanotubes appear as dark parallel lines and are seen to have breaks and disclinations that are the result of acid attack (see arrows on the figure). Several hundred tubes were observed at various locations in the sample and found to be empty. Thus, by patient collection of a large number of HRTEM images of the material in the as-purified condition and after 225 °C anneal, it is apparent that peapods do not exist in the material following these processing steps. The CA material was also examined in the as-purified condition. In this material, no surfactant was used and therefore the SWNTs were clean, but damaged. No peapods were found in this material.

Annealing experiments were carried out at various temperatures on a number of specimens in order to determine the formation mechanism of the peapods. In Figure 5, the same sample as in Figures 3 and 4 is shown after annealing at 450 °C for two hours in vacuum. The walls of the SWNTs are partially healed. Many peapods are seen to be present indicating that the formation of peapods occurred during the annealing treatment following acid purification and the 225 °C cleaning anneal. These results indicate that C_{60} is entering the nanotubes during these relatively low temperature anneals. It is known from a number of analyses that acid purified PLV material contains residual crystallites of C_{60}. It is therefore likely that C_{60} enters the nanotube by transporting to open ends or sidewall defects by surface diffusion or in the gas phase.

In order to test this hypothesis, in-situ annealing experiments were conducted in the FEG-TEM at temperatures between 225 °C and 375 °C on a specimen that was prepared from PLV material and subsequently cleaned by baking for 24 hours at 225° C. The temperature was equilibrated at 25° increments for observation. The image of Figure 6 was recorded at 350°C. C_{60} molecules have adsorbed to the surface of the nanotubes, which were originally clean as in Figure 4. The molecules adsorb to the surface and are present for only short times before vanishing. The process is quite dynamic with many molecules appearing, briefly residing on the surface and then disappearing so quickly as to be undetectable to the eye. Considering that the interior motion of C_{60} demonstrated in Figure 2 also occurs rapidly, this result cannot be used as to determine that sublimation or surface diffusion is the dominant C_{60} transport mechanism.

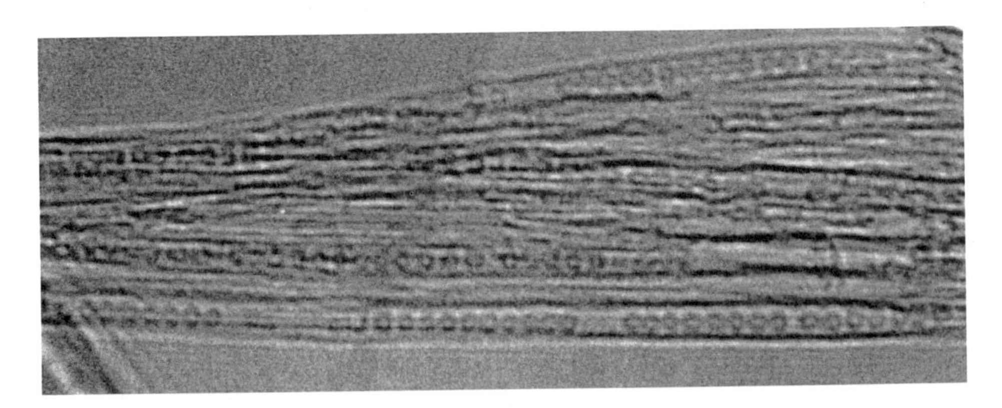

Figure 5 Formation of peapods. A rope of PLV material is shown after further annealing treatment for one hour at 450 °C. Many of the SWNTs are now filled with chains of C_{60} molecules.

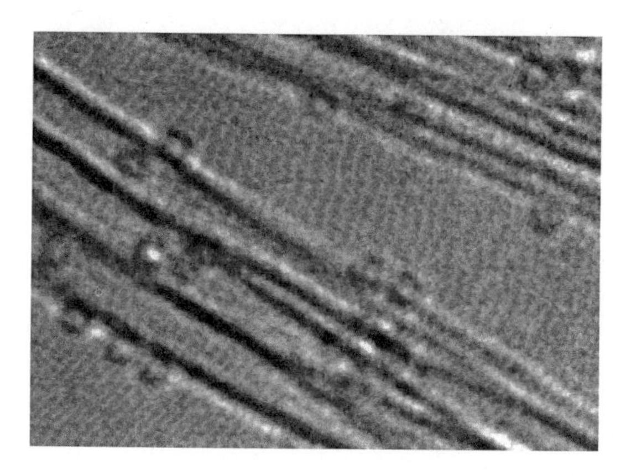

Figure 6 An in-situ HRTEM image at 350 °C showing the adsorption of C_{60} molecules onto the surface of SWNTs.

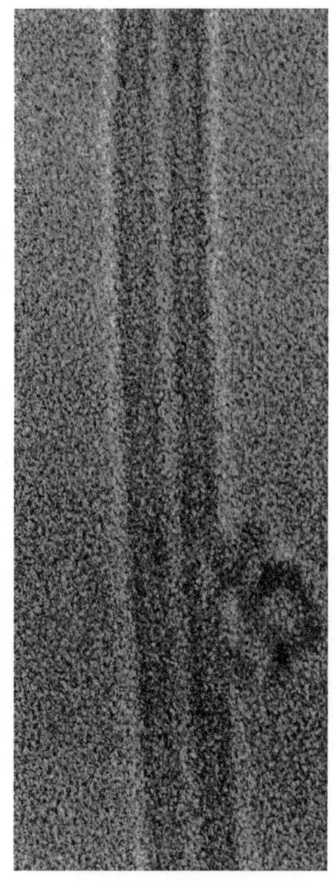

Figure 7 After annealing peapods at 1200 °C, capsules and co-axial tubes (CATs) are formed. In the image, a section of a CAT is shown. The inner tube of 0.7 nm diameter is centered within the surrounding 1.4 nm diameter nanotube. The external feature at the lower

The rapid activity discussed above was seen at 350 °C but not at 325 °C. Thus the onset temperature lies between these limits. If the mechanism of C60 transport is via sublimation and via the vapor phase, then the onset temperature should be a strong function of vacuum system pressure. On the other hand, a surface diffusion mechanism should be less sensitive to pressure. This remains to be tested. Once the fullerene arrives on the nanotube surface, the large amount of filling that is seen argues for surface diffusion to open end or sidewall defects. Under a vapor phase transport mechanism, as the temperature is increased, the residence time of C_{60} on the surface of the SWNT, and thus the probability of entering the SWNT will decrease. At high temperatures, the defects in the SWNTs will anneal thereby blocking the entrance of C_{60} molecules into the tubes. Thus it would appear that there will be a critical temperature window within which peapods can be efficiently formed.

A final test of the peapod formation mechanism was conducted using the CA material, which contains very little residual C_{60} after synthesis and acid purification. To one specimen, C_{60} suspended in di-methyl formamide was applied to the nanotube material and allowed to dry. Another specimen was kept in the as-received condition. Both specimens were annealed at 400 °C for 1 hour. Consistent with the large difference in the quantity of C_{60} in the two specimens, peapods were found to be abundant in the former and sparse in the latter specimen.

In Figure 7, an image of a double tube is shown. Due to the special nature of these tubes, only being found as pairs with an inner 0.7 nm diameter nanotube encased in an outer 1.4 nm diameter nanotube, they have been named co-axial tubes (CAT). It is interesting to note that these features have been found in the 1100 °C annealed PLV material, but were never found in any specimen annealed at the lower temperatures of the experiments described above. In order to determine the formation mechanism, as-received PLV material was annealed in vacuum at 450 °C for two hours to form peapods and then immediately annealed at 1200 °C for 24 hours. In this specimen only CATs and long capsules were found with no residual peapods. Thus it appears that CATs form through the coalescence of C_{60} molecules at high temperatures. The surrounding SWNT acts as a reaction chamber insuring that the C_{60} molecules collide at near-zero impact parameter and templating the reaction product.

In addition to the above features, we have found that nanoscopic hybrid materials form with SWNTs of different diameters, with larger fullerenes encapsulated within larger SWNTs. However, the most abundant of these materials have a SWNT of 1.4 nm diameter. The hybrid structures presented above as well as others not shown are members of a complete structural set of hybrid materials containing hemispherically-capped cylinders of varying length and these 1.4 nm diameter SWNTs. The diameter of the encapsulated cylinders is that of C_{60}, 0.7 nm. The complete structural set can be written as $\{[C_{(60+10n:\ n>=0)}]_{(0.7\ nm\text{-}\phi)}\}$ @ $\{SWNT_{(1.4\ nm\text{-}\phi)}\}$. The end member hybrid structure with n=0 is the SWNT containing C_{60} molecules (now called peapods); the hybrid structure with n=1 is the SWNT containing C_{70} molecules, etc. At n=∞ (actually the maximum n commensurate with the length of the surrounding SWNT), the other end member hybrid material is a two-walled nanotube (now called a co-axial tube or CAT). In contrast to multiwall nanotubes that typically contain many graphene layers and a large inner core diameter, CATs always contain two walls with an outer diameter of 1.4 nm.

SUMMARY

The mechanism for the formation of encapsulated C_{60} chains is now apparent, and we have successfully synthesized them in large fractions of the tubes that comprise our samples. In contrast to prior studies showing the uptake of liquid into large diameter

MWNTs, the mechanism involves transport in the vapor phase or the surface diffusion of individual molecules. SWNTs with small interior diameters are completely filled over long lengths. The extent of filling argues for the entrance of C_{60} through sidewall defects as well as open ends. A minimum temperature must be achieved in order to promote exterior C_{60} to enter the tubes. Annealing at too high a temperature will limit the residence time of C_{60} on the SWNT as well as cause the healing of the nanotubes walls, thereby eliminating access to the nanotube interiors. There is thus a critical temperature window required for the formation of bucky-peapods. The ability to synthesize these unique hybrid materials in bulk will now permit scaled-up empirical study. Interestingly, there is no obvious physical reason why SWNTs could not be filled with any appropriately-sized molecule with an affinity for nanotubes, and present in the gas phase or highly mobile at temperatures within the critical temperature window. Thus, the present work opens the possibility for the general functionalizing of the particular diameter SWNTs produced at highest concentration by PLV and CA methods.

ACKNOWLEDGEMENTS.

We thank Profs. D.T. Colbert and R.E. Smalley for the acid purified and anealed PLV materials and Prof. P. Bernier and Dr. P. Petit for the acid-purified CA material. This work benefited from discussions with colleagues Profs. J.E. Fischer, L.A. Girifalco and V.Vitek. This work was funded by the National Science Foundation with central facility support from the U. Penn MRSEC.

REFERENCES

1. Ajayan, P.M., *et al.*, *Opening Carbon Nanotubes with Oxygen and Implications for Filling.* Nature, 1993. **362**: p. 522.
2. Ajayan, P.M. and S. Iijima, *Capillarity-Induced Filling of Carbon Nanotubes.* Nature, 1993. **361**: p. 333.
3. Ajayan, P.M., *et al.*, *Carbon Nanotubes as Removable Templates for Metal Oxide Nanocomposites and Nanostructures.* Nature, 1995. **375**: p. 564.
4. Davis, J.J., *et al.*, *The Immobilisation of Proteins in Carbon Nanotubes.* Inorganica Chimica Acta, 1998. **272**: p. 261.
5. Dujardin, E., *et al.*, *Capillarity and Wetting of Carbon Nanotubes.* Science, 1994. **265**: p. 1850.
6. Sloan, J., *et al.*, *Selective Deposition of UCl_4 and $(KCl)_x(UCl4)_y$ inside Carbon Nanotubes Using Eutectic and Noneutectic Mixtures of UCl_4 and KCl.* J. Solid State Chem., 1998. **140**: p. 83.
7. Tsang, S.C., P.J.F. Harris, and M.L.H. Green, *Thining and Opening of Carbon Nanotubes by Oxidation using Carbon Dioxide.* Nature, 1993. **362**: p. 520.
8. Tsang, S.C., *et al.*, *A Simple Chemical Method of Opening and Filling Carbon Nanotbues.* Nature, 1994. **372**: p. 159.
9. Ugarte, D., *et al.*, *Filling Carbon Nanotubes.* Appl. Phys. A, 1998. **67**: p. 101.
10. Thess, A., *et al.*, *Crystalline Ropes of Metallic Carbon Nanotubes.* Science, 1996. **273**: p. 483.
11. Nikolaev, P., *et al.*, *Diameter Doubling of Single-Wall Nanotubes.* Chem. Phys. Lett., 1997. **266**: p. 422.
12. Smith, B.W., M. Monthioux, and D.E. Luzzi, *Encapsulated C_{60} in Carbon Nanotubes.* Nature, 1998. **396**: p. 323.

13. Rinzler, A.G., *et al.*, *Large Scale Purification of Single Wall Carbon Nanotubes: Process, Product and Characterization.* Applied Physics A, 1998. **67**: p. 29.
14. Journet, C., *et al.*, Nature, 1997. **388**: p. 756.
15. Petit, P., in *private communication*. 1999.
16. Monthioux, M., B.W. Smith, and D.E. Luzzi, unpublished data, 1998.
17. Burteaux, B., *et al.*, *Abundance of Encapsulated C60 in Single-Wall Carbon Nanotubes.* Chem. Phys. Lett., 1999(in press).
18. Smith, B.W., M. Monthioux, and D.E. Luzzi, *Carbon Nanotube Encapsulated Fullerenes: A Unique Class of Hybrid Materials.* Chem. Phys. Lett., 1999(in press).

LINEAR AUGMENTED CYLINDRICAL WAVE METHOD FOR NANOTUBES: BAND STRUCTURE OF $[Cu@C_{20}]_{\infty}$

Pavel N. D'yachkov and Oleg M. Kepp

Institute for General and Inorganic Chemistry,
Russian Academy of Sciences,
Leninskii pr. 31,
117907 Moscow, Russia

INTRODUCTION

Preparation of the fullerene carbon nanotubes[1-7] offers strong possibilities of manufacturing the nanowires with intriguing mechanical, electronic, and magnetic properties[4, 8-13]. The inner cavity of these tubules can be filled with a variety of substances including the transition metals[2-8]. We are now being confronted with a problem of predicting structural properties of nanotubes intercalated with transition metals. Here we present a linear augmented cylindrical wave (LACW) method and its application to band structure of the tubes doped with 3d-metals.

FORMALISM

As usually in the muffin-tin (MT) approximation, we suggest that the effective one-electron potential is spherically symmetric inside MT spheres and is constant in the interstitial region. In nanowires the infinite motion of electrons is possible in one direction (along the z axis) giving rise to electronic bands. We suggest that a system is confined between the impenetrable cylindrical potential barrier for the other two dimensions, so that the wave functions reduce to zero on the boundary of the cylinder and in the region beyond it. A radius a of the cylinder is chosen in such a way that the essential part of the electronic is located inside the cylinder with the πa^2 cross section.

In order to obtain the basis, one has to augment eigenfunctions (cylindrical waves) in the interstitial region with functions constructed from the solutions inside the MT-spheres. The resulting base functions (LACWs) should be continuous with continuous first derivatives on the MT-spheres. In the interstitial region the base functions are solutions of the wave equation for a free electron in the cylinder. The base function is a product[14]

Science and Application of Nanotubes, edited by Tománek and Enbody
Kluwer Academic / Plenum Publishers, New York, 2000

$\psi_P(\mathbf{Z})\psi_M(\Phi)\psi_{|M|,N}(R)$. The $\psi_P(\mathbf{Z})$ function corresponds to a free electron movement along z axis

$$\psi_P(\mathbf{Z})=(1/\sqrt{c})\exp[i(\boldsymbol{k}+\boldsymbol{k}_P)\mathbf{Z}],\ k_P=(2\pi/c)P,\ P=0,\pm1,\pm2,\ ... \tag{1}$$

Where c is the lattice constant for z direction, $-\pi/c \leq \boldsymbol{k} \leq \pi/c$. The function ψ_M is:

$$\psi_M(\Phi)=(1/\sqrt{2\pi})\exp(iM\Phi),\ M=0,\pm1,\pm2,\ ... \tag{2}$$

Finally, the function $\psi_{M,N}(R)$ is the radial solution of the part of the Schrodinger equation:

$$[-\frac{1}{R}\frac{d}{dR}R\frac{d}{dR}+\frac{M^2}{R^2}]\psi_{M,N}(R)+U(R)\psi_{M,N}(R)=E_{|M|,N}\psi_{M,N}(R) \tag{3}$$

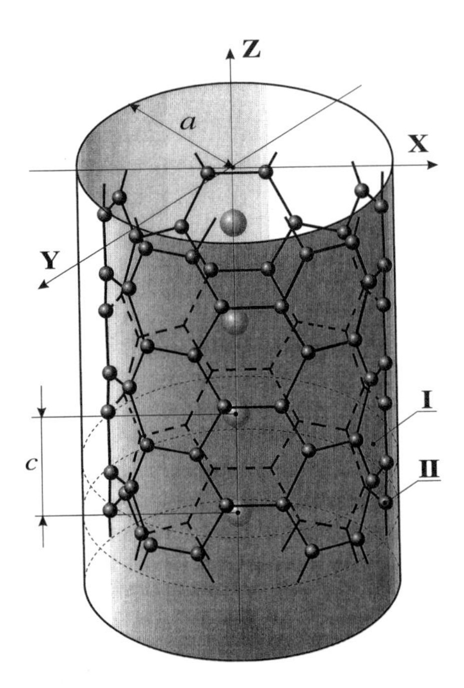

Figure. 1. Metal doped nanotubes $[M@C_{20}]_\infty$ inside a cylindrical potential

Here $U(R)=U$ for $R<a$ and $U(R)=0$ for $R>a$; $E_{|M|,N}$ is the energy spectrum which depends on two quantum numbers N and M. The electronic energy corresponding to cylindrical wave $\Psi(\mathbf{Z},\Phi,R)$ is:

$$E=(k+k_P)^2+E_{|M|,N}. \tag{4}$$

For $R \leq a$ Eq (3) has the form:

$$[\frac{d^2}{dR^2}+\frac{1}{R}\frac{d}{dR}+\kappa^2_{|M|N}-\frac{M^2}{R^2}]\psi_{M,N}(R)=0, \tag{5}$$

where $\kappa_{|M|,N}=\{E_{|M|,N}\}^{1/2}$. This is the well known Bessel equation[15-16]. The solution is $\psi_{M,N}(R) = CJ_M(\kappa_{|M|,N}R)$, where J_M is the cylindrical Bessel function of the order M and C is a constant.

The wave function equals to zero on the impenetrable barrier $\psi_{M,N}(a) = J_M(\kappa_{|M|,N}a) = 0$; thus, the energy spectrum $E_{|M|,N}$ is determined by the roots of Bessel function as

$$E_{|M|,N}=(\alpha_{|M|,N})^2/a^2. \tag{6}$$

Here $\alpha_{|M|,N}$ is the root number N (N=1, 2, ...) of the cylindrical Bessel function of the order M. The constant C is obtained from the normalisation condition for $\psi_{M,N}(R)$[15-16]:

$$C = \sqrt{2}/\{a\,|J'_M(\kappa_{|M|,N}a)|\}, \tag{7}$$

where J'_M is the derivative of the Bessel function. Finally, in the cylindrical coordinate system the base wave function in the interstitial region Ω_{II} is:

$$\Psi_{II}(Z,\Phi,R|k,P,M,N)=\{\sqrt{\Omega}\,|J'_M(\kappa_{|M|,N}a)|\}^{-1}\exp\{i(K_PZ+M\Phi)\}\,J_M(\kappa_{|M|,N}R). \tag{8}$$

Here $\Omega = \pi ca^2$ is the unit cell volume and $\boldsymbol{K_P}=\boldsymbol{k}+\boldsymbol{k_P}$.

Inside the α-th MT-sphere, the base function is expanded in terms of spherical harmonics[17, 18]:

$$\Psi_{I\alpha}(r,\theta,\varphi|k,P,M,N)=\sum_{l=0}^{\infty}\sum_{m=-l}^{l}\{A_{lm}u_l(r,E_l)+B_{lm}\dot{u}_l(r,E_l)\}\,Y_{lm}(\theta,\varphi), \tag{9}$$

where $u_{l\alpha}$ is the solution of the radial Schrodinger equation in the α-th MT sphere for the energy E_l, $\dot{u}_{l\alpha}=[\partial u/\partial l]_{E_l}$, and r, θ, φ are the polar coordinates of the vector ρ_α of the local coordinate system of the MT sphere α. The coefficients A_{lm} and B_{lm} are obtained under the condition that the wave function (8), (9) is continuous with the first continuous derivative. These coefficients are readily calculated taking into account the addition theorem for the cylindrical Bessel functions[15, 16].

Using these base functions, which we call the linear augmented cylindrical waves (LACW), one can calculate[18,20] the overlap $<P_2M_2N_2|P_1M_1N_1>$ and Hamiltonian matrix

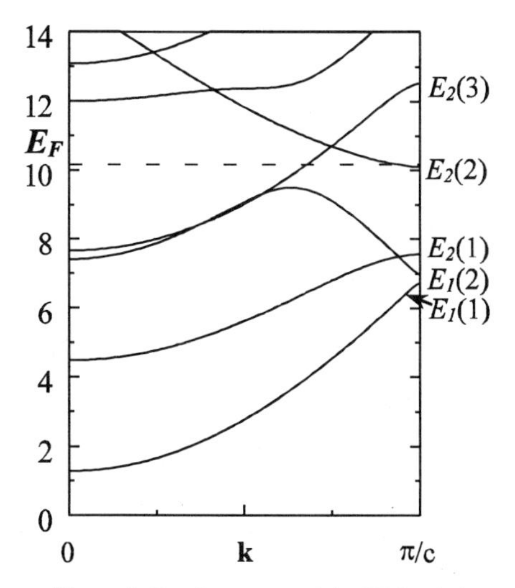

Figure 2. Band structure of the $[Cu]_\infty$ chain

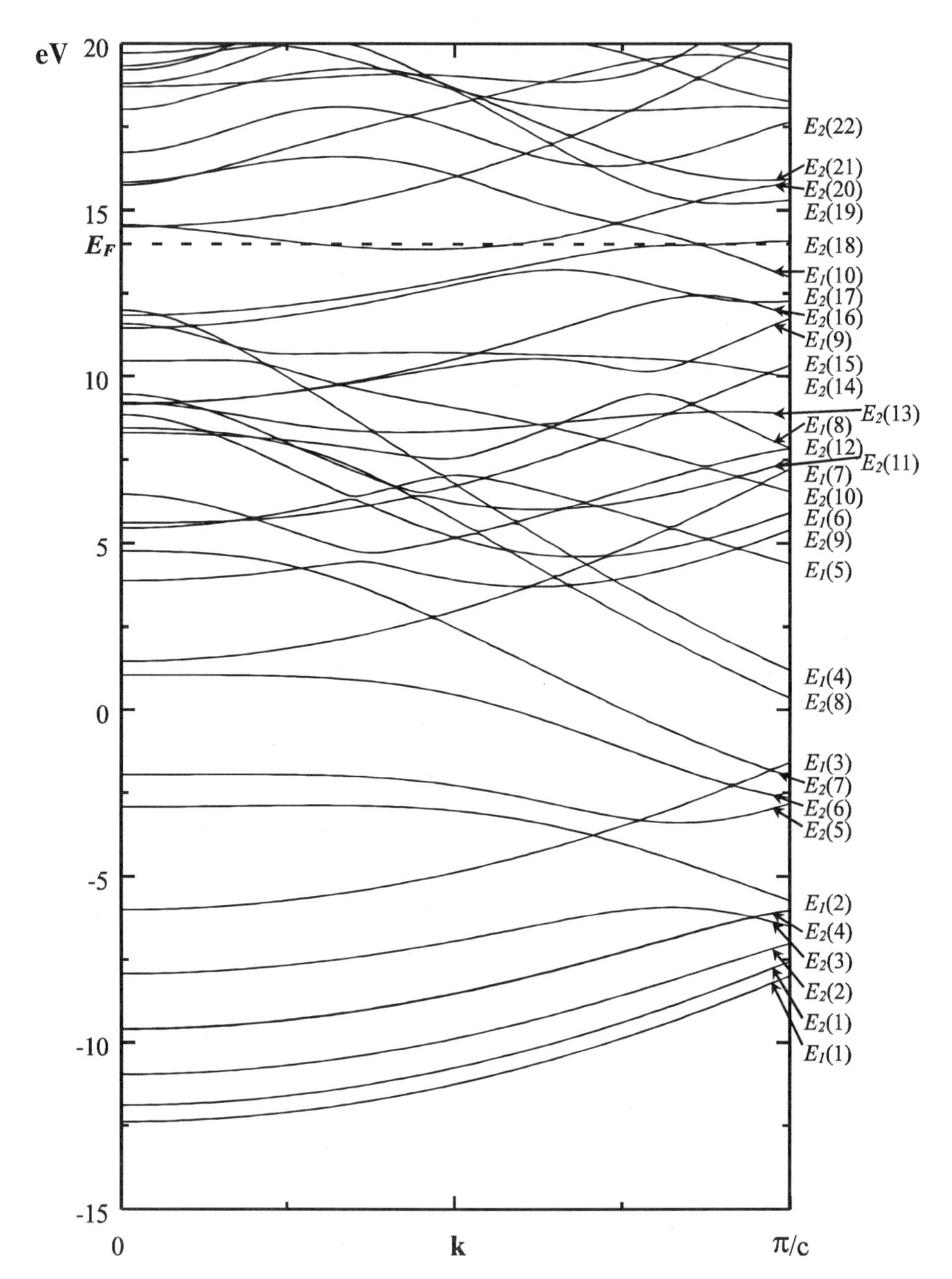

Figure. 3. Band structure of the $[Cu@C_{20}]_\infty$ tube

elements $<P_2M_2N_2|H|P_1M_1N_1>$. Finally, using the secular equation:

$$\det \| <P_2M_2N_2|H|P_1M_1N_1> - E(\boldsymbol{k}) <P_2M_2N_2|P_1M_1N_1> \| =0 \qquad (17)$$

we calculate dispersion curves $E(\boldsymbol{k})$.

APPLICATIONS

The developed method was implemented as the program in FORTRAN and applied to study the band structure of linear metallic chains $[M]_\infty$ with interatomic distances equal to the doubled atomic radii and the band structure of metal doped carbon nanotubes $[M@C_{20}]_\infty$ with armchair geometry (Fig. 1) of the carbon system[19-21]. In the former case a potential cylinder radius a was supposed to be equal to 3.5Å, which may be considered as radius of the host carbon nanotubes. One-electron potential was supposed to coincide with the atomic one inside the MT-spheres and zero in the intersphere region. In the case of metal doped carbon nanotubes $[M@C_{20}]_\infty$ this radius was taken to be equal to the van der Waals radius of the system.

The dispersion curves of $3d$-transition metals are generally analogous. As a typical example, Fig. 2 shows the band structures on the $[Cu]_\infty$ and chain. The character of the low-lying states can be interpreted in terms of tight binding method. The nondegenerate $E_1(1)$ and $E_3(1)$ bands are σ-bonding and antibonding combinations of $4s$- and $3d(z^2)$-functions. The doubly degenerate $E_2(2)$ band located in a gap between $E_1(1)$ and $E_3(1)$ states is contributed by $3d(x^2-y^2)$ and $3d(xy)$ functions, whose orientation results in bonding interactions of the δ-type. The next is doubly degenerate π-type antibonding band formed by $3d(xz)$ and $3d(yz)$ functions.

Figure 3 shows the band structure of the $[Cu@C_{20}]_\infty$ tube, whose structure is shown in the Fig. 1. Comparison with calculation results of the $[C_{20}]_\infty$ tube[22] indicates that the bands $E_1(1)$- $E_8(1)$ and $E_1(2)$-$E_{16}(2)$ are similar to those of the pure carbon tube. The bands $E_9(1)$-

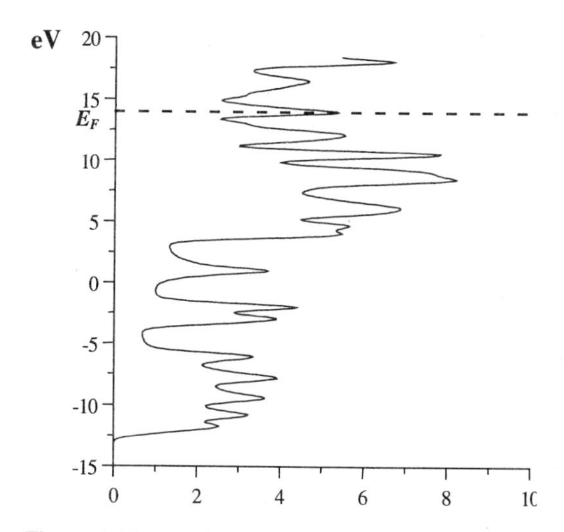

Figure. 4. Electron density of states of the $[Cu@C_{20}]_\infty$ tube

$E_{10}(1)$ and $E_{17}(2)$-$E_{18}(2)$ are similar 4s and 3d bands of $[Cu]_\infty$. In the vicinity of Fermi level, the branches of the carbon and metal subsystems are intersected, which point to their considerable interaction and joint participation in conductivity. In the pure carbon armchair tube, the Fermi level is known to be located in dip of electron density of states. Intercalation of the transition metal fills this dip (Fig. 4), which results in a growth of electron density of states in the Fermi level and consequently in a considerable increase in conductivity of the armchair carbon nanotubes.

ACKNOWLEDGEMENT

This work has been supported by Russian Basic Researches Foundation, grant 97-03-33729.

REFERENCES

1. S. Iijima, *Nature*, 351: 56 (1991).
2. S. Iijima and T. Ichihashi, *Nature*, 363: 603 (1993).
3. W. Kratschmar, *Nuovo Cimento*, 107: 1077 (1994).
4. T. W. Ebbesen, *Physics Today*, 49: 26 (1996).
5. Y. K. Tsang, Y. K. Chen, P. J. F. Harris, and M. L. H. Green, *Nature*, 372: 152 (1994).
6. O. Zhou, R. M. Fleming, D. W. Murphy, and C. H. Chen, *Science*, 263: 1744 (1994).
7. R. Beyers, C.-H. Kiang, R. D. Johnson, and J. R. Salem, M. S. de Vries, C. S. Yannoni, D. S. Bethune, H. C. Dorn, P. Burbank, K. Harich, and S. Stevenson, *Nature*, 370: 196 (1994).
8. T. Yamabe, *Synth. Metals*, 70: 1511 (1995).
9. L. Lou, P. Nordlander, and R. E. Smalley, *Phys. Rev.* B., 52: 1429 (1995).
10. L. Langer, V. Bayot, E. Grivei, J.-P. Issi, J. P. Heremans, C. H. Olk, L.Stockman, C. Van Haesendonck, and Y. Bruynseraede, *Phys. Rev. Lett.* 76: 479 (1996).
11. M. Kosaka, T. W. Ebbesen. H. Hiura, and K. Tanigaki, *Chem.Rev.Lett.*, 233: 47 (1995).
12. W. A. de Heer, A. Chatelain, and D. Ugarte, *Science*, 270: 1179 (1995).
13. L. A. Chernozatonskii, Yu. V. Gulyaev, Z. Ja. Kosakovskaja, *Chem. Phys. Lett.*, 233: 47 (1995).
14. V.M.Galitskii, B.M.Karnakov, V.I.Kogan., *Problems of Quantum Mechanics*, Moscow, Nauka (1984) (in Russian).
15. G.A.Korn, T.M.Korn, *Mathematical Handbook for Scientists and Engineers*, New York, McGrow-Hill, (1961).
16. G.N.Watson, *A Treatise on the Theory of Bessel Functions,* (1945).
17. L. D. Landau and E. M. Lifchitz, *Quantum Mechanics*, Moscow, Nauka (1974) (in Russian).
18. V. V. Nemoschkalenko and V. N. Antonov, *Methods of Computational Physics in Solid State Theory. Band Structure of Metals*, Kiev, Naukova Dumka (1985) (in Russian).
19. P. N. D'yachkov, O.M.Kepp, and A. V. Nikolaev, *Dokl. Akad. Nauk*, 365: 215 (1999).
20. O.M.Kepp, and P. N. D'yachkov, *Dokl. Akad. Nauk*, 365:350. (1999)
21. P. N. D'yachkov, O.M.Kepp, and A. V. Nikolaev, *Macromol. Symp.* 136:17 (1998).
22. J.W.Mintmire,B.I.Dunlap, and C.T.White, *Phys. Rev. Lett.* 68:631 (1992).

COMPARATIVE STUDY OF A COILED CARBON NANOTUBE BY ATOMIC FORCE MICROSCOPY AND SCANNING ELECTRON MICROSCOPY

P. Simonis[a], A. Volodin[b], E. Seynaeve[b], Ph. Lambin[c] and Ch. Van Haesendonck[b],

[a]Laboratoire de Spectroscopie Moléculaire de Surface, University of Namur, rue de Bruxelles 61, B-5000 Namur, Belgium;
[b]Laboratorium voor Vaste-Stoffysica en Magnetisme, K. U. Leuven, Celestijnenlaan 200D, B-3001 Leuven, Belgium;
[c]Laboratoire de Physique du Solide, University of Namur, rue de Bruxelles 61, B-5000 Namur, Belgium

ABSTRACT

Depending on the way they are produced, carbon nanotubes can present different morphologies. For example, coiled carbon nanotubes can be found in the nanotube materials produced by catalytic decomposition of acetylene. We have used atomic force microscopy (AFM) in non-contact mode to study this particular form of nanotube material. A coiled carbon nanotube has been investigated by AFM and by scanning electron microscopy (SEM). By so doing, geometrical and structural characteristics have been examined. These experimental work have been compared to previous results, giving us a way to determine wether the coils are structures made up of elastically deformed tubes or structures containing topological defects like pentagons or heptagons made during the catalytic growth process.

INTRODUCTION

Since their discovery by Iijima in 1991[1], as a by-product collected on the electrodes used for production of the C_{60} by the arc discharge method, carbon nanotubes have generated much interest among the scientific community. The publications dealing with carbon nanotubes clearly outnumber today the ones related to fullerenes. Two reasons account probably for this success. Firstly, although fullerenes still lack of demonstrated economically viable applications, several promising breakthroughs have already been realised with carbon nanotubes: they are used as AFM tips[2], they have been incorporated in flat panel displays[3], they have been used as electron emitters[4], and more interesting realisations are still to come. Secondly, because of their serendipitous electronic structure, carbon nanotubes, and especially single wall ones have allowed to discover or re-discover the very foundations of solid state physics. Not to mention the astonishing dependence of their electronic versus atomic structures, single wall carbon nanotubes are ideal (however real) objects for demonstrating the existence of the one-dimensional "particle on a string" system that is often taught before the 3-dimensional "particle in a box" in every introductory solid state physics textbook. More sophisticated physics has been recently extracted from experiments with single wall carbon nanotubes: Demonstration of the weak localization[5], of the Aharonov-Bohn effect[6] and their behaviour as a Luttinger liquid[7].

Depending on the way they are produced, carbon nanotubes exhibit different morphologies. The arc discharge method can give multiwall or single wall nanotubes, sometimes arranged in ropes but mainly straight ones because they are pulled by the electric field between the electrodes. On the other hand, HRTEM images of the nanotubes grown by the catalytic decomposition of acetylene often show predominantly multiwalled, but also single walled and helicoïdal nanotubes regularly wound as rings or coils. The coiling process during the growth is not yet fully explained. Some people think that coiling arises during the catalytic growth, because of the anisotropic extrusion velocities[8] related to the different facets exhibited by the catalyst nanoparticle on which the nanotube is growing. It might also be caused by some steric hindrance effect during the growth. Although such coiled nanotubes have been frequently seen by Transmission Electron Microscopy (TEM), very few STM or AFM pictures have been published, however, by any means atomic resolution pictures could not yet be recorded on these objects.

We believe that because of their remarkable shape, coiled nanotubes will reveal themselves as very interesting objects in the future, from both theoretical and applied points of view. Indeed, some theoretical study[9] has already shown that coils are metallic or semimetallic depending on their characteristics. On the other hand, we can think of coiled single walled nanotubes as the archetype of the nanospring or of the nanosolenoïd to be used for nano-mechanical or nano-electrical applications. We anticipate that the magnetic properties of coiled nanotubes shall be of special interest.

A helicoïdal nanotube can be described by its inner and outer radii and its pitch (i.e. the distance between two adjacent loops of the helix). The coiling can be achieved in two ways: either after growth or during growth. First, after growth, by applying an external mechanical stress and elastically bending a straight nanotube in such a way that it closes upon itself. It will stabilise with the van der Waals attraction between the contacting loops. Because they are made of a limited number of adjacents loops in close contact to each other and because the minimal diameter for this kind of tubes is about 200 nm their aspect ratio is more comparable to a ring than to a helix. Another possibility is linked to the inclusion of pentagon-heptagon pair-defects during growth in the hexagonal lattice. A pentagon will induce a positive curvature and a heptagon a negative one. Each defect pair can twist the tube by an angle between 0° and 36°. This kind of helicoïdal structure is composed of straight tubes segments connected by defects in order to induce the coiling. TEM images of coils seem to suggest that the last case is more widespread, because they often show polygonized multiwall tubes.

Recently Avouris et al.[10] have reported a simple and ingenious process for producing nanotube rings from nanotubes ropes produced by laser ablation. The tubes are then shortened and twisted by using an acidic treatment in an ultrasonic bath. After the sonication, SEM reveals the presence of numerous rings having diameters between 250 and 500 nm. Their explanation is that the rings or coils are formed by bending straight nanotubes or ropes during the ultrasonication process and that the structures are stabilised by van der Waals attraction between adjacent loops provided that the overlap is long enough. They expect to achieve kinetic monitoring of the ring formation.

In this paper we report on the observation by two different techniques: SEM and AFM in air of one catalytically grown coiled carbon nanotube The observed coil has a length of

3.5 µm and a small thickness. AFM has been chosen because it is a powerful technique to see the topography of coils and to measure their elasticity. In this case, we first tried modulation force microscopy (MMF) measurements in order to deduce the Young Modulus of the tube. The kind of tip is decisive for the MMF measurements. A very fine and soft tip is needed in order to get some resolution and avoid displacing nanotubes. We did not get any results in this case because the tip was either too big or too rigid. The observation of such a small tube by AFM and SEM will however be helpful to discuss the possible origins of the coils.

EXPERIMENTAL

The carbon nanotubes were produced in the NMR laboratory of the University of Namur by catalytic decomposition of acetylene at 700°C using nitrogen as the carrier gas in a fixed bed flow reactor at atmospheric pressure. After the growth, the catalyst Co/zeolithe NaY prepared by impregnation was removed from the carbon material by dissolution in fluorhydric acid. In order to remove the catalyst, the residual material was dissolved in $KMNO_4$. The purified nanotubes were diluted in isopropanol, sonicated for half an hour and then deposited on a passivated silicon substrate. This latter had been preliminary patterned by e-beam lithography. A pattern of gold crosses had been made in order to trace back the nanotubes in AFM after SEM measurements.

The AFM images were taken in the VSM laboratory of the K.U. Leuven with an Autoprobe CP microscope (Park Scientific Instruments) operated in the tapping mode. The tapping mode minimizes the lateral forces and avoids to drag the tube. The silicon tip had an apex radius of about 10 nm. The softness of the tip is of type B (rigidity constant of 3.2 N/m). With this kind of tip, it is possible to avoid displacing the nanotubes and coils during the scanning. The tip moved from 2 to 5 nm from the surface with a resonance frequency of the cantilever of 90 kHz. The SEM study was carried out at the same place using a Philips XL-30FEG microscope. The electrons were accelerated at 5kV and a magnification from 10 000 to 100 000 x allowed us to see the details of the coil.

RESULTS

Figure 1 shows an AFM image recorded in tapping mode of one regular coiled nanotube of 3.5 µm length. One can clearly see the effect of a double tip that produces a ghost structure at 0.5 µm left from the real coil. The coil has two different parts characterised by different pitches and orientations. Between both parts, one can see three loops sticking together. They are two possibilities for tentatively explaining such a configuration. A first hypothesis could be that we have two separate coils and that these two different coils are intertwined, with the first loop of one coil screwed inside the first two loops of the other one. This situation is unlikely to occur because it requires that the two tubes have the very same diameter and pitch. This is not the case here as we shall demonstrate later. The other possibility is that, due to a bending, the tube takes two different orientations. Between both straight segments, a couple of loops are compressed in order to accommodate the deformation energy just like a regular spring will do.

Figure 2 shows a closer view of the upper part of the tube. We observe that on this part, the pitch of the helix is approximately constant and that it can be measured with the help of longitudinal height profiles. The corresponding line cut (A) is shown in Figure 3 from which a pitch value of about 125 nm can be deduced, by averaging over four successive loops. This value is in reasonable agreement with the value of 132 nm determined by spatially Fourier Transforming the AFM data.

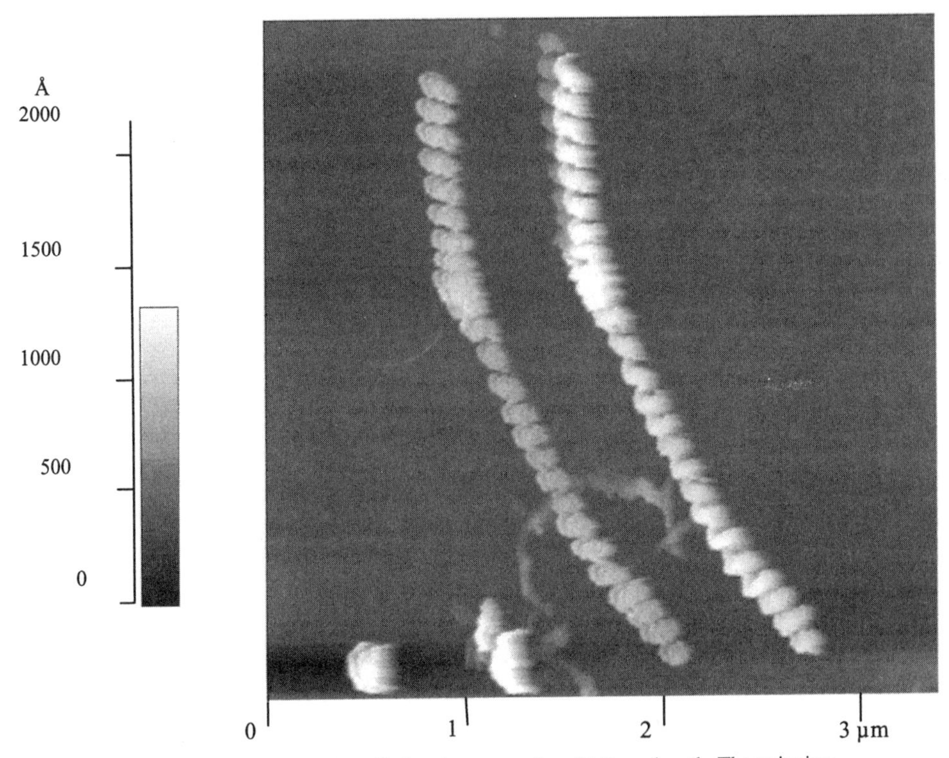

Figure 1. AFM image of a coiled carbon nanotube of 3.5 μm length. The twinning is a double tip effect. There is a deformation in the middle of the tube and thre loops sticking together

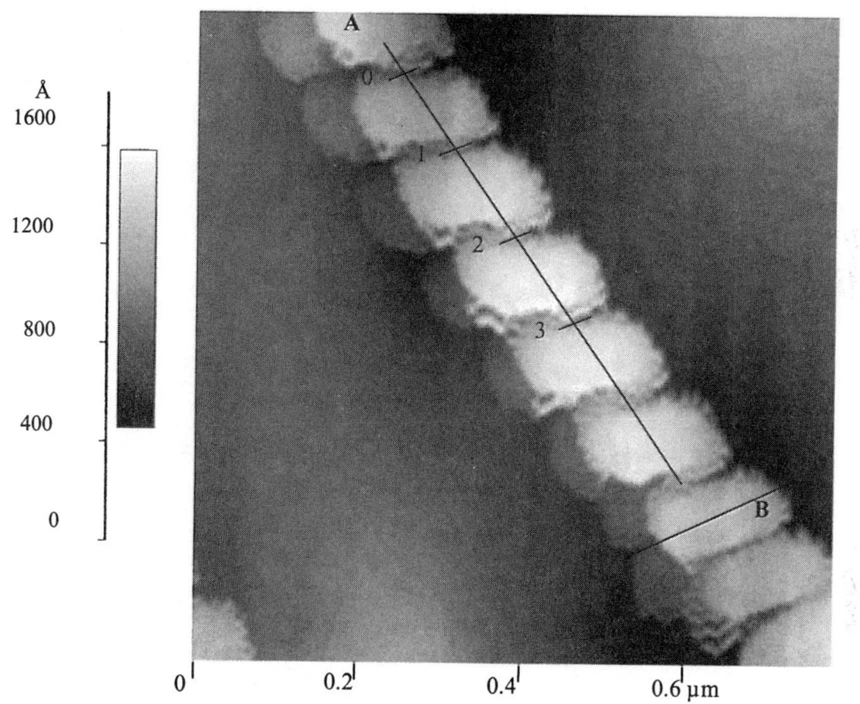

Figure 2. AFM image of the upper part of the tube showed on figure 1. One can see that the pitch is constant on this section.

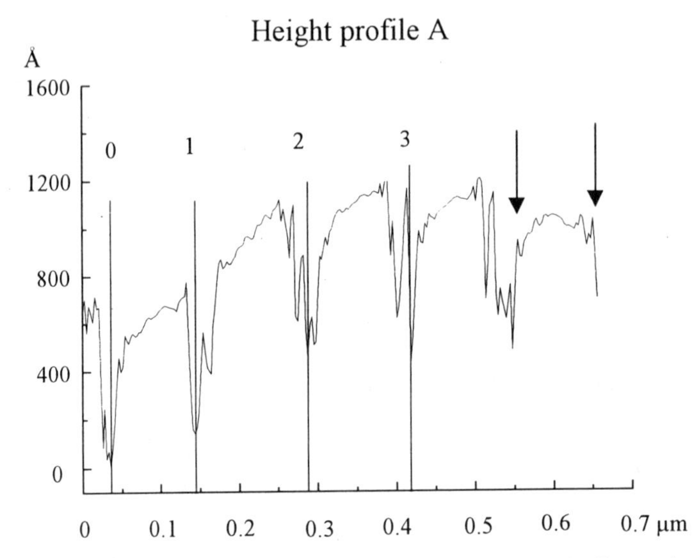

Figure 3. Height profile of the coiled nanotube corresponding to the line cut A on figure 2. The periodicity of the coil is well seen.

Table 1. Distance between different loops of the coil corresponding to the line cut showed on figure 3.

	markers	Distance between markers (nm)
height profile A	0-1	125.0
	1-2	122.5
	2-3	127.5

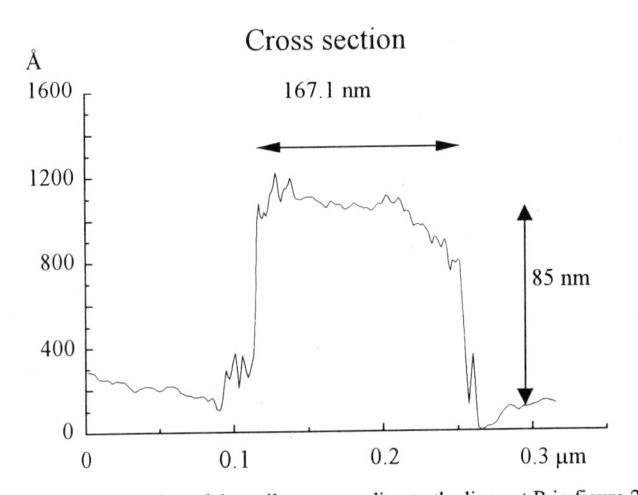

Figure 4. Cross section of the coil corresponding to the line cut B in figure 2.

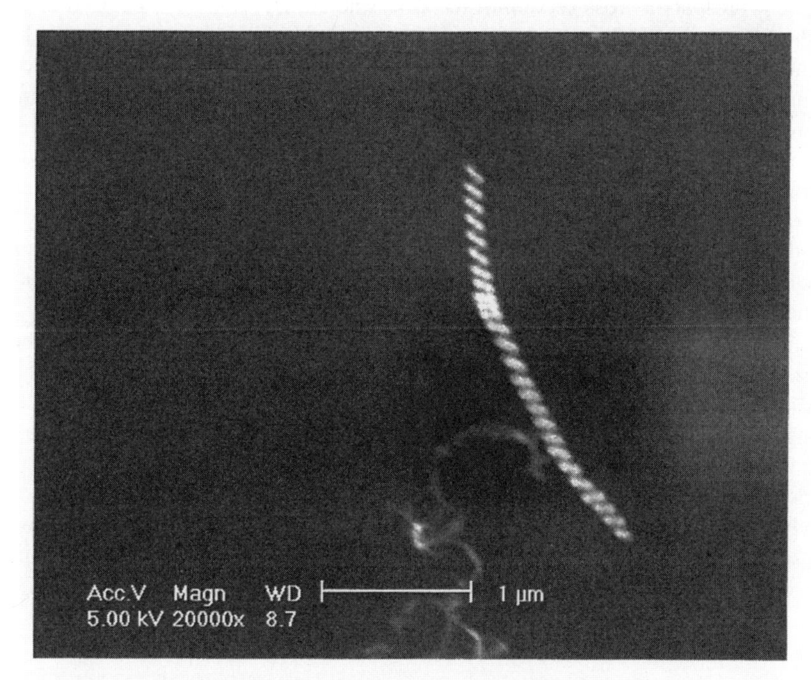

Figure 5. SEM image of the same coiled carbon nanotube. One can see the same bending in the middle of the tube with the three loops closer.

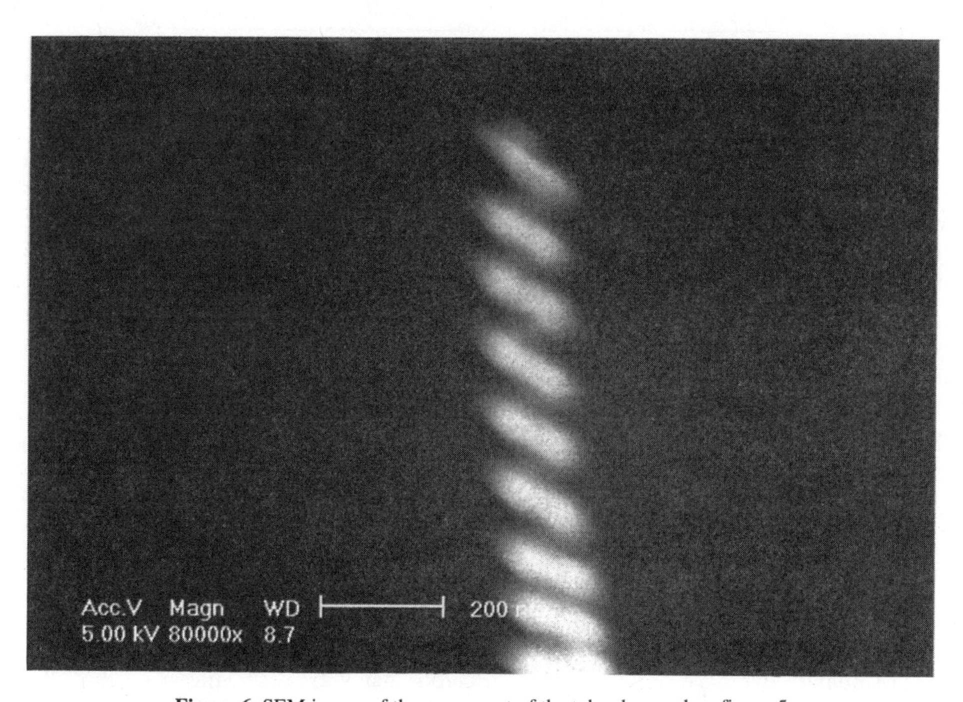

Figure 6. SEM image of the upper part of the tube showned on figure 5.

Figure 4 presents a cross section of the coil (line cut B) which can be used to estimate the width of the coil at 167 nm and its height at about 85.8 nm. Such a discrepancy between height and width, which normally should both reflect the same value of the diameter of the helix, can be explained by the effect of convolution with the tip and by the AFM mode used for this measurement. Indeed, with the tapping mode, the tip pushes the coil towards the substrate and squashes it. Therefore, the measured height will not reflect the exact diameter of the coil that is more precisely determined from the diameter of the helix. The effect of the convolution of the tip with the tube is only important at the very beginning of both sides of the tube because the tip is very fine (10 nm radius). On the other hand, the tip size does not affect the measurement of the pitch because the tip radius has no effect on the spatial periodicity.

Figure 5 shows a SEM image taken on the same object. One can clearly see there that it consists of only one single coil. The pitch is also different in the upper and the lower sections and we can distinctly see the three closer loops in the middle section of the tube. This image is in favour of the interpretation of one single coiled nanotube acting as a string instead of two separate coils fitted together.

A closer look at the upper part of the tube shown on the SEM image of figure 5 can be used to check the diameter of the coil. One finds a value of 153 nm. This is very close to the one measured by AFM. We can also measure the outer diameter of the multiwall nanotube forming the coil: 46 nm in comparison to about 80 nm determined by AFM from the

Table 2. Comparison between AFM and SEM measurements for the same coiled carbon nanotube.

	diameter of the coil	outer diameter	pitch
AFM	167 nm	80 nm	125 nm
SEM	153 nm	46 nm	113 nm

distance between the two arrows of Figure 3. The pitch is 113 nm and is also close to the AFM value. The difference between the outer diameters of the nanotubes can be explained by the fact that the AFM tip is too large to penetrate between two loops. The sides of the tip will rapidly interact with the sides of the loops and consequently, the outer diameter will be larger when measured by AFM.

The study of this nanotube makes us believe that it did not result from the sonication process. Indeed, it is very unlikely that such a coil with so small a pitch and a diameter can be produced by sonication.

These results describe a coiled object rather different from the nanotubes rings produced by Avouris[10] from the sonication of straight tubes. We must emphasize that the coil analysed here is a multiwalled nanotube and not a rope of single walled one as in the Avouris experiment. It is most probable that this tube contains a lot of defects as pentagon-heptagon pairs and is not bent elastically.

CONCLUSION

In summary, we have observed one coil carbon nanotube with SEM and AFM. The coil has a diameter of 153 nm that is the smallest one eve reported by AFM to our knowledge. We have compared our results with previous study[10] of coils and rings and suggested an intrinsically growth process for coiled carbon nanotubes material. A more detailed study should be done on helicoïdal structures made by sonication in order to analyse whether such long coils are possible or not and to conclude about the coil formation.

ACKNOWLEDGMENTS

This work was supported by the Belgian Interuniversity programme on "Reduced Dimensionality Systems" (PAI-IUAP 4/10) sponsored by the Belgian Office for Scientific, Cultural and Technical Affairs.

We thank A. Fonseca and J. B.Nagy from the RMN Laboratory of Namur for providing us with the nanotube material.

P. S. acknowledges financial support from FRIA.

REFERENCES

[1] S. Iijima, *Nature* 354:p 56 (1991)
[2] Ch. M. Lieber, *Solid State Comm.* 107:607 (1998).
[3] Q. H. Wang et al., *Appl. Phys. Lett.* 72:2912 (1998).
[4] Y. Saito, S. Uemura, K. Hamaguchi, *J. Appl. Phys.* 37:L346 (1998).
[5] L. Langer et al. *Phys. Rev. Lett.*, 76:479 (1996).
[6] A. Bachtold et al. *Nature*, 397:673 (1999).
[7] M. Bockrath et al. *Nature*, 397:598 (1999).
[8] S. Amelinckx et al., *Science* 265:635 (1994)
[9] K. Akagi et al., *Phys. Rev. Lett.* 74:2307 (1995)
[10] R. Martel, R. Shea and P. Avouris, *Nature* 398:299 (1999)

INVESTIGATION OF THE DEFORMATION OF CARBON NANOTUBE COMPOSITES THROUGH THE USE OF RAMAN SPECTROSCOPY

C.A.Cooper[1] and R.J.Young[1]

[1]Manchester Materials Science Centre
UMIST / University of Manchestér
Grosvenor Street
Manchester M1 7HS

ABSTRACT

Since the discovery of carbon nanotubes in 1991, the study of these new carbon materials has undergone rapid development in this fast moving research area. The deformation micromechanics of single-walled carbon nanotube (SWNT) and multi-walled carbon nanotube (MWNT) particulate nanocomposites have been studied using Raman spectroscopy. SWNTs and MWNTs prepared by two different methods (pulsed-laser and arc-discharge) have been used as reinforcement for a polymer matrix nanocomposite. A four-point bend test has been employed to deform samples of a dispersion of SWNT and MWNTs in epoxy resin to follow the deformation of the nanotubes with Raman spectroscopy. Both SWNT and MWNTs exhibit well-defined Raman peaks. It has been found that for all samples deformed, the Raman peak position chosen shifted to a lower wavenumber upon application of a tensile strain. In this way it has been possible to demonstrate that the effective modulus of SWNTs dispersed in a composite is in the range 100–500 GPa. In addition, it has been found that the bands in the Raman spectrum of SWNTs shift to a higher wavenumber when dispersed in ethanol and stays at this higher position even when the solvent is evaporated.

INTRODUCTION

Composites are formed when two or more materials are joined to give a combination of properties that cannot be attained in the original materials. Composite materials are selected to give desirable improvement of properties such as stiffness, strength and weight and may be placed into three categories: particulate, fibre and laminar, based on the shape of the reinforcing phase. In general it is found[1] that the highest levels of reinforcement are obtained through the use of continuous fibres although such composites are often difficult to fabricate. Composites consisting of reinforcing particles as a dispersed phase are considerably easier to fabricate and therefore used widely, although the levels of reinforcement obtained are not particularly high.

Science and Application of Nanotubes, edited by Tománek and Enbody
Kluwer Academic / Plenum Publishers, New York, 2000

Over recent years there has been some important developments in the use of Raman spectroscopy to follow the mechanisms of deformation in fibre-reinforced composites[2]. Stress transfer from the matrix to the fibres has been demonstrated from significant Raman band shifts in the reinforcing fibres. This article considers the extension of the Raman technique to composite systems where the reinforcing phase is in the form of particles (i.e. nanotubes) rather than fibres and therefore assess the reinforcement ability of the nanotubes.

In addition to the spherical structure of C_{60}, it is possible to synthesize carbon nanotubes and nested concentric carbon nanotubes[3]. A nanotube can be considered as a hexagonal arrangement of carbon atoms arranged in sheet form, which has been rolled up to form a tube. The tubes can either be open ended or have caps formed from half a C_{60} molecule at either end. Carbon nanotubes are thought to have remarkable mechanical properties with theoretical Young's modulus being quoted as between 1 TPa[4] and 5 TPa[5]. The calculated theoretical tensile strength of carbon nanotubes is about 200 GPa[4]. These high values of mechanical properties make them ideal as reinforcements in nanocomposites. Raman spectroscopy has been used to characterize single-walled carbon nanotubes and multi-walled carbon nanotubes. It has also been possible to follow particle deformation in the particulate composite systems from stress-induced Raman band shifts.

EXPERIMENTAL

Materials and Preparation

The SWNTs were supplied by Rice University, USA (prepared by pulsed laser vaporization process)[6] and Sussex university, UK (arc-discharge prepared with Ni / Y catalyst)[6]. The Rice University SWNTs were supplied suspended in a basic (NaOH, pH 10) non-ionic surfactant/water solution which were re-suspended in ethanol by diluting and centrifuging several times. The MWNTs were supplied by Dynamic Enterprise Ltd., UK (arc-discharge method), and Sussex University, USA (arc-discharge method). A two-part cold-curing epoxy resin consisting of 100 parts by weight of Araldite LY5052 resin to 38 parts by weight of HY5052 hardener was employed both as the composite matrix and substrate for the beam specimens. The resin was cured at room temperature for 7 days in a square 'picture frame' mould then cut into strips measuring $3 \times 10 \times 60 \pm 0.2$ mm. A carbon nanotube / ethanol mixture was added to uncured epoxy resin and sonicated for two hours before leaving in a vacuum oven overnight to evaporate the solvent. The hardener was added and the mixture stirred well to distribute the nanotubes and again placed in a vacuum oven for about 30 min to remove any air bubbles. The epoxy resin / nanotube mix was applied to the surface of the beam and was cured at room temperature for 7 days before testing.

Raman Spectroscopy

The spectra were obtained using a Renishaw 1000 Raman system using the 633 nm red line of a 25mW He-Ne laser. The intensity of the laser beam on the sample surface using a ×50 microscope objective lens was about 1.3 mW. The peak values were derived by using Lorentzian routines fitted to the raw data obtained from the CCD detector.

The beam samples were inserted into a four-point bending rig and placed on the Raman microscope stage. The configuration of the beams under tensile loading can be seen in Fig 1. The surface strain was measured using a resistance strain gauge, of gauge factor 2.09, bonded to the surface of the samples using a cyanoacrylate adhesive. Strain was applied successively and several Raman spectra were acquired at 0.25% strain intervals with the polarisation of the laser beam parallel to the direction of tensile strain.

RESULTS AND DISCUSSION

Single-Walled Nanotubes

SWNTs were obtained from two different sources, both using different methods of preparation, one using a pulsed-laser vaporization process, (SWNT-*P*), and the other an arc-discharge method (SWNT-*A*). In addition, the pulsed-laser prepared SWNTs were transported in an aqueous suspension containing a surfactant to ensure sufficient wetting. Raman spectroscopy was used both to characterize and follow the deformation behaviour of the SWNTs obtained from each source.

Fig 2 shows the Raman spectrum of SWNT-*A*. The strong Raman bands at 1532 and 1569 cm^{-1} (G band) are a combination of the A_{1g}, E_{1g} and E_{2g} vibrational modes[7]. The band

Tensile Deformation

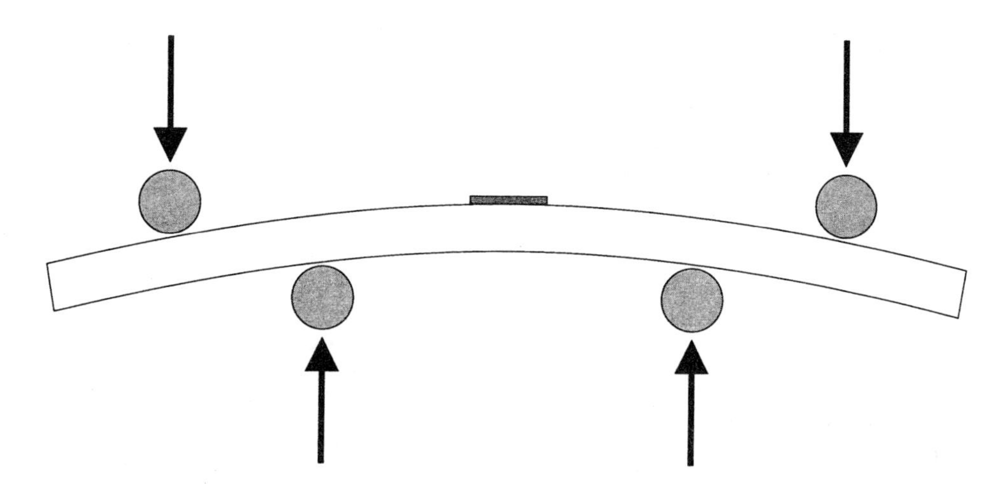

Figure 1. Composite beam under four-point bending to give tensile loading on the top surface

at 1308 cm^{-1} (D band) arises from the disorder-induced mode A_{1g}, and the band at 2610 cm^{-1} (D' band) is an overtone of the D band and corresponds to the vibrational breathing mode A_{1g}[8].

Fig 3 shows the Raman spectrum of SWNT-*P* which is similar to that obtained from SWNT-*A*. It can be seen that all of the Raman band peak positions for SWNT-*P* are at a higher Raman band position.

The band positions were upshifted for the SWNT-*P* as shown in Fig 4, with the D' band being shifted by almost 30 cm^{-1} to 2639 cm^{-1}. The reason for this upshift is unclear but may be a result of the different production methods used or because of the SWNTs re-suspension in a solvent. The response of SWNTs Raman spectra after being immersed in a solvent was assessed by comparing the Raman peak positions of SWNT-*A* in air, immersed in ethanol, after evaporation and after 7 days evaporation.

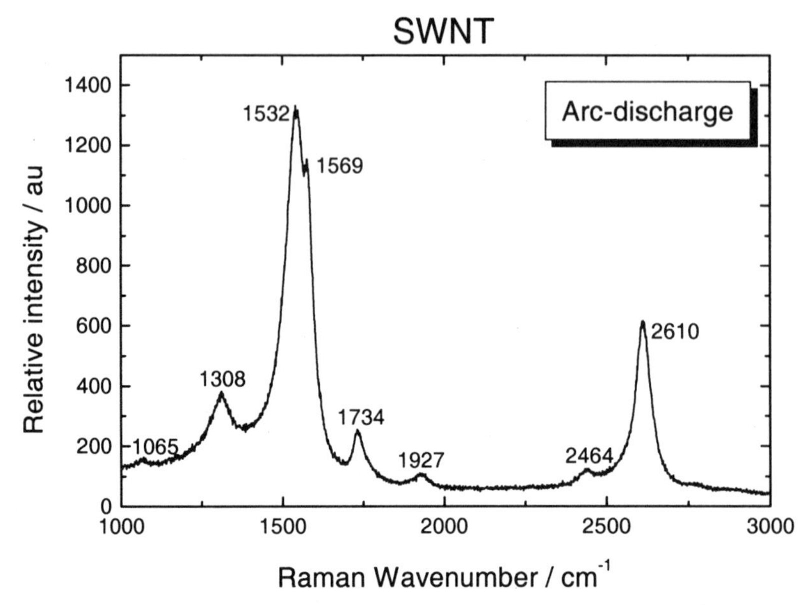

Figure 2. Raman spectrum of SWNT-*A* in the range 1000 - 3000 cm^{-1}

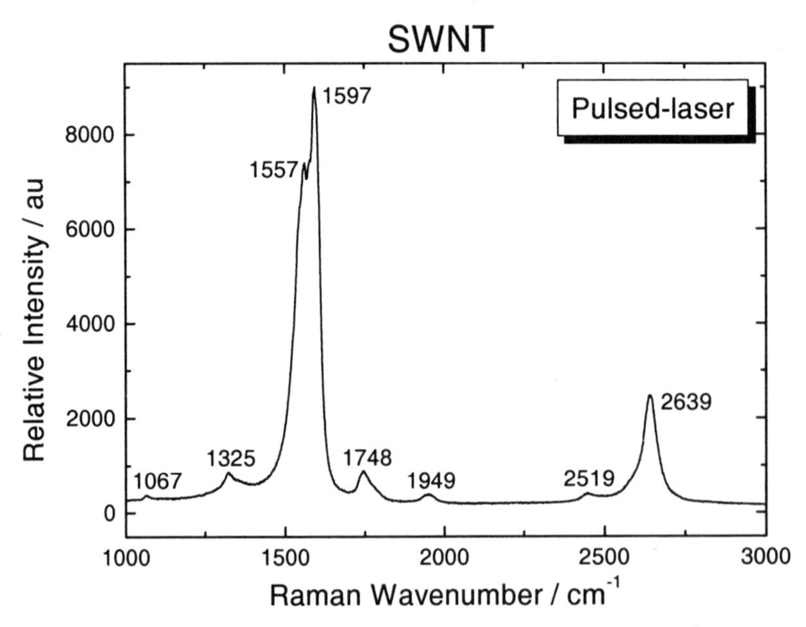

Figure 3. Raman spectrum of SWNT-*P* in the range 1000 - 3000 cm^{-1}

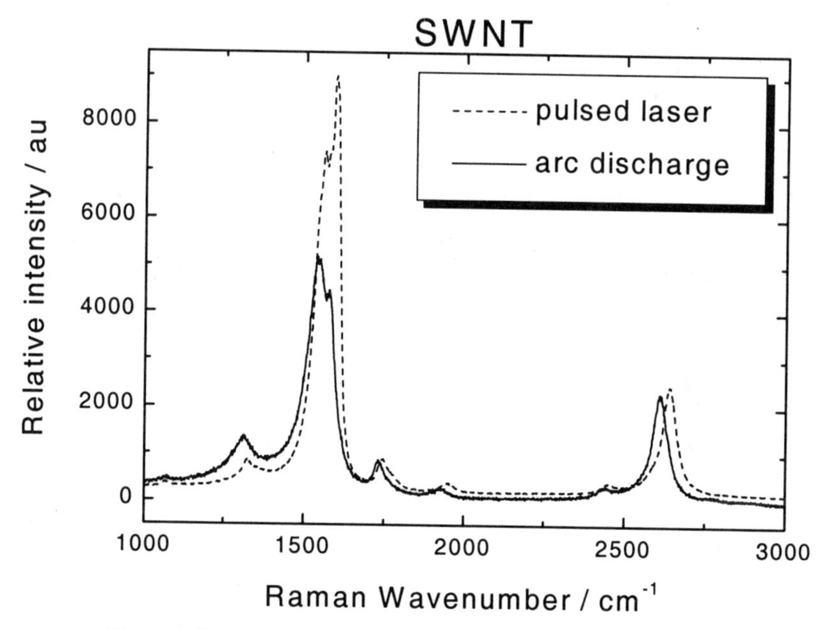

Figure 4. Comparison of spectra obtained from SWNT-*P* and SWNT-*A*

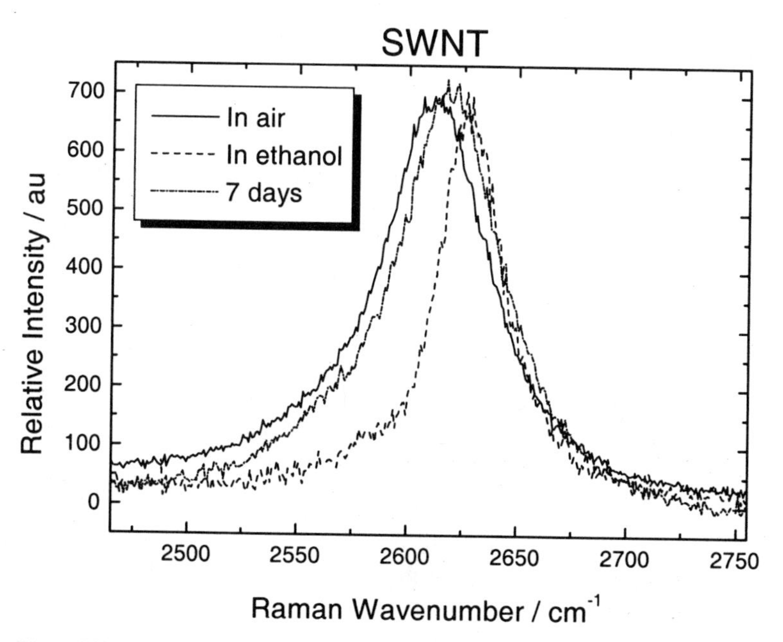

Figure 5 Variation of the SWNT-*A* D' peak with presence or absence of solvent (ethanol)

It can be seen from Fig 5 that the SWNT-*A* shifts to a higher position when the nanotubes are immersed in ethanol. The response of the SWNT-*A* D' Raman peak position to being immersed in ethanol is summarised in Table 1.

It can be seen from Table 1 that the SWNT-*A* D' Raman peak position shifts to a higher wavenumber after being immersed in ethanol and does not go back to its original position even after evaporation indicating a permanent change. Wood and Wagner [9] have followed the response of the SWNT D' Raman band when the nanotubes are immersed in different liquids. They found that the D' band shifted to a higher wavenumber; the amount depending on the liquid used.

Care was taken not to introduce thermal residual stresses into the nanotubes when fabricating the epoxy resin composites by allowing them to cure for 7 days at room temperature rather than using a thermal cure. Table 2 summarises the D' Raman peak positions for the SWNTs after their dispersion in an epoxy resin matrix

Table 1. Variation of the SWNT-*A* D' Raman peak position to the presence or absence of ethanol

Experimental Conditions for SWNT-*A*	SWNT-*A* D' Raman Peak Position / cm^{-1}
In air	2610
In ethanol	2628
After immediate evaporation	2619
After 4 days evaporation	2619
After 7 days evaporation	2620

Table 2. Response of SWNT D' Raman peak position after dispersion in epoxy resin

Experimental Conditions for SWNTs	SWNT-*A*	SWNT-*P*
In air	2610	2640
In ethanol (after 7 days evaporation)	2620	2639
In epoxy resin matrix	2625 ± 1.3	2637 ± 1.0

± = standard deviations

It can be seen that the D' Raman band position for the SWNT-*P* has not increased after being dispersed in the epoxy resin. However, the D' Raman band position for the SWNT-*A* has moved to a higher wavenumber. This may be because the SWNT-*A* interact further with the epoxy chemicals.

Deformation of SWNT Nanocomposites

The SWNTs were distributed dilutely into an epoxy matrix resulting in a low volume fraction (< 0.1 wt %) nanocomposite. The nanocomposites were subjected to tensile deformation in the four-point bending rig. The variation of the D' Raman peak position with strain for the SWNT-*A* composite can be seen in Fig. 6.

The D' Raman peak position moves to a lower wavenumber with tensile strain indicating that the macroscopic stress on the composite deforms the SWNTs. This is in agreement with the deformation behaviour of other carbon materials and in particular, carbon fibres[10]. Fig 7 shows the strain-induced Raman band shift for the SWNT-*P* reinforced composite. It can be seen that the peak position of the 2640 cm^{-1} Raman band shifts significantly (about –8 cm^{-1} / % strain) to a lower wavenumber on application of a tensile strain. This demonstrates the reinforcement of epoxy resin with the SWNTs which is comparable with the behaviour of carbon fibre reinforced composites[10].

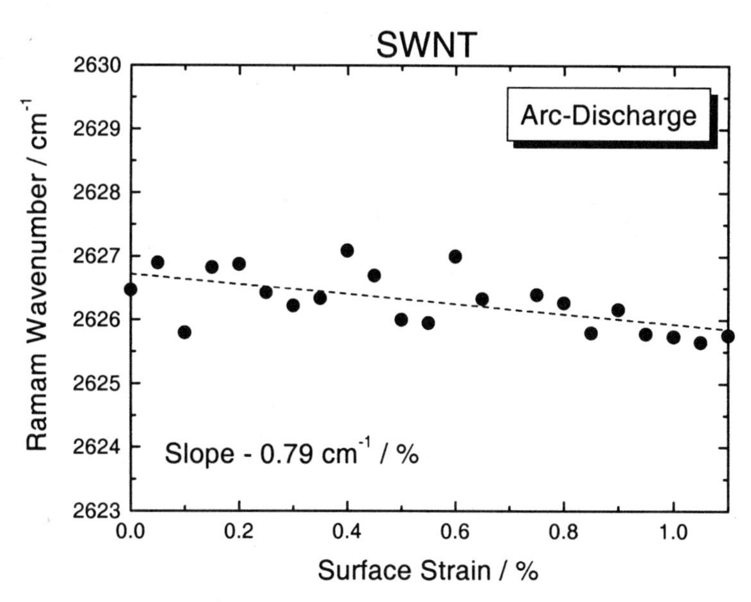

Figure 6 Variation of the 2610 cm^{-1} Raman peak position with tensile strain for SWNT-*A* dispersed in epoxy resin

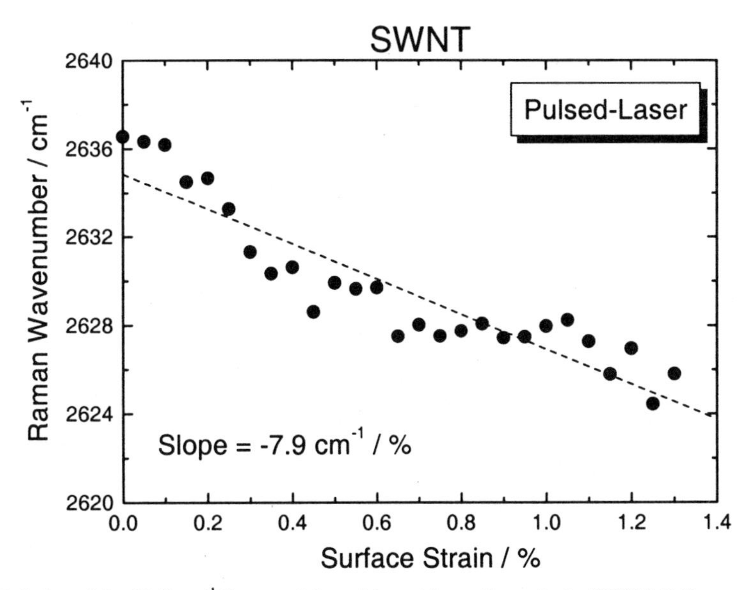

Figure 7. Variation of the 2640 cm^{-1} Raman peak position with tensile strain for SWNT-*P* dispersed in epoxy resin

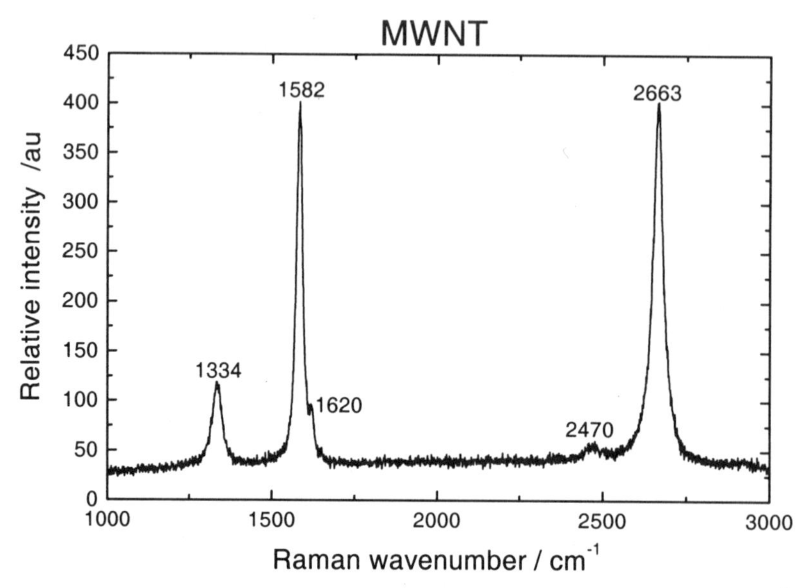

Figure 8. Raman spectrum of multi-walled carbon nanotubes in the range 1000 - 3000 cm^{-1}

Multi-Walled Nanotubes

The Raman spectrum and deformation behaviour of the MWNTs obtained from the two sources were very similar; both were produced by the arc-discharge method.

Fig 8 shows the Raman spectrum of the MWNTs. It should be noted that the Raman spectrum for MWNTs differs from that for the SWNTs. The spectrum resembles closely that of carbon fibres[12], and graphite[13]. A sharp peak is seen at 1582 cm^{-1} that corresponds to one of the E_{2g} modes[7]. A band is seen at 1334 cm^{-1} (D band) which has been attributed to the breakdown of translational symmetry produced by the microcrystalline structure[13]. The Raman band at 2663 cm^{-1} is the overtone of the D band.

Fig 9 shows the strain-induced Raman band shift for the MWNT-reinforced particulate composite. It can be seen that the peak position of the 2663 cm^{-1} Raman band shifts to a lower wavenumber on application of a tensile strain. The Raman band shift rate for the MWNTs is lower than that obtained for SWNT-*P* but higher than that obtained for SWNT-*A*.

Reinforcement of Epoxy Resin with Carbon Nanotubes

The results of this study show that the maximum Raman band shift rate is –8 $cm^{-1}/\%$ (for SWNT-*P*). Stress transfer from the epoxy resin to the nanotubes has been demonstrated from the stress-induced bandshift; therefore, the carbon nanotubes must provide some reinforcement. The D' Raman band of 400 GPa modulus carbon fibres (Thornel T50-U, untreated) at 2660 cm^{-1} is –32 $cm^{-1}/\%$. The Raman bandshift rate per unit strain is proportional to the fibre modulus[12]. Young showed that the bandshift rate increased as the fibre modulus increased. The bandshift rate for SWNT-*P* is –8 $cm^{-1}/\%$ which is approximately ¼ of that for the T50-U carbon fibres which implies that the nanotube modulus is at least 100 GPa. The carbon nanotubes however, have a 3-dimensional random distribution within the epoxy matrix which contrasts with the unidirectional alignment of the tested carbon fibres.

An orientation efficiency factor, η_o has been proposed by Cox [1], to take into account fibre orientation distributions and assumes that the matrix and fibre deform elastically. Values for η_o have been calculated by Krenchel [1], for different fibre orientation distributions and for three-dimensional random orientations, $\eta_o = 1/5$. Therefore, if this random orientation is taken into account, the carbon SWNT modulus could be as high as 500 GPa.

It is of interest to consider why the rates of band shift for the SWNT-*A* and MWNT are lower than for the SWNT-*P* material. There are a number of possibilities. They may have lower modulus values or poorer dispersion in the resin. The adhesion to the epoxy matrix may be worse than for the SWNT-*P* material or the aspect ratio of the nanotubes may be different. It is impossible to say at this stage which of these factors is the most important. It is significant, however, that the levels of reinforcement are different for the different nanotubes and it may be possible in the future to produce materials that given even higher levels of reinforcement.

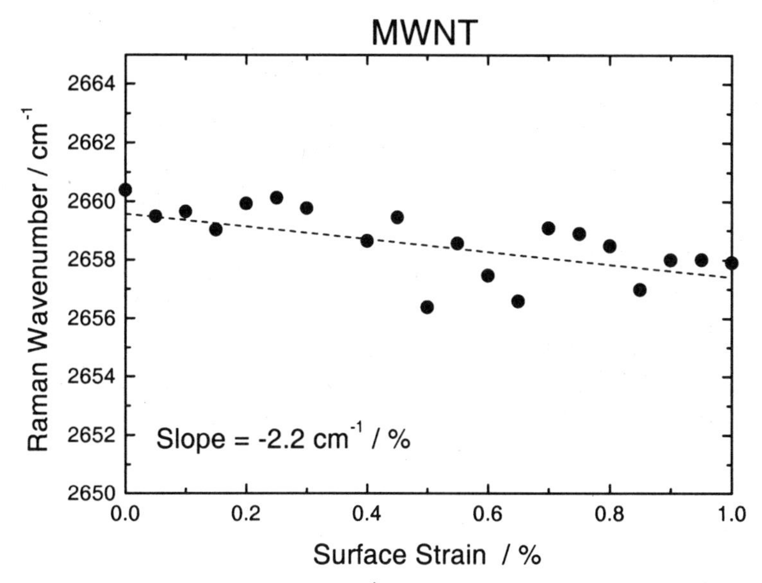

Figure 9. Variation of the 2663 cm^{-1} Raman peak position with tensile strain

CONCLUSIONS

It has been demonstrated that Raman spectroscopy can be used to characterise and follow the tensile deformation of a dilute dispersion of SWNT and MWNTs in a particulate composite system. Well-defined Raman spectra have been obtained from the dispersed phase and stress transfer between the different phases has been demonstrated from stress-induced Raman band shifts. Using this method, the effective modulus of the nanotubes in the composite can be determined. Possible reasons why the effective modulus measured for the nanotubes is less than theoretical predictions or experimental estimates are as follows:

- The nanotubes may not have been well-dispersed in the epoxy resin matrix.
- There was poor adhesion between the nanotubes and matrix.
- The aspect ratio of the nanotubes may have been too small to get optimum reinforcement
- The previous estimates of nanotube modulus may be too high.

It is impossible at this stage to say which of these factors is the most important but it is clear that the Raman technique has the potential to determine the efficiency of reinforcement by carbon nanotubes in any composite systems that may be developed in the future.

ACKNOWLEDGEMENTS

This work is part of a larger programme supported by the EPSRC and one author, C.A. Cooper is grateful for the EPSRC studentship.

REFERENCES

1. Hull, D. and Clyne, T.W., *An Introduction To Composite Materials,* 2nd Edition, Cambridge University Press, 1996.
2. Andrews, M.C., Bannister, D.J. and Young, R.J., *J. Mat. Sci.*, **31** (1996) 3893.
3. Rao, A.M., Richter, E., Bandow ,S., Chase, B., Ecklund, P.C., Williams, K.A.,Fang, S., Subbaswamy, K.R., Menton, M., Thess, A., Smalley, R.E., Dresselhaus, G., *Science,* **275**, (1997) 187.
4. Meyyappan, M. and Han, J., *Proto. Tech. Int.*, **6** (1998) 14.
5. Lourie, O. and Wagner, H.D., *J. Mat. Res.*, **13** (1998) 2418
6. Birkett, P.R. and Terrones, M., *Chem. in Brit.*, (1999) 45.
7. Nemanich, R.J. and Solin, S.A., *Sol. St. Comm.*, **23** (1977) 417.
8. Tuinstra, F. and Koenig, J.L., *J. Chem. Phys.* **53**, (1970) 1126.
9. Wood, J.R and Wagner, H.D., Private communication
10. Huang Y. and Young R.J., *Composites*, **26** (1995) 541.
11. Cooper, C. and Young, R.J., to be published.
12. Young, R.J., *J. Tex. Inst.*, **86** (1995) 360.
13. Nemanich, R.J. and Solin, S.A., *Phys. Rev. B*, **20** (1979) 392.

ELECTRONIC STATES, CONDUCTANCE AND LOCALIZATION IN CARBON NANOTUBES WITH DEFECTS

T. Kostyrko[1], M.Bartkowiak[1,2] and G.D. Mahan[2]

[1]*Institute of Physics, A. Mickiewicz University, ul. Umultowska 85, 61-614 Poznań, Poland*
E-mail: tkos@phys.amu.edu.pl
[2]*Solid State Division, Oak Ridge National Laboratory, Oak Ridge, TN 37831-6030, and*
Department of Physics and Astronomy, University of Tennessee, Knoxville, TN,37996-1200

INTRODUCTION

The discovery of carbon nanotubes (NT)[1] (see [2, 3] for reviews) provided the first genuine case of a one dimensional (1-d) system of almost macroscopic size. Many electronic and transport properties of 1-d systems are determined by the effects of disorder, which leads to localization of all electronic states (see [4] for a recent review). In nanotubes, the role of disorder remains a controversial issue. The experiments which show almost perfect ballistic conductance both in single wall[5] and multiwall[6] NTs imply that the localization length greatly exceeds the sizes of the studied NTs (typically 1-10 μm). On the other hand measurements often show a gradual decrease of the conductance with lowering of temperature. This, together with the dependence of the conductance on magnetic field may be explained[7] in terms of transition to a weakly localized regime.

On theoretical ground, one may expect noticeable dependence of the sensitivity to disorder on the NT's radius, with thicker NTs (individual multiwall NTs and ropes) resembling more 2-d or 3-d systems. Also, a change of chemical potential by means of doping or by a change of a gate voltage introduces more energy bands at the Fermi level thereby making the system more like a 2-d one.

The purpose of the present article is to discuss possible role of the disorder effects on the basis of a simple, yet still fairly realistic, model of single wall carbon nanotubes. The model is described by a tight binding Hamiltonian with one π-orbital per carbon atom. The disorder is created by randomly positioned point like defects of strength E, i.e., we assume that a site energy may take a value 0 or E with probability $1-c$ and c, respectively, where c denotes the concentration of defects.

The main advantage of our approach is that it allows us to obtain simple (sometimes even analytical) criteria for estimation of the role of disorder as a function of defects strength and concentration, as a function of NT radius and of the position of the chemical potential.

First we discuss the general formalism used in our computations. It combines a transfer matrix approach and a Green function method to describe scattering of electrons by barriers. A barrier is defined as a sequence of unit cells containing defects. We show that the transmission and reflection matrices, obtained from this approach can be directly related to a conductance calculated from a linear response theory, i.e. Kubo formula. As an interesting by-product of this analysis we show how a one particle lattice Green function for a perfect NT can be easily calculated using the matrix approach.

Next, we present results of the numerical calculation of the conductance from a single defect and show its scaling with NT radius and defect strength. The main conclusion of this part is that the point-like defect leads to appearance of a quasibound (resonant) state close to the Fermi energy of an undoped NT. The quasibound state significantly reduces transmission, cutting off one of the available conducting channels. We compare numerically the scattering from a point-like defect with a so-called 5-77-5 defect, obtained by rotation (by an angle of $\pi/2$) of one of the C-C bonds. We also present the results for systems with a small finite number of randomly distributed defects and explain how the conductance may be approximately interpreted semi-classically in terms of scattering from a single defect.

The subsequent section contains an analysis which confirms interpretation of the results for a single defect in terms of the scattering by the quasibound state. We present the dependence of the quasiparticle lifetime (for scattering from defects) on the energy. The inverse of the quasiparticle lifetime is shown to peak near the maximum in the one–particle density of states which originates from the quasibound state. These results are supported by exact computations using finite chains (of 3000 unit cells) with small concentration of defects. For weak defects we estimate the mean free path ℓ_{mfp}, and find it to be equal to the scattering length obtained from a semi-classical analysis of the reflection from a number of defects.

The last but one section is devoted to analysis of the electron localization due to the disorder. We show how one can use transfer matrix to calculate localization length. We present numerical results showing the dependence of the localization length on the energy, defect strength and defect concentration. The computations lead to a simple proportionality relation between the mean free path and the localization length in the limit of weak defect and small concentration: $\xi(0) \approx 3\ell_{\mathrm{mfp}}$.

Finally, in the last section we discuss the limitations of the model, and compare our results to the work of other groups. We speculate on the applicability of our results for the real systems and try to answer the question to what extent our results can be modified by effects neglected in the model.

SCATTERING THEORY APPROACH TO CONDUCTANCE

In the analysis of transport properties of 1-d systems one can start from a simple model of the experimental arrangement, which can be represented as follows:

```
(lead)(device)(lead)
```

We assume "leads" to be perfect and semi-infinite. Then, by use of a generalized version of a Landauer formula[8], one can find the electronic conductance G of the system by calculating a transmission T through the "device"

$$G = n_c \frac{e^2}{h} T = n_c \frac{e^2}{h} (1 - R) \tag{1}$$

Above, R denotes a reflection coefficient and n_c is a number of conducting channels (including spin). In our case, "device" will represent the disordered fragment of a

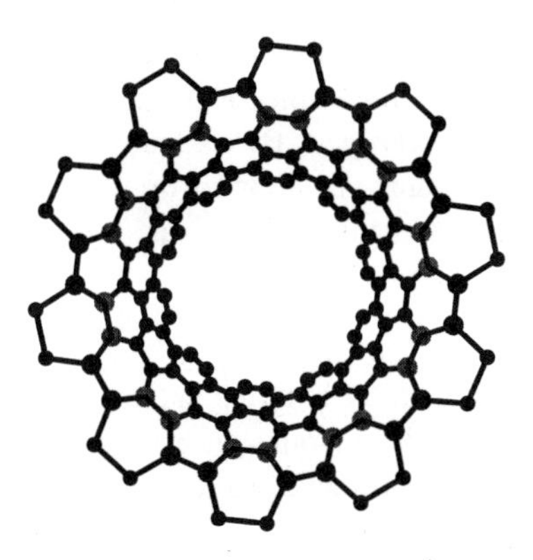

Figure 1: (10,10) Armchair nanotube with a and b atoms showed by light gray and dark gray filled circles, respectively.

nanotube, whereas "leads" will be taken as semi-infinite nanotubes of the same radius and helicity. By doing this, we separate the problem of coupling of the NT to realistic metallic contact from the effects of disorder in the "device".

Transfer Matrix Approach

In order to find the transmission, we formulate the scattering problem using a transfer matrix method[9]. For this purpose it is convenient to write the tight binding Hamiltonian for the NT in a form[10]

$$\mathcal{H} = \sum_j \left[a_{j+1}^\dagger t_j b_j + b_j^\dagger t_j^\dagger a_{j+1} + a_j^\dagger H_j^a a_j + b_j^\dagger H_j^b b_j + a_j^\dagger W_j b_j + b_j^\dagger W_j^\dagger a_j\right] \tag{2}$$

where a_j, b_j denote n_b-component electron operators for the a and b orbitals of j-th cell (see Fig.1). The first two terms in Eq.(2) describe electron hopping between neighboring cells, and t_j is a matrix of the corresponding hopping parameters. The next two terms, including hermitian matrices $H_j^{a,b}$, describe intra cell hopping processes as well as interaction with on site impurities. The last two terms describe intra-cell hopping between a and b orbitals. In zigzag nanotubes, in addition to a and b orbitals one can have orbitals not connected directly to either of the neighboring cells. They can be eliminated at a later stage of the theory at the cost of modifying the intra-cell hopping matrix W_j[10]. In Eq.(2) and in the rest of our paper we concentrate, for the sake of simplicity, on the properties of the armchair nanotubes.

A wave function of the Hamiltonian for an eigenenergy ω, can be generally written as

$$\Psi = \mathrm{e}^{-i\omega t} \sum_j \left(a_j^\dagger x_j + b_j^\dagger y_j\right) |0\rangle \tag{3}$$

Inserting this wave function into the Schrödinger equation with our Hamiltonian, Eq.(2), leads to a system of recurrence equations for the vector coefficients x_j, y_j. The recurrence equations may be written using a transfer matrix T_j as follows

$$\begin{pmatrix} x_{j+1} \\ y_j \end{pmatrix} = T_j \begin{pmatrix} x_j \\ y_{j-1} \end{pmatrix} , \quad T_j = \begin{pmatrix} t_j^{-1\dagger} B_j W_j^{-1} A_j - t_j^{-1\dagger} W_j^\dagger & t_j^{-1\dagger} B_j W_j^{-1} t_{j-1} \\ -W_j^{-1} A_j & -W_j^{-1} t_{j-1} \end{pmatrix} \tag{4}$$

where $A_j = \omega - H_j^a$ and $B_j = \omega - H_j^b$. A diagonal representation of the transfer matrix T_{perfect} for a perfect lattice may be written as

$$T_D = \begin{pmatrix} \Lambda & \mathbf{0} \\ \mathbf{0} & \tilde{\Lambda} \end{pmatrix} = S^{-1} T_{\text{perfect}} S \ , \quad \Lambda\tilde{\Lambda} = I \tag{5}$$

The complex eigenvalues of T_{perfect} with modulus equal to unity correspond to running waves. Besides, there are in general complex eigenvalues of T_{perfect} which correspond to decaying solutions for the given energy ω. By convention, we include in Λ all eigenvalues of T_{perfect} which correspond to solutions allowed by a causality condition (see below).

A current in the state defined by Eq.(3) reads:

$$iJ_j = x_{j+1}^\dagger t_j y_j - y_j^\dagger t_j^\dagger x_{j+1} \tag{6}$$

Because A_j and B_j are hermitian, J_j is independent of the site j. Using the transfer matrix T_{perfect} and the matrix S, a current conservation law for the perfect lattice can be written in the block-matrix representation as:[11]

$$T_D^\dagger S^\dagger \begin{pmatrix} \mathbf{0} & t \\ -t^\dagger & \mathbf{0} \end{pmatrix} S T_D = S^\dagger \begin{pmatrix} \mathbf{0} & t \\ -t^\dagger & \mathbf{0} \end{pmatrix} S = \begin{pmatrix} iV & -\gamma \\ \gamma^\dagger & i\tilde{V} \end{pmatrix} \tag{7}$$

In the last equation we introduced a velocity matrix V, after Molinari[11].

Let us consider now a particle approaching our device from infinity. The particle is represented by a running wave which can be written as a linear combination of all possible solutions of a Schrödinger equation for the given energy ω. In a perfect (N_a, N_a) armchair nanotube for $\omega \approx 0$, there are always two conducting channels per spin, but there are at least six channels per spin for $|\omega/t| > \pi/N_a$ (for $N_a = 10$, this is relevant for doping as small as one dopant per 100 carbon atoms).

The particle incident at the barrier splits into reflected and transmitted parts. Both these parts can now have contributions corresponding to evanescent waves with amplitudes exponentially decreasing with distance from a barrier. The amplitude of the reflected part is determined by the reflection matrix ρ and the amplitude of the transmitted part is determined by the transmission matrix τ. Relations between these two matrices can be obtained from the requirement that the transmitted wave does not include an exponentially growing contributions as well as any part corresponding to a wave running towards the barrier (a causality condition). Comparing the current, Eq.(6), on the two sides of the barrier and using Eq.(7) leads then to the relation

$$\begin{aligned} \tau^\dagger V \tau + \rho^\dagger \left(-\tilde{V}\right) \rho &= V - \rho^\dagger \gamma^\dagger + \gamma \rho \\ \rho &= Q_{22}^{-1}\, Q_{21} \\ \tau &= Q_{11} - Q_{12}\, Q_{22}^{-1}\, Q_{21} \end{aligned} \tag{8}$$

where the matrices $Q_{\mu\nu}$ $(\mu, \nu = 1, 2)$ are block submatrices of a $n_b \times n_b$ matrix Q representing a $M + 1$-cell barrier beginning at site L:

$$Q = S^{-1}\, T_{L+M} \ldots T_{L+1} T_L\, S \tag{9}$$

(We restrict the discussion to real transfer matrices assuming zero magnetic field, in this case $\tilde{V} = -V$[11]).

While Eq.(8) still includes the solutions decaying exponentially with distance from the barrier and is a direct consequence of the current conservation, far from the barrier the decaying solutions are not important. With a help of a projection operator P_c having the property that $P_c \tilde{\Lambda} = \tilde{\Lambda} P_c = P_c \Lambda^\dagger$, we simplify matrices τ, ρ and V, by setting to zero all their rows and columns related to the decaying solutions: $\tau_r = P_c\, \tau\, P_c$. From Eq.(8) we obtain then the equation for the reduced matrices τ_r, ρ_r, V_r:

$$\tau_r^\dagger V_r \tau_r + \rho_r^\dagger V_r \rho_r = V_r \tag{10}$$

Eq.(10) can be rewritten in the convenient form:

$$\tilde{\tau}^{\dagger}\tilde{\tau} + \tilde{\rho}^{\dagger}\tilde{\rho} = I \ , \quad \tilde{\tau} = v\tau_r\tilde{v} \ , \quad \tilde{\rho} = v\rho_r\tilde{v} \tag{11}$$

where v and $\tilde{v}$ are defined by the relations $v^{\dagger}v = V_r$ and $v\,\tilde{v} = P_c$. Eq.(11) is a multichannel reflection law, describing the relation between the unit fluxes of reflected and transmitted electrons far from the barrier. The transmission coefficient is a non-negative number defined here as

$$T = \frac{1}{n_c}\mathrm{tr}\left(\tilde{\tau}^{\dagger}\tilde{\tau}\right)$$

and related to a reflection coefficient $R = \frac{1}{n_c}\mathrm{tr}\left(\tilde{\rho}^{\dagger}\tilde{\rho}\right)$ by: $T = 1 - R$.

Landauer Conductance and the Kubo formula

So far we implicitly assumed that the conductance given by Eq.(1), and directly related to a reflection R, is identical to the quantity measured in a typical two-probe measurement. This fact is well established in a number of works where the equivalence between the Landauer method and the more general approach based on the linear response theory was shown. The existing proofs[12], however, were given only for the cases of simple structures of the leads (free-electron or tight-binding electron spectrum with one kind of atoms in unit cell). In the present case, we deal with two groups of atoms per unit cell of the perfect leads, which formally resembles the situation in an alternating (e.g. dimerized) lattice. One can ask, whether the equivalence of the linear response theory (the Kubo formula) and the Landauer type approach still holds. Although the expected answer is yes[13], we outline here the main points of the proof, because it allows us to reformulate the problem of scattering in another useful way.

We use the standard scattering theory approach which starts from the following general relation between amplitudes of incoming $(\alpha_{n\nu}, \beta_{n\nu})$ and the scattered $(x_{n\nu}, y_{n\nu})$ waves

$$\begin{pmatrix} x_{n\nu} \\ y_{n\nu} \end{pmatrix} = \begin{pmatrix} \alpha_{n\nu} \\ \beta_{n\nu} \end{pmatrix} + \sum_m G_{0,nm}\Delta_m \begin{pmatrix} x_{m\nu} \\ y_{m\nu} \end{pmatrix} \tag{12}$$

$$= \begin{pmatrix} \alpha_{n\nu} \\ \beta_{n\nu} \end{pmatrix} + \sum_m G_{nm}\Delta_m \begin{pmatrix} \alpha_{m\nu} \\ \beta_{m\nu} \end{pmatrix} . \tag{13}$$

In Eqs.(12,13), G_0 and G denote Green functions for the perfect alternating lattice and for the alternating lattice with disorder, respectively. Δ_m describes the modulation of the perfect lattice values of $H^{a,b}$ due to disorder at site m (for simplicity we restrict ourselves to a diagonal disorder only). α and x (β and y) are amplitudes of incoming waves for a (b) atoms, n, m number unit cells and ν is the solution index. Our aim is to find a transmission matrix which defines the relation between the incoming and transmitted wave: $x_{n\nu} = \sum_{\mu} \tau_{r,\mu\nu}\alpha_{n\mu}$.

The components of the Green functions can be obtained by solving the system of the equations of motion in real space using the transfer matrix. In this case the causality condition mentioned above leads to the result:

$$G_{0,nm} = \begin{pmatrix} S_{11}\Lambda^{n-m-1}\tilde{G}^{aa} & S_{11}\Lambda^{n-m-1}\tilde{G}^{ab} \\ S_{21}\Lambda^{n-m}\tilde{G}^{aa} & S_{21}\Lambda^{n-m}\tilde{G}^{ab} \end{pmatrix}, \quad \tilde{G} = S^{-1}\begin{pmatrix} G^{aa}_{0,m+1m} & G^{ab}_{0,m+1m} \\ G^{ba}_{0,mm} & G^{bb}_{0,mm} \end{pmatrix} \tag{14}$$

The matrices $S_{\mu\nu}$ $(\mu, \nu = 1, 2)$ are block submatrices of the matrix S diagonalizing T_{perfect}. The proof of equivalence of the Landauer and Kubo formula rests on use of Eq.(14) together with the Dyson equation

$$G = G_0 + G_0 \Delta G = G_0 + G \Delta G_0$$

to eliminate the matrices Δ_m from Eq.(12). The final result for the transmission matrix is

$$\tau_r = P_c \tilde{\Lambda}^{j-1} \mathcal{S}^{-1} G^{aa}_{jm} \left(\tilde{G}^{aa} \right)^{-1} \Lambda^m P_c \tag{15}$$

where: $\mathcal{S} = S_{11}$. Using the causality condition we can relate the elements of the perfect lattice Green function to the velocity matrix V

$$\left(\tilde{\Lambda} \tilde{G}^{aa} \mathcal{S}^{-1\dagger} \right)^{-1} P_c = -iV P_c \tag{16}$$

which leads to the following formula for conductance in the Landauer type approach

$$G = \frac{e^2}{h} \text{tr} \tilde{\tau} \tilde{\tau}^\dagger = \frac{e^2}{h} \text{tr} \left(V \mathcal{S}^{-1} G^{aa}_{jm} \mathcal{S}^{-1\dagger} V \mathcal{S}^{-1} G^{aa\dagger}_{jm} \mathcal{S}^{-1\dagger} \right) \tag{17}$$

The above formula is equivalent[14] to a result of the Kubo linear response theory. It is also completely equivalent to the formula obtained using transfer matrix approach in the previous paragraph.

For the case of single defect, the formula for the transmission matrix, Eq.(15), can be rewritten into a more transparent form as follows. The Dyson equation can be rewritten with a help of a $\mathcal{T}$-matrix (see, e.g. [15]) as

$$G = G_0 + G_0 \mathcal{T} G_0, \quad \text{where: } \mathcal{T} = \Delta (I - G_0 \Delta)^{-1} \tag{18}$$

Let us consider a defect in b-subsystem at the site L. Using Eqs.(14,15,16) we obtain from Eq.(18) a simple relation between the transmission matrix and the $\mathcal{T}$-matrix

$$\tilde{\tau} = P_c - iV^{-\frac{1}{2}} \mathcal{T}_S V^{-\frac{1}{2}} \tag{19}$$

where

$$\mathcal{T}_S = P_c \mathcal{S}^\dagger \mathcal{T}^{bb}_L \mathcal{S} P_c, \quad \mathcal{T}^{bb}_L = \Delta^b_L (1 - G^{bb}_{0,LL} \Delta^b_L)^{-1} \tag{20}$$

The advantage of Eq.(19) is that it allows to see clearly the general features of the single defect scattering. For a weak defect the transmission drops rapidly close to band edges where the velocity vanishes and the second term in Eq.(19) becomes important. For a strong defect one may have a resonant state in a multiband system[17]. The resonant state manifests itself as an effective enhancement of the defect strength which leads to a reduction of the transmission.

In order to use Eq.(19), first one has to know the matrix S defined in Eq.(5). Besides, one has to know the perfect lattice Green function, $G^{bb}_{0,LL}$, which again can be determined[13] using the causality conditions. The result is

$$G^{bb}_{0,LL} = -t^{-1} A \mathcal{S} \left(\Lambda - \tilde{\Lambda} \right)^{-1} \mathcal{S}^{-1} t^{-1\dagger} \tag{21}$$

The above results is equivalent to the Fourier transform of a corresponding formula available in a k-space.

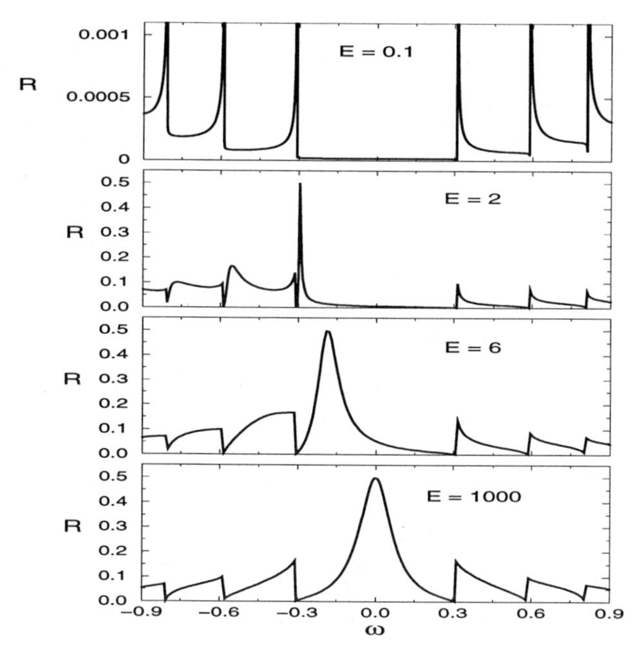

Figure 2: Reflection coefficient for a point defect of $E/t = 0.1, 2, 6, 1000$ for armchair (10,10) NT.

SCATTERING FROM A SIMPLE BARRIER IN A NANOTUBE

In this section we present results of our numerical computations using previously derived formula for the reflection R. We chose to discuss reflection rather than conductance, because we found that it is reflection which follows a very simple scaling law with NT radius. While the general principles underlying our computations were discussed in the previous section, we refer to our papers [10, 16] for the further details.

Single Defect Scattering

We start from a discussion of a reflection from a barrier created by a simple point-like defect which may be considered as a zero order approximation for all defects with spatial extension much smaller than a NT radius. This includes not only substitutional defects (like N or B atoms substituting a C atom) but also, as will be seen later, more complex defects, such as 5-77-5 defect.

In Fig.2, we present evolution of the ω-dependence of the reflection coefficient R with a strength of defect E for the armchair (10,10) nanotube. For small value of E, $E \ll 1$, the reflection R shows a behavior closely resembling the one of the density of states (DOS) for the same nanotube. With increase of the defect strength the reflection grows initially as E^2 and the peaks in ω dependence become more smooth. At the same time the overall picture becomes more asymmetric with respect to $\omega = 0$. The latter, unlike in a single band case, may have consequences also for other transport properties of the system, such as the thermoelectric power (TEP).

For large enough E, $E > 1$, we find that the peak which for small E is located at $\omega/t = \sin(\pi/N_a)$ smoothes and moves towards higher values of ω as E increases. This behavior of R suggests appearance of a resonance state[17]. The condition for the appearance of the resonant or quasibound solutions are given by zeros of the real part of denominator of the $\mathcal{T}$-matrix. In this case the zero matches the position of the

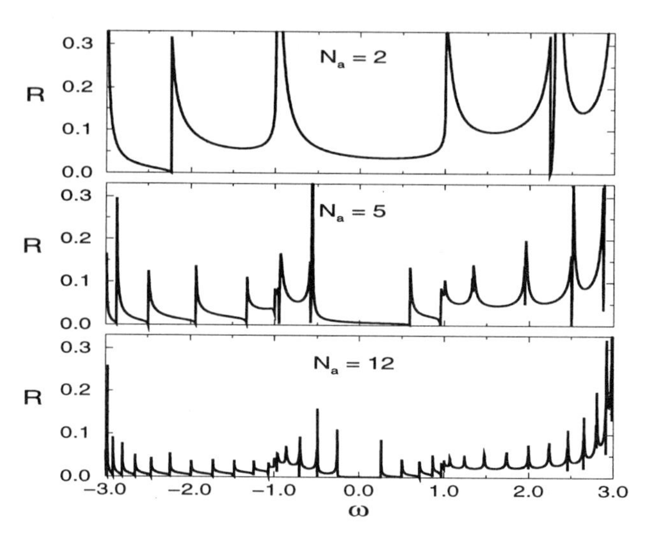

Figure 3: Reflection coefficient for a point defect $E = t$ for several armchair NTs.

maximum in R, $R = 1/2$. The position of the zero approaches $\omega = 0$ for very strong E. One might then interpret the round maximum of R near $\omega = 0$ in Fig.2 for large E as related to strongly overdamped resonance.

The value of the reflection coefficient at the points of resonance can be understood as follows[10]. Suppose ω_0 is the exact eigenvalue of the Hamiltonian $\mathcal{H}$ and satisfies $\det[\mathrm{Re}(I - G_0\Delta)] = 0$. Furthermore, assume that $\varepsilon_0 = \varepsilon_{k_1,1} = \ldots = \varepsilon_{k_g,g} = \omega_0 + \mathcal{O}(1/N)$, where ε_0 is the eigenvalue of the Hamiltonian with no defect, $\mathcal{H}_0$, with degeneracy g (we consider running waves only), ν is the band index, and N is the number of unit cells in the system. Then one can find $g - 1$ independent linear combinations of the eigenstates of $\mathcal{H}_0$, $\phi_{k_\nu,\nu}$, $\nu = 1..g$, of the form $\psi_\mu = \sum_{\nu=1..g} \alpha_{\mu\nu}\phi_{k_\nu,\nu}$, with a property that: $\mathcal{H}\psi_\mu = \varepsilon_0\psi_\mu$ and $\Delta\psi_\mu = 0$. These states will not scatter from the defect and contributions from them to $\mathrm{Tr}\left(\rho^\dagger\rho\right)$ will vanish. A wave corresponding to a resonant (or quasi–localized) solution, orthogonal to states ψ_μ, will be perfectly reflected and will contribute 1 to $\mathrm{Tr}\left(\rho^\dagger\rho\right)$. As a result the reflection coefficient at the energy of a resonance will be equal to $1/n_c$ and have a local maximum at this point. This property seems to hold to a good approximation also for any defect of extension much smaller than the nanotube radius as can be exemplified by the 5-77-5 defect for large enough N_a (see below and Fig.5).

The evolution of the ω-dependence of the reflection with the diameter for $E = 1$ is presented in Figs.3. One can see a deepening of the transmission window around $\omega = 0$ with increase of N_a and a relatively slow decrease of R in the rest of ω range. A similar tendency is found for the zigzag NTs.

In order to make a more general conclusion concerning dependence of the reflection coefficient on N_a and E, we plotted the reflection coefficient calculated for $\omega = 0$ as a function of N_a for several values of E in the log-log scale in the Fig.4. The curves obtained for weak E values show that R scales like

$$R \sim s\frac{E^2}{N_a^2} \tag{22}$$

with $s \approx 1/6$ and $s \approx 1/2$ for the armchair and zigzag nanotubes respectively. This scaling holds in the case of larger defect strength as well, but then one has to go to

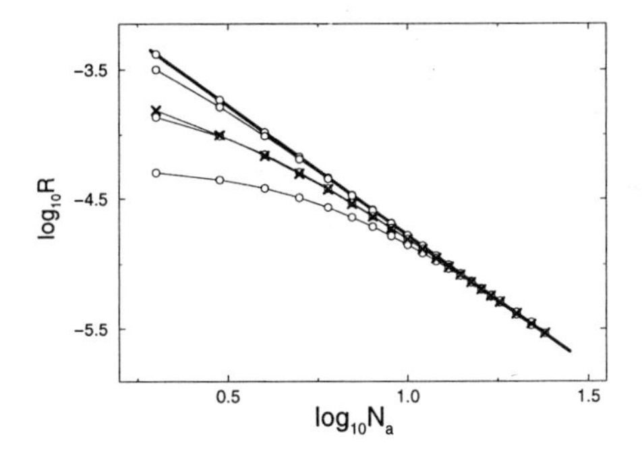

Figure 4: Reflection coefficient at $\omega = 0$ for a point defects $E/t = 0.1, 2, 5, 10$ for armchair NTs. Crosses: 5-77-5 defect. Heavy line: $1/N_a^2$ dependence.

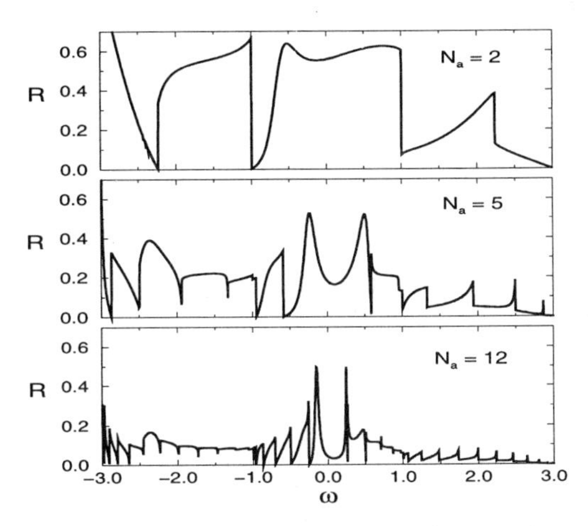

Figure 5: Reflection coefficient for a 5-77-5 defect for several armchair NTs.

larger values of N_a to see it[18].

The scaling behavior can be understood by considering low electron excitations near $\omega = 0$ (i.e. half filled band). Independent on the nanotube radius, there are only two bands which intersect the Fermi surface. It is therefore possible to disregard ("project out"[19]) the other bands, provided that the defect strength is small enough ($E \ll 1$). In this case, the initial Hamiltonian can be reduced to a two band model with distance dependent hopping and an on-site point defect of an effective strength $\tilde{E} = E/4N_a$. With a change of the nanotube radius, only the effective defect strength changes, while the hopping integrals do not depend on N_a. For small enough $\tilde{E}$ the reflectivity will go like square of a defect strength, because $R = \mathrm{Tr}(\rho^\dagger \rho)/n_c$ (with $\rho \sim \Delta \sim \tilde{E}$) while $n_c = 2$ for any N_a. This gives rise to the $\frac{1}{N_a^2}$ dependence of R.

As the defect strength becomes larger, the effects of the band structure in R are less pronounced (see Fig.2 for $E = 1000$), because the numerator and the denominator in $\mathcal{T}$ are comparable. In this region of E, we obtained a very good quantitative agreement with the results of Chico *et al.*[20], showing a relatively smooth and symmetric with respect to $\omega = 0$ dependence of the conductance for (4,4) nanotube with a vacancy.

Let us now turn to so called 5-77-5 defect in an armchair nanotube. This defect can be obtained by rotating one of the C-C bonds by $\pi/2$ resulting in the transforma-

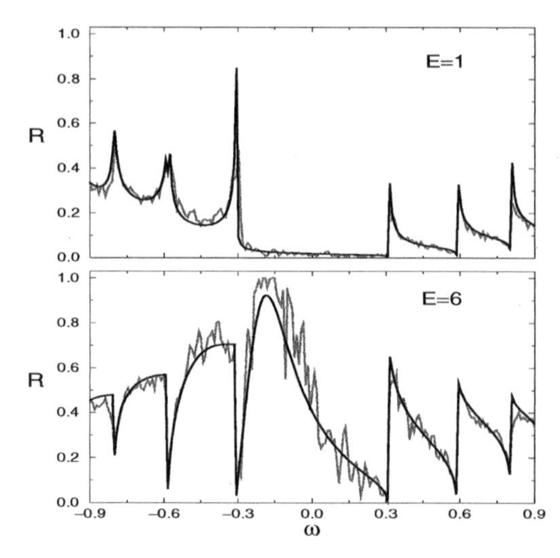

Figure 6: Gray line: conductance fluctuations for 12 defects placed at random in a chain of 100 unit cells of (10,10) nanotube. Black line is the semiclassical approximation: $R_n = \frac{nR_1}{1+(n-1)R_1}, n = 12$

tion of four of nearby hexagons into a pair of heptagons having a common side and separating two pentagons (in this case the barrier matrix is made of a product of the two consecutive transfer matrices). The details of the ω-dependence of the reflection coefficient for the 5-77-5 defect show some similarities with those for point defects with a very high value of E. In fact, we found the same scaling ($R \sim 6/N_a^2$) with N_a as in the case of the point defect with energy $E \approx 6$. This is illustrated in Fig.4.

Interference Effects and Semiclassical Approach

We consider now a barrier formed by a random arrangement of a number of point like defects. Typical result for R in the case of 12 defects distributed randomly on 100 unit cells of (10,10) NT is presented in the Fig.6. The reflection shows rapid oscillations as a function of energy of the incident electron (note that similar oscillations have been recently observed experimentally at low temperatures[21]). They depend on the particular arrangements of the defects. Apart from these oscillations, the overall energy dependence of R is related to the single defect results. In general, the bigger the reflection R_1 from a single defect, the bigger is R for a sequence of the defects. Noticeable enhancements of R correspond to boundaries of the energy regions with different number of conducting channels: the increase of R is due to the reduced velocity of electrons near the top or bottom of consecutive bands[22].

The fluctuations of R, characteristic for a given defect arrangement, are quantum effects related to interference of the electron waves scattered from different defects. Let us neglect for a moment these quantum effects and treat the scattering electrons as classical particles reflected from a point defect with a probability R_1[14]. We then get an estimation of the reflection from N identical defects:

$$R_N = \frac{N R_1}{1 + (N-1)R_1} \tag{23}$$

This is plotted in the Fig.6 against the exact result. One can see that Eq.(23) explains quite well the energy dependence of reflection from a small number of defects in terms of that from a single defect, except that it does not show the oscillations.

To obtain analytical estimation of the conductance for weak defects we substitute for R_1 the result expressed by the previously established scaling law, Eq.(22). If the system is large enough, we can define a defect concentration c per carbon atom: $c = N/4N_aL$ (L is the NT length in units of the lattice constant $a = 0.14 \times \sqrt{3}/2$ nm). We then find, using Eq.(23), an estimation of conductance neglecting interference effects Γ[14] for $|E/tN_a|$, $c \ll 1$:

$$\frac{\Gamma_N}{\Gamma_0} \approx \frac{\ell_e}{\ell_e + aL}\,, \quad \ell_e = \frac{a}{4cN_aR_1} \overset{\omega \to 0}{\to} \frac{a}{4s}\frac{N_a}{c}\left(\frac{t}{E}\right)^2 \tag{24}$$

which defines a scattering length ℓ_e. For $\ell_e \gg La$ the conductance is almost equal to the ideal limit Γ_0 — this is the ballistic range. In the opposite limit, when $\ell_e \ll La$, the conductance goes as inverse of L — this is the limit of validity of Ohm's law.

QUASIPARTICLE LIFETIME AND DENSITY OF STATES

In this section we discuss the effect of disorder on quasiparticle self energy and the density of states (DOS). We present results obtained using Green function approach and the exact diagonalizations of the Hamiltonian for finite chains.

Green Function of Nanotube with a Dilute Disorder

Here we are interested in the self energy of the Green function due to scattering on the point-like defects. We assume that the concentration of the defects, c, is small enough that we may restrict ourselves to linear term with respect to c.

The Green function may be obtained formally using an Averaged T-Matrix Approximation (ATA)[15]. Although the validity of the approximation is limited to small concentration of defects, it has no restriction for the defect strength. The self energy in ATA is given by, up to terms linear in c, by

$$\Sigma = \frac{c\,E}{1 - E\,F(\omega)} \tag{25}$$

where

$$F(\omega) = \frac{1}{4LN_a}\sum_{k\nu}\frac{1}{\omega + i0^+ - \varepsilon_{k\nu}} \tag{26}$$

$\varepsilon_{k\nu}$ denotes the ν-th band of the perfect NT[23]. Function $F(\omega)$ is in fact identical to a diagonal element of the unperturbed lattice Green function obtained previously, $G^{bb}_{0,LL}$, and given in a more convenient form by Eq.(21). Note that in the case of point defects, Σ does not depend on the band index and on the wave vector.

The imaginary part of $F(\omega)$, which gives the density of states of a perfect NT, can be obtained explicitly for any N_a. It is an even function of ω, $\mathrm{Im}F(-\omega) = \mathrm{Im}F(\omega)$. As a result, the real part of F vanishes at $\omega = 0$. The imaginary part of the self energy gives the lifetime of the quasiparticle due to the scattering by defects. This leads to the simple expression for the lifetime for $\omega = 0$

$$\hbar\tau^{-1}(\omega = 0) = \frac{cE^2}{\sqrt{3}N_a|t|}\left[1 + \frac{E^2}{3N_a^2t^2}\right]^{-1}\,, \tag{27}$$

exact up to linear term in c. This result is a generalization of the one obtained in a paper of White and Todorov[19] to the case of arbitrary strong defects. In fact, keeping

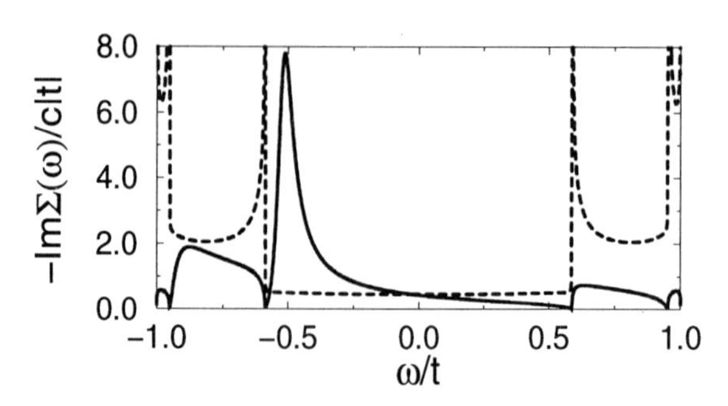

Figure 7: Energy dependence of $-\mathrm{Im}\Sigma/c|t|$ for the (5,5) NT and point defects of strength $E/t = 2$. Heavy line is obtained from Eq.(25), and dashed line from the Born approximation.

only terms linear with respect to E^2 in Eq.(27) makes a good approximation (exact to within 10%) for $|E|$ as large as ≈ 15 eV for (10,10) NT.

The energy dependence of the lifetime may be important in estimation of more subtle transport properties of the system such as the thermoelectric power. We find that the result depends considerably on the defect strength (see Fig.7). In the case of a very weak defect, the self energy is almost symmetrical with respect to ω. In this case, it is quite well described by the Born approximation, i.e. second order calculation with respect to the defect strength[19] and proportional to the unperturbed density of states.

With increase of E, the denominator of the $\mathcal{T}$ (see also Eq.(27)) becomes more important and the asymmetry in $\tau(\omega)$ develops. $\tau(\omega)^{-1}$ starts to exhibit a maximum in the central region of energy spectrum. The maximum shifts towards $\omega = 0$ with increase of E, resembling the behavior found for the reflection in the case of single defect scattering. The position of the maximum is close to the maximum of R. We conclude, that the origin of the maximum is appearance of the quasibound state. Indeed, zeros of the real part of the denominator of Eq.(27), define positions of the resonance states.

Using the results for the quasiparticle lifetime, we find an explicit expression for the mean-free path at $\omega = 0$ in the limit of weak defect strength and small c. For the two bands intersecting at the Fermi wave vector $k_F = 2\pi/3$, we have:

$$\varepsilon_{k\pm} = \pm t\left(1 - 2\cos\frac{k}{2}\right)\,, \quad v_F = \frac{a|t|}{2\hbar}\sqrt{3} \tag{28}$$

and the mean-free path $\ell_{\mathrm{mfp}} = v_F\tau$ reads

$$\ell_{\mathrm{mfp}} = \frac{3aN_a}{2c}\frac{t^2}{E^2} \tag{29}$$

This coincides with the expression for the scattering length ℓ_e, derived above. The agreement is due to the neglect of interference effects from scattering from different defects both in Eq.(25) and in Eq.(24).

Density of States

The Green function obtained in the previous paragraph can be used to determine the change in the density of states (DOS) due to the defects. We first calculate the result

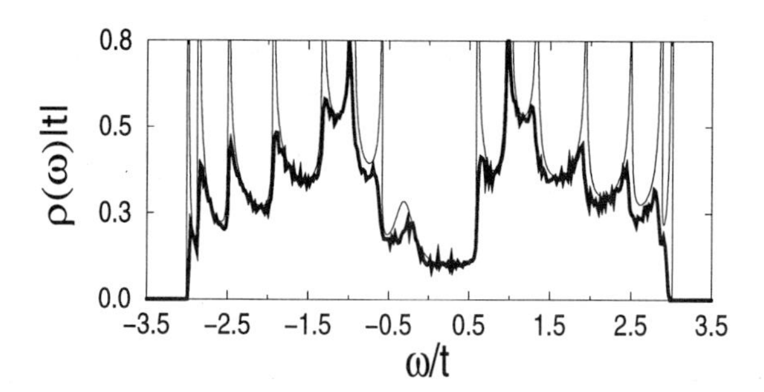

Figure 8: DOS for (5,5) NT with concentration $c = 5\%$ of point defects having strength $E/t = 6$. Heavy solid line — exact diagonalization of a chain of $L = 3000$ unit cells. The thin solid line is obtained using a single defect solution. Bands of localized states, existing for $5 < \omega/t < 7$, are not shown here.

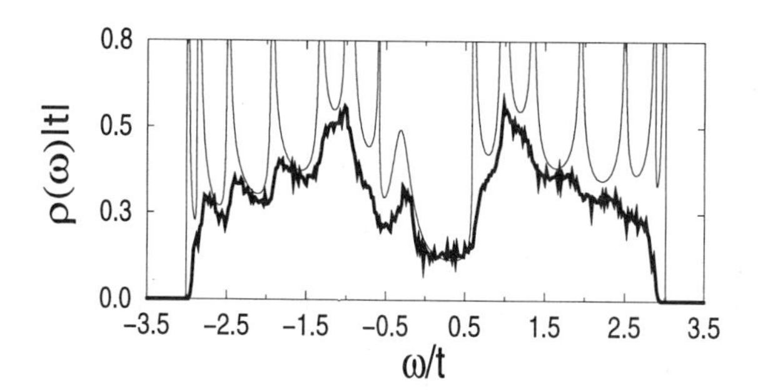

Figure 9: As above, for $c = 10\%$. Note washing out of the van Hove singularities with increase of the defect concentration. At the same time some part of density of states shifts into the region of localized states.

for the single defect and then multiply it by the defect concentration

$$\pi\delta\rho(\omega) = cE\,\mathrm{Im}\frac{\frac{dF(\omega)}{d\omega}}{1 - EF(\omega)} \tag{30}$$

The change in DOS peaks near zero of the denominator of Eq.(30), the real part of which defines the energy of resonance state.

The above approximate estimation of DOS can be compared with exact results obtained by means of diagonalization of the Hamiltonian (2) for finite chain for a specific (generated by the computer) random configurations of point defects. The approximate calculations based on Eq.(30) compare fairly well with the diagonalization for the (5,5) NT for $c \leq 5\%$. We conclude that for the small c much of change of DOS can be interpreted as being due to the quasibound states. At the same time in the diagonalization results we can observe much of reduction of DOS near the position of van Hove singularities, which is not accounted for by the independent defects approximation.

With increase of the NT radius (for constant E, c) the resonance state maximum in the exact DOS becomes more clearly defined, which is due to the decrease of the DOS in the two band region as $1/N_a$. Also the van Hove singularities in wider NTs are more easily obscured by the disorder.[24]

LOCALIZATION LENGTH

In this section we examine quantitatively the role of interference effects in scattering from defects. Interference is a reason that all electron states in 1-d systems are localized, i.e. restricted to a finite region. The spatial extension of this region is determined by the *localization length.*

In real systems the localization length may be sometimes much larger than the size of the system itself — in this case we have a ballistic conductor with the conductance approaching maximum possible value of $n_c G_0$ and only weakly dependent on the system size. In the opposite case of a strongly localized system the localization length is much smaller than the size of the system. In this case conductance decreases exponentially with the system size, with the exponent determined by the inverse of the localization length. Numerical computations presented below show how the transition between the two limiting cases can come about.

Lyapunov Exponents and Localization Length

The transfer matrix method presented in the previous sections may be viewed as a way to find the eigenstates for a given arrangement of defects. There is, however, a convenient way to get information about the spatial extension of a wave function directly from a product of the transfer matrices. Let us first define a product of the transfer matrices

$$P_{L+1} = T_L T_{L-1} ... T_0 \ , \quad T_0 = I$$

then the limit

$$\mathcal{L} = \lim_{L\to\infty} \left(P_L^T P_L\right)^{1/2L}$$

can be shown to exist.[25] The logarithms of the eigenvalues of the matrix $\mathcal{L}$ are called *Lyapunov exponents* and are denoted by λ_ν ($\nu = 1..2n_b$). In actual computations we deal with the L-dependent exponents, $\lambda_{L\nu}$. They are related to the product $P_L^T P_L$, and in the limit $L \to \infty$ approach the Lyapunov exponents ($\lambda_{L\nu}$ are also called Lyapunov exponents for brevity). The smallest positive Lyapunov exponent, λ_L^{min}, determines the rate of decay of the wave function for the given energy. It is related to a localization length by

$$\xi^{-1} = \lim_{L\to\infty} \lambda_L^{min}$$

In calculation of the Lyapunov exponents one has to overcome problems with the accuracy of calculation of the product of the matrices. This problem may be solved either by QR decomposition or by singular value decomposition at each iteration. In our computations we used the methods described in a monograph by Crisanti *et al.*[26], which allowed us to study localization in a chain of several million of unit cells in a reasonable time.

Numerical Results for Localization Length

For a given concentration of defects, the dependence of Lyapunov exponents on the nanotube length allows us to see a gradual transition from a ballistic to a localized limit. In Fig.10 we present the values of the smallest Lyapunov exponent obtained from 50 samples with different realizations of disorder for (5,5) NT. With the increase of the system length, the dependence on the defect distribution gradually decreases and all values of λ_L begin to group around their mean value. Note that the mean value did not change much with L, unlike its mean square deviation.

In Fig.11, we present the results for the localization length as a function of en-

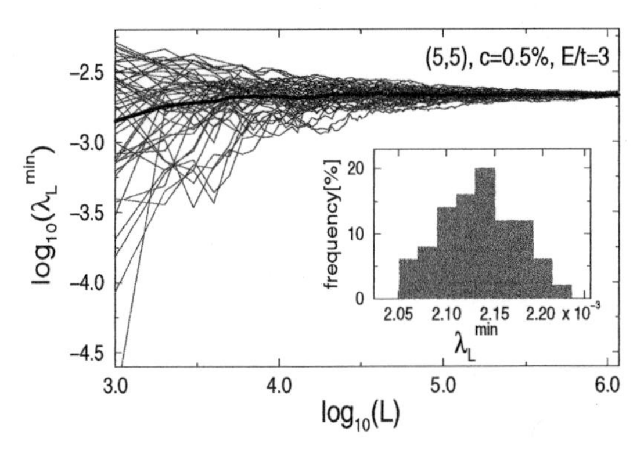

Figure 10: The logarithm of smallest positive Lyapunov exponent for different configurations of defects in (5,5) NT, $E = 3t$ and $c = 0.5\%$ as a function of the system's length (thin lines). The thick line is the average over 50 values of λ_L^{min}. Inset shows the distribution of the values of λ_L^{min} for $L = 10^6$.

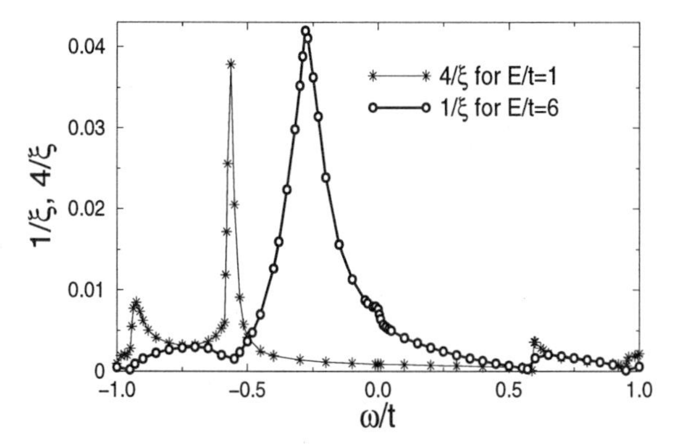

Figure 11: Energy dependence of inverse of the localization length for (5,5) NTs with the concentration $c = 0.5\%$ of point defects of strengths $E/t = 1$ and $E/t = 6$.

ergy. They were obtained for systems of length of $10^5 \ldots 10^7$ unit cells, depending on parameters c, E. Further increase of L did not noticeably change ξ_L. The presented results were calculated by averaging over 50 different realizations of the disorder. The accuracy of ξ, estimated from the standard deviation, was of order of few percent.

The energy dependence of the inverse of the localization length in general follows the corresponding dependence of the reflection R. In particular, the regions of sharp increase of ξ^{-1} are due to the enhanced scattering near the band top or bottom. The pronounced maximum in ξ^{-1} near $\omega = -0.2t$ for $E/t = 6$ is related to the existence of the quasibound state. The localization length for weak defects is a smooth function of ω near $\omega = 0$. With increase of E, a tiny maximum in $\xi(\omega)$ emerges near $\omega = 0$, suggesting appearance of more complex quasibound states, possibly due to neighboring defects. This seems to be supported by the plot the ratio of $\ell_{\rm mfp}/\xi(0)$ as a function of the defect strength for concentration $c = 1\%$, Fig.12. The dependence is clearly non-monotonous, showing a significant decrease of $\xi(0)$ for $E \approx 3t$.

It is interesting to compare the value of the localization length with the calculated mean-free path $\ell_{\rm mfp}$ for $\omega = 0$. In Fig.13, the value of $\ell_{\rm mfp}/\xi(0)$ is plotted as function of the concentration c. For $|E/tN_a| < 1$ the points corresponding to different values

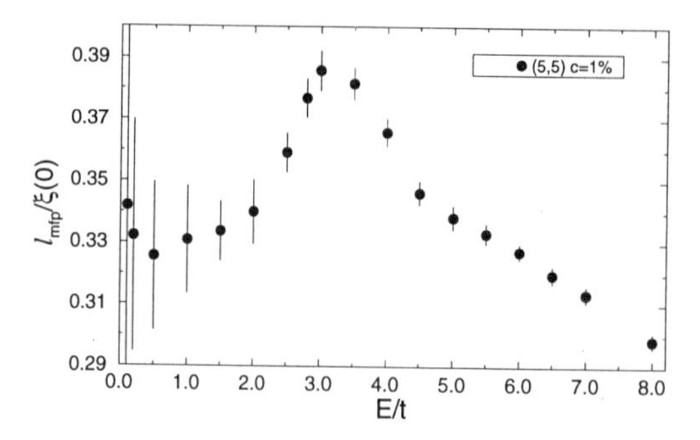

Figure 12: Ratio $\ell_{\text{mfp}}/\xi(0)$ vs. defect strength for the concentration $c = 1\%$ in (5,5) NT. Vertical bars show the values of the standard deviation of $\xi(0)$.

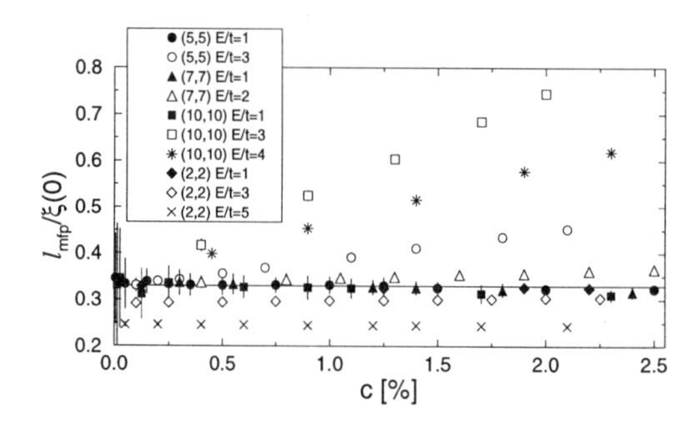

Figure 13: Ratio $\ell_{\text{mfp}}/\xi(0)$ vs. concentration c of defects of various strengths in NTs of various radii. Solid line corresponds to the relation $\ell_{\text{mfp}} = 0.33\xi(0)$.

of E and N_a fall onto lines which all converge to a common point at $c = 0$, where $\ell_{\text{mfp}}/\xi(0) \approx 0.33$. The $c = 0$ limit of the ratio $\ell_{\text{mfp}}/\xi(0)$ is similar to that for the one-band Anderson model,[27] where $\ell_{\text{mfp}}/\xi(0) = 0.5$.

Increase of a ratio $E/N_a t$ spoils the simple form of scaling of the localization length. The ratio $\ell_{\text{mfp}}/\xi(0)$ depends noticeably on the defect strength for $|E/N_a t| \gg 1$. This is exemplified in the Fig.13 by the results obtained for (2,2) nanotube. Note, however, that in the typical case of (10,10) NT the scaling law is well obeyed for E as large as $E \approx 10t \approx 30$ eV.

Our result can be applied to determine if a given NT sample should exhibit properties characteristic for ballistic, ohmic or localized regime.[14] In particular case of the 5-77-5 defect corresponding to $E/t \approx 6$, with equilibrium defect concentration $c \approx 1.5 \times 10^{-5}$ we find $\xi(0) \approx 10$ μm for (10,10) NT. This shows that a NT of length 2 μm (see, e.g. Tans *et al.*[5]) can be well in the ballistic limit if no other strong defects are present.

CONCLUDING REMARKS

In this paper we discussed the influence of disorder on electron states and conductance of single wall carbon nanotubes. Our studies were based on a simple tight binding hamiltonian, with point like defects (with an important exception of the 5-77-5 defect). We disregarded the effects of electron-electron and electron-phonon interactions, as well as possible interactions between the nanotubes forming ropes. We now try to assess the role of some of the neglected effects and validity of the simplifications made.

One of deficiencies of the point like defect approximation is inability to observe the Friedel sum rule with only one value of the defect potential. This problem can be remedied by considering finite extension of a potential[28,29] which can be relatively easily implemented using the transfer matrix method. On the other hand we may also compare our results with the ones obtained using an *ab initio* approach by Choi and Ihm[30]. The qualitative feature of the substitutional defects of N and B obtained in the *ab initio* work (the appearance of the quasibound state) are quite well reproduced using point like defects. The same holds true for 5-77-5 defect, with the difference that the tight binding calculation seem to overestimate the reflection from this defect near $\omega = 0$.

The neglect of Coulomb interaction may be considered the most serious deficiency of our approach in the light of many works done using the Luttinger model. The presence of just a single barrier is known to reduce conductance completely at zero temperature in the Luttinger model, independently on the barrier strength[31]. With increase of temperature the power law decrease of the conductance was predicted. This conclusion, obtained for a one band model, may not be however valid in a many band system, like a nanotube. We also point out that the two-band approach is not sufficient in the case of the nanotubes with strong defects: one cannot obtain a quasibound state in the reduced two-band Hamiltonian.

Other conclusions, based on the exact calculation of the zero temperature Kubo conductivity were obtained for a model of interacting electrons with disorder on a small 2D cylinder[32]. These results suggest, that the conductivity may only weakly depend on the interaction in the weak interaction limit.

On experimental side, there are some indications of the behavior typical for the Luttinger model. The temperature dependence of the conductance of small ropes of single wall NTs[33] was found to be consistent with a power law. In certain samples, however, the conductance at 4.2 K was found to *exceed* its value at 300 K for some values of the gate voltage[21] (with a conductance value close to the ballistic limit). Also, so far unexplained conductance fluctuations as a function of the gate voltage was observed at low temperatures.

While we consider a question of a role of Coulomb interactions in NT far from being settled we expect that some interplay of the effects of disorder and electron correlations may be a key to fully account for transport experiments in NTs.

ACKNOWLEDGMENT

Authors are grateful to M.P. Anantram for a helpful discussion. Research support is acknowledged from the University of Tennessee, from Oak Ridge National Laboratory managed by Lockheed Martin Energy Research Corp. for the U.S. Department of Energy under contract DE-AC05-96OR22464, and from a Research Grant No. N00014-97-1-0565 from the Applied Research Projects Agency managed by the Office of Naval Research. T.K. and M.B. acknowledge also support from K.B.N Poland, Project Nos. 2 P03B 056 14 and 2 P03B 037 17.

REFERENCES

1. S. Iijima, *Nature* **354**, 56 (1991).

2. R. Saito, G. Dresselhaus and M.S. Dresselhaus, *Physical Properties of Carbon Nanotubes* (Imperial College Press, London, 1998).
3. P.M. Ajayan and T.W. Ebbesen, *Rep. Prog. Phys.* **60**, 1025 (1997).
4. B. Kramer and A. MacKinnon, *Rep. Prog. Phys.* **56**, 1469 (1993).
5. S.J. Tans *et al.*, *Nature* **386**, 474 (1997).
6. S. Frank *et al.*, *Science* **280**, 1744 (1998).
7. L. Langer *et al.*, *Phys. Rev. Lett.* **76**, 479 (1996);
   H.R. Shea, R. Martel, and Ph. Avouris, *preprint.*
8. R. Landauer,*Philos. Mag.* **21**, 863 (1970).
9. J.M. Ziman, *Models of Disorder* (Cambridge University Press, London, 1979).
10. T. Kostyrko, M. Bartkowiak and G.D. Mahan, *Phys. Rev.* B **59**, 3241 (1999).
11. L. Molinari, *J. Phys.* A **30**, 983 (1997).
12. D.S. Fisher and P.A. Lee, *Phys. Rev.* B **23**, 6851 (1981)
13. T. Kostyrko, in preparation.
14. S. Datta, *Electronic Transport in Mesoscopic Systems* (Cambridge, 1995).
15. R.J. Elliott, J.A. Krumhansl and P.L. Leath, *Rev. Mod. Phys.* **46**, 465 (1974).
16. T. Kostyrko, M. Bartkowiak and G.D. Mahan,*Phys. Rev.* B **60**, (Nr.15) (1999).
17. P.F. Bagwell, *Phys. Rev.* B **41**, 10354 (1990).
18. Quite recently this scaling behaviour was confirmed analytically in a paper by H.J. Choi and J. Ihm, *Sol. St. Commun.* **111**, 385 (1999).
19. C.T. White and T.N. Todorov, *Nature* **393**, 240 (1998);
    T. Ando and T. Nakanishi, *J. Phys. Soc. Jap.* **67**, 1704 (1998).
20. L. Chico *et al.*, *Phys. Rev.* B **54**, 2600 (1996).
21. D.H. Cobden *et al.*, preprint cond-mat:9904179.
22. Similar results for the strong, point like defects were obtained by M.P. Anantram and T.R. Govindan, *Phys. Rev.* B **58**, 4882 (1998).
23. R. Saito, G. Dresselhaus and M.S. Dresselhaus, *Chem. Phys. Lett.* **195**, 537 (1992).
24. This was also observed for continuous distributions of disorder by S. Roche and R. Saito, *Phys. Rev.* B **59**, 5242 (1999), and K. Harigaya, *Phys. Rev.* B **60**, 1452 (1999).
25. V.I. Oseledec, *Trans. Moscow Math. Soc.* **19**, 197 (1968).
26. A. Crisanti, G. Paladin and A. Vulpiani, *Products of Random Matrices in Statistical physics* (Springer Verlag, Berlin, 1993).
27. D.J. Thouless, *J. Phys.* C **6**, L49 (1973).
28. J. Rudnick and E.A. Stern, *Phys. Rev.* B **7**, 5062 (1973).
29. G.D. Mahan, Int. J. Mod. Phys. B **9**, 1313 (1995).
30. H.J. Choi and J. Ihm, *Phys. Rev.* B **59**, 2267 (1999) and APS'99 Meeting, talk IC02.02.
31. C.L. Kane and M.P.A. Fischer, *Phys. Rev.* B **46**, 15233 (1992).
32. R. Berkovits and Y. Avishai, *Phys. Rev. Lett.* **76**, 291 (1996).
33. M. Bockrath *et al.*, *Nature* **397**, 598 (1999).

PHYSICS OF THE METAL-CARBON NANOTUBE INTERFACES: CHARGE TRANSFERS, FERMI-LEVEL "PINNING" AND APPLICATION TO THE SCANNING TUNNELING SPECTROSCOPY

Yongqiang Xue[1] and Supriyo Datta[2]

[1]*School of Electrical and Computer Engineering, Purdue University, West Lafayette, IN 47907, USA*
Email: yxue@ecn.purdue.edu
[2]*School of Electrical and Computer Engineering, Purdue University, West Lafayette, IN 47907, USA*
Email: datta@ecn.purdue.edu

INTRODUCTION

After its discovery in 1991,[1] carbon nanotube has rapidly emerged as the most promising candidate for molecular electronics due to its quasi-one dimensional structure and the unique characterization of its electronic structure in terms of two simple geometric indices.[2] Besides its huge technological potential, carbon nanotube also serves as the artificial laboratory in which one-dimensional transport can be investigated,[3] similar to the semiconductor quantum wire.[4] However, unlike its semiconductor cousin where transport is mostly ballistic, the study of transport in carbon nanotube has been distressed by the difficulty of making low resistance contact to the measuring electrodes. The high resistances reported in various two- and three-terminal measurements[5] have led Tersoff[6] (and also independently by one of the authors[7]) to suggest that wavevector conservation at the metal-carbon nanotube contact may play an important role in explaining the high contact resistance.[8] The complexity and importance of the metal-carbon nanotube interface makes it an immediate challenge to both theorists and experimentalists.

The single most important property of the metal-carbon nanotube interface (and in general, of any interface involving metal) is the position of the energy bands (or energy levels) of the nanotube relative to the Fermi-level of the metal which manifests itself in the electronic transport property of the interface. Depending on the contact geometry, transport can occur in the direction parallel to the nanotube axis, in the case of the nanotube field-effect-transistor (FET),[5,9] or perpendicular to it, in the case of the STS measurement.[10,11] In the STS measurement, the Fermi-level is found to have shifted to the valence band edge of the semiconducting nanotube.[10] Such observed Fermi-level "pinning" has been used to explain the operation of the nanotube FETs with high-resistance contacts,[5] where the measured two-terminal resistance for metallic nanotube is $\sim 1M\Omega$. Recently low resistance contacts with two-terminal resistance as low as $20k\Omega$ have been obtained.[9] However, low temperature transport measurements using these low resistance contacts show that the Fermi-level is located between the valence and the conductance band of the semiconducting nanotube, instead of being "pinned" at the valence band edge. This conflict raises the important question of whether the Fermi-level positioning may depend on the contact geometry and the interface coupling.

In this paper we present a theory of the scanning tunneling spectroscopy of a single-wall carbon nanotube (SWNT) supported on the Au(111) substrate. The central idea is that the work function difference between the gold substrate and the nanotube leads to charge transfers across the interface, which induce a local electrostatic potential perturbation on the nanotube side. This atomic-scale interfacial potential perturbation shifts the energy level of the nanotube relative to the gold Fermi-level, and gives rise to the observed Fermi-level shift in the STS current-voltage characteristics. However, for transport in the direction parallel to the nanotube axis, as in the case of nanotube transistors, such local potential perturbation at the interface is not important in determining the Fermi-level position *if the coupling between the metal and the nanotube is strong* (i. e. , low resistance contact). In this case, the metal-induced gap states (MIGS) model provides a good starting point for determining the Fermi-level position. Based on this model, we expect that any discrepancy between the metal Fermi-level and the nanotube "charge-neutrality level" should be rapidly screened out by the metal-induced gap states in the nanotube side,[12–14] leading to the "pinning" of the Fermi-level. Another important feature in our theory is that we have taken into account the localized 5d orbitals of the platinum tip in our treatment of the STS which can have significant effects on the interpretation of the STS data.[15] Our discussion is restricted to the low temperature regime, in correspondence with the experimental works.[9,10]

METHODS

A convenient way of characterizing the band lineup problem at any interface is to find a reference level, the role of which is to put all materials forming the interface on a common absolute energy scale.[13,14] If the position of the reference level depends only on the bulk property, then the relative position of the energy bands at the interface can be determined trivially by merely lining up the reference levels. This is the elegant idea of "charge-neutrality level",[12] which has been applied with impressive success by Tersoff[13,14] to various metal-semiconductor junctions and semiconductor heterojunctions. For metal, the reference level is the Fermi-level E_F, while for semiconductor, it is the so called "charge-neutrality level" which can be taken as the energy where the gap states cross over from valence- to conduction-band character.[14]

This approach greatly simplifies the band lineup problem and gives quantitatively accurate prediction of the Schottky barrier height when applied to the metal-semiconductor interface.[13] The success of this model relies on the fact that there exists a continuum of gap states around E_F at the semiconductor side of the metal-semiconductor interface due to the tails of the metal wavefunction decaying into the semiconductor, which can have significant amplitude a few atomic layers away from the interface.[12] Any deviation from the local charge neutrality condition in the interface region will be screened out rapidly by these metal-induced gap states (MIGS). In this way, the local charge and potential perturbations right at the interface are not important in determining the barrier height observed in the transport characteristics since the range of this local perturbation is only a few atomic layers, and the charge carriers can easily tunnel through this region.[13,14]

This is no longer true for the interface formed when a single-wall nanotube (SWNT) is deposited onto the gold substrate, and the transport characteristic is measured using the scanning tunneling microscope (STM). In this case, transport occurs in the direc-

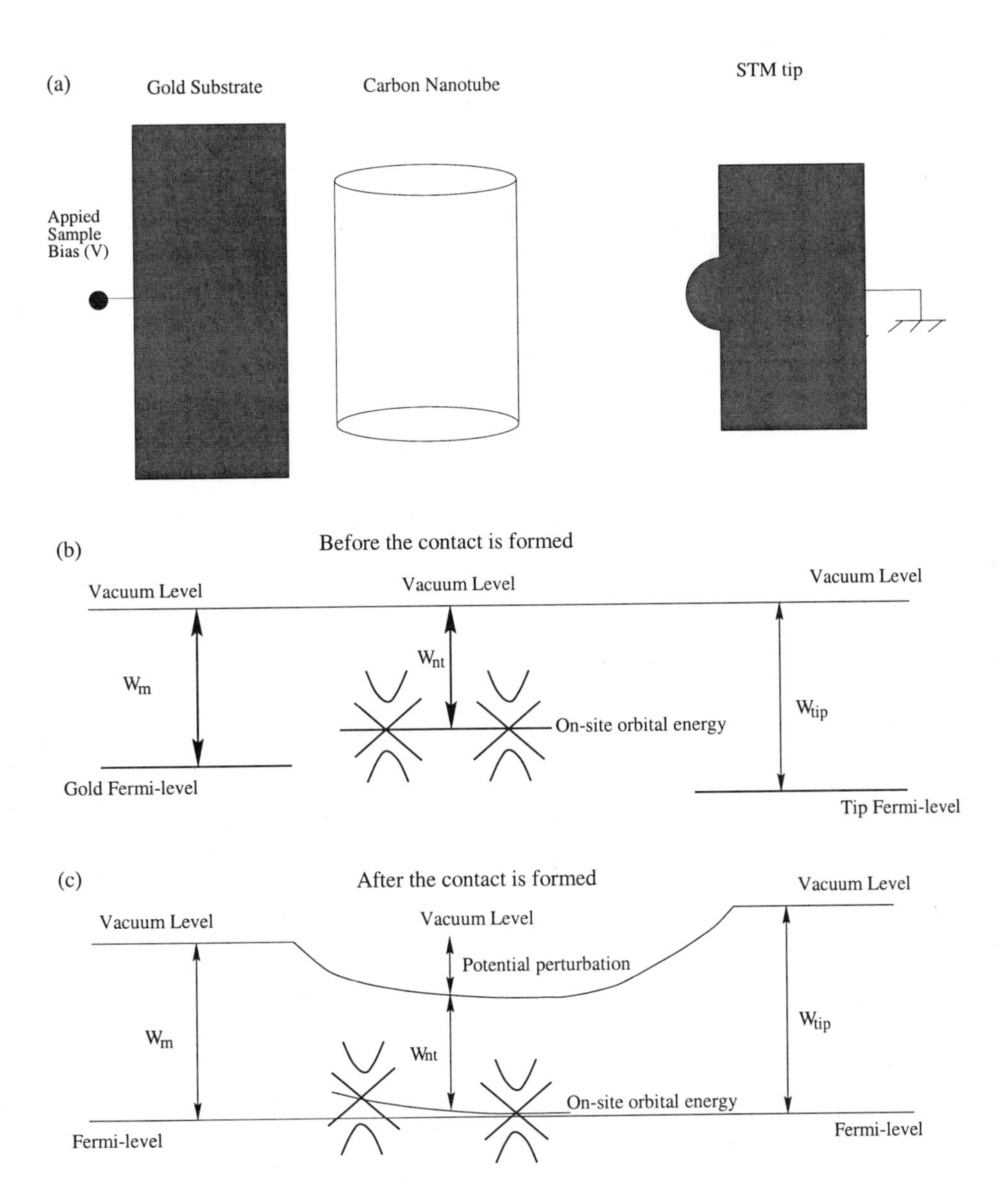

Figure 1: Illustration of the formation of contact. (a) SWNT deposited on the gold substrate, also shown is the STM tip. (b) Energy band diagram of the substrate-SWNT-tip system before the formation of contact, here only the case of a metallic SWNT is shown. Note the lining up of the vacuum levels. The work function of the substrate W_m is $5.3(eV)$, while that of the nanotube W_{nt} and the tip W_{tip} (made of platinum) are $4.5(eV)$ and $5.7(eV)$ respectively. (c) Energy band diagram after the formation of contact. Note that the shifts of the vacuum levels equal the interfacial potential perturbations. After equilibrium is reached, the Fermi-level is constant throughout the system.

tion perpendicular to the nanotube axis and the characteristic length scale involved is the diameter of the nanotube. Since the diameter of the nanotube used in the STS measurement is only of the nanometer range which can be comparable to the range of the interfacial potential perturbation,[10] the detailed potential variations in this dimension will be important in determining the STS current-voltage characteristics, similar to the case of molecular adsorbates on metal surfaces.[16,17]

Fig. 1 illustrates schematically the local electrostatic potential profile at the tip-nanotube-substrate heterojunction. Unlike ordinary metal-semiconductor interfaces, a better reference level for studying the interface formed between the gold substrate and the supported nanotube is the vacuum level (Fig. 1 (b)). We consider the formation of the interface by moving the nanotube toward the gold surface. If the charge distributions on both sides don't change when interface is formed, then the vacuum levels just line up. However, due to the difference of work functions (the work function of the nanotube W_{nt} is $4.5(eV)$, while that of gold W_m is $5.3(eV)$[10]), electrons will transfer from the nanotube to the gold substrate, which also changes the electrostatic potential at the interface whose value $\delta\phi$ should be determined self-consistently. Since the perturbation due to the tip is much weaker than the perturbation associated with the formation of the nanotube-substrate interface, we will neglect the tip when treating the nanotube-substrate interface, the only role of which is to make sure that after equilibrium is reached, the Fermi-level should be constant throughout the system.

We assume an ideal substrate-carbon nanotube interface and describe the electronic structure of the interface using tight-binding (TB) methods. We use the π-electron tight-binding model of the nanotube,[18] which provides a good description of the electronic structure except for those nanotubes with very small diameters.[19] In this model, the on-site π orbital energy coincides with the Fermi-level (metallic) or the mid-gap level (semiconducting) of the isolated SWNT. We take the Fermi-level of the gold as the energy reference, then the initial π orbital energy at each carbon atom of the nanotube is $W_m - W_{nt} = 0.8(eV)$. The final on-site π orbital energy is the superposition of the initial value and the change in the electrostatic potential $\delta\phi$ which changes as one moves away from the gold substrate (Fig. 1(c)). For the gold substrate we use the TB parameters of Papaconstantopoulos[20] which include the $5d6s6p$ orbitals of gold and up to second-nearest-neighbor interactions. For the coupling between the nanotube and the gold surface, we use the values obtained from the Extended Hückel Theory (EHT).[21] This gives a crude but reasonable estimate of the interface coupling, especially since the exact atomic geometry of the interface is not known. We have included only the couplings between nearest neighbors and involving those carbon atoms closest to the gold surface.

Since the carbon nanotube has periodic symmetry along its axis, only one unit cell needs to be considered. We use the Green's function method to calculate the electron population of each carbon from the expression:

$$n_i = -\frac{2}{\pi} Imag\{\int_{-\infty}^{E_F} G_{i,i}(E)dE\} \tag{1}$$

where $G(E)$ is the projection of the Green's function onto one unit cell of the nanotube and $G_{i,i}$ is the ith diagonal matrix element corresponding to atom i in the unit cell. We calculate $G(E)$ by projecting the Hamiltonian of the whole interface onto one unit cell of the nanotube using the decimation technique,[22] thus reducing the whole Hamiltonian into an effective one in which the interactions between the given unit cell and the rest of the interface system are incorporated into the corresponding self-energy operators (for detail see ch. 3 of Datta[23]).

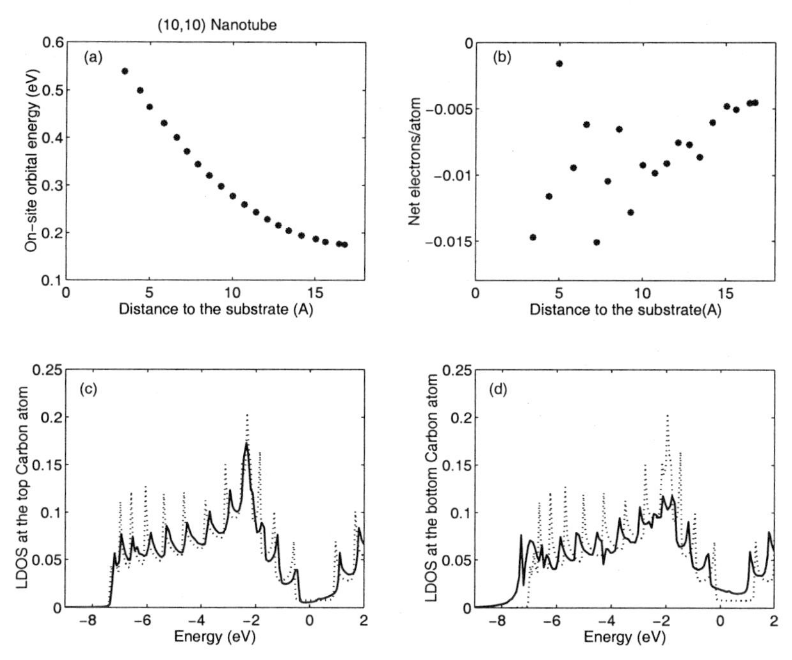

Figure 2: The calculated self-consistent potential and charge perturbations of (10,10) nanotube for nanotube-substrate distance of 3.2Å. (a) The potential perturbations are characterized by the final on-site π orbital energies which are the superposition of the initial values $0.8(eV)$ and the induced diagonal perturbations $\delta\phi_i$. (b) The charge perturbations are characterized by the net electron population on each carbon atom $n_i - n_b$ which are negative due to the electron transfer from the nanotube to the substrate. (c) and (d) show the LDOS at the top (furthest to the substrate) and the bottom (closest to the substrate) carbon atom. Dotted lines are that of the isolated SWNT which have been horizontally shifted by the corresponding potential perturbations $\delta\phi_i$.

The charges thus obtained from the initial Hamiltonian (i. e. , the on-site π orbital energies equal $0.8(eV)$) are different from those in the bulk of the nanotube n_b, the difference of which represent the charge perturbations induced by the formation of the interface. Since we are interested in the electrostatic potential variation as a function of the distance to the gold surface, we group the carbon atoms into equivalent layers according to their distance to the gold substrate. Within tight-binding theory, the self-consistency is achieved by introducing perturbations in the diagonal elements of the Hamiltonian and imposing Hartree consistency between the potential perturbations and the charge perturbations $\delta n_i = n_i - n_b$. We use the self-consistent scheme developed by Flores and coworkers[24] which has been used successfully to study the metal-semiconductor junctions and semiconductor heterojunctions (different but equivalent method has also been used by Harrison to study semiconductor heterojunctions[25]). In this scheme, the diagonal perturbations at each layer $\delta\phi_k$ are related to the induced surface charge densities at each layer ρ_k by the following equations:

$$\begin{aligned}
\delta\phi_{metal} &= 0, \\
\delta\phi_1 &= 4\pi d_{mnt}\rho_{metal}, \\
\delta\phi_2 &= \delta\phi_1 + 4\pi d(\rho_{metal} + \rho_1), \\
&\cdots \\
\delta\phi_n &= \delta\phi_{n-1} + 4\pi d(\rho_{metal} + \textstyle\sum_{k=1}^{n-1} \rho_k)
\end{aligned} \tag{2}$$

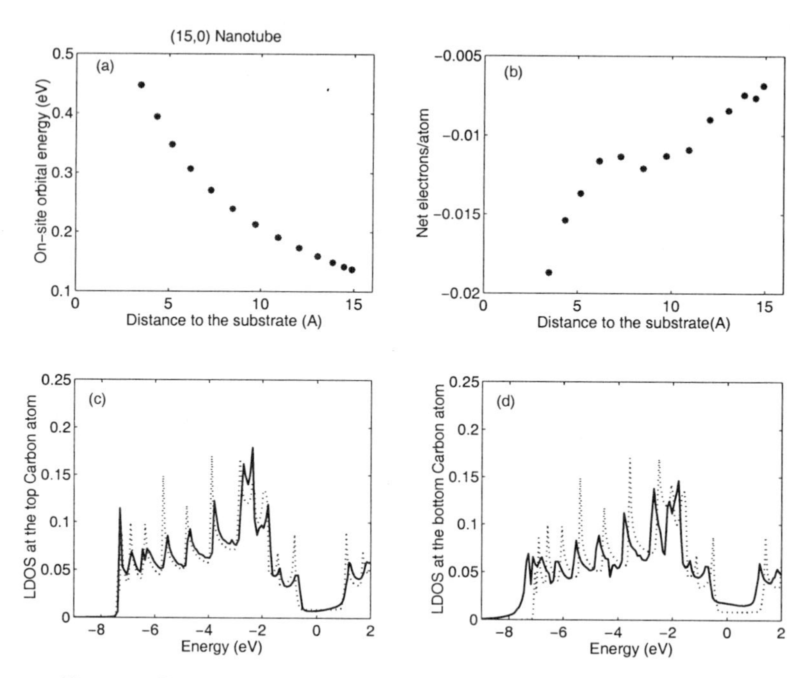

Figure 3: Calculated results for (15,0) nanotube, otherwise same as Fig. 2.

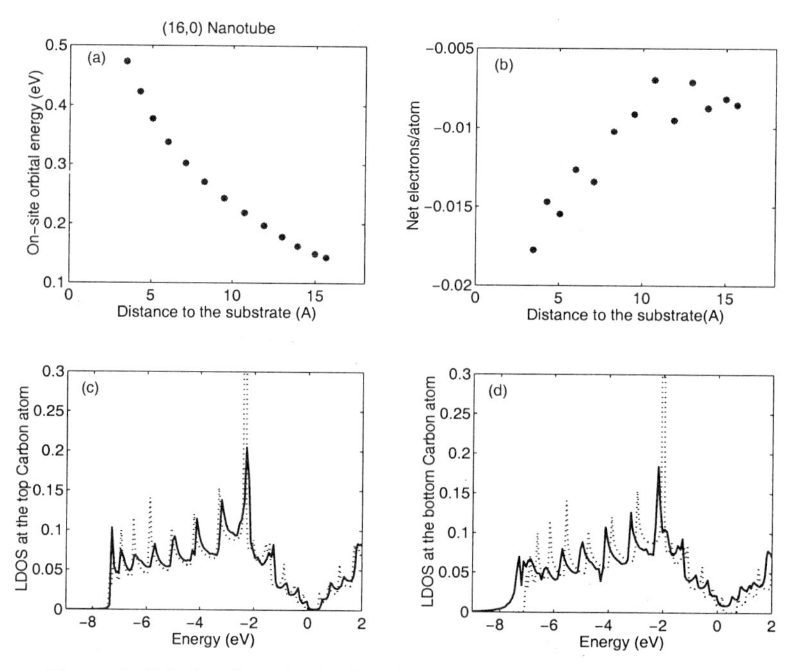

Figure 4: Calculated results for (16,0) nanotube, otherwise same as Fig. 2.

where d_{mnt} is the substrate-SWNT distance, d is the interlayer spacing within the SWNT, and n is the number of layers within the SWNT. We have assumed that the charge transfers into the substrate are localized within the surface layer. Once self-consistency is achieved, the resulting potential perturbation is not sensitive to how we group the carbon atoms into layers.

RESULTS

Fig. 2 - Fig. 5 show the results for four different nanotubes: (10,10), (15,0), (16,0) and (14,0) with diameters of 1.36, 1.17, 1.25 and 1.10 (nm) respectively, close to those measured in Wildöer et al. .[10] All nanotubes show similar behavior. We have not studied chiral nanotubes due to the large size of their unit cell. Since the electronic structure of the SWNTs depend only on their metallicity and radius,[26] we believe similar conclusions can be reached for the chiral tubes. The distance between the nanotube and the gold substrate for this calculation is 3.2 (Å). At this distance, the coupling across the interface is weaker than the carbon-carbon coupling or the gold-gold coupling. This is appropriate since experiments show that the coupling between the nanotube and the substrate is not strong chemical bonding, but rather van der Waals bonding.[27]

Since the π orbital energy coincides with the position of the Fermi-level (mid-gap level) of the isolated metallic (semiconducting) SWNT, then the Fermi-level shift in the STS measurement should correspond to the on-site π orbital energy of the carbon atom closest to the STM tip *if only this atom is coupled to the tip.* However, considering the cylindrical shape of the SWNT, more carbon atoms could be coupled to the tip and the Fermi-level shift then corresponds to the average value of the on-site orbital energies of the carbon atoms within the coupling range. From the plotted values, we expect Fermi-level shifts of $\sim 0.2(eV)$ for all the nanotubes studied here, close to the measured values.[10] The total potential drop across the nanotube is comparable to that across the vacuum region between the nanotube and the gold surface, meaning the net electric field within the nanotube is much smaller than that within the vacuum region, in agreement with Benedict et al. .[28] The similarity between the metallic and the semiconducting nanotube shown here can also be understood from Benedict et al. ,[28] which shows that the dielectric response of the nanotube in the direction perpendicular to the axis doesn't depend on the metallicity, only on the diameter.

Also shown are the charge transfers at each carbon atom (defined as $n_i - n_b$), and the local density of states (LDOS) at the carbon atoms closest (bottom) and furthest (top) to the gold substrate. The charge transfers per atom are small and mainly localized on the carbon atoms close to the gold surface, in agreement with recent *ab initio* calculations.[29] Due to the hybridization with the gold surface atomic orbitals, the peak structures in the LDOS of the bottom carbon atom corresponding to the Van Hove singularities of typical 1-d system are broadened. Their positions also change, which may be understood from the bonding-antibonding splitting resulting from the hybridization of the nanotube molecular orbitals and the gold orbitals. This is actually not surprising since in the direction perpendicular to the nanotube axis, the wavefunctions of the nanotube are just a discrete set of molecular orbitals confined along the circumference and the situation is not much different from the molecular adsorbate on noble and transition metal surfaces.[30] Also notable is the enhancement of the density of states in the gap at the expense of the valence band, reminding us of the Levinson theorem which states the total number of states should be conserved in the presence of perturbation, be it due to impurity[31] or due to surface.[32] In contrast,

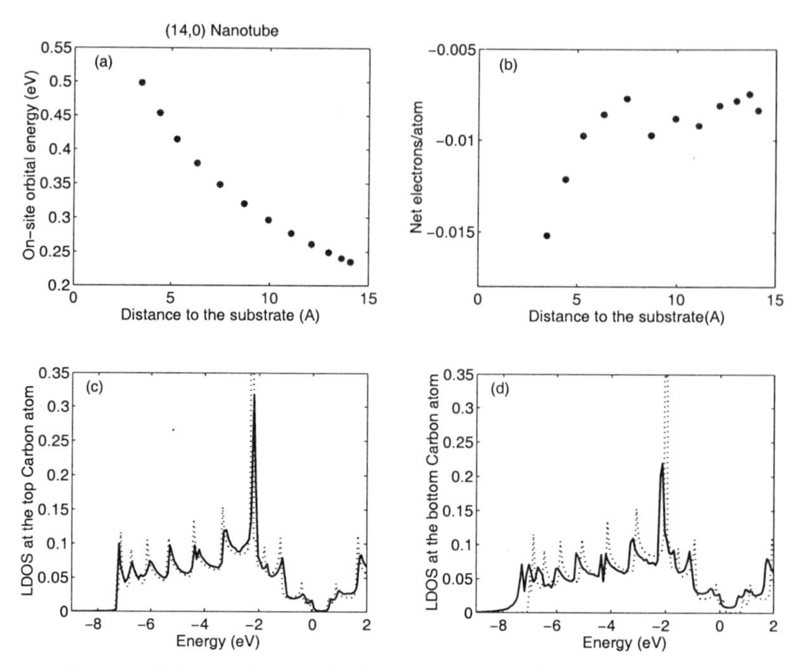

Figure 5: Calculated results for (14,0) nanotube, otherwise same as Fig. 2.

the perturbation to the LDOS at the carbon atom furthest to the gold surface is much weaker. Since the coupling to the substrate is rather weak, the overall shape of the density of states stay relatively unperturbed, especially so for the carbon atom at the top.

We can also increase the interface coupling to see how the interface electronic structure changes with the coupling strength. In EHT, this is achieved by moving the nanotube closer to the substrate. We have made sample calculations for natotube-gold distance of 1.9 (Å). At this distance, the coupling across the interface is much stronger than the bulk correspondents and the on-site π orbital energies of the carbon atoms can be below the gold Fermi-level. The LDOS at the bottom carbon atom is strongly distorted, resulting in much larger charge transfers to the gold substrate and large density of states in the gap. The peak structures are completely smeared out. The shifting of the carbon π orbital energy to lower than the gold Fermi-level has actually been observed in the STS measurement of Wildöer et al. [10] where for one nanotube, the Fermi-level position is found to have shifted to the conduction band edge. The authors attribute it to the anomalously low work function of the gold substrate, while our sample calculation points out an alternative way of thinking. However, the way we obtain strong interface coupling is rather artificial and we have assumed an ideal substrate-nanotube interface. In reality, when deposited onto the gold substrate, the nanotubes can be flattened due to the tube-surface interaction, thus increasing the contact area. When there are imperfections on the substrate such as steps and kinks, the nanotubes tend to follow the surface morphology and can be bent or even be broken. All these can lead to stronger coupling across the interface as having been documented in a series of work by Avouris and coworkers.[27] Further analysis is needed before conclusions can be made about the interface electronic structure in the presence of all the nonidealities.

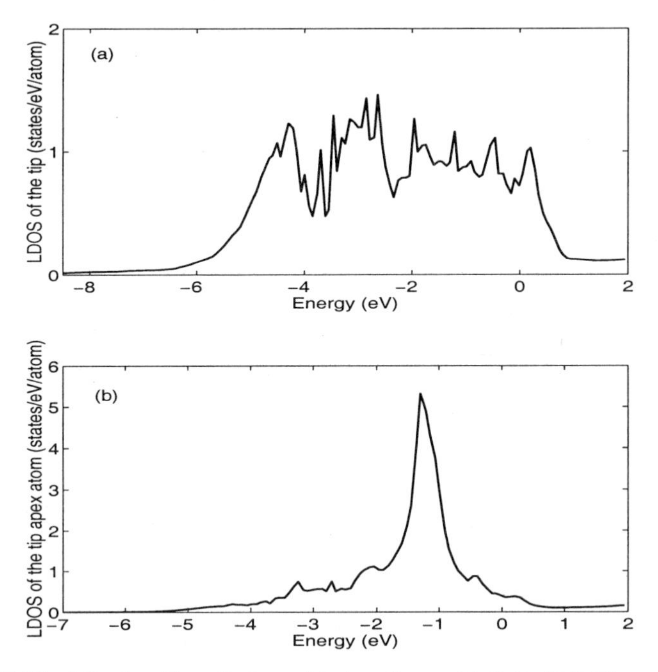

Figure 6: (a) LDOS of the tip. Here tip is modeled as the semi-infinite Pt(111) crystal. (b) LDOS of the tip apex atom. Here tip is modeled as a Pt atom stacked on the surface of the semi-infinite Pt(111) crystal.

Application to the scanning tunneling spectroscopy. The differential conductance dI/dV (or the normalized one $d\ln I/d\ln V$) obtained from the STS measurement is often interpreted as to reflect the local density of states of the sample, based on the s-wave model of the tip.[33] Although this allows an easy interpretation of the experimental results, first-principles calculations have shown that this model is inadequate for tips made from transition metals, where small clusters tend to form at the tip surface giving rise to localized d-type tip states.[15] As a result, the tip electronic structure can have a profound effect on the interpretation of the STS measurement.[17,34]

We can calculate the current using the standard expression from scattering theory (see ch. 3 of Datta[23]):

$$\begin{aligned} I &= \frac{2e}{h}\int_{-\infty}^{+\infty} dET(E,V)[f(E-eV)-f(E)] \\ &\simeq \frac{2e}{h}\int_{0}^{eV} dET(E,V) \end{aligned} \tag{3}$$

However, considering the large uncertainties in the atomic geometry of the tip and the coupling parameters between the tip and the sample, we have taken a simpler approach instead, aiming only to illustrate how the tip electronic structure may affect the transport characteristics. Since the coupling across the nanotube-tip interface is weak, the tunneling Hamiltonian theory may be invoked to write the current crudely as:

$$I \propto \int_{0}^{eV} \rho_{nt}(E-\eta eV)\rho_{tip}(E-eV)dE \tag{4}$$

where ρ_{nt} and ρ_{tip} are the density of states of the SWNT and the platinum tip respectively, η is the voltage division factor which characterizes the additional voltage drop across the nanotube when the external bias V is applied.[16,17]

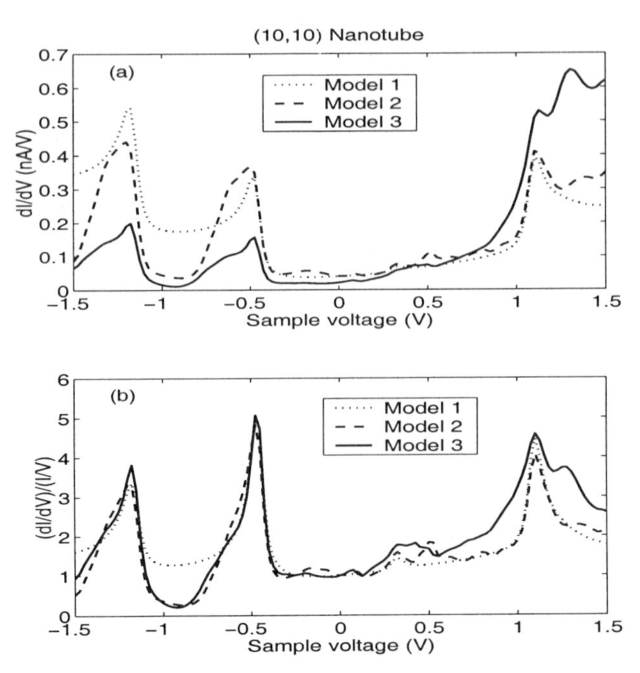

Figure 7: STS conductance-voltage characteristics of (10,10) nanotube calculated using the tunneling Hamiltonian theory (Eq. 4) for nanotube-substrate distance of 3.2Å. Model 1: constant ρ_{tip}; Model 2: tip modeled as a semi-infinite Pt(111) crystal; Model 3: tip modeled as a Pt atom stacked on the surface of the semi-infinite Pt(111) crystal. The bias voltage is applied to the sample. (a) dI/dV vs V. (b) $d\ln I/d\ln V$ vs V.

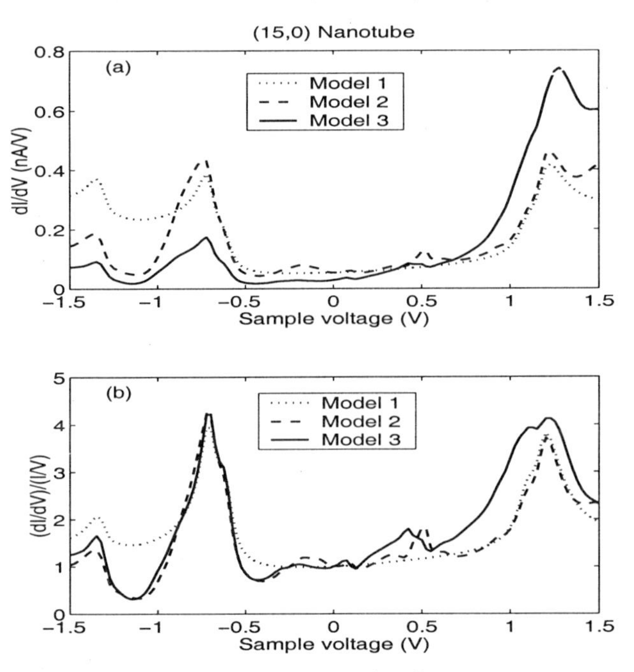

Figure 8: STS conductance-voltage characteristics of (15,0) nanotube, otherwise same as Fig. 7.

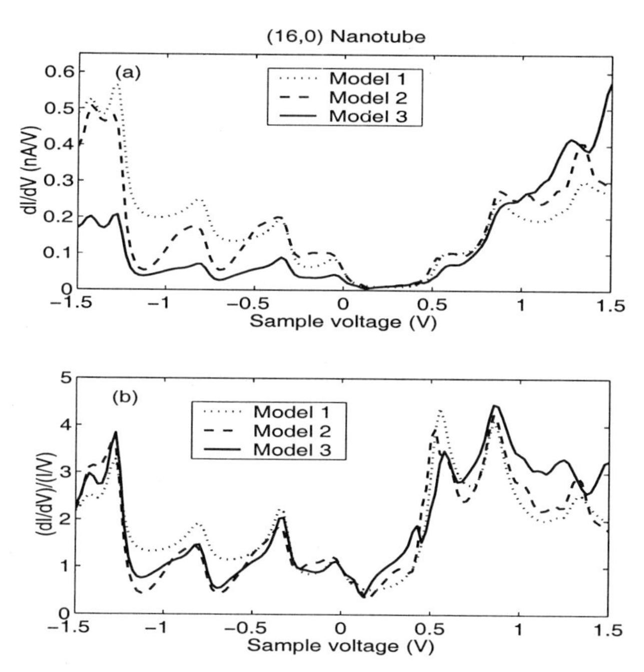

Figure 9: STS conductance-voltage characteristics of (16,0) nanotube, otherwise same as Fig. 7.

The differential conductance thus obtained then reflects the energy convolution of the density of states of the nanotube and the tip. Since the tip-nanotube distance is generally larger than the nanotube-substrate distance and the net field within the nanotube is small,[28] we can take $\eta = 0$ for low applied bias. If ρ_{tip} is constant within the range of the integral, we recover the usual expression $dI/dV \propto \rho_{nt}$. Note ρ_{nt} is calculated using the Hamiltonian with the on-site perturbations and the coupling to the gold substrate included (we use the LDOS of the carbon atom closest to the tip for ρ_{nt} here) .

We have used two models of the tip: (1) Tip is modeled as a semi-infinite Pt(111) crystal; (2) Tip is modeled as a Pt atom stacked on the surface of the semi-infinite Pt(111) crystal.[17,34] The TB parameters are those of Papaconstantopoulos.[20] Fig. 6 shows the calculated LDOS of the surface and the tip. For the second model of the tip, the on-site Hamiltonian matrix elements at the tip apex atom have been chosen so that local charge neutrality is satisfied (for details see Cerda et al. [35]). Fig. 7 - Fig. 10 show the calculated STS conductance-voltage characteristics for different models of the tip. Nanotube-substrate distance is the same as that used in the calculation of Fig. 2 - 5. As can be seen from the plots, additional fine structures are introduced between the peak structures of ρ_{nt} by taking into account the electronic structure of the tip.

DISCUSSIONS AND CONCLUSIONS

Recent experiments on the current-voltage characteristics of carbon nanotube field effect transistors show distinct behavior regarding the Fermi-level position at the metal-carbon nanotube interface. Measurements with high resistance contact[5] have been

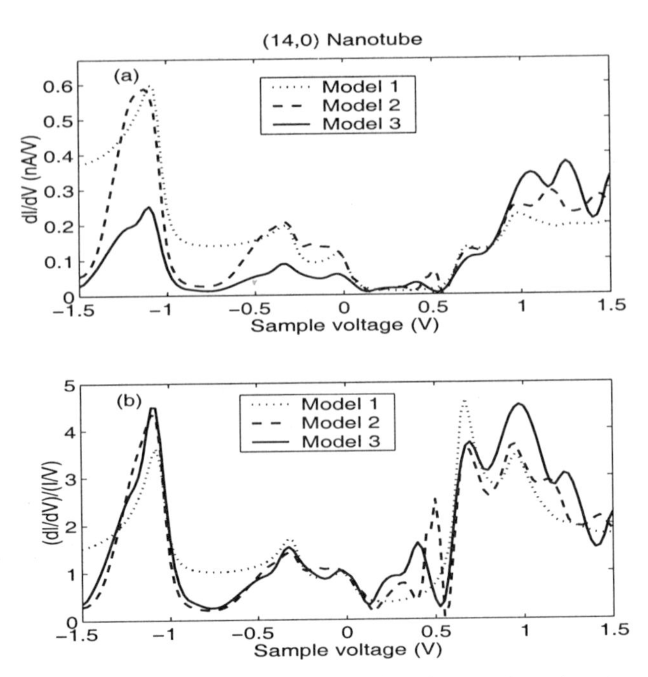

Figure 10: STS conductance-voltage characteristics of (14,0) nanotube, otherwise same as Fig. 7.

explained by assuming that the Fermi-level is "pinned" at the valence band edge of the semiconducting nanotube which is deduced from the STS studies. With the advancement of new techniques of making electric contact to the nanotube,[9] low resistance contacts with two-terminal conductance close to the conductance quantum have been obtained. Measured current-voltage characteristics at low temperature using these new technique show that the Fermi-level is located in the gap of the semiconducting nanotube. In these experiments,[9] SWNTs are grown from the patterned catalyst islands on the surface of an oxidized silicon wafer using the chemical vapor deposition (CVD) method,[36] Au/Ti contact pads are then placed on the catalyst islands fully covering the islands and extending over their edge.

Since the SWNTs thus grown are mostly capped,[36] the coupling between the SWNT and the electrode is presumably similar to that of fullerene where it is well known that fullerene forms strong chemical bond with the noble and transition metal surfaces.[37] The large contact area between the SWNT and the metal will make the coupling across the interface even stronger. This will allow the metal wavefunction to penetrate deep into the nanotube side and we expect that the formation of the Schottky barrier will be dominated by the contribution from the metallic screening by MIGS. Translating into the language of the MIGS model of metal-semiconductor junction,[12,13] this means that the density of the MIGS at the interface will be sufficiently large[a] to screen out any discrepancy from the local charge neutrality condition, resulting in the lining up of the metal Fermi-level and the "charge neutrality level" of the nanotube.

[a]A local density of only 0.02 state/atom eV can give a screening length of about $3\AA$ from the Thomas-Fermi model of screening, see Tersoff.[13]

If the energy band of the nanotube is exactly symmetric (e. g. , within the π-electron model), the "charge neutrality level" will at the mid-gap, although different values can be obtained using more accurate models of the electronic structure. Our point here, however, is not to give a quantitative estimate of the barrier height, but rather to show that the MIGS model provides the conceptual framework of understanding in the limit of strong interface coupling.

The situation gets complicated for measurements using high resistance contacts, where the nanotube is side-contacted and the coupling across the interface is weak.[6,38] Due to the weak coupling across the interface, the metal wavefuncion cannot penetrate into the nanotube side and the MIGS model is no longer applicable. In this case, the interface defects and bending of the nanotube at the edge of the contact[38] can induce localized states at the interface region which will accommodate additional charges and affect the formation of Schottky barrier.[39] Therefore, we expect the final Fermi-level position will depend on the detailed contact conditions and *may or may not be located at the valence band edge.* We believe detailed *ab initio* analysis may be mandatory to clarify the various mechanisms involved.

To conclude, we have used a simple self-consistent tight-binding method to study the charge and potential perturbations at the interface between the supported single-wall carbon nanotube and the gold substrate. The Fermi-level shift deduced from the STS current-voltage characteristics is explained in terms of this local potential perturbation, rather than being "pinned" at the valence band edge of the SWNT, therefore resolving the confusion about the interface Fermi-level position arising from conflicting results obtained in recent transport measurements on carbon nanotube transistors. The major approximation in our approach is the use of the empirical tight-binding Hamiltonian and a simple self-consistent scheme to describe the electronic structure of the interface. The applicability of the π-electron model of the SWNT to STS studies seems to having been justified by recent *ab initio* calculations.[29] More serious approximation involves the use of Extended Hückel Theory to obtain the interface coupling which is at best semi-quantitative. This can be justified partly by the inevitable uncertainties in the exact atomic geometry of the SWNT-metal substrate interface which make the quantification of the interface coupling difficult, partly by the fact that our results are insensitive to small changes in the coupling parameters. Similar tight-binding methods have been used to study the metal-semiconductor junctions and semiconductor hetero-junctions[24,25] and shown to be at least qualitatively correct. In any case, first principle methods (currently under investigation) are required in order to obtain more accurate knowledge of the interface electronic structure, but we believe the tight-binding analysis here provides the correct starting physical picture which one should have in mind before devoting to the fully self-consistent calculations.

ACKNOWLEDGMENT

This work is jointly supported by NSF and ARO through grant number 9708107-DMR. We are indebted to M. P. Anantram for drawing our attention to this important topic.

REFERENCES

1. S. Ijima, Nature **354**, 56 (1991).

2. R. Saito, G. Dresselhaus and M. S. Dresselhaus, *Physical Properties of Carbon Nanotubes* (Imperial College Press, London, 1998).
3. C. Dekker, Phys. Today **52** (5), 22 (1999); D. H. Cobden, Nature **397**, 648 (1999).
4. D. K. Ferry, *Transport in Nanostructures* (Cambridge University Press, Cambridge, 1997).
5. S. J. Tans, A. R. M. Verschueren and C. Dekker, Nature **393**, 49 (1998); R. Martel et al. , Appl. Phys. Lett. **73**, 2447 (1998).
6. J. Tersoff, Appl. Phys. Lett. **74**, 2122 (1999).
7. S. Datta, unpublished.
8. M. P. Anantram, S. Datta and Yongqiang Xue, e-print cond-mat/9907357.
9. H. T. Soh et al. , Appl. Phys. Lett. **75**, 627 (1999).
10. J. W. G. Wildöer et al. , Nature **391**, 59 (1998).
11. T. W. Odom et al. , Nature **391**, 62 (1998).
12. V. Heine, Phys. Rev. A **138**, 1689 (1965); C. Tejedor, F. Flores and E. Louis, J. Phys. C **10**, 2163 (1977).
13. J. Tersoff, Phys. Rev. Lett. **52**, 465 (1984); J. Tersoff, Phys. Rev. B **30**, 4874 (1984); **32**, 6928 (1985).
14. J. Tersoff, in *Heterojunction Band Discontinuities: Physics and Device Applications*, edited by F. Capasso and G. Margaritondo (North-Holland, Amsterdam, 1987).
15. *Scanning Tunneling Microscopy III*, 2nd edition, edited by R. Wiesendanger and H.-J. Güntherodt (Springer-Verlag, Berlin, 1996).
16. S. Datta et al. , Phys. Rev. Lett. **79**, 2530 (1997) and references thererein.
17. Yongqiang Xue et al. , Phys. Rev. B **59**, 7852 (1999) and references thererein.
18. J. W. Mintmire, B. I. Dunlap and C. T. White, Phys. Rev. Lett. **68**, 631 (1992); R. Saito et al. , Phys. Rev. B **46**, 1804 (1992).
19. X. Blase et al. , Phys. Rev. Lett. **72**, 1878 (1994).
20. D. A. Papaconstantopoulos, *Handbook of the Band Structure of Elemental Solids* (Plenum Press, New York, 1986).
21. R. Hoffmann, J. Chem. Phys. **39**, 1937 (1963).
22. F. Guinea et al. , Phys. Rev. B **28**, 4397 (1983).
23. S. Datta, *Electron Transport in Mesoscopic Systems* (Cambridge University Press, Cambridge, 1995).
24. F. Guinea, J. Sanchez-Dehesa and F. Flores, J. Phys. C **16**, 6499 (1983); A. Muñoz, J. Sanchez-Dehesa and F. Flores, Phys. Rev. B **35**, 6468 (1987); J. C. Durán et al. , *ibid.* **36**, 5920 (1987).
25. W. A. Harrison and J. E. Klepeis, Phys. Rev. B **37**, 864 (1988).
26. J. W. Mintmire and C. T. White, Phys. Rev. Lett. **81**, 2506 (1998); J.-C. Charlier and Ph. Lambin, Phys. Rev. B **57**, 15037 (1998).
27. T. Hertel, R. Martel and Ph. Avouris, J. Phys. Chem. B **102**, 910 (1998); A. Rochefort, D. R. Salahub and Ph. Avouris, Chem. Phys. Lett. **297**, 45 (1998).
28. L. X. Benedict, S. G. Louie and M. L. Cohen, Phys. Rev. B **52**, 8541 (1995).
29. A. Rubio et al. , Phys. Rev. Lett. **82**, 3520 (1999); A. Rubio, e-print cond-mat/9903215.
30. R. Hoffmann, Rev. Mod. Phys. **60**, 601 (1988).
31. R. Rennie, Adv. Phys. **26**, 285 (1977).
32. J. A. Appelbaum and D. R. Hamann, Phys. Rev. B **10**, 4973 (1974); C. Kallin and B. I. Halperin, *ibid.* **29**, 2175 (1984).
33. J. Tersoff and D.R. Hamann, Phys. Rev. B **31**, 805 (1985); N.D. Lang, *ibid.* **34**, 5947 (1986).

34. M. Tsukada et al. , Surf. Sci. Rep. **13**, 265 (1991).
35. J. Cerda et al. , Phys. Rev. B **56**, 15885 (1997); J. Cerda et al. , *ibid.* **56**, 15900 (1997).
36. J. Kong et al. , Nature **395**, 878 (1998); J. Kong, A. M. Cassell and H. Dai, Chem. Phys. Lett. **292**, 567 (1998).
37. M. S. Dresselhaus, G. Dresselhaus and P. C. Eklund, *Science of Fullerenes and Carbon Nanotubes* (Academic Press, San Diego, 1996), ch. 17.
38. A. Rochefort et al. , e-print cond-mat/9904083.
39. W. Mönch, Rep. Prog. Phys. **53**, 221 (1990); see also the articles in *Metallization and Metal-Semiconductor Interfaces*, edited by I. P. Batra (Plenum Press, New York, 1989).

SINGLE PARTICLE TRANSPORT THROUGH CARBON NANOTUBE WIRES: EFFECT OF DEFECTS AND POLYHEDRAL CAP

M. P. Anantram and T. R. Govindan

NASA Ames Research Center, Mail Stop: T27A-1,
Moffett Field, CA 94035-1000, U. S. A.

INTRODUCTION

The ability to manipulate carbon nanotubes with increasing precision has enabled a large number of successful electron transport experiments. These studies have primarily focussed on characterizing transport through both metallic and semiconducting wires[1–6]. Reference [1] demonstrated ballistic transport in single-wall nanotubes for the first time, although the experimental configuration incurred large contact resistance. Subsequently, methods of producing low contact resistances have been developed and two terminal conductances smaller than 50kΩ have been repeatably demonstrated in single-wall and multi-wall nanotubes. In multi-wall nanotubes, reference [5] demonstrated a resistance of approximately $h/2e^2$ in a configuration where the outermost layer made contact to a liquid metal. This was followed by the work of reference [6] where a resistance of approximately $\hbar/27e^2$ (478Ω) was measured in a configuration where electrical contact was made to many layers of a multi-wall nanotube. References [5] and [6] note that each conducting layer contributes a conductance of only $2e^2/h$, instead of the $4e^2/h$ that a single particle mode counting picture yields. These small resistances have been obtained in microns long nanotubes, making them the best conducting molecular wires to date. The large conductance of nanotube wires stems from the fact that the crossing bands of nanotubes are robust to defect scattering[7–9].

In STM experiments, electrons tunnel from the substrate being probed to the tip of a capped or open nanotube. Experiments in molecular electronics may also use a nanotube tethered to metal or other parts of the circuit. For these applications, understanding the physics of electron transport through capped nanotubes is relevant. To gain this understanding, references [10–13] analyzed the wave function and density of states of capped nanotubes, and reference [14] recently studied the electron transmission probability through polyhedral capped nanotubes.

The outline of this article is as follows. In section we discuss a method to calculate the transmission through nanotubes with an arbitrary number of leads. In section , we review the role of defects in determining the conductance of nanotubes. Two types of defects are discussed here. The first is weak uniform disorder and the second is strong isolated defects. The transmission of electrons at the band center is reasonably insensitive to the first type of defect but is greatly diminished by the second. Transmission through caps is discussed in section . Our discussion is for polyhedral caps. The transmission is found to correspond to the local density of states (LDOS) at all energies

Science and Application of Nanotubes, edited by Tománek and Enbody
Kluwer Academic / Plenum Publishers, New York, 2000

except those corresponding to the localized states. Hybridization between substrate and carbon nanotube states leads to interference paths that yield strong transmission antiresonances. The presence of defects in the nanotube can transform these antiresonances to resonances. The current carrying capacity of these resonances, however, depends on the location of the defects. Conclusions are presented in section .

METHOD

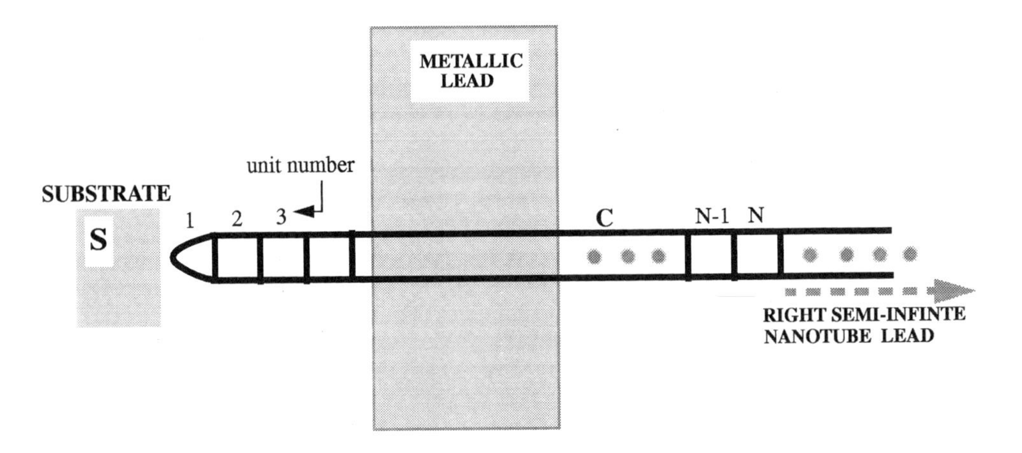

Figure 1: Carbon nanotube in contact with three leads: a semi-infinite carbon nanotube on the right, and metallic leads in the center and in close proximity to the cap at the left.

In this section, we present the model and method used to analyze transport through CNT. Fig1. shows a nanotube (denoted by C) with three leads, a semi-infinite carbon nanotube on the right, a metallic lead in the center and a substrate (denoted by S) which is the metallic lead in close proximity to the cap at the left. Placing the metallic lead at the center of the nanotube is akin to the experiments in reference [1], where the metal is side-contacted to the nanotube. The transmission and LDOS are calculated from the Green's function of the nanotube connected to the leads. The leads are by definition objects that are capable of injecting electrons into the device, and so are typically modeled as either metals[15] or semi-infinite carbon nanotubes[9,16]. The Green's function is calculated by the procedure given in references [9, 17]. The equation governing the Green's function is:

$$\left[EI - H_C - \sum_{\alpha} \Sigma^r_\alpha\right] G^r_C(E) = I, \tag{1}$$

where H_C is the Hamiltonian of the isolated nanotube and I is the identity matrix of size equal to the number of atoms in C. Σ^r_α is the retarded self energy arising due to connection with lead α, and is given by,

$$\Sigma^r_\alpha(E) = V_{C-\alpha} g^r_\alpha(E) V_{\alpha-C} \, . \tag{2}$$

$V_{C-\alpha}$ ($V_{\alpha-C}$) is the Hamiltonian representing the interaction between C (lead α) and lead α (C), and g^r_α is the Green's function of the isolated lead. Σ^r_α is a square matrices of size equal to the number of atoms in C. Note that only sub-matrices of Σ^r_α that correspond to atoms in C coupled to lead α are non-zero. Further, only those elements of g^r_α that couple atoms (nodes) in α to C are required. The real part of the diagonal component of Σ^r_α represents the change in on-site potential of the C atoms, and the imaginary part is responsible for injection of electrons from α to C. In general, the off-diagonal elements of Σ^r_α cannot be neglected, since they play an important role in determining transport characteristics.

The single particle LDOS at site i [$N_i(E)$] and transmission probability between leads α and β [$T_{\alpha\beta}(E)$] at energy E are obtained by solving Eq. (1) for the diagonal element G^r_{ii} and the off-diagonal sub-matrix of G^r in C that couple α and β:

$$N_i(E) = -\frac{1}{\pi} Im[G^r_{ii}(E)] \tag{3}$$

$$T_{\alpha\beta}(E) = Trace[\Gamma_\alpha G^r \Gamma_\beta G^a] \ . \tag{4}$$

$\Gamma_\alpha = 2\pi V_{C-\alpha}\rho_\alpha(E) V_{\alpha-C}$, where $\rho_\alpha(E) = -\frac{1}{\pi} Im[g^r_\alpha(E)]$ is the surface density of states of terminal α (Im extracts the imaginary part).

For computational efficiency and to calculate the transmission through a long nanotube region, we divide the structure C into N smaller units with each unit typically representing a few rings of atoms along the circumference of the tube (Fig. 1). The units need not necessarily be of the same size, but the main idea is to order atoms in the units such that $[EI - H_C - \sum_\alpha \Sigma^r_\alpha]$ is a tridiagonal block matrix,

$$\begin{pmatrix} A_1 & B_{12} & O & O & O & O & O \\ B_{21} & A_2 & B_{23} & O & O & O & O \\ O & B_{32} & \bullet & \bullet & O & O & O \\ O & O & \bullet & \bullet & \bullet & O & O \\ O & O & O & \bullet & \bullet & \bullet & O \\ O & O & O & O & \bullet & \bullet & B_{N-1,N} \\ O & O & O & O & O & B_{N,N-1} & A_N \end{pmatrix} \begin{pmatrix} G^r_{11} \\ G^r_{21} \\ \bullet \\ \bullet \\ \bullet \\ G^r_{N-11} \\ G_{N1} \end{pmatrix} = \begin{pmatrix} 1 \\ O \\ \bullet \\ \bullet \\ \bullet \\ O \\ O \end{pmatrix} . \tag{5}$$

The advantage of casting the matrix in the block tridiagonal form is the efficiency of matrix inversion compared to a non-tridiagonal sparse matrix[18]. If the ith unit contains N_i atoms, then the diagonal block, $A_i = [EI - H_C - \sum_\alpha \Sigma^r_\alpha]_{ith\ block}$ is a N_i by N_i square matrix. The off diagonal sub-matrix B_{ij} represents coupling between units i and j, and is a N_i by N_j rectangular matrix, where N_j is the number of atoms in unit j. Also, note that B_{ij} is non zero only when $|i-j| = 1$. In Eq. (4), computing $T_{\alpha\beta}$ only requires solving for the off-diagonal rectangular matrix $G_{i_\alpha i_\beta}$ between units i_α and i_β. i_α and i_β are units representing carbon atoms connected to leads α and β respectively.

Finally, the single-particle Hamiltonian of the nanotube is[19],

$$H_C = \sum_i \epsilon_i c^\dagger_i c_i + \sum_{i,j} \left[t_{ij} c^\dagger_i c_j + c.c \right] \ , \tag{6}$$

where, ϵ_i is the on-site potential and t_{ij} is the hopping parameter between lattice sites i and j. $\{c^\dagger_i,\ c_i\}$ are the creation and annihilation operators at site i. The Hamiltonian between a nanotube (C) and metal (M) is,

$$H_{CM} = \sum_{i \in C\ x \in M} \left[\tau_{ix} c^\dagger_i d_x + c.c \right] \ , \tag{7}$$

where, τ_{ix} is the hopping parameter between lattice sites i of the nanotube and x of the metal. $\{d^\dagger_x,\ d_x\}$ are the creation and annihilation operators at site x in the metal. The Hamiltonian of the metal is similar to Eq. 6, except that the on-site potential, hopping parameters and lattice coordination corresponds to that of the metal.

DEFECTS

Having described the system, model and method, we now consider the effect of two types of defects on transport. The effect of weak uniform disorder on the localization length of nanotubes was first considered in reference [7], within the context of a two band model by calculating the elastic mean free path and relating the mean free path to the localization length. Reference [9] independently calculated the transmission probability through disordered nanotubes of varying lengths connected to disorder free semi-infinite carbon nanotube leads. Weak uniform disorder is herein modeled by randomly changing the on-site potential of the atoms, $\epsilon_i \rightarrow \epsilon_i + \delta\epsilon_i$ in Eq. (6), where $\delta\epsilon_i$ is randomly chosen from the interval $\pm|\epsilon_{random}|$ at all lattice points. A larger $|\epsilon_{random}|$ implies larger disorder.

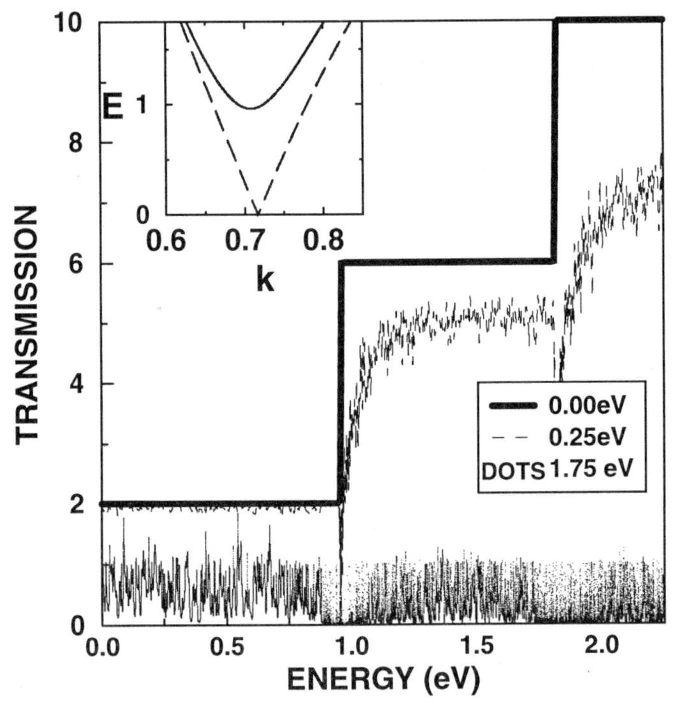

Figure 2: Transmission probability versus energy of a (10,10) nanotube with disorder distributed over a length of 1000 Å. The significant features here are the robustness of the transmission probability around the zero of energy as the strength of disorder is increased and the dip in transmission probability at energies close to the beginning of the second sub-band. The inset shows energy versus wave vector for the first (solid) and the second sub-band (dashed); the velocity of electrons at the minima of the solid line is zero. (From reference [9].)

Fig. 2 is a plot of the transmission probability versus energy for three different values of disorder. There are three main points that we want to convey here. The first is the insensitivity of the transmission probability to disorder at energies close to the band center, where only two sub-bands are present. $|\epsilon_{random}| = 0.25eV$ results in an insignificant reflection probability. A large value of $|\epsilon_{random}| = 1.75eV$ causes more reflection but insufficient to change the transmission drastically at the band center. The reason for this robustness is that electrons in a CNT can find transmission paths around defects because of the large number of atoms in a cross section. And, importantly they result in only two modes at the band center into which electrons can reflect. If this intuitive explanation is correct, we would expect that for a given $|\epsilon_{random}|$, the transmission should depend on the radius of the tube. That is, with increasing number of atoms along the circumference of the tube, the transmission should increase. In Fig. 3, we show that this is indeed the case by comparing the transmission of (10,10), (5,5) and (12,0)-zigzag tubes, each with a 1000 $\mathring{A}$ disordered region. The diameters of these tubes are 13.4 $\mathring{A}$, 9.4 $\mathring{A}$ and 6.7 $\mathring{A}$ respectively. For the (10,10) and (5,5) tubes, the band structure at energies close to the Fermi energy are similar[19]. However, the number of atoms in a unit cell of a (5,5) tube is only half of that of a (10,10) tube (20 and 40 atoms, respectively). The important point here is that although the transmission of disorder-free (10,10) and (5,5) tubes at energies around the band center are identical, transmission is smaller for the (5,5) tube in the presence of disorder. The (12,0) zigzag tube has a diameter between that of the (10,10) and (5,5) tubes. Correspondingly, the average transmission is between the values of the (10,10) and (5,5) tubes. Reference [8], using analytical expressions addresses the reason for small back scattering.

The second point we wish to make about Fig. 2 concerns localization. We have calculated conductance as a function of the length of the disordered region. For disordered regions larger than the localization length (L_0), the conductance of any quasi one dimensional structure has been predicted to decrease exponentially with length, $g = g_0 \, exp(-L/L_0)$, in the phase coherent limit[20]. For lengths shorter than the localization length, the decrease in conductance is not given by this relation. We observe this to be the case in Fig. 4. The inset of Fig. 4 shows that the conductance does change exponentially with length for disordered regions larger than L_0. The value of L_0 corresponding to disorder strengths of 1eV and 1.75eV are 3353 $\mathring{A}$ and 1383 $\mathring{A}$ respectively. We note that in Fig. 4 the conductance of a micron long (10,10) nanotube with $|\epsilon_{random}| = 0.25eV$ is *large*, approximately $1.5\frac{e^2}{h}$.

The third point we want to convey is the large decrease in transmission probability at energies where many sub-bands are present, and especially the dip in transmission probability corresponding to the opening of a new sub-band. The origin of this dip in transmission probability is the low velocity electrons in the second sub-band (inset of Fig. 2) and can be explained as follows. In a perfect lattice, the velocity (dE/dk) of electrons in the second sub-band and with an energy close to the minimum is nearly zero. These low-velocity electrons are easily reflected by the smallest of disorders. Disorder causes mixing of the first and second sub-bands. As a result, electrons incident in either sub-band at these energies have a large reflection coefficient (in comparison to electrons with energies close to the band center). Increasing the disorder strength results in further reduction of the conductance and also results in the broadening of the dip. A dip in transmission probability at the formation of new sub-bands was also found in reference [21], in their study of quasi one dimensional wires. At energies away from the band center, many sub-bands exist. As a result, an electron incident at these energies has many sub-bands into which it can reflect. This accompanied by the large reflection due to low velocity states at sub-band openings leads to a transmission probability that is greatly diminshed in comparison to that at band-center energies, where there are only two sub-bands (Fig. 2). For example, in Fig. 2, the transmission probability at $E = 2eV$ is equal to 10 in a disorder-free (10,10) nanotube, while even a small $|\epsilon_{random}| = 0.25eV$ reduces this to a about 6 or 7.

Although transmission through CNTs is relatively insensitive to weak-disorder type defects, there may be other type of defects that are capable of changing the transmission probability more dramatically at the band center. We consider one such defect type below.

Strong isolated defects (defined to be lattice locations onto which an electron

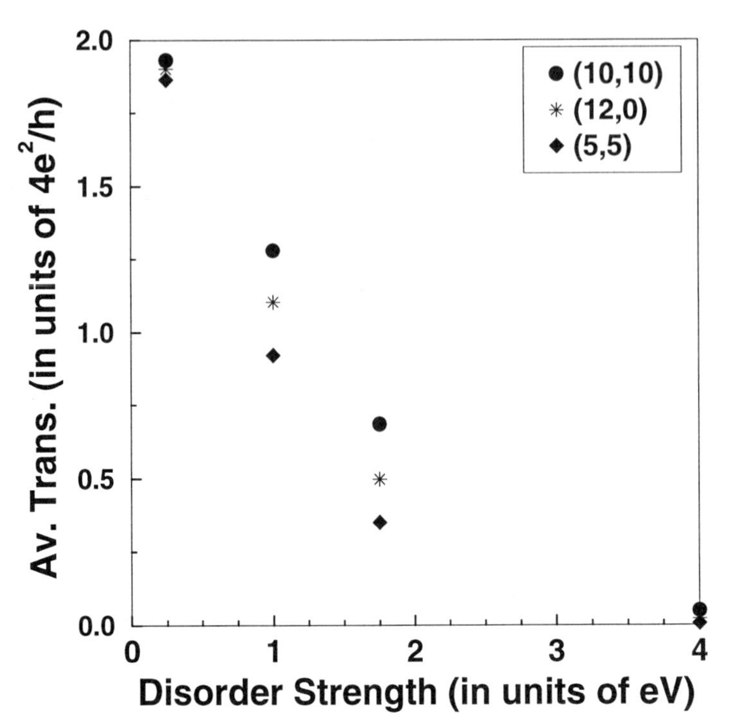

Figure 3: Average transmission probability at the band center versus disorder strength for wires of different diameter. The transmission probability has been averaged over a thousand different realizations of the disorder. The main feature here is that the average transmission probability decreases with decrease in the number of atoms along the circumference of the wire. (From reference [9].)

cannot hop) are isolated due to a either a large mismatch in the on-site potential or weak bonds with neighbors. It was shown in reference [16] that scattering from a single such defect causes a maximum reduction in the transmission probability at the band center E=0. The transmission probability of a 1000 $\mathring{A}$ long (10,10) nanotube with ten defects scattered randomly along the length is plotted in Fig. 5. The main point is that independent of the exact location of these defects, a *transmission gap* opens at the center of the band as the number of defects is increased. The width of the transmission gap increases with the increase in defect density[9]. The transmission probability also decreases sharply at energies corresponding to the opening of the second sub-band, although this effect is weak compared to that due to weak uniform disorder. While there is as yet no experimental evidence for such strong scattering, we expect that when the Fermi energy is close to $E = 0$, the low bias conductance will be small in the presence of such defects[9].

TRANSPORT THROUGH A POLYHEDRAL CAP

The discussion of transport through capped nanotubes is restricted to armchair tubes with polyhedral caps. The discussion closely follows our work in reference[14]. The main issues we address here are: (i) the relationship between the LDOS and the transmission probability through cap atoms in a defect free CNT, (ii) the effect of the localized discrete energy levels in the cap, and (iii) the effect of defects on tunnel current/transmission.

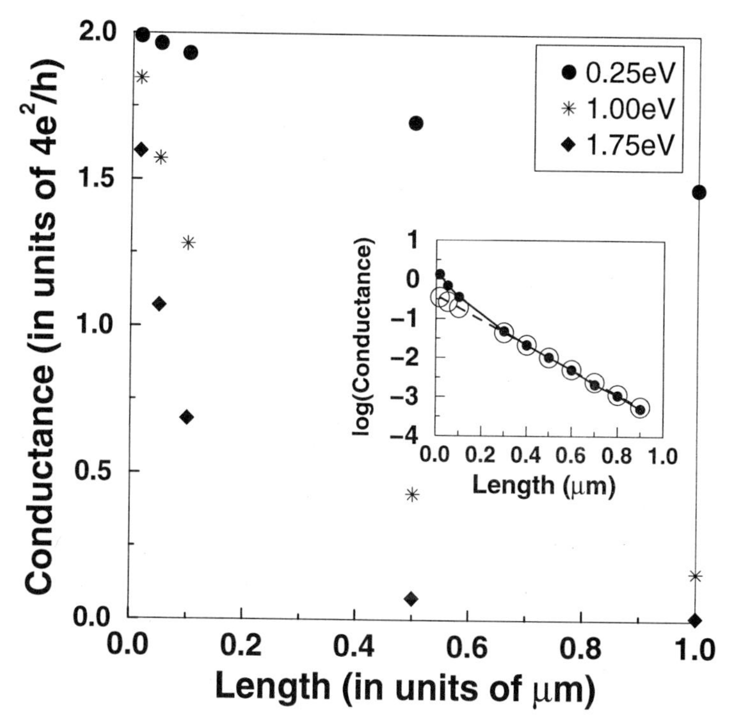

Figure 4: Conductance versus length of the (10,10) CNT. While for the large disorder strengths, the conductance is significantly affected by disorder, the conductance is reasonably large for the smaller values of disorder. This demonstrates the robustness of these wires to weak uniform disorder at energies close to the band center. Inset: log(Conductance) versus length for disorder strength of 1.75eV, in a (5,5) CNT. The solid line/filled circle corresponds to the simulation and the dashed line/empty circle corresponds to that obtained using $g = g_0 exp(-L/L_0)$. (From reference [9].)

The transmission probability and LDOS are calculated for a (10,10) nanotube with a polyhedral cap (Fig. 6). This cap has a five-fold symmetry, with one pentagon at the cap center and five pentagons placed symmetrically around it. The geometry is the same as in Fig. 1 but with the metal contact at the center absent. The transmission probability is calculated from the substrate to the semi-infinite nanotube lead at the right hand side of Fig. 1. We assume only a single atom in the cap makes contact with the substrate. The atom making contact is shown circled in Fig. 6. The contact is modeled by an energy independent self-energy [in Eqs. (1) and (2)], Σ_S^r. Within the context of a single parameter tight binding Hamiltonian, the polyhedral cap has localized states (with an infinite life-time) that decay into the nanotube[10]. Coupling of the cap to the substrate causes hybridization between the substrate states, nanotube continuum states and localized states of the polyhedral cap. This hybridization transforms the localized states into quasi-localized states (finite life-time).

The density of states averaged over the cap and over unit cells at various distances away from the cap is shown in Fig. 7. The distance is measured in terms of the number of unit cell lengths of the (10,10) nanotube. In Fig. 7, there are two localized states in the energy range considered, one around 0.25 eV and the other around -1.5 eV. The transmission probability (Fig. 8) corresponds directly to the LDOS at most energies. The major difference is near the localized state energy, where *the LDOS peaks corresponds to transmission zeroes.* The solid line in Fig. 8 is for $\Sigma_S^r = i \cdot 0.25$eV, a purely imaginary number, which corresponds to a smooth antiresonance. However,

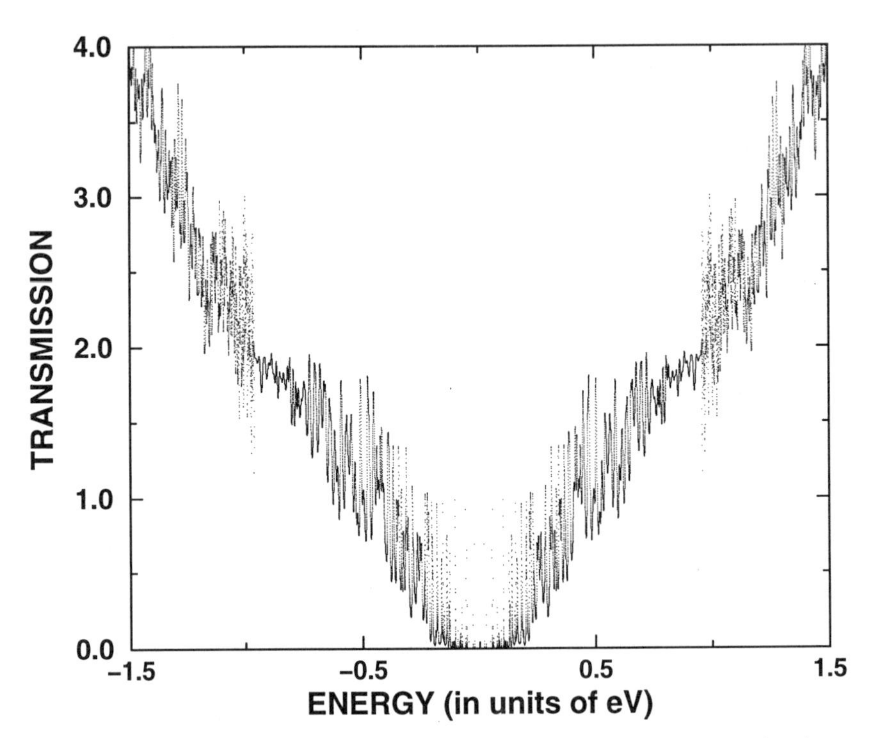

Figure 5: Transmission probability versus energy of a (10,10) CNT with ten strong isolated scatterers sprinkled randomly along a length of 1000 Å. The main prediction here is the opening of a gap in the transmission probability around the zero of energy. The transmission probability of a defect-free tube is shown in Fig. 2. (From reference [9].)

when the cap makes contact to the substrate Σ_S^r would in general also have a real part. The dashed curve in Fig. 8 corresponds to $\Sigma_S^r = 0.25 + i \cdot 0.25$eV. This curve retains the anti-resonance feature, but the shape has changed as a result of the non zero real part, which is consistent with earlier calculations in quantum wires with stubs[22]. The transmission dip arises from hybridization of localized and continuum states via coupling to the substrate, in a manner akin to Fano resonances[23]. States in the CNT cap comprise localized (ϕ_L) and continuum (ϕ_C) states that are uncoupled to each other. Bringing the substrate in close proximity to the cap couples ϕ_L and ϕ_C to the substrate states (ϕ_S). As a result, electrons can be transmitted from ϕ_S to ϕ_C either directly ($\phi_S \rightarrow \phi_C$), through the localized states ($\phi_S \rightarrow \phi_L \rightarrow \phi_S \rightarrow \phi_C$), and through higher order interactions. The interference between these paths gives rise to the transmission antiresonances.

We now consider changes to the antiresonance picture due to defects in a tube. A defect locally mediates mixing/hybridization of localized and continuum states. This leads to transmission paths similar to a double barrier resonant tunneling structure, where the two scattering centers provide coupling between the substrate, the nanotube localized state and the defect. In addition, the paths leading to the transmission antiresonance discussed previously also exist, and are accounted in the model.

To demonstrate the effects of defects on transmission through a polyhedral cap, we consider a topological bond rotation defect (box in Fig. 6)[24]. The LDOS remains similar to Fig. 7 but in comparison to Fig. 8, the transmission probability has changed significantly around the localized energy levels as shown in Fig. 9. Resonant peaks

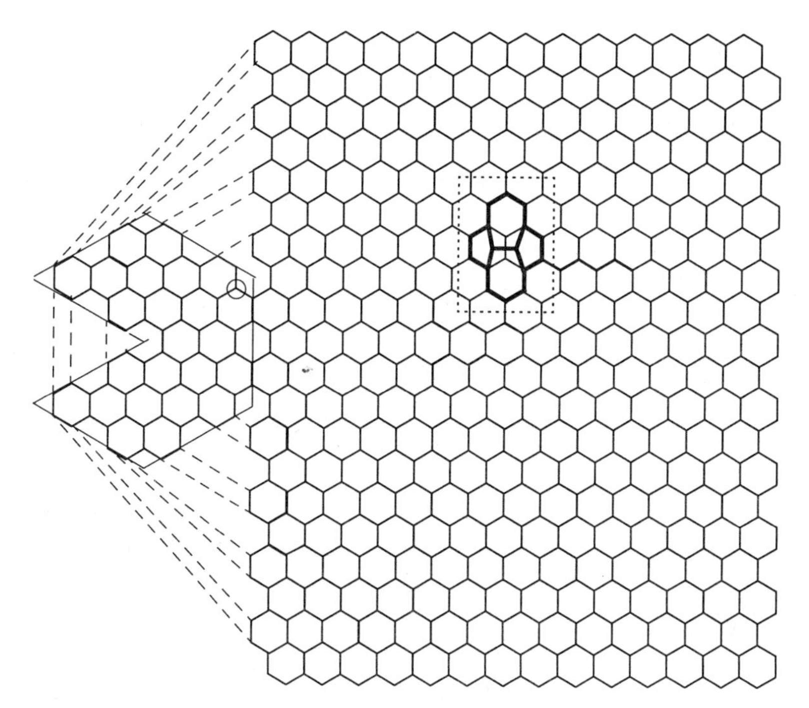

Figure 6: (10,10) carbon nanotube with a polyhedral cap. The dashed lines connect equivalent sites of the cap and nanotube in this two dimensional representation. The dashed box shows a bond rotation defect (From reference [14].)

appear in the transmission probability. The resonance width here is determined by two contributions: hybridization of the localized and continuum cap states due to the substrate and the distance of defect from the cap. In particular, the second contribution depends on $| < \phi_C | H_{defect} | \phi_L > |$, where H_{defect} is Hamiltonian of the defect. $|\phi_L|^2$ (or the density of states of the localized state) decays with distance away from the cap. As a result, the width of the transmission resonance depends on location of the defect. Fig. 9 shows the transmission for two different distances of the defect from the cap (L_D): $L_D = 7$ and $L_D = 15$, where L_D is in units of the one dimensional unit cell length of armchair tubes. The main feature in Fig. 9 is that the *transmission resonance width becomes smaller as distance of the defect from the cap increases.* This can be understood from the fact that the strength of hybridization between continuum and localized states in the cap caused by the defect ($| < \phi_C | H_{defect} | \phi_L > |$) decreases as distance of the defect increases from the cap. In terms of the current carrying capacity, clearly from Fig. 9, the resonant state is capable of carrying a large current per unit energy compared to that of the background energies.[a]

[a] In addition to the transmission resonance around $0.25eV$, there is a new narrow resonance around $0.5eV$ in Fig. 6. This new resonance is due to quasi-localized states associated with the bond rotation defect. This is discussed further in reference[14].

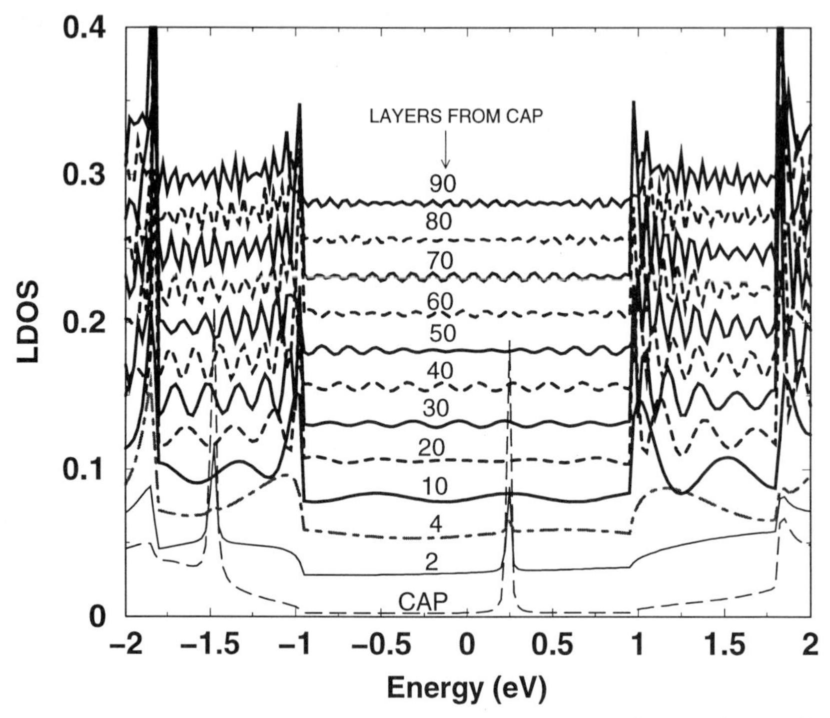

Figure 7: The bottom-most curve is the DOS averaged over all atoms in the cap. The quasi-localized level is strongly peaked. Other curves represent the DOS averaged over a unit cell of the armchair tube at varying distances from the cap. The curves have been displaced by multiples of 0.025 along the y-axis. The circled site in Fig. 6 makes contact to the substrate with coupling strength $\Sigma_S^r = i \cdot 0.25eV$

CONCLUSIONS

Two aspects of single particle transport through carbon nanotube wires were reviewed. In section , calculations of the transmission probability through carbon nanotube wires with simple defect models were discussed[9]. *It was shown that carbon nanotube wires with weak uniform disorder are relatively insensitive to back scattering at energies where only the crossing bands (close to E=0) contribute to transport.* The localization length of nanotube wires calculated in references [7, 9, 25] is large, implying that nanotubes are good molecular wires. The reflection probability at energies away from the band center (where many sub-bands coexist) is much larger. At energies corresponding to the opening of new sub-bands, coupling of high and low velocity states causes a dip in the transmission, which should be observable in experiments where the Fermi energy can be tuned. The transmission probability through nanotube wires with strong scatterers were also calculated. These structures show a transmission gap around E=0, whose width increases with the number of defects.

Transport experiments have demonstrated that carbon nanotubes are excellent metallic molecular wires, having allayed to a large extent the concerns of large contact resistance and the effect of deformation due to coupling with the substrate[5,6,26,27]. Apart from their promise in applications, carbon nanotubes are also a good test bed for experiments in both molecular electronics and the basic study of physics in quasi one dimensional systems with electron-electron interactions[27–29]. These comments would be incomplete without pointing out that references [5] and [6] measure conductances that are very close to integer multiples of $\frac{2e^2}{h}$, instead of integer multiples of $\frac{4e^2}{h}$ that a simple mode counting picture would yield. An open question is if this phenomenon is particular only to these experiments[5,6] or is it more universal. In any case, a simple and clear explanation of this does not exist at present in the literature.

The second aspect, disussed in section dealt with transport through armchair nanotube wires with polyhedral caps. Localized and continuum energy levels have been shown to coexist[10]. In comparison, asymmetric caps in general lead to quasi-localized states as opposed to localized states, as shown in references [11–13]. The transmission probabilty from the substrate to an atom in the cap and into the nanotube continuum

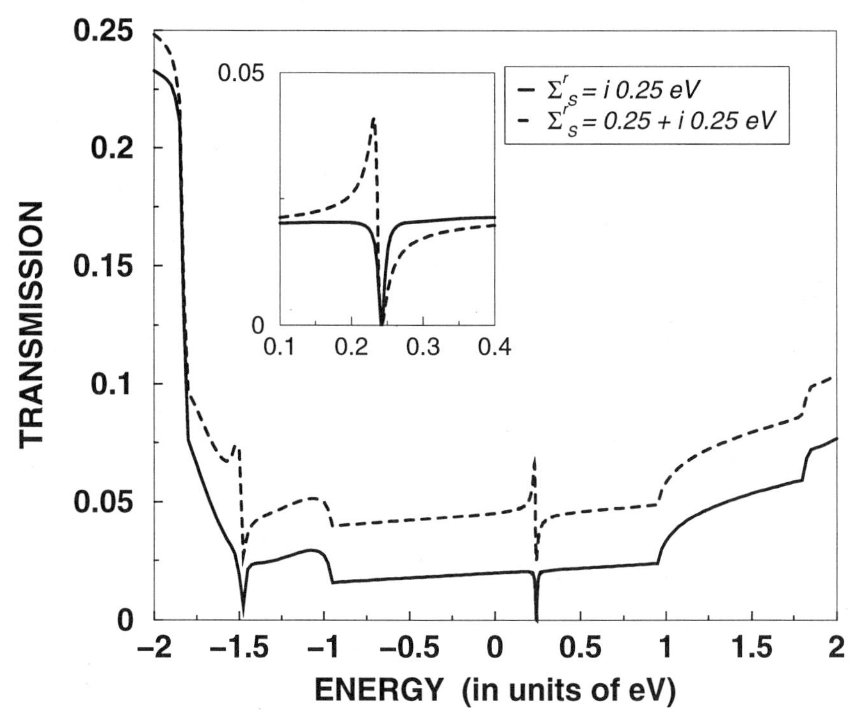

Figure 8: The transmission probability corresponds to the LDOS at most energies except at the peaks in Fig. 7. Around these energies, there is a transmission antiresonance.

states follows the LDOS of the atom. The one exception to this is at the energy of the localized states. At this energy, while the LDOS is peaked, the transmission probability exhibits an antiresonance with minimum equal to zero[14]. A defect in the tube causes hybridization between the continuum and localized levels. As a result, scattering due to defects in the tube can transform this sharp antiresonance to a transmission resonance due to new interference paths, similar to the behavior of a structure with two scattering centers. The current carrying capacity of this resonance is however sensitive to the position of the defect in the tube. This makes sense because the hybridization between the continuum and localized levels becomes weaker with an increase in the distance of the defect from the cap.

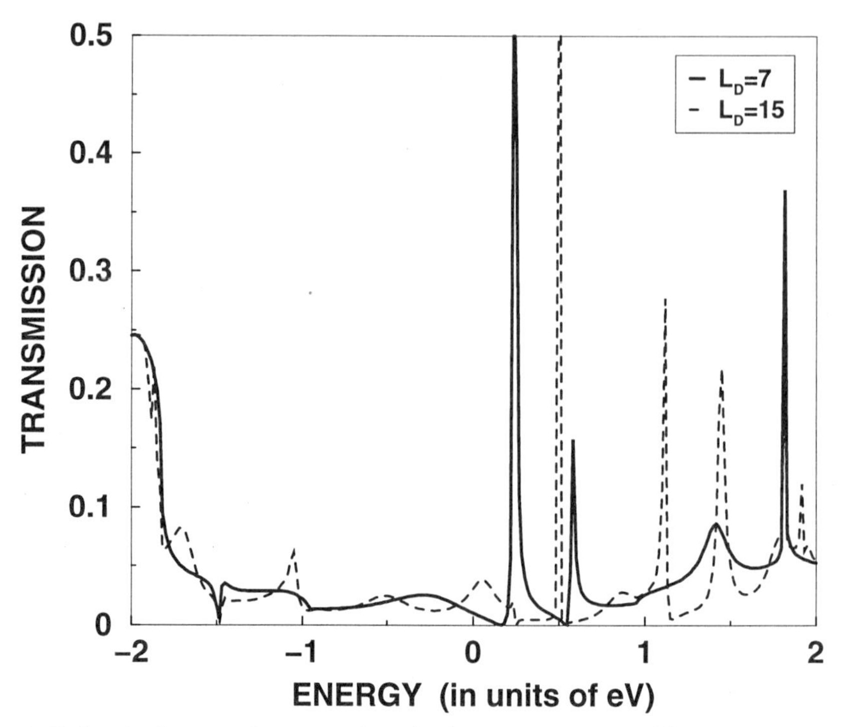

Figure 9: Defects in the nanotube can tranform the sharp antiresonances of Fig. 8 to resonances. The current carrying capacity of the resonance however depends on the location of the defect, as shown by comparing two defect locations.

REFERENCES

1. S. J. Tans, M. Devoret, H. Dai, A. Thess, R.E. Smalley, L.J. Geerligs and C. Dekker, Nature **386**, 474 (1997).
2. P. G. Collins, A. Zettl, H. Bando, A. Thess and R.E. Smalley, Science **278**, 100 (1997).
3. D.H. Cobden, M. Bockrath, N. Chopra, A. Zettl, P.L. McEuen, A. Rinzler, A. Thess, and R.E. Smalley, Physica B **249-251**, 132 (1998).
4. A. Bezryadin, A. R. M. Verschreren, S. J. Tans, and C. Dekker Phys. Rev. Lett. **80**, 4036 (1998).
5. S. Frank, P. Poncharal, Z. L. Wang and W. A. de Heer, Science **280**, 1744 (1998);
6. P. J. de Pablo, E. Graugnard, B. Walsh, R. P. Andres, S. Datta and R. Reifenberger, Appl. Phys. Lett. **74**, 323 (1999)
7. C. T. White and T. N. Todorov Nature , **393**, 240 (1998)
8. T. Ando and T. Nakanishi, J. Phys. Soc. Jpn **67**, 1704 (1998)
9. M. P. Anantram and T. R. Govindan, Phys. Rev. B **58**, 4882 (1998)
10. Ryo Tamura and Masaru Tsukada, Phys. Rev. B **52**, 6015 (1995);
11. D. L. Carroll, P. Redlich, P. M. Ajayan, J. C. Charlier, X. Blase, A. De Vita and R. Car, Phys. Rev. Lett. **78**, 2811 (1997)
12. P. Kim, T. W. Odom, J.-L. Huang and C. M. Lieber, Phys. Rev. Lett. **82**, 1225 (1999)
13. A. De Vita, J.-Ch. Charlier, X. Blase and R. Car, Appl. Phys. A **68**, 283 (1999)
14. M. P. Anantram and T. R. Govindan, *Transmission Through Carbon Nanotubes With Polyhedral Caps*, cond-mat/9907020 (1999)

15. M. P. Anantram, S. Datta and Yongqiang Xue, *Coupling of carbon nanotubes to metallic contacts*, cond-mat/9907357 (1999)
16. L. Chico, V.H. Crespi, L.X. Benedict, S.G. Louie and M.L. Cohen, Phys. Rev. Lett. **76**, 971 (1996).
17. S. Datta, Electronic Transport in Mesoscopic Systems, Cambridge University Press, Cambridge, U.K (1995); C. Caroli, R. Combescot, P. Nozieres and D. Saint-James, J. Phys. C: Solid St. Phys. **4**, 916 (1971); Y. Meir and N.S. Wingreen, Phys. Rev. Lett. **68**, 2512 (1992)
18. W. H. Press, S. A. Teukolsky, W. T. Vetterling and B. P. Flannery, Numerical Recipes in C, The Art of Scientific Computing, Second Edition, Cambridge University Press (1992).
19. M. S. Dresselhaus, G. Dresselhaus and P. C. Eklund, Chap. 19 of Science of Fullerenes and Carbon Nanotubes, Academic Press, New York (1996).
20. C.W.J. Beenakker, Rev. Mod. Phys. **69**, 731 (1997) and references there in.
21. A. Kumar and P. F. Bagwell, disorder Phys. Rev. B **44**, 1747 (1991).
22. P. F. Bagwell and R. K. Lake, Phys. Rev. B **46**, 15329 (1992); Z. Shao, W. Porod and C. S. Lent, Phys. Rev. B **49**, 7453 (1994); P. Singha Deo and A. M. Jayannavar, Phys. Rev. B **50**, 11629 (1994);
23. U. Fano, Phys. Rev. **124**, 1866 (1961).
24. B. I. Yakobson, App. Phys. Lett. **72**, 918 (1998); M. B. Nardelli et al., Phys. Rev. B **57** 4277 (1998)
25. T. Kortyrko, M. Bartkowiak and G. D. Mahan, Phys. Rev. B **59**, (1999) to appear
26. H. T. Soh, A. F. Morpurgo, J. Kong, C. M. Marcus, C. F. Quate and H. Dai, Preprint (1999).
27. Z. Yao, H. W. Ch. Postma and C. Dekker, Preprint
28. C. Kane, L. Balents, M. P. A. Fisher, Phys. Rev. Lett. **79** 5086 (1997)
29. M. Bockrath, D. H. Cobden, J. Lu, A. G. Rinzler, R. E. Smalley, L. Balents, P. L. Mceuen, *Luttinger Liquid Behavior in Carbon Nanotubes*, cond-mat/9812233

CARBON NANOTUBES FROM OXIDE SOLID SOLUTION : A WAY TO COMPOSITE POWDERS, COMPOSITE MATERIALS AND ISOLATED NANOTUBES

Christophe Laurent, Alain Peigney, Emmanuel Flahaut, Revathi Bacsa, and Abel Rousset

Laboratoire de Chimie des Matériaux Inorganiques, ESA CNRS 5070, Université Paul-Sabatier, 31062 Toulouse cedex 4, France

INTRODUCTION

Most synthesis methods of carbon nanotubes (C_{NTs}) (see by Journet and Bernier[1] for a review) are based on the sublimation of carbon in an inert atmosphere, such as the electric-arc-discharge process, the laser ablation method and the solar technique, but chemicals methods such as the catalytic decomposition of hydrocarbons, the electrolysis in a molten ionic salt, the heat-treatment of polymers, the low-temperature solid pyrolysis and the in-situ catalysis are also used. Several of these methods involve the use of nanometric metal particles. Laurent et al.[2] have reviewed the various mechanisms proposed for nanotube nucleation and growth from such particles and the micro/nanostructure of the materials obtained by the different methods. The chemical methods are promising owing to great quantities and low cost. In particular, the catalytic methods using the decomposition of hydrocarbons or the disproportionation of CO on metal particles (generally based on Fe, Co or Ni) leads to Iijima-type C_{NTs} when the catalyst particles are sufficiently small. However, the main difficulty is to obtain nanometric particles, i.e. active particles, at the relatively high temperature (usually higher than 800°C) required for the formation of C_{NTs} by catalytic decomposition of hydrocarbons. Ivanov et al.[3] and Hernadi et al.[4] used a zeolite-supported Co catalyst and obtained C_{NTs} only 4 nm in diameter as well as 60 μm long tubes, but they point out that the longest tubes are also the thickest. Dai et al.[5] reported the formation, albeit

Science and Application of Nanotubes, edited by Tománek and Enbody
Kluwer Academic / Plenum Publishers, New York, 2000

in low yield, of single-wall carbon nanotubes (SWNTs) together with a small amount of double-wall nanotubes (DWNTs) by disproportionation of CO on alumina-supported Mo particles at 1200°C. Kong et al.[6] reported the synthesis of SWNTs by chemical vapor deposition of CH_4 at 1000°C on impregnated Fe_2O_3 /Al_2O_3 catalyst but they used a very low catalyst loading and they furthermore pointed out that a quantitative measure of the SWNTs yield in their materials is lacking. Cheng et al.[7] prepared ropes of SWNT bundles by catalytic decomposition of hydrocarbons over metal particles derived from the decomposition of ferrocene and report the presence of impurities such as carbon nanoparticles, catalyst particles and carbon blacks on the surfaces of or among the bundles, requiring some purification. Hafner et al.[8] have synthesized SWNTs and DWNTs by catalytic decomposition of both CO and C_2H_4 over a supported metal (Mo and Fe:Mo) catalyst. Using calculations of the energies of curved graphene sheets for the nanotubes and capsule forms of carbon, these authors show that the nanotube energy is lower than the capsule energy for diameters below ca. 3 nm, in good agreement with the experimental results.

Works by the present laboratory[9-15] have shown that composite powders containing enormous amounts of SWNTs and small multi-wall carbon nanotubes (MWNTs) are advantageously prepared by selective reduction of oxide solid solutions in H_2-CH_4 gas mixtures. The reduction of solid solutions between a non-reducible oxide such as Al_2O_3, $MgAl_2O_4$ or MgO and one or more transition metal oxide(s) produces very small transition metal (Fe, Co, Ni and their alloys) nanoparticles at a relatively high temperature (usually higher than 800°C). The decomposition of CH_4 over the freshly formed metal nanoparticles prevents their further growth and thus results in the very strong proportion of SWNTs and small MWNTs compared to other forms of carbon.

This paper presents the highlights of the synthesis of the C_{NTs}-metal-oxide composite powders and reports results obtained on the extraction and purification of the C_{NTs} on the one hand and on the preparation, microstructure and properties of dense ceramic-matrix composites prepared from the composite powders on the other hand.

EXPERIMENTAL

Powders

Oxide Solid Solutions. Amorphous $Al_{2-2x}Fe_{2x}O_3$ solid solutions were prepared by decomposition in air at 400°C for 2 h of the corresponding mixed oxalate as described elsewhere.[16] A further calcination in air at 800-1000°C produced the cubic form η-$Al_{2-2x}Fe_{2x}O_3$ and the stable rhombohedral form α-$Al_{2-2x}Fe_{2x}O_3$ was obtained after a treatment above 1000°C. The $MgAl_2O_4$- and MgO-based solid solutions with FeO, CoO or

NiO were prepared by the combustion synthesis route proposed by Patil.[17] Briefly, the appropriate amounts of the desired metal nitrates (Mg, Al, Fe, Co, Ni) are mixed in stoichiometric proportions with urea and dissolved in a minimum amount of water in a Pyrex dish, which is then placed in a furnace pre-heated at 600 or 650°C. Within a couple of minutes, the highly exothermic redox reaction occurs, producing the desired oxide. One combustion batch yields about 6 g of oxide powder. The $MgAl_2O_4$-based combustion products were attrition-milled (2000 rpm, 30 min) in an aqueous solution of dispersant using alumina balls and a nylon rotor in a nylon vessel. The obtained product was passed through a sieve using ethanol to wash the alumina balls and the vessel. Excess ethanol was removed by evaporation at 60°C in an oven for 24 hours. The oxide powders were calcinated in air at 500°C for 30 min in order to burn the contamination caused by the erosion of nylon during milling.

C_{NTs}-Metal-Oxide Powders. In a typical experiment, 1.5 g of the oxide powder was placed in a furnace under a flowing H_2-CH_4 mixture (flow rate 250 sccm) dried on P_2O_5 and was heated at a temperature in the range 900-1070°C, which was maintained for times ranging from 6 to 240 min. Heating and cooling rates are equal to 5°C/min.

Powder Characterization. The powders were studied using scanning and high - resolution transmission electron microscopy (SEM and HRTEM), X-Ray diffraction (XRD), thermogravimetric analysis (TGA) and room temperature ^{57}Fe Mössbauer spectroscopy. Parts of the composite powders were oxidized in air at 850 or 900°C in order to eliminate all the carbon, as required for the specific surface area study. The specific surface areas of the oxide solid solution powder (S_{ss}), of the powders obtained after reduction (S_r) and of the specimens oxidized at 850 or 900°C (S_o) were measured by the BET method using N_2 adsorption at liquid N_2 temperature. The difference $\Delta S = S_r - S_o$ represents the surface of carbon by gram of composite powder. Since most of the carbon is in the form of nanotubes and since these species have a very high surface area, it has been proposed elsewhere[9, 10] that ΔS essentially represents the quantity of nanotube bundles in the composite powder. The increase in specific surface area by gram of carbon, $\Delta S/C_n$, represents the quality of carbon, a higher figure for $\Delta S/C_n$ denoting more carbon in tubular form and/or a smaller average tube diameter and/or fewer walls,[9, 10] which we consider a higher quality. We used a Micromeritics FlowSorb II 2300 apparatus that gives a specific surface area value from one point (i.e. one adsorbat pressure) and requires calibration. We determined that the reproducibility of the results is in the ± 3% range. The carbon content was determined by flash combustion with an accuracy of ± 2%.

Dense composites

The composite powders were uniaxially hot-pressed in graphite dies, in a primary vacuum, at different temperatures in the 1300-1550°C range depending on the nature of the oxide matrix and the metal content. The dwell time at the appropriate temperature was in the range 15-90 min. The dense specimens (20 mm in diameter and 2 mm thick) for mechanical tests were ground to a finish better than 6 μm with diamond suspensions.

Surfaces polished to an optical finish and fracture profiles were observed by scanning electron microscopy (SEM). Densities were calculated from the mass and dimensions of the dense composites or by the Archimedes method. The materials were studied by XRD. The transverse fracture strength (σ_f) was determined by the three-point-bending test on parallelepipedic specimens (1.6 x 1.6 x 18 mm^3) machined with a diamond blade. The fracture toughness (K_{Ic}) was measured by the SENB method on similar specimens notched using a diamond blade 0.3 mm in width. The calibration factor proposed by Brown and Srawley[18] was used to calculate the SENB toughness from the experimental results. Cross-head speed was fixed at 0.1 mm/min. The values given for σ_f and K_{Ic} are the average of measures on 7 and 6 specimens, respectively. The electrical properties were measured at room temperature on similar specimens. Fracture surfaces were coated with a Ag paste. The current intensity was measured versus the voltage which was varied between 0.1 and 0.28 V.

RESULTS AND DISCUSSION

Powders

Alumina-based powders. After a preliminary study[9] demonstrating the possibility to prepare C_{NTS}-Fe-Al_2O_3 powders containing enormous amounts of C_{NTS} (Fig. 1) by reduction of an α-$Al_{1.8}Fe_{0.2}O_3$ solid solution, it was attempted[10] to increase the active-to-inactive catalyst particle ratio by using powders with a high specific surface area such as poorly crystallized or uncrystallized $Al_{1.8}Fe_{0.2}O_3$ solid solutions.

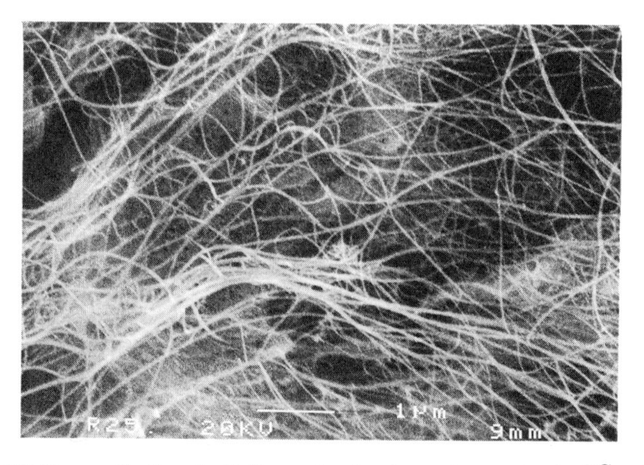

Figure 1. SEM image of a C_{NTS}-Fe-Al_2O_3 powder showing a dense network of C_{NTs} bundles.

The amorphous, η (cubic) and the stable α (corundum) forms of $Al_{1.8}Fe_{0.2}O_3$ were prepared and reduced in a H_2-CH_4 gas mixture at 900 or at 1000°C, giving rise to composite powders containing alumina, α- and γ-Fe, Fe_3C (as revealed by XRD and Mössbauer spectroscopy), and several forms of carbon, including C_{NTs}, thick tubes and clusters of graphitic carbon nanoparticles (SEM, HREM).

The powders have also been studied using a combination of chemical analysis and specific surface area measurements. In particular, we have made use (Table1) of the specific surface area of carbon (ΔS) as a representation of the quantity of carbon nanotubes in a powder and we have considered the ratio of this value to the carbon content ($\Delta S/C_n$) as a quality figure, a higher figure for $\Delta S/C_n$ denoting more carbon in tubular form and/or a smaller average tube diameter and/or fewer walls. This method allows to characterize the composite powders at a macroscopical scale, thus producing results more representative of the material than local techniques do.

The composite powders prepared from the amorphous or η solid solutions contain important quantities of non-tubular carbon (notably thick tubes and clusters of graphitic carbon nanoparticles, Fig. 2a), resulting in a poor value of $\Delta S/C_n$. Moreover, TGA revealed that some carbon is entrapped within the oxide grains upon the crystallization of alumina during the reduction step. This entrapped carbon is more stable than the C_{NTs} with respect to oxidation in air and probably, more generally, with respect to most thermal treatments, which may prove detrimental if those powders were to be used for the production of dense materials by sintering. In contrast, the composite powders prepared from the stable α-solid solution contain carbon essentially in the form of C_{NTs} (Fig. 2b), in line with a higher value of $\Delta S/C_n$.

Table 1. Specific surface area (S_{ss} : oxide solid solutions, S_r : composite powders , S_o : powders oxidized at 850°C, $\Delta S = S_r - S_o$) and carbon contents (C_n : nanocomposites, C_{on} : nanocomposites oxidized at 850°C) of the powders. Am. : amorphous.

Oxide	S_{ss} (m^2/g)	Reduction Temperature (°C)	S_r (m^2/g)	S_o (m^2/g)	C_n (wt%)	C_{on} (wt%)	ΔS (m^2/g)	$\Delta S/C_n$ (m^2/g)
Am.	79.5	900	23.5	20.0	13.4	0.65	3.5	26
		1000	16.0	11.0	20.1	0.58	5.0	25
η	30.5	900	44.1	39.3	8.9	0.30	4.8	54
		1000	31.0	25.1	18.3	0.21	5.9	32
α	3.1	900	6.7	4.1	1.65	0.00	2.6	155
		1000	8.4	3.5	6.2	0.00	4.9	79

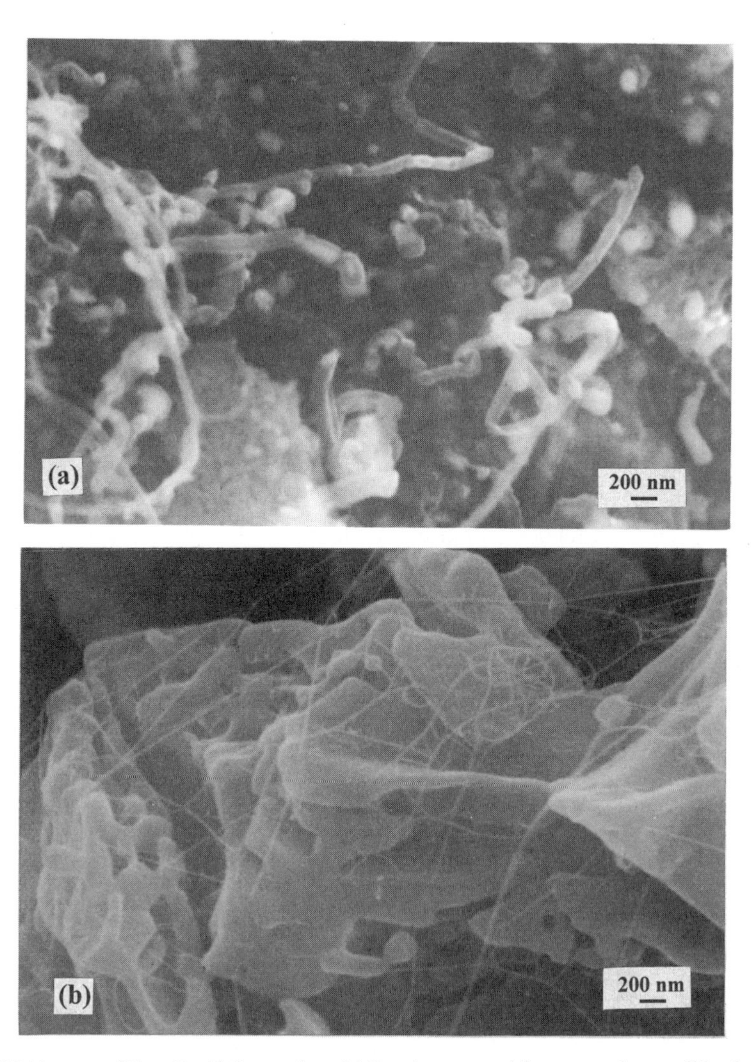

Figure 2. SEM images of C_{NTS}-Fe-Al_2O_3 powders. (a) Powder prepared from amorphous solid solution : thick tubes and carbon nanoparticles most of which are filled with Fe metal/carbide particles. (b) Powder prepared from α solid solution : bundles of C_{NTS}.

This could partly result from the fact that the catalyst particles formed upon reduction are smaller than when using amorphous or η solid solutions. In these ex-α powders, most C_{NTs} are less than 10 nm in diameter and they are arranged in bundles several tens of micrometers long. Starting from an α-solid solution, the increase in reduction temperature from 900 to 1000°C produces an increase in the amount of C_{NTs}, probably owing to the higher CH_4 sursaturation level in the reduction atmosphere. However, a concomitant decrease in carbon quality is also observed.

The aim of a further study[11] was to determine the influence of the Fe content in the starting alumina-hematite compounds on the amount of C_{NTs} and on the quality of carbon in composite powders prepared by reduction in H_2-CH_4 at 900 or 1000°C (6 mol. % CH_4, 240 min). Oxides based on α-alumina and containing various amounts of Fe (2, 5, 10, 15 and 20 cat.%) were prepared by decomposition and calcination of the corresponding mixed-oxalates. XRD revealed that specimens in which the Fe content is below the solubility limit of hematite in α-alumina (≤ 10 cat.% Fe) are monophase solid solutions of general formula α-$Al_{2-2x}Fe_{2x}O_3$ ($0 \leq x \leq 0.1$), whereas those containing more Fe rather consist in a very intimate mixture of an Al_2O_3-rich oxide and an Fe_2O_3-rich phase. Whatever the reduction temperature (900 or 1000°C), it is preferable to reduce the monophase oxides rather than the mixtures with respect to the formation of C_{NTs}. Indeed, the highest quantity of C_{NTs} is obtained using α-$Al_{1.8}Fe_{0.2}O_3$ as starting compound, i.e. the maximum Fe concentration allowing to retain the monophase solid solution (Fig. 3a). A further increase in Fe content provokes the formation of an Fe_2O_3-rich phase which upon reduction produces too large Fe particles. The higher carbon quality (Fig. 3b) is obtained with only 5 cat.% Fe (α-$Al_{1.9}Fe_{0.1}O_3$), probably because the surface Fe/Fe_3C nanoparticles formed upon reduction are slightly smaller, being less numerous and therefore less prone to coalescence than those formed from α-$Al_{1.8}Fe_{0.2}O_3$, thereby allowing the formation of C_{NTs} with a smaller diameter.

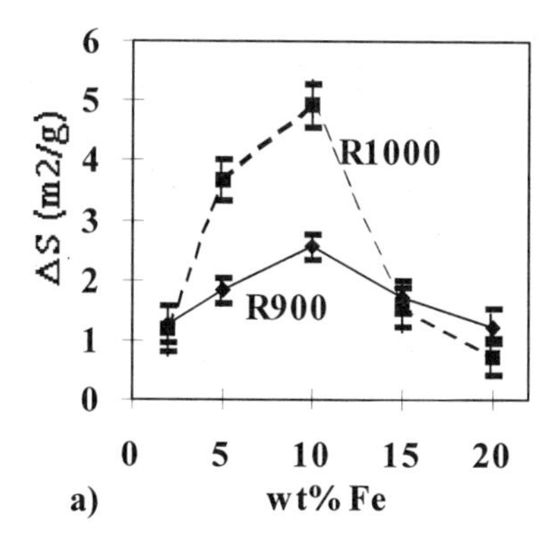

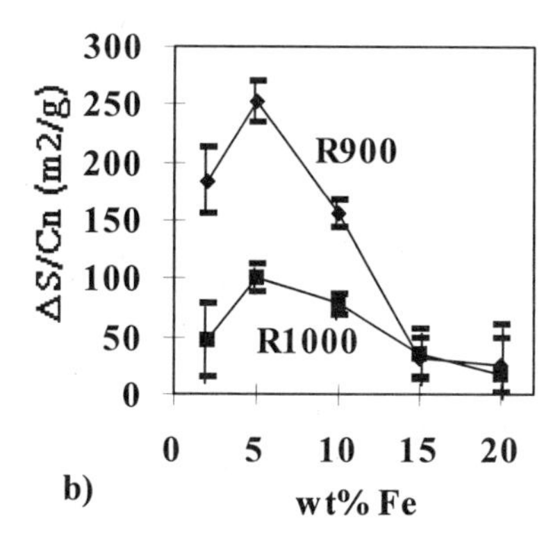

Figure 3. ΔS and $\Delta S/C_n$ (see text) versus the Fe content in the C_{NTS}-Fe-Al_2O_3 powders.

For a given Fe content, lower than 10 cat.%, the increase in reduction temperature from 900 to 1000°C increases the quantity of C_{NTs}, because the CH_4 sursaturation level in the reducing atmosphere is higher at 1000 than at 900°C. However, the simultaneous decrease in carbon quality points out that a higher reduction temperature also favors tube thickening and/or the deposit of much non-tubular carbon species in these experimental conditions.

A next work[13] was aiming at studying the influence of the CH_4 content in the H_2-CH_4 gas mixture on the synthesis of C_{NTs}- Fe/Fe_3C-Al_2O_3 nanocomposite powders. The stable, monophase, α-$Al_{1.9}Fe_{0.1}O_3$ solid solution, which has previously given the better results in terms of carbon quality was used as a starting compound. C_{NTs}-Fe/Fe_3C-Al_2O_3 powders were prepared by selective reduction of α-$Al_{1.9}Fe_{0.1}O_3$ in different H_2-CH_4 gas mixtures (0, 1.5, 3, 4.5, 6, 9, 12, 14, 16, 18, 24, 30 and 45 mol % CH_4). Mössbauer spectroscopy showed that the reduction of the Fe^{3+} ions to metallic Fe is tremendously favored by the presence of CH_4 in the reduction atmosphere, notably for low CH_4 contents. However, the subsequent differences existing from one powder to the other seem to mainly concern intragranular particles inactive for the formation of C_{NTs}. The α-Fe and Fe_3C particles located at the surface of the matrix grains represent *ca.* 26 % of all the iron species whatever the CH_4 content and no clear evolution of their respective proportion could be observed. The trend is that there is more Fe_3C than α-Fe for CH_4 contents higher than 4.5 mol %. More carbon is deposited when the CH_4 concentration is increased, whereas the quantity of C_{NTs} (Fig. 4a) is maximum for 24 mol % CH_4 and the carbon quality (Fig. 4b) is maximum for the 9 - 18 mol % CH_4 range. For specimens prepared using a CH_4 content lower than 9 mol %, it is proposed that carbon is preferentially engaged in species such as Fe-C alloys, Fe_3C and graphenic layers at the surface of the catalyst particles rather than in C_{NTs}.

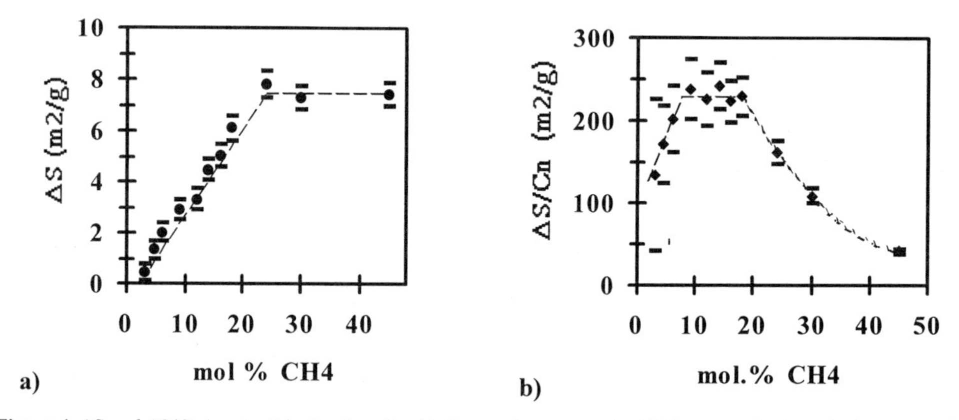

Figure 4. ΔS and $\Delta S/C_n$ (see text) in the C_{NTS}-Fe-Al_2O_3 powders versus the CH4 content in the reducing atmosphere.

However, the proportion of C_{NTs} increases upon the increase in CH_4 content. For CH_4 concentrations ranging between 9 and 18 mol %, it is proposed that the extra CH_4 content in the atmosphere produces carbon solely contributing to the formation of C_{NTs} and that the maximum quality achievable in these experimental condition is obtained. For higher CH_4 contents, the quantity of C_{NTs} is constant and the formation of spheroidal carbon particles causes a decrease in quality.

The C_{NTs} are arranged in very long bundles homogeneously dispersed in the composite powder. SWNTs have been observed (Fig. 5a) but most C_{NTs} have 2, 3 or 4 walls (Fig. 5b). Almost all the C_{NTs} observed in HREM are free of pyrolytic or amorphous carbon deposits and have an inner diameter in the 1-6 nm range, which could indicate that the catalyst particles active for C_{NTs} formation are in this size range. The exact nature of these particles (Fe, Fe-C alloy or Fe_3C) remains an open question.

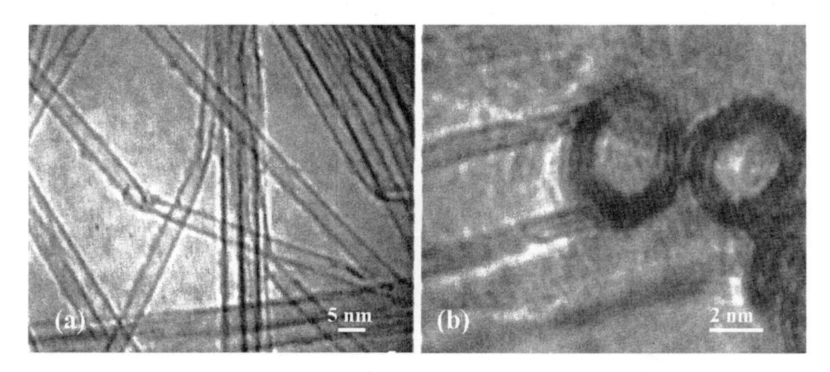

Figure 5. HREM images of C_{NTs} in a C_{NTs}-Fe-Al_2O_3 powder.

$MgAl_2O_4$-based powders. The spinel $MgAl_2O_4$ has the advantage over Al_2O_3 that it forms total solid solutions with $FeAl_2O_4$, $CoAl_2O_4$ and $NiAl_2O_4$. Thus, work has been undertaken[12] to determine the influence of the nature and content of the transition metal (Fe, Co, Ni) on the amount and quality of the C_{NTs} in the spinel-matrix composite powder. C_{NTs}-metal-spinel powders were prepared by selective reduction in H_2-CH_4 of $Mg_{1-x}M_xAl_2O_4$ (M = Fe, Co or Ni, x = 0.1, 0.2, 0.3 or 0.4) solid solutions. XRD revealed that in the case of Fe, formation of Fe_3C particles is observed in addition to metallic Fe. SEM and HREM (Fig. 6) observations reveal that the metal-oxide grains are uniformly covered by a web-like network of C_{NTs} bundles, several tens of micrometers long, most of which are made up of SWNTs with a diameter close to 4 nm. Fig. 6a shows a capped SWNT with no catalyst particle at the tip. Fig. 6b shows a DWNT. The C_{NTs} thus have a very high aspect ratio and appear to be very flexible. HREM observations suggest that the mechanisms for nanotube formation are distinct from those proposed for the synthesis of hollow carbon fibers[19] and that several mechanisms such as the yarmulke[5] and the template growth[20] could be active. Only the smallest metal particles (< ca. 5 nm) may be connected with the formation of C_{NTs}.

The macroscopical characterization method (Table 2) shows that the increase in transition metal content yields more C_{NTs} up to a metal content of 10.0 wt% (x = 0.3) but provokes a continuous decrease in carbon quality, suggesting a compromise which could be found for specimens with only 6.7 wt% of metal (x = 0.2). Co gives superior results with respect to both the quantity and quality parameters. In the case of Fe, the quality of the obtained carbon is notably hampered by the formation of Fe_3C particles. The observed differences between the Co- and Ni-specimens point towards an intrinsic effect of the chemical nature of the metals.

These results prompted another study[14] to investigate the influence of alloying Fe, Co and Ni. C_{NTs}-alloy-spinel powders were prepared by selective reduction in H_2-CH_4 of $Mg_{0.8}M_yM'_zAl_2O_4$ (M, M' = Fe, Co or Ni, y + z = 0.20 and y or z = 0, 0.05, 0.10, 0.15 and 0.20) solid solutions. The composition of the Fe/Co, Fe/Ni and Co/Ni alloy nanoparticles in

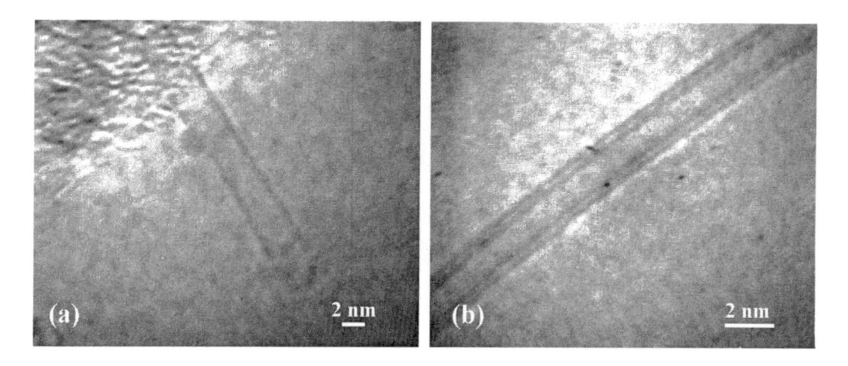

Figure 6. HREM images of C_{NTs} in a C_{NTS}-Fe-$MgAl_2O_4$ powder.

Table 2. ΔS and $\Delta S/C_n$ (see text) for C_{NTS}-M-$MgAl_2O_4$ (M = Fe, Co, Ni) powders containing different amounts of metal. The metal content is noted as x in $Mg_{1-x}FexAl_2O_4$.

Specimen	$Fe_{0.1}$	$Fe_{0.2}$	$Fe_{0.3}$	$Fe_{0.4}$	$Co_{0.1}$	$Co_{0.2}$	$Co_{0.3}$	$Co_{0.4}$	$Ni_{0.1}$	$Ni_{0.2}$	$Ni_{0.3}$	$Ni_{0.4}$
ΔS (m^2/g)	3.0	7.7	8.7	8.7	9.2	12.8	13.7	13.6	2.3	4.3	6.6	6.4
$\Delta S/C_n$ (m^2/g)	167	133	95	74	354	337	269	192	192	215	200	123

the composite powders strongly affects the conversion of CH_4 into carbon species. Compared to pure Fe, alloying with Co or Ni hampers the formation of Fe_3C and reduces the metal particles size and size distribution, thus favoring carbon deposition and the formation of C_{NTs}. In particular, the $Fe_{0.75}Co_{0.25}$ and $Fe_{0.50}Co_{0.50}$ compositions appear to be the most efficient for the formation of C_{NTs} (Table 3). Alloying Ni with Co produces specimens with a quality as high as when using pure Co but decreases the quantity of C_{NTs}. Electron microscopy observations shows that the metal-oxide grains are uniformly covered by a web-like network of C_{NTs} bundles. Most of the observed C_{NTs} (ca. 80 %) are SWNTs, with a diameter in the 0.8-5 nm range. MWNTs have a diameter smaller than 10 nm and generally have 2 walls. C_{NTs} with 3-5 walls are very rarely observed.

Table 3. ΔS and $\Delta S/C_n$ (see text) for C_{NTS}-M/M'-$MgAl_2O_4$ (M/M' = Fe/Co and Co/Ni) powders containing different amounts of metal. The metal content is 6.7 wt% (M0.2 in Table 2).

Specimen	$Fe_{0.75}Co_{0.25}$	$Fe_{0.5}Co_{0.5}$	$Fe_{0.25}Co_{0.75}$	$Co_{0.75}Ni_{0.25}$	$Co_{0.5}Ni_{0.5}$	$Co_{0.25}Ni_{0.75}$
ΔS (m^2/g)	14.9	19.9	14.9	7.9	4.5	6.7
$\Delta S/C_n$ (m^2/g)	210	284	265	328	259	338

MgO-based powders. Extraction and purification of the C_{NTs}. The solid solution method has been extended to MgO-based solid solutions and since the best results have previously been obtained when using cobalt as a catalyst, a C_{NTs}-Co-MgO composite powder was prepared by reduction of $Mg_{0.9}Co_{0.1}O$ in H_2-CH_4 (18 mol.% CH_4, 1000°C, 6 min). The carbon content (C_n) in the composite powder was found equal to 5.98 wt%. XRD analysis revealed the presence of ε-Co in addition to MgO. Peaks corresponding to carbonaceous material were not detected. SEM observations revealed that the metal-oxides grains, 100-300 nm in size, are uniformly entrapped in a web-like network of very long (tens or hundreds of micrometers), flexible filaments smaller than 20 nm in diameter. TEM studies of similar materials have revealed that such filaments are bundles of carbon nanotubes. ΔS was found equal to 24.1 m^2/g and $\Delta S/C_n$ to 403 m^2/g, which compares favorably with our best previous result ($\Delta S/C_n$ = 354 m^2/g when starting from $Mg_{0.9}Co_{0.1}Al_2O_4$).

Most importantly, the MgO matrix presents the advantage over alumina and spinel that it can be readily dissolved in acids, thus allowing the extraction of the carbon nanotubes. The C_{NTs}-Co-MgO composite powder was submitted to a treatment in a HCl aqueous solution (36%, room temperature) in order to extract the carbon nanotubes by dissolution of MgO and part of Co. The resulting product was washed with de-ionized water and air dried overnight at room temperature. The XRD pattern of the obtained solid showed the peaks of ε-Co and a peak which could correspond to the distance between graphene layers (d_{002} = 0.34 nm). MgO peaks were not detected, showing that the matrix was totally dissolved along with some Co particles. The carbon content in this product was found equal to 64.4 wt% (i.e. 90 at.%). TEM observations revealed that the extracted product consists in a mixture of C_{NTs} and catalyst particles covered by graphene layers. The observed C_{NTs} were found to have not been damaged by the mild acid treatment. HREM observations were carried out on isolated C_{NTs} (Fig. 7). A bundle of 5 DWNTs is shown in Fig. 7a and a SWNT is shown in Fig. 7b. Amorphous carbon deposits are scarcely observed in the present materials and it is inferred that most of it results from the degradation of the C_{NTs} under the electron beam. It is also noteworthy that pyrolytic thickening of the C_{NTs} did not occur in the present experimental conditions.

The number of walls as well as the inner and outer diameters of 80 isolated C_{NTs} were measured on HREM images. More than 80 % of the C_{NTs} have only 1 or 2 walls, about half of these being SWNTs (Fig. 8a). This proportion is lower than that reported by Hafner et al.[8] for C_{NTs} prepared by passing C_2H_4 over supported Fe:Mo particles at 700°C (70%), but higher than the one they obtain at 850°C (30%). Most internal and external diameters are in the 0.5-5 nm range (Fig. 8b), which is a much larger distribution than those reported for C_{NTs} prepared by arc-discharge[20-22], laser-vaporization[23, 24] and hydrocarbon-ferrocene decomposition.[7] However, it is in good agreement with other results obtained with the catalysis method.[5, 6, 8] Chen et al.[25] have prepared C_{NTs} from CH_4 decomposition and CO disproportionation on a Ni-MgO powder derived from the H_2 reduction of a $Mg_{0.6}Ni_{0.4}O$ solid solution. The obtained MWNTs are 15-20 nm in diameter, in contrast with the present results. This arises owing to important differences with the present work, besides the use of cobalt instead of nickel : firstly we use a much lower amount of transition metal (x = 0.1 compared to x = 0.4), secondly the present reduction is performed in a H_2-CH_4 mixture with no prior treatment in pure H_2, thus preventing a premature growth of the catalyst particles.

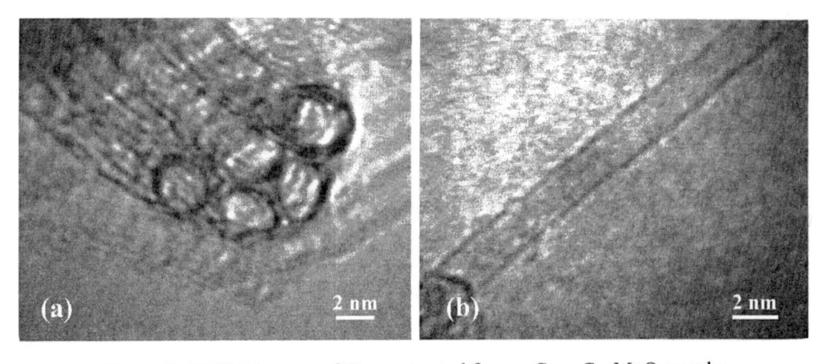

Figure 7. HREM images of C_{NTs} extracted from a C_{NTs}-Co-MgO powder.

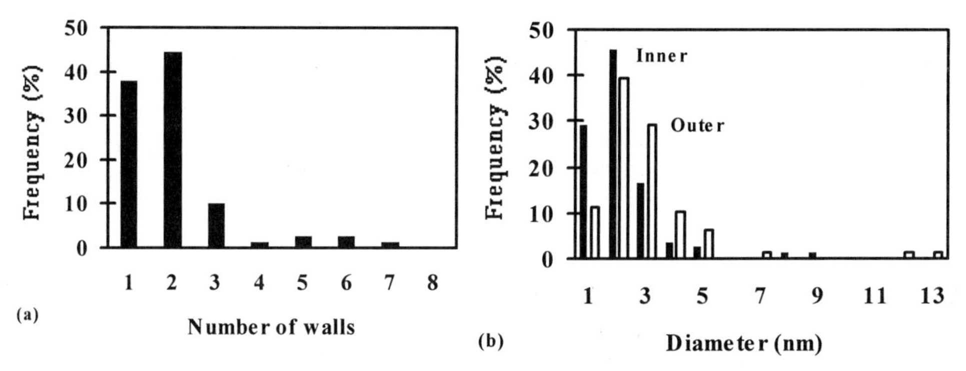

Figure 8. Number of walls (a) and inner and outer diameters (b) of C_{NTs} extracted from a C_{NTs}-Co-MgO powder.

More than 90 % of the present C_{NTs} have a diameter lower than 3 nm, which is in excellent agreement with the model proposed by Hafner et al.[8] However, SWNTs with diameters higher than 3 nm are observed, which suggests that slightly larger catalytic particles can also be active, as opposed to becoming onionated. Large diameter C_{NTs} could be interesting for applications such as hydrogen absorption.[26] In contrast to other results,[5, 8] some DWNTs have a lower inner diameter than most SWNTs. This may indicate that at least some tubes are formed by the yarmulke mechanism,[5] a characteristic of which is that the outer wall is formed first. Both particle-free and particle-containing tubes tips have been observed. These features are characteristics of a "base-growth" and a "tip-growth" mechanisms, respectively[27-29] and thus they both could be active in the present experimental conditions, assuming no modification of the sample is provoked by the preparation for TEM examination.

Attempts at improving the purification have shown that the methods developed for C_{NTs} produced by the electric-arc route or by laser ablation could not be applied since they resulted in the oxidation of the nanotubes.[30-32] In contrast, a succession of oxidation treatment in air (250°C) and dissolution in HCl was found to yield a material containing 84 wt% (97 vol%) carbon.

Dense composites

C_{NTs}-Fe-Al_2O_3 dense composites have been prepared[33] by hot-pressing composite powders which differ one from another not only by their Fe content (2, 5, 10, 15 and 20 wt%) but also by the quantity and the quality of the C_{NTs}, which both depend on the Fe content and on the reduction temperature (900 or 1000°C) used for the powder preparation.[11] C_{NTs} have been detected in the hot-pressed materials but with a decrease in quantity in comparison to the corresponding powders. This phenomenon is less pronounced for the ex-R900 than for the ex-R1000 specimen, possibly because the R900 powders contain a higher proportion of C_{NTs} with respect to the total carbon content. The presence of carbon as C_{NTs} and others species (Fe carbides, thick and short tubes, graphene layers) in the powders modifies the microstructure of the hot-pressed specimens in comparison to that of similar carbon-free nanocomposites. Densifications are lower and the matrix grains appear to be twice smaller in the present materials (ca. 1 μm versus 2 μm), possibly because carbon may inhibit some diffusion processes. Also, the metal particles are smaller, probably because the graphene layers which cover the Fe (and/or Fe-carbide) hamper the coalescence of these particles during hot-pressing.

In the dense composites, SEM observations of the fracture surface showed C_{NTs} bundles not larger than 100 nm in diameter and appearing to be remarkably flexible (Fig. 9). The SEM observations could indicate that the C_{NTs} bundles can dissipate some fracture energy. However, the additive effect that was expected from the hybridization of metal-alumina nanocomposites with very long C_{NTs} bundles is not observed. Indeed, most fracture strengths of C_{NTs}-Fe-Al_2O_3 composites are only marginally higher than that of Al_2O_3 and are generally markedly lower than those of the carbon-free Fe-Al_2O_3 composites[34] (Fig. 10a, b). Moreover, the fracture toughness values are lower than or similar to that of Al_2O_3 (Fig. 10c, d).

These relatively poor mechanical properties could notably result from a poor densification of the composites and a weak cohesion between the C_{NTs} bundles and the matrix.

The fracture strength of $MgAl_2O_4$- and MgO-based materials were also measured. They are reported in Table 4 with the relative density and the matrix grain size of the

Figure 9. SEM image of the fracture surface of a C_{NTS}-Fe-Al_2O_3 hot-pressed composite.

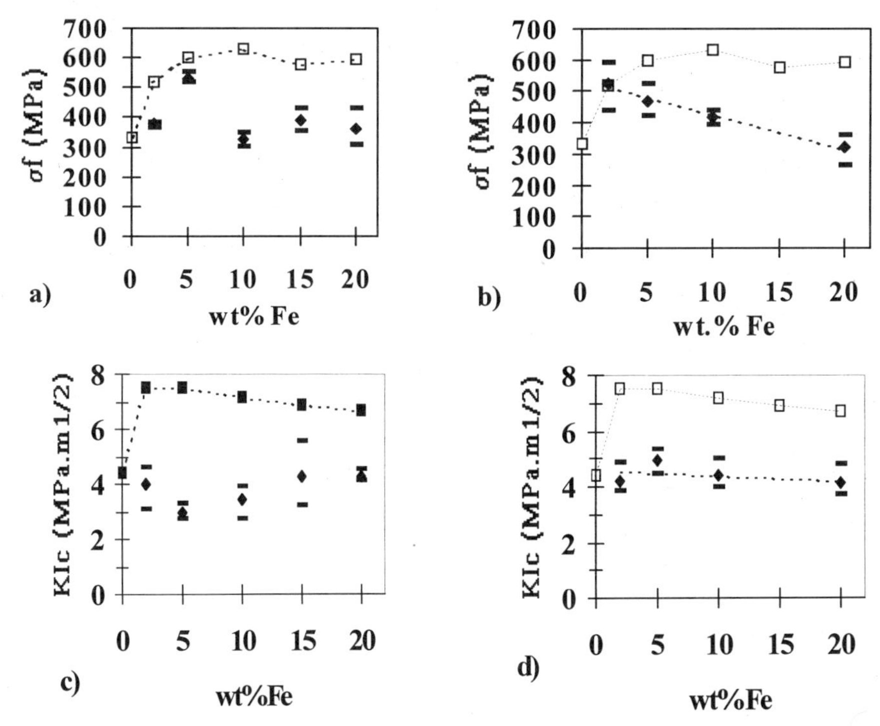

Figure 10. Mechanical properties of C_{NTS}-Fe-Al_2O_3 (filled diamonds) and Fe-Al_2O_3 (open squares) hot-pressed composites versus the metal content. The Fe-Al_2O_3 composites were prepared as described elsewhere[34]. Fracture strength of C_{NTS}-Fe-Al_2O_3 prepared from powders reduced at 900°C (a) and 1000°C (b). Fracture toughness of C_{NTS}-Fe-Al_2O_3 prepared from powders reduced at 900°C (c) and 1000°C (d).

Table 3. Relative density (d, %) matrix grain size (D, μm) and fracture strength (σ_f, MPa)of $MgAl_2O_4$- and MgO-based materials.

Specimen	d	D	σ_f
$MgAl_2O_4$	> 99	13	300
$Fe_{0.5}Co_{0.5}$-$MgAl_2O_4$	99	0.8	210
C_{NTs}-$Fe_{0.5}Co_{0.5}$-$MgAl_2O_4$	> 91	0.55	240
MgO	94	3 - 8	200
Co-MgO	97	5 - 10	280
C_{NTs} -Co-MgO	> 82	1 - 5	250

materials. The C_{NTs}-containing composites are much less densified and generally have a lower matrix grain size. Although the fracture strength are never higher than those of both the unreinforced matrix and the metal-oxide composite, they can be considered as acceptable values considering the lower density of the materials.

Interestingly, electrical measurements have evidenced that in contrast to the oxide and metal-oxide dense materials which are insulating, the C_{NTs}-metal-oxide composites show an electrical conductivity of the order of 2 S/cm. Work is in progress to get a better understanding of the electrical behavior of these composites.

CONCLUSIONS

We have demonstrated the ability to prepare C_{NTs}-metal-oxide powders that contain enormous amounts of C_{NTs}, most of which are SWNTs or small MWNTs with internal and external diameters in the 0.5 - 5 nm range. The results confirm that only the smallest metal particles (smaller than ca. 6 nm) may catalyze the formation of SWNTs and small MWNTs and underline the need that the catalyst is in the form of such nanoparticles at a temperature which is usually above 800°C in the catalysis methods. In this way, the reduction of oxide solid solutions allows to produce metal particles at a temperature which is high enough for the hydrocarbon gas to somehow interact with them so as to form the nanotubes prior to any exaggerate particle growth.

In the case of MgO-based materials, the oxide matrix and part of the Co catalyst can be dissolved by a combination of air oxidation and mild acid treatment that does not damage the C_{NTs}. The proposed method could be a real improvement in the low-cost, large-scale synthesis of C_{NTs}.

Dense materials can be prepared by hot-pressing the composite powders. These ceramic-matrix composites have acceptable mechanical properties and interestingly display an electrical conductivity owing to the dispersion of a network of C_{NTs} bundles, which could lead to some applications.

Acknowledgments

The authors would like to thank Mr. L. Datas for his assistance in the HREM observations, which have been performed at the Service Commun de Microscopie Electronique à Transmission - Université Paul-Sabatier.

REFERENCES

1. C. Journet and P. Bernier, *Appl. Phys. A* 67:1 (1998).
2. Ch. Laurent, E. Flahaut, A. Peigney and A. Rousset, *New J. Chem*. 22:1229 (1998).
3. V. Ivanov, A. Fonseca, J.B. Nagy, A. Lucas, P. Lambin, D. Bernaerts and X.B. Zhang, *Carbon* 33:1717 (1995).
4. K Hernadi, A. Fonseca, J.B. Nagy, D. Bernaerts, J. Riga and A. Lucas, *Synthetic Metals* 77:31 (1996).
5. H. Dai, A.G. Rinzler, P. Nikolaev, A. Thess, D.T. Colbert and R.E. Smalley, *Chem Phys. Lett.* 260:471 (1996).
6. J. Kong, A.M. Cassell and H. Dai, *Chem. Phys. Lett.* 292:567 (1998).
7. H.M. Cheng, F. Li, X. Sun, S.D.M.Brown, M.A. Pimenta, A. Marucci, G. Dresselhaus and M.S. Dresselhaus, *Chem. Phys. Lett.* 289:602 (1998).
8. J.H. Hafner, M.J. Bronikowski, B.K.Azamian, P. Nikolaev, A.G. Rinzler, D.T. Colbert, K.A. Smith and R.E. Smalley, *Chem. Phys. Lett.* 296:195 (1998).
9. A. Peigney, Ch. Laurent, F. Dobigeon and A. Rousset, *J. Mater. Res.* 12:613 (1997).
10. Ch. Laurent, E. Flahaut, A. Peigney and A. Rousset, *J. Mater. Chem.* 8:1263 (1998)
11. A. Peigney, Ch. Laurent, O. Dumortier and A. Rousset, *J. Eur. Ceram. Soc.* 18:1995 (1998).
12. E. Flahaut, A. Govindaraj, A. Peigney, Ch. Laurent, A. Rousset and C.N.R. Rao, *Chem. Phys. Lett.* 300:236 (1999).
13. A. Peigney, Ch. Laurent and A. Rousset, *J. Mater. Chem.* 9:1167 (1999).
14. A. Govindaraj, E. Flahaut, Ch. Laurent, A. Peigney, A. Rousset and C.N.R. Rao, *J. Mater. Res.* 14:2567 (1999).
15. E. Flahaut, A. Peigney, Ch. Laurent and A. Rousset, *Chem. Phys. Lett.* (1999) submitted for publication.
16. X. Devaux, Ch. Laurent, M. Brieu and A. Rousset, *Nanostruct. Mater.* 2:339 (1993).
17. K.C. Patil, *Bull. Mater. Sci.* 16:533 (1993).
18. W. S. Brown and J. E. Srawley, *ASTM Spec. Tech. Pub. 410*, ASTM:Philadelphia (1972).
19. N.M. Rodriguez, *J. Mater. Res.* 8:3233 (1993).
20. S. Iijima and T. Ichihashi, *Nature* 363:603 (1993).
21. S. Seraphin and D. Zhou, *Appl. Phys. Lett.* 64:2087 (1994).
22. C. Journet, W.K. Maser, P. Bernier, A. Loiseau, M. Lamy de la Chapelle, S. Lefrant, P. Deniard, R. Lee and J.E. Fisher, *Nature* 388:756 (1997).
23. A. Thess, R. Lee, P. Nikolaev, H. Dai, P. Petit, J. Robert, C. Xu, Y.H. Lee, S.G. Kim, A.G. Rinzler, D.T. Colbert, G.E. Scuseria, D. Tomanek, J.E. Fisher and R.E. Smalley, *Science* 273:483 (1996).
24. T. Guo, P. Nikolaev, A. Thess, D.T. Colbert and R.E. Smalley, *Chem. Phys. Lett.* 260:471 (1996).
25. P. Chen, H.B. Zhang, G.D. Lin, Q. Hong and K.R. Tsai, *Carbon* 35:1495 (1997).
26. A.C. Dillon, K.M. Jones, T.A. Bekkedahl, C.H. Kiang, D.S. Bethune and M.J. Heben, *Nature* 386:377 (1997).
27. G.G. Tibbetts, M.G. Devour and E.J. Rodda, *Carbon* 25:367 (1987).
28. R.T.K. Baker, *Carbon* 27:315 (1989).
29. S. Amelinckx, X.B. Zhang, D. Bernaerts, X.F. Zhang, V. Ivanov and J.B. Nagy, *Science* 265:635 (1994).

30. J. Liu, A.G. Rinzler, H. Dai, J.H. Hafner, R.K. Bradley, P.J. Boul, A. Lu, T. Iverson, K. Shelimov, C.B.Huffman, F.Rodriguez-Macias, Y-S. Shon, T.R. Lee, D.T. Colbert and R.E. Smalley, *Science* 280:1253 (1998).
31. A.G. Rinzler, J.Liu, H. Dai, P. Nikolaev, C.B. Huffman, F.J. Rodriguez-Macias, P.J. Boul, A.H. Lu, D. Heymann, D.T. Colbert, R.S. Lee, J.E. Fischer, A.M. Rao, P.C. Eklund and R.E. Smalley, *Appl.Phys. A* 69:29 (1998).
32. E. Dujardin, T.W. Ebbesen, A. Krishnan and M.M.J. Treacy, *Adv. Mater.* 10:611 (1998) .
33. Ch. Laurent, A. Peigney, O. Dumortier and A. Rousset, *J. Eur. Ceram. Soc.* 18:2005 (1998).
34. X. Devaux, Ch. Laurent, M. Brieu and A. Rousset, *C. R.. Acad. Sci. Paris Série II.* 312:1425 (1991).

IMPULSE HEATING AN INTERCALATED COMPOUND USING A 27.12 MHz ATMOSPHERIC INDUCTIVELY COUPLED ARGON PLASMA TO PRODUCE NANOTUBULAR STRUCTURES

Thomas J. Manning[1], Andrea Noel[1], Mike Mitchell[1], Angela Miller[1], William Grow[1], Greg Gaddy[1], Kim Riddle[2], Ken Taylor[2], Joseph Stach[3]

[1] Dept. of Chemistry, Valdosta State University,Valdosta, GA, USA, 31698
[2] Dept. of Biology, EM Lab, Florida State University, Tallahassee, FL, 31602
[3] Advanced Energy (formerly RF Power Products), Voorhees, NJ, USA

INTRODUCTION

Carbon nanotubes can be synthesized by several methods including the carbon arc, carbon vapor, and laser abalation[1]. In each of these methods, the starting material is either a solid graphite rod or a hydrocarbon and the atmosphere is held under low pressure in helium, argon, or an inert gas/oxygen/hydrocarbon mixture. The most popular method remains the carbon arc[2], in which two graphite rods are used as electrodes in a low-pressure inert gas environment. This system must be shut down and the electrodes scrapped and extracted in order to synthesize and purify carbon nanotubes.

Laser pyrolysis has been used to produce nanoscale particles by rapidly heating (100,000 °C/s) and cooling a gas phase reaction zone[3-11]. Haggerty[3] was the first to report a gas phase pyrolysis reaction and produced sub-100 nanometer size particles of Si, SiC, Si_3N_4. Since than additional studies of Si, SiC, Si_3N_4, ZrB_2, TiO_2 and Fe_3C have been reported in the literature[4-11]. Patents won by the Exxon Corporation involving the synthesis of nanoscale iron carbide particles demonstrate the potential commercial application of this particles[9,10]. Bi et al.[11] at the University of Kentucky (UK) followed these patents with a detailed study aimed at the physical characterization of these iron carbide particles. The UK study outlines three potential applications for the nanoscale Fe_xC_y particles: 1. They are found on the surface of iron catalysts in the Fischer-Tropsch synthesis of hydrocarbon fuels, 2. They have been found to have high saturation magnetization coupled with high hardness and oxidation resistance, and 3. They are being investigated for use as catalysts for coal liquefaction. In a previous paper[12], we showed that the covalently bonded graphite intercalation compound (GIC) fluorinated graphite (C_1F_1) forms exfoliated graphite (EG) when processed in an atmospheric pressure 27.12 MHz inductively coupled argon plasma. This EG differed from other EG produced[13] in that the ends of the graphitic sheets were rolled or formed tubes. The micron size carbon dust produced by this process has a low density (0.013 g/cm^3), high surface area (150 m^2/g), minimum impurities (99.5%+/- C), honey-combed structure (SEM pictures) and tubular endings (1345 cm^{-1}, 1590 cm^{-1}, 1620 cm^{-1}, Raman Spectroscopy)[12].

The industrial production of fluorinated-graphite (C_nF_n; n=1) compounds is carried out at high temperatures (600 °C) by reacting graphite with F2 gas for several hours (7-24 hrs)[14]. Fluorine-graphite intercalation compounds are nonconductors. It has an experimental density of 0.64 g/cm^3, a surface area of 218 m^2/g, and thermally decomposes at 450 °C. Brominated-graphite is the other common halogen based GIC and are considered good conductors[15,16]. Brominated graphite is easily synthesized by reacting graphite with Br_2 liquid or vapor at room temperature and pressure and it has a reversible intercalation-deintercalation process. When the brominated-graphite compound is heated above a critical temperature it puffs up or becomes exfoliated. When the process is reversed, there exist a residual bromine component in the graphite complex. How much bromine is left is dependent on the original particle size, temperatures, etc. For our work, we sought a final product free of most impurities and preferably 100% carbon.

Scientists have conducted extensive work with GIC compounds Br_2, ICl, and HNO_3, and demonstrated that the exfoliation or puffing of the structure takes place above 170°C, 190 °C, and 130 °C, respectively[17,18]. The Br_2-GIC and ICl-GIC collapse when cooled below 100 °C /210 °C and 190 °C, respectively. The Br_2-GIC compound collapse temperature varies with the starting material (HOPG, single crystal graphite, respectively). Mazieres et al.[19] observed that Br_2-graphite on pyrocarbons underwent first exfoliation at 200 °C and collapsed at 100 °C when cooled. Mazieres et al.[20] also observed irreversible exfoliation after heating brominated graphite on pyrocarbons to 1000 °C and cooling the compound in an argon atmosphere.

Exfoliated graphite synthesized by other methods is conducted by forming and decomposing graphite intercalation compounds (GIC) with various acids. For example EG can be produced from reactions involving HNO_3/H_2SO_4, $FeCl_3/NH_3$, and $AlCl_3$.[21-23]. Other methods not using the acids, such as the heating of potassium-graphite-tetrahydrofuran ternary compounds ($KC_{24}(THF)_1$ and $KC_{24}(THF)_2$), have also been used to produce EG[24]. These methods can be time consuming in that they require an initial synthesis of a specific graphite intercalation compound, the conversion of the GIC to the exfoliated graphite, and a residue such a metal ions or an acid removed from the EG by additional washing or heating. Exfoliated graphite (EG), which has several industrial applications, has been used in numerous chemical and nuclear plants as sealing materials under the name of *Grafoil*[25,26]. Flexibility and elasticity characteristics, high thermal resistance, high chemical stability, lightweight, etc characterize Grafoil. In figure 1 is a Scanning Electron Microscope (SEM) image of EG produced in this lab and clearly shows the puffed or exfoliated nature of the material. EG produced in this lab remains puffed for extended (months) periods of time.

Another form of pure carbon is reticulated vitreous carbon (RVC)[27,28] is another pure form of carbon and has a large, honey comb structure with 10 to 100 pores per inch. The pores in those structures are approximately 1000 times larger than those reported by this lab but have a large (500 m^2/g) surface area and a low density (0.048 g/cm^3). This material has the appearance of fused whiskers. A range of applications have been found for RVC including, electrode material, filtration, and high temperature insulation. The inductively coupled plasma (ICP) is an instrument widely used in other areas of plasma chemical synthesis, as an ion source for mass spectrometry, and as atomic and ionic emission source for spectrochemical studies. It can be operated at atmospheric or sub-atmospheric pressure. Past work by the PI has demonstrated that a low pressure ICP can be operated with all inert gases, various hydrocarbons, nitrogen, oxygen, carbon dioxide, etc and subsequently could be used as a stable light source for high resolution

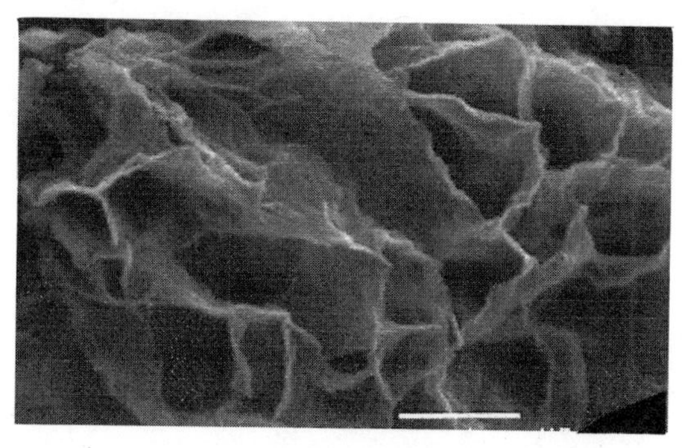

Figure 1. The sheets of graphite are ripped apart, puffed or exfoliated. The pores in the image illustrated above are 100-1000 nm in width. The walls of EG have a thickness in the 4-10 nm range.

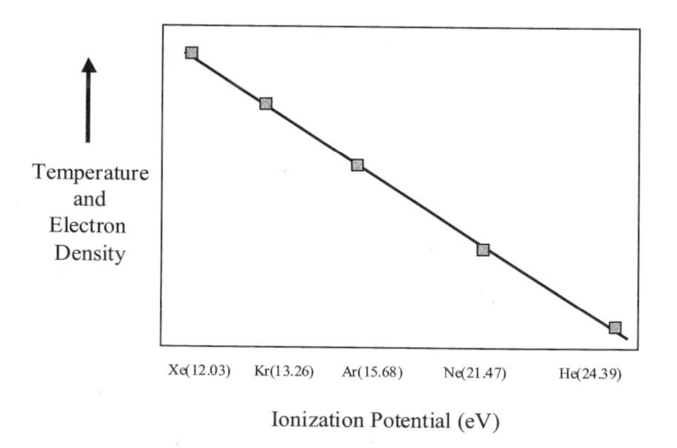

Figure 2. The inert gas content of plasma controls its relative temperature and electron density provided other parameters are held constant.

Fourier transform Spectroscopy[29,30]. ICP's at atmospheric pressure have been widely demonstrated with argon, helium, nitrogen, oxygen, and air. Varying the inert gas can be important because the temperature (electron, Doppler, vibrational, etc.) and electron density of the discharge can be altered as a function of the discharges average ionization potential. Figure 2 illustrates the relative temperature and electron density of a discharge as a function of the plasma's inert gas content. Helium, with a high ionization will produce relatively cool, low electron density plasma, while xenon, with a relatively low ionization potential will produce a relatively high temperature, high electron density plasma. The plasma's continuum background in the ultraviolet and visible can also be correlated linearly with the gases temperature and electron density. In our work, the ICP is operated with argon. The gas has a total flow of approximately 17 liters per minute that enter the quartz torch through three separate channels as illustrated in figure 3 below. The plasma flow (14 l/min) enters the torch at a 90 ^{o} angle and produces a tangential flow. The auxiliary gas (2 l/min) enters the middle channel with a tangential flow and is used to control the height of the plasma above the torch. The nebulizer gas enters the plasma through the central channel (1 l/min) with a laminar flow. In this work it carries the micrometer sized fluorinated dust into the plasma. The geometry of the torch is critical if the dust particles are to penetrate the plasma at its hottest region.

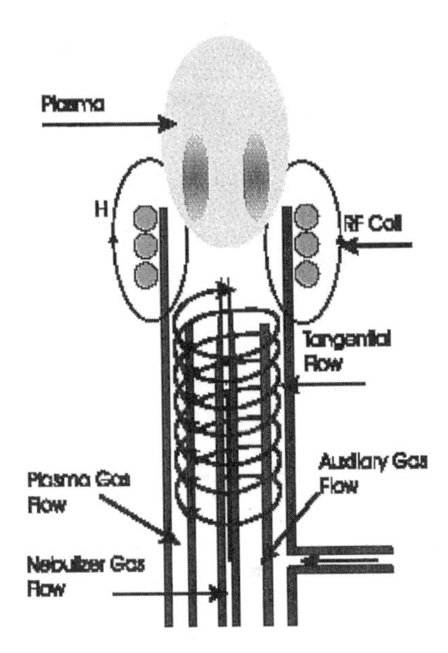

Figure 3. Schematic of the ICP quartz torch used in this work.

Also shown in figure 3, is the ¼ inch, water-cooled copper coil used to carry the radiofrequency (27.12 MHz) energy used to generate a magnetic field (H) that maintains the plasma. Not shown is the tesla coil used to ignite the discharge. We typically operate in the 1-1.5 kW range with 10-20 W reflected. The central channel is where the mixtures of Ar/O_2 to be discussed below took place. Pure argon ICP's are characterized by temperature in the 5,000 to 7,000 °C range with electron densities in the 1-3 x 10^{15} cm^{-3} range. The addition of particle matter (i.e. C_1F_1 dust) and the subsequent molecular fragments undoubtedly cools the discharge. Specifically, gas mixtures by volume from (100% Ar/0% O_2) to (50 % Ar/50% O_2) were tested.

In this work we extend our studies in three directions: *1.* Analyze the effect of intercalation ($C_1F_{0.8}$, C_1F_1, $C_1F_{1.1}$) on the final product *2.* Test several oxygen/argon ratios and its effect on the final product and *3.* Treat the EG with $FeCl_3$ and reheat at 800 °C and search for carbon nanotubes and nanoencapsulates in the soot by SEM and TEM. We also include discussion related to the basic operation of inductively coupled plasma, background on graphite intercalation compounds, and some future projections about the use of intercalated compounds and inductively coupled plasmas in synthetic nanotechnology. A brief discussion related to the production of spherical allotropes of carbon (i.e. C_{60}, C_{70}, etc.) is also included.

Before a discussion of our results begins, we should discuss why we are using graphite intercalated compounds as the starting material and the inductively coupled plasma as the medium for high temperature chemistry. Fluorinated graphite, which is one of two covalently bound IC's (oxygenated graphite is the other), presents two parameters that differentiate it from pure graphite that we find interesting. First, in the C_1F_1, the carbon sheets are further apart than in pure graphite. This changes or minimizes any steric hindrance that might take place in rolling the graphite sheets up to form nanotubes. Second, the carbon atoms in C_1F_1 are predominantly sp^3 hybridized, compared to sp^2 found in graphite. This sp^3 geometry is potentially more reactive than its trigonal pyramidal counterpart. In graphite, each C atom is held in place by three bonds (1 C=C at 611 kJ/mol, 2 C–C at 345 kJ/mol each), where as the F atom is held in place by one bond (C-F at 439 kJ/mol). Clearly, the F atom is easier to remove than a C atom. Breaking a C-F bond destabilizes the remaining carbon backbone and lowers the energy barrier needed to fragment and/or roll the carbon sheet into a nanotube. The advantages of GIC's can be extended to the carbon arc. Specifically, the end of a graphite rod could be intercalated, packed with an intercalated compound, or packed with a metal/fluorinated-graphite mixture for fullerene and/or nanotube synthesis studies.

The ICP offers several potential advantages over the arc method. It is a continuous flow method so if this technique proved to be a successful method to produce spherical or tubular allotropes of carbon, it could be operated continuously in an industrial environment. Second, because the product of the source can be continuously collected, the plasma does not have to be shut down. Fourth, the electrodes (RF coils) are nonintrusive therefore the magnetic field is homogeneous as a function of time and parameters such as field strength, temperature, electron and ion density remain fairly constant with time. Fourth, the ICP is a flexible source that can be operated over a range of pressure, temperatures, geometry's and chemical conditions.

DISCUSSION AND RESULTS

Pure graphite (C_1F_0; 1-2 um), and three fluorinated graphite compounds ($C_1F_{0.8}$, C_1F_1, $C_1F_{1.1}$) were tested. Analysis by SEM and TEM revealed a direct correlation between the degree of intercalation and the degree of exfoliation and nanotube formation. Specifically little exfoliation and no nanotubular structures were found in pure graphite dust and the largest degree of exfoliation and evidence for nanotubes was found using the $C_1F_{1.1}$ dust as the starting material. Although fluorine covalently bonds C in the intercalation compound, analyses of the final carbon soot product by a CHN analyzer consistently showed greater than 99% carbon. Illustrated in figure 4 (below) is a typical nanotubular structure found on the surface of the exfoliated graphite. We ran Ar/O_2 mixtures on each of the four dust types discussed above. Past work has shown that Ar/O_2 mixtures produced spherical fullerenes and work by shows that O_2 can be used in certain plasma's to enhance fullerene formation[5]. In our work we found that O_2 actually inhibited the formation of nanotubular structures.

Specifically, a pure Ar plasma with $C_1F_{1.1}$ produced the greatest degree of exfoliation and nanotube formation. Work by Jimenez et al.[15] produced EG using benzene-derived graphite fibers. The Raman spectra of the EG produced in their process differed from ours in spectral position (i.e. 1370 cm^{-1} vs 1345 cm^{-1}), relative spectral intensities and FWHM

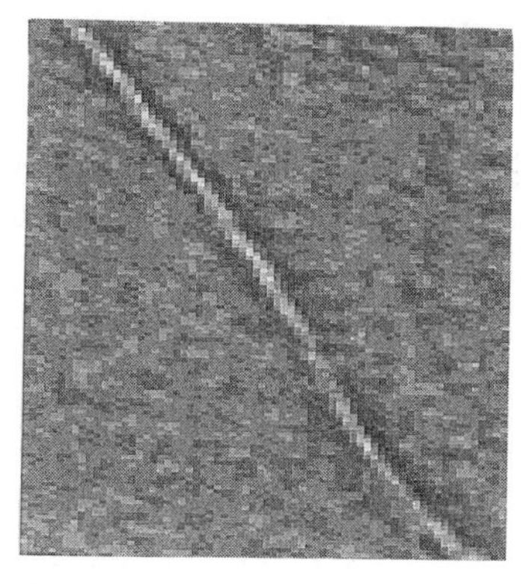

Figure 4. Carbon nanotube (2 nm wide) found on surface of exfoliated graphite.

of the spectral peaks. In previous work from this lab[12], the Raman spectra of the EG produced in this processed was shown to approximate that of carbon nanotubes. Although the majority of carbon present was in the forms of sheets, it was proposed that the terminals of the sheets were rolled accounting for the nanotubular spectra.

Figure 5 (below) illustrates the surface of EG and small tubular structures that have formed from the $C_1F_{1.1}$ precursor. The arrows point out tubular structures 2-4 nm wide produced by this process that are found throughout the processed samples. Although the untreated EG produced modest yields of these tubular structures, an important insight to the formation of these structures in an ICP is provided. The surface of the graphite sheet was fluorinated before entering the plasma.

Upon entering the plasma, fluorine is eliminated and graphite sheets are fragmented and stripped from the particle surface. Although graphitic carbons are sp^2 hybridized, fluorinated carbons are sp^3 hybridized[31] giving rise to a structure that is less ordered and potentially more reactive than pure graphite in a plasma. The gas (Ar) velocity carrying the dust to the ICP is 4700 cm/s and the estimated residence time of the dust in the plasma is ≈1-10 ms. With these conditions and the constant degradation of the dust by the robust plasma conditions, it would be difficult to argue that a nanotube could nucleate and grow on a single exposed surface in several milliseconds[32]. Because the tubular structures are only found on the carbon surface indicates that formation is the result of the degradation of the graphite sheets (i.e. large to small structural changes) and not the nucleation and growth of an individual structure from atomic and small molecular components (i.e. small to large).

From our experiments and observations it becomes obvious that there exists large quantities of disordered graphitic sheets present during the formation of the nanotubular structures. At several thousand degrees, numerous atomic, molecular and ionic species are present and provide the final building blocks for the formation of nanotubular structures from the graphitic sheets. The weak van deer Waals forces that hold the graphite sheets together are still responsible for holding the nanotubular structure to the carbon surface. Past work combining iron and graphite has been demonstrated in the synthesis of carbon nanotubes[33] and carbon whiskers[34]. Cobalt and nickel have also been used to enhance nanotube formation in the carbon arc[35].

In this work, 10 mls of a 10 ppm $FeCl_3$ (in 1% HNO_3) aqueous solution was prepared. 10 mg's of the exfoliated graphite carbon dust produced from the C_1F_1 precursor was added to the solution, stirred for 1 hour, and allowed to precipitate over 24 hours. The iron chemically absorbed on the exfoliated graphite. The EG-Fe dust was dried (110 °C) in

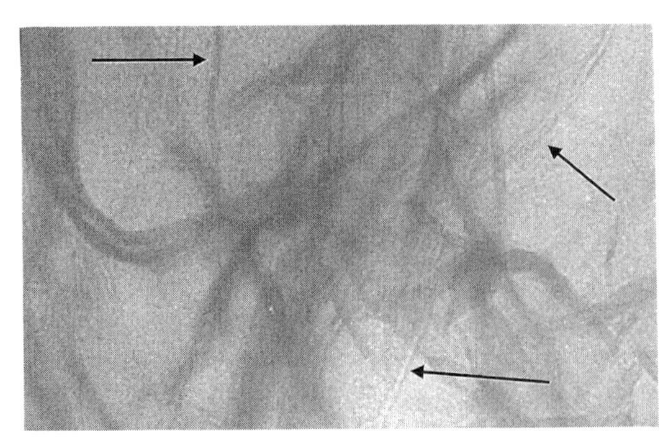

Figure 5. Nanotubes (2-3 nm wide) are found in the product of the EG produced from $C_1F_{1.1}$.

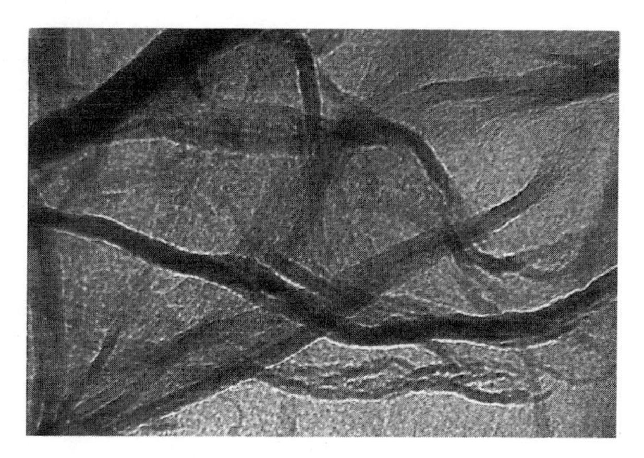

Figure 6. Partially wrapped graphite sheets produced when heated with Fe.

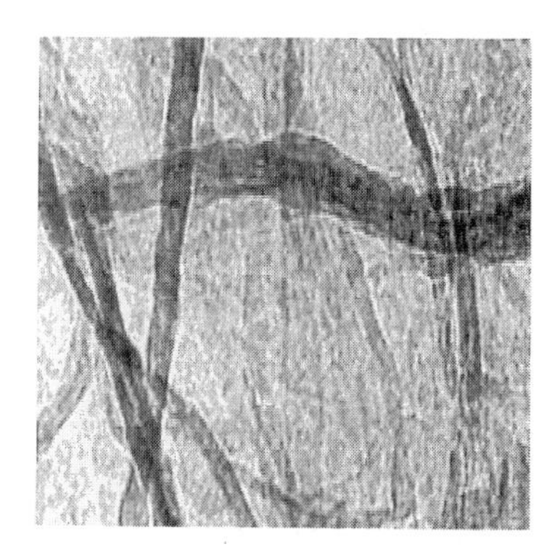

Figure 7. TEM images of tubular structures between 6-14 nm wide produced after Fe treatment of EG.

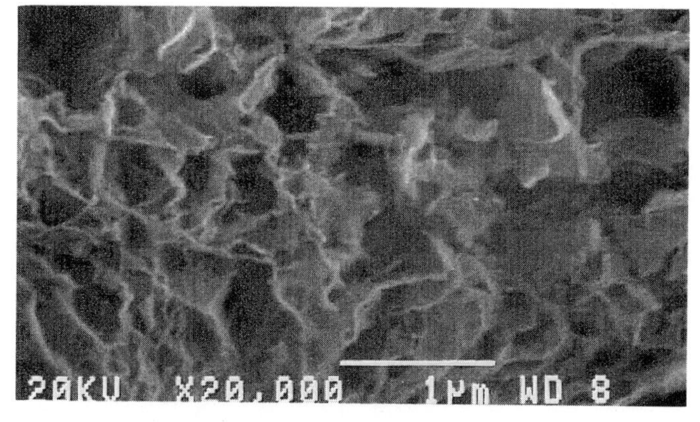

Figure 8. SEM image of EG treated by Fe and heated at 800 °C.

an inert gas environment. The dust was than heated to 800 °C for 60 minutes. At this temperature any potential salt that might form ($FeCl_3$, $Fe(NO_3)_3$, etc.) would thermally decompose leaving Fe(s). The solid Fe microparticles that result catalyze open nanotube formation. Figure 6 and 7 are TEM images of nanotubular structures is indicative of the structures found throughout the sample. It appears that the $FeCl_3$/800°C causes some graphite sheets to roll but not to form closed structures.

After treating EG with Fe and heating in a furnace, a large number of the rolled graphite sheets appear. Fe is contained within these carbon wraps. It is proposed that the disordered EG system composed of graphite sheets interacts with Fe(s) by energy transfer from the plasma to the graphite[16]. The Fe(s) couples and transfers energy from the inert gas environment to the carbon structure. Consider the following: At the oven temperature (800 °C), the robust plasma conditions needed for the generation of atomic and small molecular species do not exist. No individual micro or nanoparticles of metal were found in the EG after it was heated but were found wrapped in the rolled sheets. There is no change in the EG structure when heated in the oven without the iron for a short period of time (<10 minutes). If heated for 30-60 minutes, EG undergoes combustion and leaves no residue. The thermal conductivity of pyrolytic graphite is 21.3 W cm^{-1} K^{-1} when parallel to the layer planes and 0.0636 W cm^{-1} K^{-1} when perpendicular to the layer planes.

It appears that the iron particles present in the EG before and during heating allow the heat to be conducted to the EG from a parallel orientation. Without the iron present, heat is conducted primarily by the less efficient perpendicular mode. The EG has a relatively large surface area (150 m^2/g) and the ends of the graphite sheets are rolled essentially blocking thermal conductivity by the parallel mode thus heat can only be conducted to the graphite by the less efficient perpendicular mode. The iron serves as a heat conduit, which

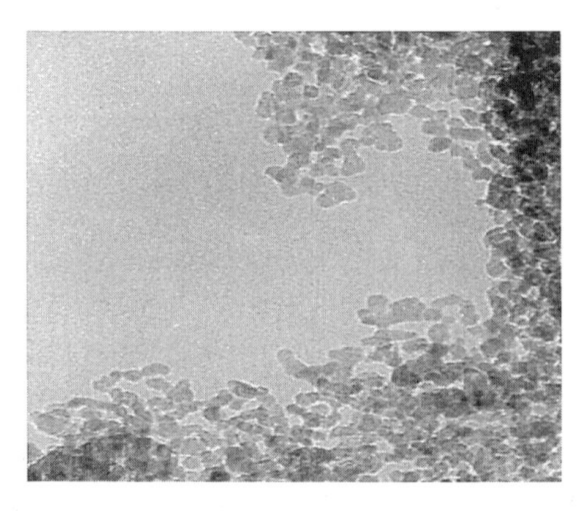

Figure 9 (above) and 10 (below). TEM image of nanoparticles that coat the Fe treated EG.

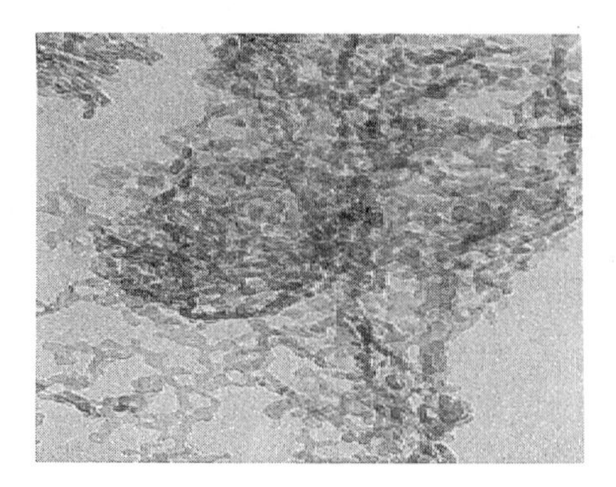

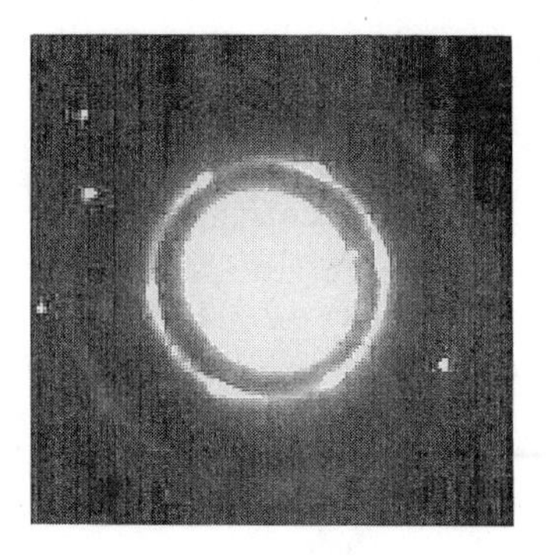

Figure 11. Electron diffraction patterns of EG treated with Fe.

leads to the structural changes. An underdetermined fraction of iron binds the graphite at its terminals, the structures most reactive area.

In examining the solid material by TEM left in the crucible after heating, we found nanoencapsulates similar to those described by Bi et al[11]. Using a TEM, we found a matrix covering the surface that consisted of nanoparticles in the 4-8 nm range. We should reemphasize that when pure EG (no Fe) was heated, it combusted and left no residue. When we treated EG with Fe and subsequently reheated, the final SEM image shows a partially collapsed exfoliated structure (see figure 8). We are in the process of conducting XRD and HRTEM studies of these particles illustrated in figure 9 and 10. We know at this stage that their formation is catalyzed by Fe and Ni and do NOT form with pure carbon/EG. Their formation mechanism is closely tied to that of the tubular structures described above and we believe they are either Fe_xC_y particles or carbon coated (graphite

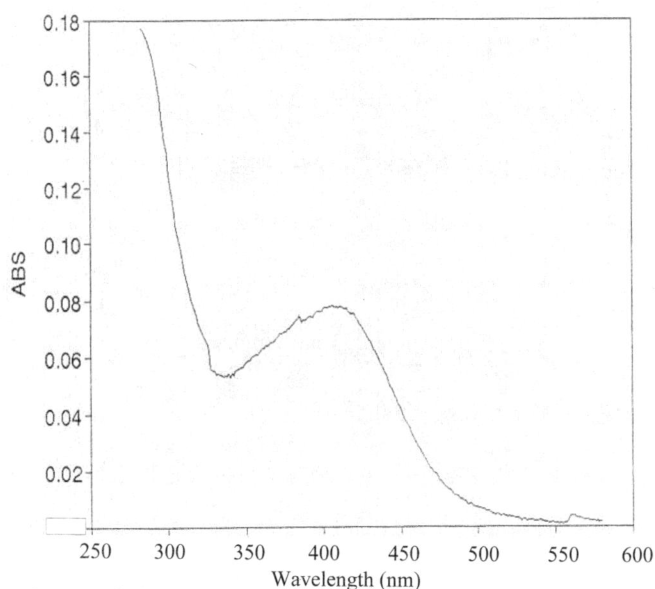

Figure 12. Some of the carbon soot samples produced a UV/VIS absorbance spectra characteristic of spherical fullerenes dissolved in a nonpolar solvent (x-axis in nm).

sheets) particles of Fe or Fe_xC_y. In figure 11, are electron diffraction patterns of the Fe treated EG material.

When this work started, we searched for spherical allotropes of carbon by ultraviolet/visible spectroscopy and, when results were encouraging, utilized an off-campus mass spectrometer. For the same reasons that we rationalized the use of an intercalated compound and the inductively coupled plasma for the synthesis of nanotubular structures, we believed this approach would be useful in the economical production of C_{60}, C_{70}, etc. Typically we collected the soot produced after C_1F_1 was passed through the ICP and extracted with toluene. On some experiments we confirmed the synthesis of C_{60} by UV/VIS spectrometry (see figure 12) and TOF-MS but much of this work was not reproducible on a consistent basis. We have no explanation for these results but may return to this endeavor shortly.

THE FUTURE?

Using intercalated compounds as the starting materials introduces three new parameters that can be tested for improved nanotube formation. These include the distance between the carbon sheets, the hybridization of the carbon atoms, and the use of a metal catalysis adsorbed on the surface of the material. At this stage I would propose three series of experiments. The experiments would involve a comparison of the carbon arc, laser ablation and ICP technique and their relative production of nanotubes and spheres on an intercalated carbon surface. The surface of a graphite rod would be intercalated would be held in a static position or slowly moved through the energy source. As nanotubes are formed on the intercalated surface, having the ability to move the rod might provide the ability to synthesize extremely long nanotubular structures. Parameters such as the type of intercalation compound (C_xF_y, C_xO_y, etc.), the degree of intercalation (i.e. $C_1F_{.5}$ to $C_1F_{1.2}$), the depth of intercalation, different metals chemically adsorbed into the intercalated compound (Fe, Ni, La, etc.) etc. could be tested on each technique. Once a favored technique is established, an attempt to optimize the technique to produce extended length tubes would be undertaken.

CONCLUSIONS

We have shown that impulse heating a covalently intercalated compound in inert gas environment yields closed nanotube structures in the exfoliated graphite. Treated with $FeCl_3$ and reheated, open nanotubular and nanoencapsulated structures are identified by TEM. This opens an exciting area of research in that a variety of intercalated compounds could be tested as precursors and several metals (Ni, Co, etc.) could be examined in the heating stage. The impulse heating of intercalated nonmetal compounds to produce nanotubular structures could also be extended to the carbon arc and laser ablation techniques.

From past work with carbon arcs, parameters such as gas pressure, voltage, current density and electrode gap were optimized to maximize fullerene and nanotube production. Other techniques have conditions that need to be optimized to produce certain types of spherical (C_{60}, C_{70}, etc.) or tubular allotropes (single wall, multi-wall, etc.) of carbon. This work shows that two other parameters, carbon hybridization (sp^2 vs. sp^3) and the distance between graphite layers in the starting material, can also be influential in the production of certain nanoscale structures.

ACKNOWLEDGEMENTS

Advanced Energy (Voorhees, NJ) is thanked for the donation of a 27.12 MHz ICP. Valdosta State University is thanked for its assistance with this work. The Florida State University Department of Biology (Kim Riddle) is thanked for SEM and TEM experiments. Dr. Ken Taylor's (FSU, EM lab) NIH grant is thanked for providing suport.

REFERENCES

1. T. Ebbesen, *Carbon Nanotubes* (CRC Press) (1997) p. 139ff
2. S. Iijima, *Nature* 354, 56, 1991
3. J.Haggerty, in Laser Induced Chemical Processes, edited by J. Steinfeld (Plenum Press, New York, 1981)
4. R. Fantoni, E. Borsella, S. Enzo SPIE 1279, 77 (1990)
5. P.Buerki, T. Troxler, S. Leutwyler,, in High Temperature Science (Humina Press Inc. 1990) vol. 27, p. 323
6. F. Curcio, G. Ghiglione, M. Musci, C. Nannetti, Appl. Surf. Sci., 36, 52-58 (1989)
7. G. Rice and R. Woodlin, J. Am. Ceram. Soc. 71, C181 (1988)
8. F. Curcio, M. Musci, N. Notaro, Appl. Surf. Sci. 46, 225-229 (1990)
9. R. Fiato, G. Rice, S. Miseo, S. Soled, United States Patent, 4,637,753 (1987)
10. G. Rice, R. Fiato, S. Soled, United States Patent 4,659,681 (1987)
11. X. Bi, B. Ganguly, G. Huffman, F. Huggins, M. Endo, P. Eklund, J. Mater. Res., vol. 8, 1666 (1993)
12. T. Manning, M.Mitchell, J. Stach, T. Vickers, *Carbon* vol. 37, 1159 (1999)
13. H. Jimenez, J. Speck, G. Roth, M. Dresselhaus, *Carbon,* 24 627-633 (1986)
14. K. Kinoshite, Carbon, Electrochemical and Physicochemical Properties, Wiley Interscience 1988, p. 207
15. K. Kinoshite, Carbon, Electrochemical and Physicochemical Properties, Wiley Interscience 1988, p. 215
16. M.S. Dresselhaus, and Dresselhaus, G., *Adv. Phys.*, 1981, 30, 139-326
17. S. Anderson, D. Chung, *Carbon*, 1984, 22, 253
18. W. Martin, Brocklehurst, J., *Carbon* 1964, 1, 133
19. C. Mazieres, G. Colin, Jegoudez and R. Setton, *Carbon* 1975, 13, 289
20. C. Mazieres, G. Colin, J. Jegoudez, and R. Setton, *Carbon*, 1976, 14, 176
21. R.E. Stevens, S. Ross, and S.Wesson, *Carbon,* , 1973, 11, 525
22. M. Dowell, *Ext. Abs. Program, 12th Bienn. Conf. Carbon*, p. 31, American Carbon Society, 1975, p. 31
23. H. Mikami, Japanese Patent No. 76 96,793. 1976
24. M. Inagaki, K. Muramastu, Y. Maeda, Maekawa, K., *Synthetic Metals*, 1983, 8, 335
25. D. Berger, J. Maire, *Mater. Sci. Eng.,* 1977, 31, 335
26. Union Carbide, *US Patent* No. 3,404,061. 1968
27. J. Wang, *Electrochimica Acta*, 1981, 26, 1721
28. V. Norvell, G. Mamantov, *Anal. Chem.* 1977, 49, 1470
29. T. J.Manning, US Patent No. #4,968,142, 1991
30. T. J.Manning, *Appl. Spec.* , 156 (1990)
31. H. Touhara, K. Kadono, M. Endo, N. Watanabe, *Proceedings of Carbon 84, Extended Abstracts* (Bordeaux, France, July2-6th) p.278 (1984)
32. T.W. Ebbesen, *Annu. Rev. Mater: Sci.* 24, 235, 1994
33. M. Jose-Yacaman, M. Miki-Yoshida, L. Rendon, Santiesteban, J.G., *Appl. Phys. Lett.* 62, 657, 1993
34. M. Endo, H. Ueno, in *Extended Abstracts of the 1984 MRS Symposium on Graphite Intercalation Compounds*, Eklund, P.C., M.S. Dresselhaus, M.S., G., Eds., Materials Research Society , Pittsburgh, PA, 1984, 177

35. S. Iijima, T. Ichihashi, *Nature* 363, 603, 1993
36. D.S. Bethune, C.H. Kiang, M.S. de Vires, G. Gorman, R. Savoy, J. Vazquez, R. Beyers, *Nature* 363, 605, 1993

THE SYNTHESIS OF SINGLE-WALLED CARBON NANOTUBES BY CVD CATALYZED WITH MESOPOROUS MCM-41 POWDER

Jun Li[1], Mawlin Foo[2], Ying Wang[2], Hou Tee Ng[2], Stephan Jaenicke[2], Guoqin Xu[2], and Sam F. Y. Li[1,2]

[1]Institute of Materials Research & Engineering
10 Lower Kent Ridge Road
Singapore 119260
[2]Department of Chemistry
National University of Singapore
Singapore 119260

INTRODUCTION

Carbon nanotubes have attracted extensive interests since their discovery in 1991 because of their extraordinary materials properties.[1-4] One of the major efforts of current research is to find a method to synthesize high purity defect-free carbon nanotubes in cheaper and better controlled ways. Arc-discharge[5] and laser ablation methods[6-9] have been the major techniques in the past. However, the large-scale synthesis of carbon nanotubes with controlled conformation, i.e. single-walled or multiwalled structures, remains challenging. In the past few years, chemical vapour deposition (CVD) has been recognized as a promising solution.[10-13]

CVD is a versatile process in which hydrocarbon gas molecules are decomposed to reactive species at elevated temperatures and then grow into carbon nanotubes. The growth of carbon nanotubes can be controlled by the selection of different transition metal catalysts such as Ni, Fe, Co, etc.[11,13] The synthesis of high purity single-walled nanotubes (SWNTs) or multiwalled nanotubes (MWNTs) has been reported by catalytic CVD method.[11-14] The selection of different types of catalyst support and hydrocarbon feedstock provides us with more control to the reaction. We report here on the growth of single-walled carbon nanotube ropes by the catalytic CVD method using MCM-41, a mesoporous molecular sieve with cylindrical channels of ~ 30 Å diameter[16,17], as catalyst support.

Typical inorganic heterogeneous catalysts can be divided into two classes, i.e. microporous (with pore diameters ≤ ~ 20 Å) and mesoporous (with pore diameters ~20-500Å) solids. The utility of these materials arises from their microstructure which allows molecules access to large internal surfaces and cavities which enhance the catalytic activity. The same principle also applies to the CVD process for carbon nanotube synthesis. However, the commonly used microporous molecular sieve catalysts such as zeolites and aluminium phosphate-based materials have channel diameters less than that of carbon nanotubes. Therefore mesoporous materials are required as catalysts for carbon nanotube synthesis by CVD methods.[12,15]

Most mesoporous materials such as silicas, transitional aluminas, or pillared clays and silicates are amorphous or paracrystalline solids. The pores in these materials are generally irregularly spaced and broadly distributed in size. MCM-41, however, has a much better defined internal structure consisting of a hexagonal array of uniform one-dimensional mesoporous channels.[16,17] The channel diameter may be controlled from 15 Å to greater than 100 Å by using surfactant templates with alkyl chains of different length.[18,19] The channel size of MCM-41 is exactly in the range of the outer diameters of carbon nanotubes. Therefore, we expect that the pore size of the catalyst support should affect the growth of carbon nanotubes. We report here our recent results of the synthesis of carbon nanotubes by CVD of methane catalyzed with Fe_2O_3 embedded in MCM-41 powder.

EXPERIMENTAL

MCM-41 was prepared with micelles of cetyltrimethylammonium bromide (CTMABr) as the templating agent. The composition of the synthesis gel followed the molar ratios given by Cheng et al[18]. The molar composition of the reactants silica, tetramethylammonium hydroxide pentahydrate (TMAOH, Fluka), CTMABr (Merck) and water are as following: 1.00 SiO_2 : 0.19 TMAOH : 0.27 CTAMBr : 40 H_2O. TMAOH and CTMABr (cationic surfactant) were added to deionized water with stirring and slightly heated to obtain a clear solution. The silica source was added to the solution with magnetic stirring for 2 hrs and then aged for 24 hrs at room temperature. The reaction mixture was placed in a Teflon-lined stainless-steel autoclave and subjected to hydrothermal synthesis at 150 °C. The reaction products were removed after two days, filtered, and washed free of bromide with a dilute solution of ammonium nitrate followed by a final rinse with deionized water. The filtrates were dried overnight in air at 80 °C and calcined in a muffle furnace (NEY 2-525) at 550 °C for 8 hours with a heating rate of 1 °C/min. The MCM-41 powder obtained in this way was characterized by nitrogen adsorption measurements (Quantachrome NOVA 2000) and powder X-ray diffraction (Siemens 5000).

The catalyst in our study is iron. We followed the preparation procedure reported by Kong at el[11]. The MCM-41 powder (0.4g) made by above method was magnetically stirred with a solution of $Fe(NO_3)_3 \cdot 9H_2O$ (0.1 g, Riedel-de-Haën) in methanol (3 ml) for 24 hrs, after which the methanol is removed by evaporation at 80 °C. The powder was then calcined at 250 °C for 4 hrs in a muffle furnace. This process presumably converted iron nitrate into catalytic active Fe_2O_3, which gave the MCM-41 powder a rust brown color.

0.5 g of the treated MCM-41 was then packed in a tubular quartz reactor (i.d. 28 mm). Helium gas (99.9% purity, Soxal) was passed through at a flow rate of 20 cm^3/min while the temperature of the reactor was increased to the required temperature of approximately 940 °C via a resistively heated electric oven. The helium flow was then stopped and CH_4 gas (99.9% purity, Soxal) was passed through at a flow rate of 50 ml/min for 12 min at 936 °C. Helium gas flow was resumed till the sample had cooled to room temperature.

SEM measurements were performed using Philips XL 30 FEG systems and TEM measurements were carried out with a Philips CM 300 FEG system. Samples were first ultrsonicated in methanol for 1 hr and then deposited on a 3 mm Cu grid covered with continuous carbon film or holey carbon films (for high-resolution measurements).

RESULTS

1. Characterization of MCM-41

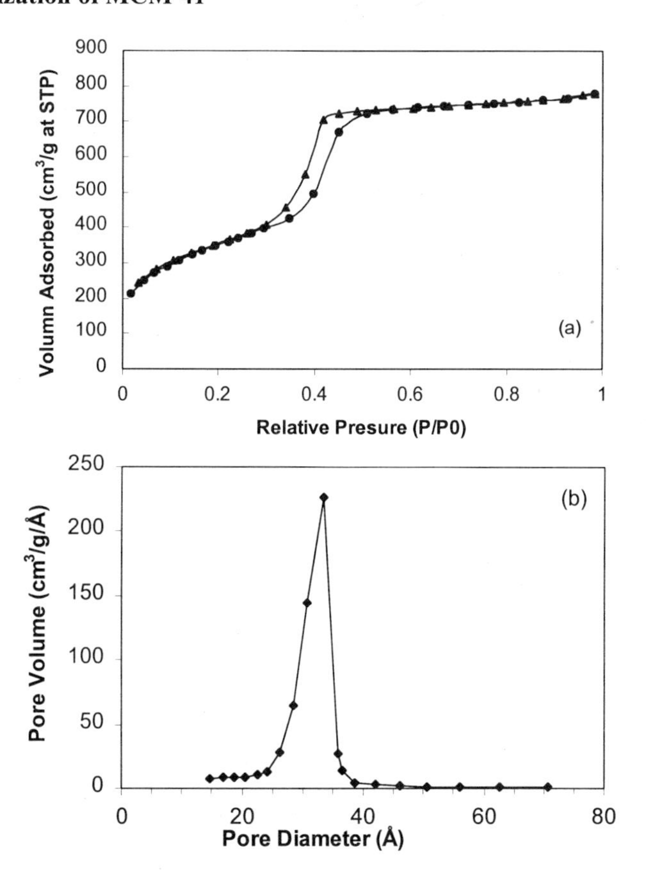

Figure 1. (a) The nitrogen adsorption isotherms and (b) pore size distribution (obtained using the BJH model with the data of adsorption branch of isotherm) of a hexagonal mesoporous MCM-41 sample.

Fig. 1 shows the nitrogen adsorption isotherm and the pore size distribution curve calculated based on the BJH model[16-19]. From the nitrogen adsorption measurements, we found that the MCM-41 sample is mesoporous with a narrowly distributed pore diameter of about ϕ = 33 Å, surface area of 1140.45 m^2/g and pore volume of 1.10 cm^3/g assuming that the pores are arrays of one-dimensional cylindrical channels. This is further confirmed by powder x-ray diffraction (XRD) as shown in Figure 2. The reflections can be indexed as (100), (110), (200), (210), and (300) diffraction from an ordered hexagonal array of one-dimensional channels. The channels are aligned perpendicular to the hexagonal plane and there is no atomic order within the channel walls. This is consistent with the fact that only in-plane diffractions (hk0) are observed. We can calculate the hexagonal lattice constant as a = 52.4 ± 0.9 Å. The discrepancy between the XRD lattice constant a (52.4 Å) and the pore diameter ϕ (33 Å) derived from N_2 adsorption measurements is due to the wall thickness. Assuming the wall thickness is uniform as shown by the schematic model in the inset of Figure 2, the thickness can be calculated as t = a-ϕ = 19.4 Å. Therefore, MCM-41 has a very large internal surface area, making it an idea catalyst support. The measured pore size is about the same as that of a SWNT. This should be a good substrate to control the synthesis of carbon nanotubes.

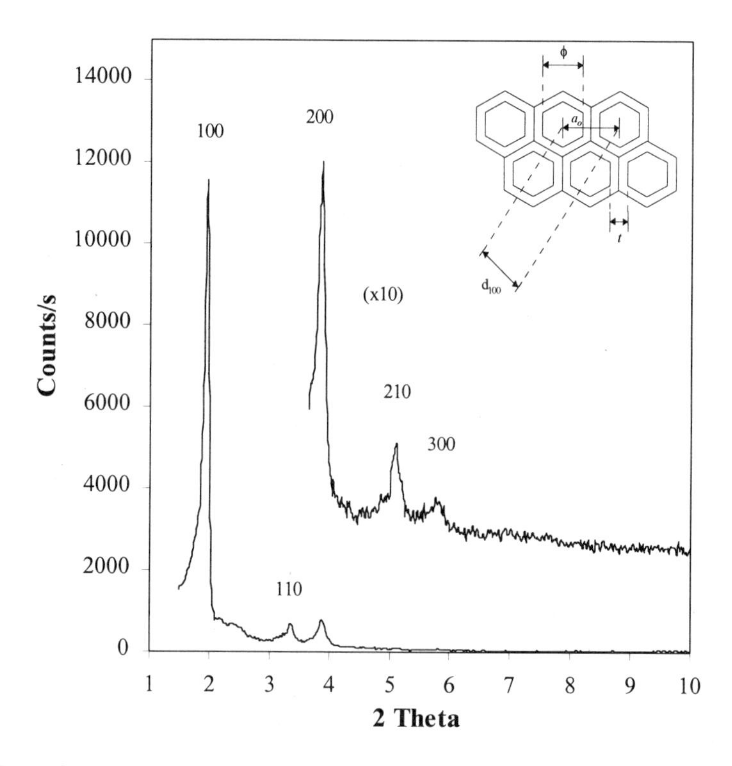

Figure 2. X-ray diffraction from MCM-41 powders. The inset shows the schematic model of hexagonally packed one-dimensional channels.

Typically, MCM-41 is formed as powder with the particle size of one to several microns. Figure 3 shows two SEM images of a single MCM-41 grain at different magnifications. The figures reveal a one-dimensionally stretched anisotropic morphology. As reported before, [20] the whole grain is like a single crystal with all the one-dimensional channels aligned in the same direction and stacked together like cylinders. The channels run from one end to the other through the whole "crystal". There also appears a small amount of tiny particles on the surface of the grain. Attempts to resolve the hexagonal structure by TEM were not successful, probably due to charging and beam damage. However, other studies[17-20] have confirmed the reported pore size by TEM and found it to be consistent with XRD results. Thus we did not go further to characterize our sample by TEM.

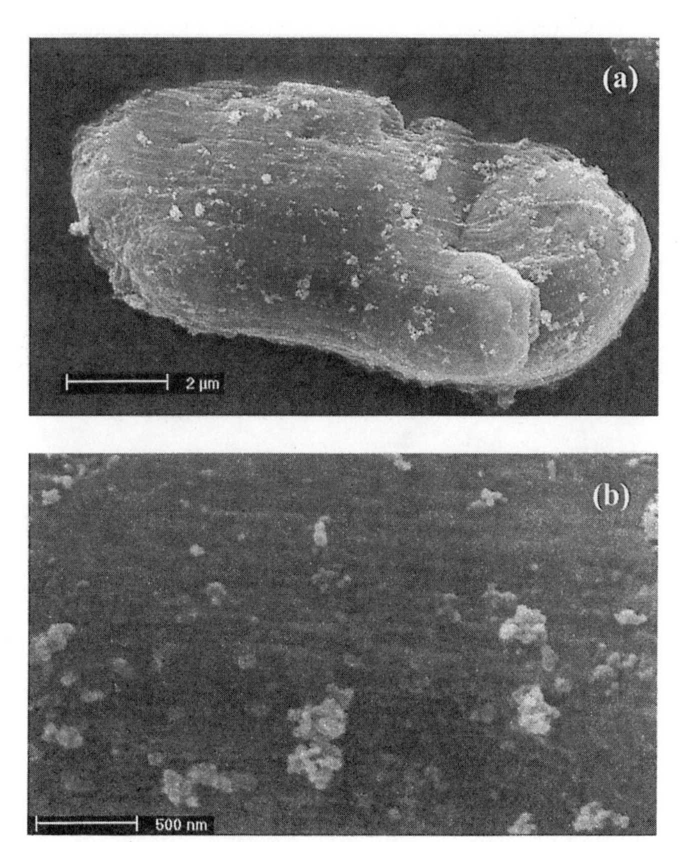

Figure 3. Scanning electron micrographs of a single MCM-41 grain.

It has been reported that Fe_2O_3 can catalyze the growth of either SWNTs[11,14] or MWNTs[12,15] depending on the type of catalyst support and preparation condition of the catalyst. It would be interesting to see whether the well-defined mesoporous channels can provide us with some control over the catalytic reaction. $Fe(NO_3)_3 \cdot 9H_2O$ was embedded into the internal channels of MCM-41 from the methanol solution and was assumed to fill up the pores. After evaporation of methanol, we expected that $Fe(NO_3)_3 \cdot 9H_2O$ solids were deposited as nanoparticles with controlled size inside the internal channels of MCM-41. Further calcination will decompose $Fe(NO_3)_3 \cdot 9H_2O$ into small iron oxide clusters.

Thus prepared catalyst is expected to possess large porosity to allow reactive hydrocarbon species access to the Fe surface. The wall of the channel should limit the lateral growth of the carbon nanotubes, and perhaps lead to an aligned growth.

2. The Scanning Electron Micrograph of Carbon Nanotubes Formed by CVD

Figure 4 shows SEM images at two different spots on a MCM-41 grain after CVD. Clearly, a dense network of carbon nanotubes formed on the surface. The average diameter can be measured as about 20 nm. These carbon nanotubes are at least several microns in length and either extend all the way out of the image frame or join into another tube. Experimentally, we can not clearly see both ends of any single carbon nanotube. Some of

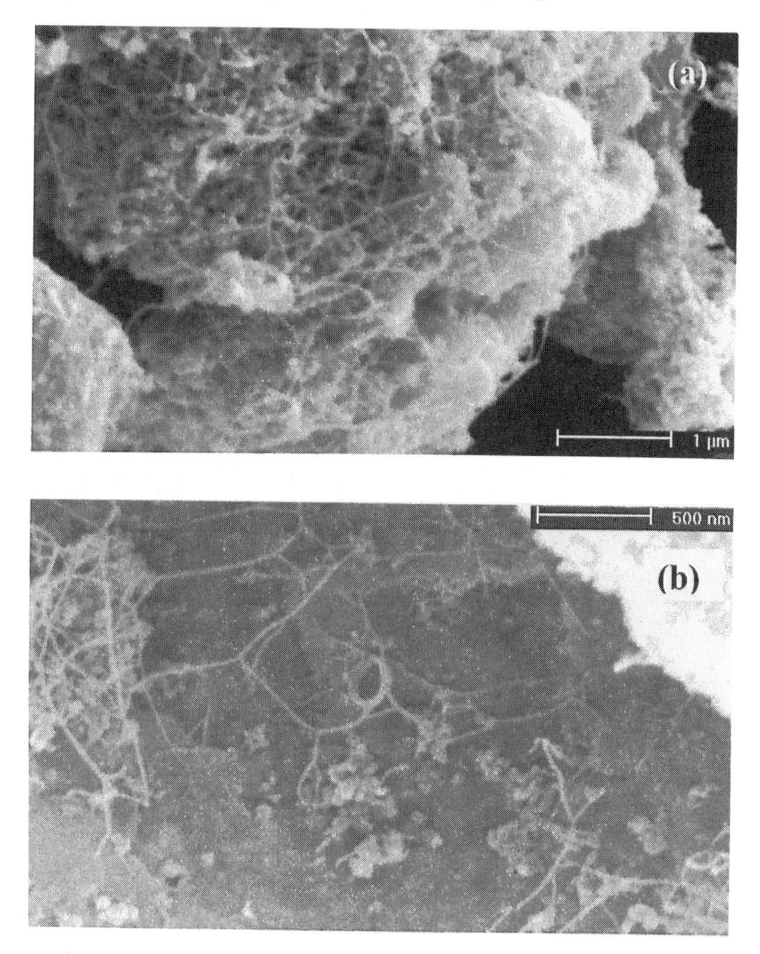

Figure 4. Scanning electron micrograph of the surface of Fe_2O_3 embedded MCM-41 powder after chemical vapor deposition of carbon nanotubes.

the carbon nanotubes seem grown out from the surface. However, within the resolution of the SEM, we cannot with certainty locate the origin of the carbon nanotubes in the mesoporous channels in MCM-41. There are also some small particles on the MCM-41 surface, as seen clearly in Fig. 4(b). These particles are similar to those shown in Fig. 3. There is no evidence that we have deposited significant amount of amorphous carbon particles onto the surface. Interestingly, some carbon nanotubes seem to end at the small particles, indicating that these particles are likely small MCM-41 grains.

After CVD, we observed that the rust brown color of MCM-41 powder was turned into black. Large amounts of carbon seem deposited onto MCM-41. It is likely that a certain percentage of them are deposited as amorphous carbon instead of nanotubes. However, the sample has significant charging problem during SEM measurements if we don't coat it with a gold film to increase the surface conductivity. The low conductivity indicates that there should not be much graphitic carbon deposited on the sample surface. The majority of the carbon was likely deposited in the form of nanotubes. This indicates the possibility of synthesizing high purity nanotubes by catalytic CVD with MCM-41.

Further inspection of Fig. 4(a) and Fig. 4(b) reveals that carbon nanotubes have quite different features at these two spots. Figure 4(a) was taken from an angle looking into the hexagonal channels and Fig. 4(b) was from the side similar to Fig. 3. Clearly, Fig. 4(a) shows a much higher density of carbon nanotubes which are interwoven into a thick layer. This indicates that the mesoporous channels indeed make the catalytic activity much higher compared to the less porous surface at the side of the grain.

A control experiment was also carried out with undoped MCM-41 powder under the same reaction condition. As shown in Fig. 5, only small carbon particles formed on the MCM-41 surface. These particles have an average diameter of about 100 nm. Apparently, no carbon nanotubes formed. The Fe_2O_3 catalyst is obviously essential in our experiments to induce the growth of carbon nanotubes during CVD.

Figure 5. SEM image of a bare MCM-41 after treated in the same condition of CVD.

3. The High-Resolution Image of Carbon Nanotubes by TEM

The diameter of carbon nanotubes shown in Fig. 4 is quite uniform with a value of about 20 nm. This is much larger than the diameter of SWNTs (~1-3 nm). On the other hand, it is hard to imagine that MWNTs can have such uniform diameters. To answer this question, high resolution TEM measurements were carried out with the sample ultrasonicated in methanol for 1 hr.

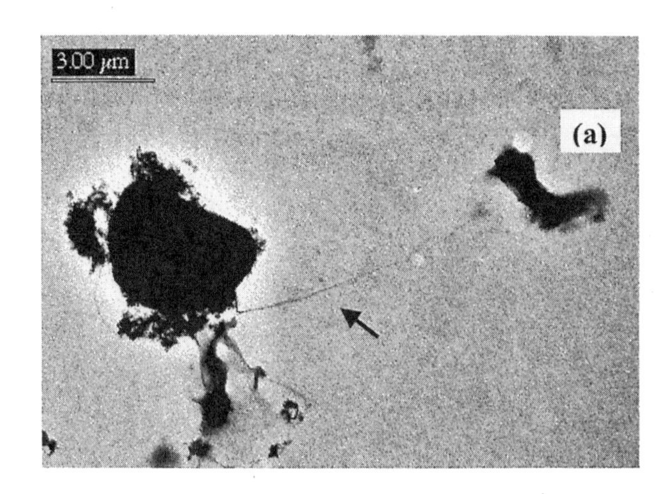

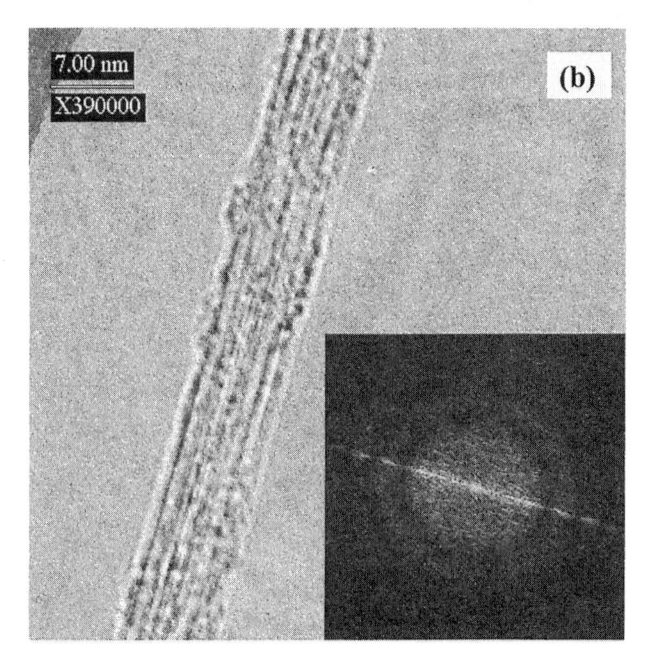

Figure 6. Transmission electron micrographs of carbon nanotubes formed by CVD of methane with Fe catalysts embedded in MCM-41. The inset in (b) is the Fast Fourier Transformation of the obtained TEM image.

Figure 6(a) shows a TEM image of two MCM-41 grains. The grains themselves are featureless because their thickness is too big for the electron beam to penetrate through. Interestingly, between these two grains, a single nanotube strand is clearly observed (as indicated by the arrow). The length of this carbon nanotube is over 8 μm. This is consistent with above-mentioned SEM images in which we can not clearly see both ends of any nanotube. It is likely that these two grains were broken apart from a single one by ultrasonication during TEM sample preparation. The nanotube has shown incredible mechanical strength since it was not broken by the violent ultrasonication for as long as one hour.

A high-resolution TEM image of a nanotube is shown in Fig. 6(b). It shows quite different features from the normal image of MWNTs. At this resolution, MWNTs typically show a clear hollow channel and are normally straight over several microns. The nanotubes synthesized with MCM-41, however, don't show the hollow channel and they are curved around (as shown in Fig. 4). The object seen in Fig. 6(b) resembles more the single-walled nanotube ropes synthesized by the laser ablation method[7]. Such structures can be attributed to about 20 to 30 single-walled carbon nanotubes stacked together to form a rope of about 7 nm in diameter. In the figure, about six nanotubes are stacked side-by-side across the 7 nm wide rope, presumably hold together by van de Waals interactions. The outer surface of the rope seems to be covered with a small amount of amorphous carbon. We tried different sample tilt but found that these SWNTs generally don't stack as orderly as reported by Thess et al[7]. Each nanotube is about 1.2 nm in diameter, in good consistency with the size of SWNTs but smaller than the mesoporous channel diameter (~3.3 nm). After Fe_2O_3 was embedded into the channel, it is not clear whether it was deposited uniformly on the channel wall making pore size smaller or if it formed nanoparticles which block parts of the channels.

The inset shows the Fast Fourier Transformation (FFT) of Fig. 6(b). It consists of two prominent features. First, there appears some diffuse intensity at the central part near the origin. Further out, a continuous ring or halo is clearly observed. This ring is due to the amorphous carbon on the carbon nanotube rope and in the thin support film on the TEM grid. More interestingly, along the direction perpendicular to the nanotubes rope, i.e. ~ 30° with respect to the x-axis, there are some broadened but discrete intensities corresponding to diffractions from the stack of SWNTs. The first order diffraction is at the position about the 1/4 to 1/5 from the origin of the radius of the halo from the amorphous carbon. This indicates that the nanotube diameter is about 4 or 5 times of the size of carbon atoms. If the nanotubes are MWNTs, we should expect a strong diffraction to appear at about the same position of the ring, e.g. corresponding to the distance between atomic layers. This is a further evidence that the carbon nanotubes synthesized by CVD with a Fe_2O_3 catalyst in MCM-41 are ropes of SWNTs.

4. CVD of Carbon Nanotubes with Fe_2O_3 Catalysts Deposited on Flat Si Surface

We have demonstrated that Fe_2O_3 embedded in MCM-41 can be used for large-scale synthesis of SWNTs. It is to our surprise that the carbon nanotubes assembled into quite uniform ropes. To understand whether MCM-41 has any effect to the growth of the nanotube ropes, we have also carried out some CVD experiments where $Fe(NO_3)_3 \cdot 9H_2O$ was deposited onto a piece of planar Si wafer instead of in mesoporous MCM-41 powder. All the other conditions for CVD were kept the same.

Fig. 7 shows SEM images of the sample surface after CVD. Clearly, there too appears a thick layer of carbon nanotubes on the surface. The nanotubes curve around with the length varying from a few microns to over 20 microns. Higher magnification image in Fig. 7(b) shows that there is a thick layer in which many nanotubes deeply stacked together. Interestingly, some of the nanotubes form helixes, which are not observed with the MCM-41 samples. The size of the nanotubes is also quite uniform, with the diameter of about 70 to 80 nm. This value, however, is about three or four times bigger than that of the nanotubes made with MCM-41. The helical structure indicates that these nanotubes are likely MWNTs. However, due to the small sample quantity, we did not have a chance to carry out further TEM measurements so far.

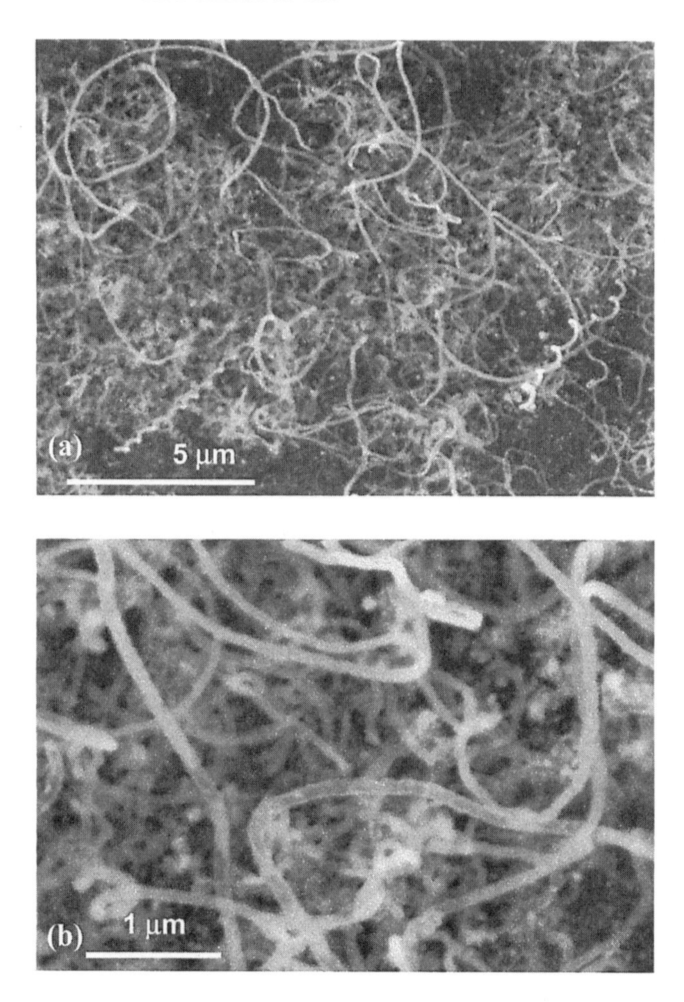

Figure 7. SEM images of carbon nanotubes grown by CVD of methane on Fe catalysts deposited on flat Si surface.

5. Discussion of the Growth Mechanism

Clearly, CVD processes under the same condition but with and without MCM-41 produce carbon nanotubes with quite different conformations, indicating that dramatic different mechanisms are involved. So far several models have been proposed to explain the growth mechanism.[21-23] It is generally accepted that the tube growth is initiated by a seed metal cluster. The outer diameter of the tube is defined by the size of this catalyst. Decomposed hydrocarbon species are dissolved in the catalyst particle resulting the oversaturation of carbon at one part of the surface. The catalyst particle can promote either tip growth or base growth, depending on the contact force between the catalyst particles and the substrate.[22] Experimental studies so far show quite diverse results which strongly depend on the nature of catalyst supports. Thess et al[7] reported that a tip growth catalyzed by a chemisorbed metal atom at the open end of the tube could explain their laser ablation experiments. Kong et al[11,14] found that CVD of methane with Fe_2O_3 catalysts embedded in alumina nanocrystals and amorphous silica produced SWNTs with closed ends, which strongly indicate the base-growth mechanism. CVD of ethylene with Fe_2O_3 embedded in porous Si film produces MWNTs also following base-growth mechanism.[15] Li et al[12] found that CVD of acetylene with Fe nanoparticles embedded in mesoporous silica could follow both mechanisms.

In our studies, the nanotube growth from Fe_2O_3 on flat Si surface seems to follow the base-growth mechanism. The formation of helical tubes could be explained by the base-growth mechanism proposed by Fonseca et al[24]. For the nanotubes grown out from MCM-41, the mesoporous channel serves as a good template to produce catalyst particles with the right size (1 to 4 nm). SWNTs with uniform sizes can quickly grow out from the channels by base-growth mechanism and then self-organize into close-packed long ropes. The packing distance (5.2 nm) of MCM-41 mesoporous channels are larger than the SWNT diameter (~1.2 nm), Therefore, only a limited number of nanotubes from neighboring channels can reach each other to form a rope. Thus, the ordered structure of the support results in a uniform rope diameter.

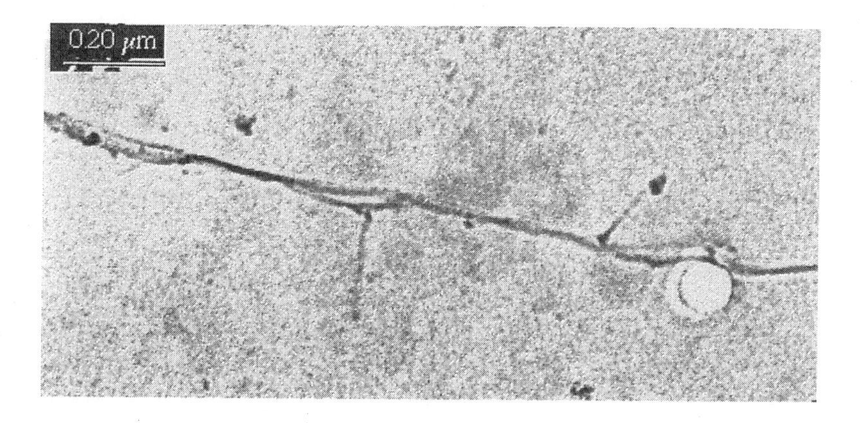

Figure 8. A TEM image of some small branches of carbon nanotubes ended with catalyst particles at the tip.

The evidence that the nanotube ropes can freely join together or separate apart (as shown in Fig. 4) and, that the ends have strong interaction with the substrate indicate that, in this case, the base-growth mechanism dominates. However, as shown in Fig. 8, we did observe that, at the end of some small branches off from the long rope shown in Fig. 6(a), there are a few catalyst particles with the size about that of the nanotubes ropes. This does not exclude that a small portion of nanotubes was grown from the tip, which is probably the reason that they quickly ended after breaking off from the main branch. However, the long rope in Fig. 6(a) indicates that both ends of the rope are strongly attached to the MCM-41 particle, which is not readily explained by either tip-growth or base-growth models.

CONCLUSIONS

We have demonstrated that the mesoporous MCM-41 can be applied as catalyst support for CVD growth of carbon nanotubes. Fe_2O_3embedded in the mesoporous channels of MCM-41 shows good catalytic properties for the growth of SWNTs. The carbon nanotubes grown by this method can self-assemble into ropes consisting of 20 to 30 SWNTs. These ropes are typically more than 10 microns long and frequently interwoven into a dense layer. The CVD method is simple and could becomes an economical solution for large-scale synthesis of high purity SWNTs.

It is not clear at this stage how the nanotubes grow out from the mesoporous channels. Neither TEM nor SEM provided us with direct information about the relation of carbon nanotubes with the MCM-41 channels. However, our experiments did show that the average diameter of the carbon nanotube ropes synthesized with MCM-41 (~ 10 to 20 nm) is much smaller than those grown from Fe on Si wafers (~70 to 80 nm). The confined size of the mesoporous channels in MCM-41 likely plays an important role.

One of the attractive properties of MCM-41 is that its pore size can be controlled precisely by selecting different surfactants as the template in sample preparation. We are in the progress to study the synthesis of carbon nanotubes with a series of MCM-41 with different pore sizes. Different catalysts such as Fe, Co, and Ni have been reported to quite different catalytic properties. It would be interesting to investigate how the size of the mesoporous supports and the different combination of metal catalysts affect the properties of the carbon nanotube material. This may provide us with some insight into the mechanism of carbon nanotube growth during CVD.

ACKNOWLEDGEMENT

We would like to thank the Institute of Materials Research & Engineering and National University of Singapore for the support of this work. We also thank Dr. Jiyan Dai, Dr. Kun Li, and Tiejun Zhang for the help with TEM experiments, and Dr. Meisheng Zhou and C. Ramasamy for their help with some SEM experiments.

REFERENCE

1. S. Iijima, *Nature*, 354, 56(1991).
2. M. Endo, S. Iijima, and M. S. Dresselhaus, Ed., *Carbon Nanotubes*, Elsevier Sciences, Oxford (1996).

3. T. W. Ebbesen, Ed., *Carbon Nanotubes: Preparation and Properties*, CRC, New York (1997).
4. S. Subramoney, *Adv. Mater.* 10 (15), 1157(1998).
5. C. Journet, W. K. Maser, and P. Bernier, *Nature*, 388, 756(1997).
6. T. Guo, P. Nikolaev, and A. Thess, *Chem. Phys. Lett.* 243, 49(1995).
7. A. Thess, R. Lee, P. Nikolaev, and H. Dai, *Science*, 273, 483(1996).
8. A. G. Rinzler, J. Liu, and H. Dai, *Appl. Phys. A* 67, 29(1998).
9. W. K. Maser, E. Munoz, and A. M. Benito, *Chem. Phys. Lett.*, 292, 587(1998).
10. H. Dai, A. G. Rinzler, and P. Nikolaev, *Chem. Phys. Lett.* 260, 471 (1996).
11. J. Kong, A. M. Cassell, and H. Dai, *Chem. Phys. Lett.* 292, 567(1998).
12. W. Z. Li, S. S. Xie, and L. X. Qian, *Science* 274, 1701(1996).
13. G. Che, B. B. Lakshmi, and C. R. Martin, *Chem. Mater.* 10, 260 (1998).
14. J. Kong, H. T. Soh, and A. M. Cassell, *Nature* 395, 878(1998).
15. S. Fan, M. G. Chapline, and N. R. Franklin, *Science* 283, 512(1999).
16. C. T. Kresge, M. E. Leonowicz, and W. J. Roth, *Nature* 359, 710(1992).
17. J. S. Beck, J. C. Vartuli, and W. J. Roth, *J. Am. Chem. Soc.* 114(27), 10834(1992).
18. C.-F. Cheng, W. Zhou, and D. H. Park, *J. Chem. Soc., Faraday Trans.* 93(2), 359(1997).
19. A. Sayari, M. Kruk, and M. Jaroniec, *Adv. Mater.* 10(16), 1376(1998).
20. D. Zhao, J. Feng, and Q. Huo, *Science*, 279, 548(1998).
21. S. Amelinckx, *Science* 265, 635(1994).
22. R. T. K. Baker, *Carbon* 27, 315(1989).
23. G. G. Tibetts, M. G. Devour, and E. J. Rodda, *Carbon* 25, 367(1987).
24. A. Fonseca, K. Hernadi, J. B. Nagy, P. Lambin, and A. A. Lucas, in: *Carbon Nanotubes,* M. Endo, S. Iijima, and M. S. Dresselhaus, ed., Elsevier Sciences, Oxford (1996).

MECHANICAL PROPERTIES AND ELECTRONIC TRANSPORT IN CARBON NANOTUBES

J. Bernholc, M. Buongiorno Nardelli, J.-L. Fattebert, D. Orlikowski, C. Roland and Q. Zhao

NC State University,
Raleigh, NC 27695-8202, U.S.A.
E-mail: bernholc@ncsu.edu

INTRODUCTION

The field of carbon nanotubes is undergoing an explosive growth due to both the intrinsic interest in these molecular structures and their technological promise in, e.g., high strength, light weight materials, superstrong fibers, novel nanometer scale electronic and mechanical devices, catalysts, and energy storage. Despite the potential impact that nanotubes could have in many areas of science and industry, the characterization of their mechanical and electrical properties is still incomplete. We show that nanotubes under high strain conditions can undergo a variety of atomic transformations, often occurring via successive bond rotations. The barrier for the rotation is dramatically lowered by strain. While very high strain rates must lead to breakage, (n,m) nanotubes with n, m < 14 can display plastic flow under suitable conditions. This occurs through the formation of a 5-7-7-5 defect, which then splits into two 5-7 pairs. The index of the tube changes between the 5-7 pairs, potentially leading to metal-semiconductor junctions. We have also computed quantum conductances of strained tubes, defects, and nanotube junctions, since these deformations are likely to occur when nanotubes are used to form nanoscale electronic devices. The results show that the defect density and the contacts play key roles in reducing the conductance at the Fermi energy, while bending and mechanical deformations affect differently the conductance of achiral and chiral nanotubes. Our results are in good agreement with recent experimental data.

MECHANICAL PROPERTIES

A number of important applications of carbon nanotubes are likely to take advantage of their outstanding mechanical properties, namely their extreme flexibility and strength

Science and Application of Nanotubes, edited by Tománek and Enbody
Kluwer Academic / Plenum Publishers, New York, 2000

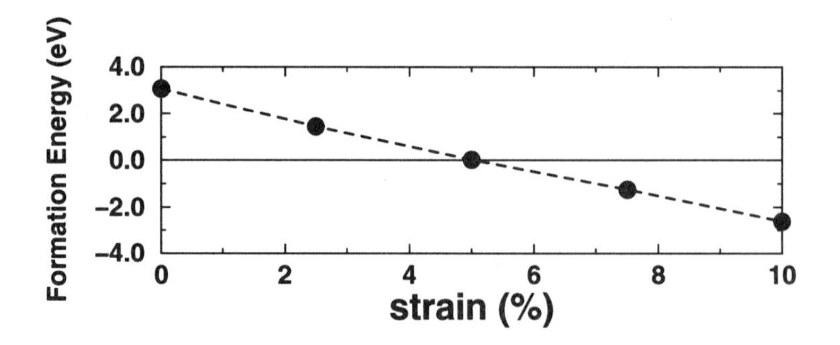

Figure 1: Formation energy of the (5-7-7-5) defect for the (5,5) tube at different strains.

at one-sixth the weight of steel. Nanotubes have already demonstrated exceptional mechanical properties: the excellent resistance of carbon nanotubes to bending has been observed experimentally and studied theoretically.[1,2,3] Their high stiffness combines with resilience and the ability to reversibly buckle and collapse: even largely distorted configurations (axially compressed, twisted) can also be due to elastic deformations with no atomic defects involved.[1,2,4,5] We have also investigated the mechanism of strain release in carbon nanotubes under uniaxial tension, in an effort to help evaluate their ultimate strength, and to determine the mechanisms of their plastic deformations and failure.

Axial strain in clean nanotubes

We have identified the first stages of the mechanical yield of carbon nanotubes and studied their mechanical failure under a tensile load. Our study, based on extensive ab-initio and classical molecular dynamics simulations, shows that an armchair nanotube releases its excess strain via spontaneous formation of a so-called (5-7-7-5) defect. A detailed knowledge of the formation and activation energies of these defects is needed for the understanding of the thermodynamics and kinetics of defect-induced processes in both the mechanical response and growth of carbon nanotubes. The values of the formation energies of a (5-7-7-5) defect in a (5,5) armchair tube are summarized in Fig. 1, as obtained in *ab initio* calculations. Calculations of the corresponding activation energies are in progress.

Fig. 1 shows that beyond about 5% tension, an armchair nanotube under axial tension can release its excess strain via spontaneous formation of topological defects.[6] This tension is partially relieved by a rotation of the C-C bond perpendicular to it (the so called Stone-Wales transformation[7]), which produces two pentagons and two heptagons coupled in pairs (5-7-7-5).[6] The appearance of a (5-7-7-5) defect can be interpreted as a nucleation of a degenerate dislocation loop in the hexagonal network of the graphite sheet. The configuration of this primary dipole is a (5-7) core attached to an inverted (7-5) core. The (5-7) defect behaves thus as a single edge dislocation in the graphitic plane. Once nucleated, the (5-7-7-5) dislocation loop can ease further relaxation by separating the two dislocation cores, which glide through successive Stone-

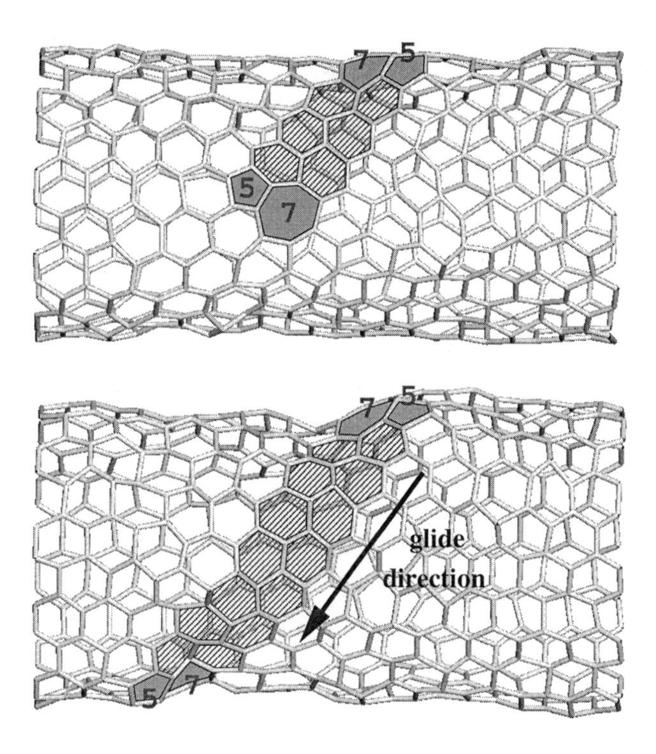

Figure 2: Plastic deformation of a (10,10) tube under axial tension at 3000 K after 2.5 ns simulation. The shaded area indicates the migration path of the edge dislocation.

Wales bond rotations.[6] This corresponds to a plastic flow of dislocations and gives rise to ductile behavior, as shown in Fig. 2. In contrast, in the case of a zig-zag nanotube, the same C-C bond will be parallel to the applied tension, which is already the minimum energy configuration for the strained bond. The formation of a Stone-Wales defect is then limited to rotation of the bonds oriented 120° with respect to the tube axis. Our calculations show that the formation energies of these defects are strongly dependent on curvature, *i.e.*, on the diameter of the tube. This gives rise to a wide variety of behaviors in the brittle-*vs.*-ductile map of stress response of carbon nanotubes.

Our calculations have identified a substantial variety of elastic responses in strained carbon nanotubes, In particular, under high strain and low temperature conditions, all tubes are brittle. If, on the contrary, external conditions favor plastic flow, such as low strain and high temperature, (n,m) tubes with n,m < 14 can be completely ductile, while larger tubes are moderately or completely brittle depending on their symmetry. These results are summarized in Fig. 3, where a map of ductile *vs.* brittle behavior of a general (n,m) carbon nanotube under an axial tensile load is presented. There are four regions indicated by different shadings. The small hatched area near the origin is the region of complete ductile behavior, where the formation of (5-7-7-5) defects is always favored under sufficiently large strain.

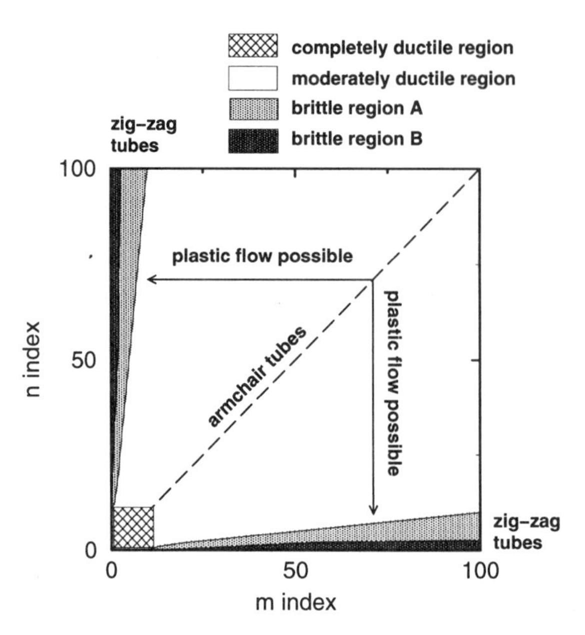

Figure 3: Ductile-brittle domain map for carbon nanotubes with diameters up to 13 nm. Different shaded areas correspond to different possible behaviors (see text).

Plastic flow, if possible, will transform the tube section between the dislocation cores along paths parallel to the axes of Fig. 3. During the transformations, the symmetry will oscillate between the armchair and the zig-zag type. The same transformations will occur in the larger (white) moderately ductile region. Tubes with indices in this area are ductile but the plastic behavior is limited by the brittle regions near the axes. Tubes in the last two regions will always follow a brittle fracture path with formation of disordered cracks and large open rings under high tensile strain conditions.

Axial strain in nanotubes with addimers

We have discovered a natural route for the formation of nanotube-based quantum dots that uses addimer-induced transformations in strained carbon nanotubes. Addimers are likely to be present in small amounts on as-grown nanotubes, or they may be deposited there with an STM tip or other methods.

The addimers induce the formation of a new set of defects, consisting of rotated hexagons separated by (5-7) pairs from the rest of the tube, as shown in Fig. 4. This defect undergoes substantial further evolution under the appropriate strain conditions. Under such strain, more hexagons are added to the initial defect via successive rotations of C-C bonds. If this process of adding hexagons were to continue, the defect structure would eventually wrap itself completely about the circumference of the tube, forming a short segment of a nanotube with a different helicity. The formation of quantum dots with addimers is particularly favorable for the (n,0) zigzag tubes, which are otherwise brittle. The winding of the defect about the nanotube suggests that the combination of addimers plus strain greater than 5% may be a natural way to produce different

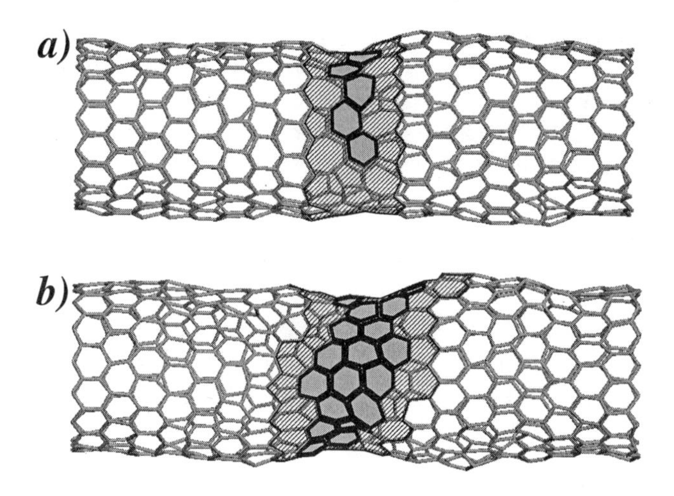

Figure 4: Time evolution of a (17,0) tube with an addimer under 7.5% strain at 3000 K, illustrating the spontaneous winding of the defect about the tube: (a) the initial configuration consisting of a single turn; (b) the final configuration, corresponding to about 3 turns after 1 ns.

electronic heterojunctions, thereby leading to the formation of nanotube-based quantum dots. Fig. 4b shows that such dots indeed form during annealing in molecular dynamics simulations.[8]

TRANSPORT PROPERTIES

Recently, there has been a substantial amount of interest in quantum transport in carbon nanotubes. Their electronic and transmission properties have been studied both experimentally and theoretically.[9,10,11,12,13,14,15,16,17,18,19] In particular, the sensitivity of nanotube's electronic properties to their geometry makes them truly unique in offering the possibility of studying quantum transport in a very tunable environment. The general problem of calculating the conductance in carbon nanotubes has been addressed with a variety of techniques that reflect the various approaches in the theory of quantum transport in ballistic systems. Most of the existing calculations derive the electronic structure of carbon nanotubes from a simple π-orbital tight-binding Hamiltonian that describes the bands of the graphitic network via a single nearest-neighbor hopping parameter. Since the electronic properties of carbon nanotubes are basically determined by the sp^2 π-orbitals, the model gives a reasonably good qualitative description of their behavior and, given its simplicity, it has become the model of choice in a number of theoretical investigations. However, although qualitatively useful to interpret experimental results, this simple Hamiltonian lacks the accuracy that more sophisticated methods are able to provide. In particular, the π-orbital model is limited to studies of topological changes, and cannot address the effects of topology-preserving distortions, such as bending, on conductance.

Recently, we have presented a general scheme for calculating quantum conductance that is particularly suitable for realistic calculations of electronic transport properties in extended systems.[20] This method is based on the Surface Green's Function Matching formalism and efficiently combines the iterative calculation of transfer matrices with the Landauer formula for coherent conductance. It is very flexible and applicable to any system described by a Hamiltonian with a localized orbital basis. The only quantities that enter are the matrix elements of the Hamiltonian operator, with no need for the explicit knowledge of the electron wave functions for the multichannel expansion. The last fact makes the numerical calculations efficient also for systems described by multi-orbital Hamiltonians. We linked this approach with a real-space, optimized local-orbital method,[21] and can now compute quantum conductance of large systems from first principles.

As the first application, we evaluated the conductance of a (5,5) tube, pristine and with a (5-7-7-5) defect, using both ab initio and tight-binding Hamiltonians, see Fig. 5. The agreement between the two sets of calculation is very good, confirming the reliability of tight-binding methods in predicting quantum conductance of carbon nanotube structures. Furthermore, our results for the (5,5) tube with a single (5-7-7-5) defect compare very well with a recent *ab initio* calculation[18] for a larger (10,10) tube.

Using the above approach and a realistic tight-binding Hamiltonian[22] we have investigated quantum conductances of a variety of strained tubes. Our results show that the defect density plays a key role in reducing the conductance at the Fermi energy. While the presence of a single defect in the nanotube wall cannot lead to a large change in the conductance spectrum, as shown above for the case of a (5,5) tube, large defective regions likely to be present in highly strained nanotubes will strongly modify their transport properties. In Fig. 6 we show the conductance spectrum of a deformed nanotube whose geometry has been obtained in one of our molecular dynamics simulations. The conductance is severely modified by the defective region and in particular a sharp decrease is observed at the Fermi level. A full description and analysis of the results will be published elsewhere.[23]

SUMMARY AND CONCLUSIONS

In summary, we have shown that carbon nanotubes at high strain conditions can undergo a variety of atomic transformations, often occurring via successive bond rotations. The barrier for the rotation is dramatically lowered by strain, and ab initio results for its strain dependence have been presented. While very high strain rates must lead to breakage, (n,m) nanotubes with n, m $<$ 14 can display plastic flow under suitable conditions. This occurs through the formation of 5-7-7-5 defects, which split into 5-7 pairs. The index of the tube changes between the 5-7 pairs, potentially leading to metal-semiconductor junctions. A different way to induce transformations is through addimers, which lead to the formation of metallic quantum dots embedded in a semiconducting nanotube.

We have also computed quantum conductances of strained nanotubes and defects. The results show that the defect density plays a key role in reducing the conductance at the Fermi energy, and that disorder strongly affects the electrical response of carbon nanotubes.

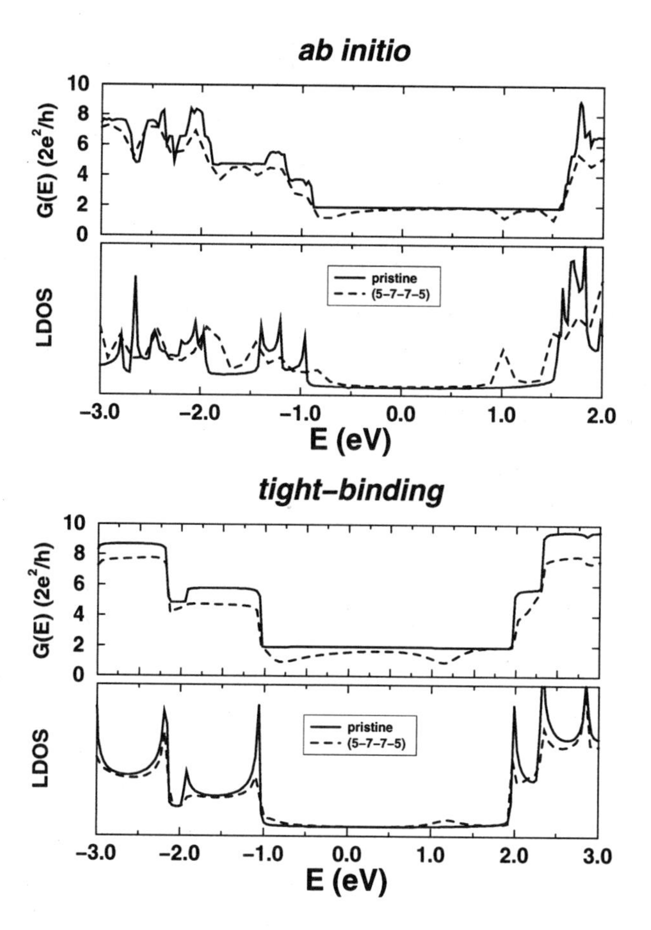

Figure 5: Upper panel: *Ab initio* conductance and local density of states for a pristine (5,5) nanotube, and with a (5-7-7-5) defect. Lower panel: Same as above using a tight-binding Hamiltonian.

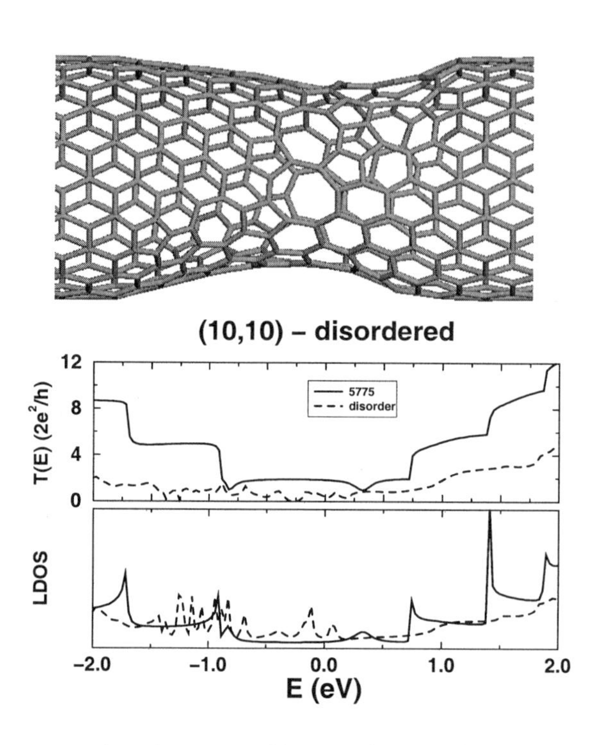

Figure 6: Upper panel: the (10,10) tube at 10% strain after annealing at 1500 K for 1.0 ns. Lower panel: Quantum conductance for the above system.

REFERENCES

1. S. Iijima, C. Brabec, A. Maiti and J. Bernholc, J. Chem. Phys. **104**, 2089 (1996).
2. B.I. Yakobson, C. J. Brabec, and J. Bernholc, Phys. Rev. Lett. **76**, 2511 (1996).
3. M.R. Falvo, G.J. Clary, R.M. Taylor II, V. Chi, F.P. Brooks Jr, S. Washburn and R. Superfine, Nature, **389**, 582 (1997).
4. J. Despres, E. Daguerre and K. Lafdi, Carbon **33**, 87 (1995).
5. N. Chopra, L. Benedict, V. Crespi, M.L. Cohen, S.G. Louie and A. Zettl, Nature **377**, 135 (1995); R. Ruoff and D. Lorents, Bull. Am. Phys. Soc. **40**, 173 (1995).
6. M. Buongiorno Nardelli, B.I. Yakobson and J. Bernholc, Phys. Rev. B **57**, R4277 (1998); Phys. Rev. Lett. **81**, 4656 (1998).
7. A.J. Stone and D.J. Wales, Chem. Phys. Lett. **128**, 501 (1986).
8. D. Orlikowski, M. Buongiorno Nardelli, J. Bernholc and C. Roland, to be published.
9. P.G. Collins, A. Zettl, H. Bando, A. Thess and R. Smalley, Science **278**, 100 (1996); S.N. Song, X.K. Wang, R.P.H. Chang and J.B. Ketterson, Phys. Rev. Lett. **72**, 697 (1994); L.Langer, L. Stockman, J.P. Heremans, V. Bayot, C.H. Olk, C. Van Haesendonck, Y. Bruynseraede and J.-P. Issi, J. Mater. Res. **9**, 927 (1994); L.Langer, V. Bayot, E. Grivei, J.-P. Issi, J.P. Heremans, C.H. Olk, L. Stockman, C. Van Haesendonck and Y. Bruynseraede, Phys. Rev. Lett. **76**, 479 (1996); S.J. Tans, M.H. Devoret, H. Dai, A. Thess, R.E. Smalley, L.J. Georliga and C. Dekker, Nature **386**, 474 (1997); A. Bachtold, C. Strunk, J.-P. Salvetat, J.-M. Bonnard, L. Forró, T. Nussbaumer and C. Schönenberger, Nature **397**, 673 (1999).
10. A. Bezryadin, A.R.M. Verschueren, S.J. Tans and C. Dekker, Phys. Rev. Lett. **80**, 4036 (1998).
11. S. Paulson, M.R. Falvo, N. Snider, A. Helser, T. Hudson, A. Seeger, R.M. Taylor II, R. Superfine and S. Washburn, http://xxx.lanl.gov/abs/cond-mat/9905304, preprint (1999).
12. W. Tian and S. Datta, Phys. Rev. B **49**, 5097 (1994).
13. R. Saito, G. Dresselhaus, M.S. Dresselhaus, Phys. Rev. B **53**, 2044 (1996).
14. L. Chico, L.X. Benedict, S.G. Louie and M.L. Cohen, Phys. Rev. B **54**, 2600 (1996).
15. R. Tamura and M. Tsukada, Phys. Rev. B **55**, 4991 (1997); *ibid*, **58**, 8120 (1998).
16. M.P. Anantran and T.R. Govindan, Phys. Rev. B **58**, 4882 (1998).
17. A.A. Farajian, K. Esfarjani and Y. Kawazoe, Phys. Rev. Lett. **82**, 5084 (1999).
18. H.J. Choi and J. Ihm, Phys. Rev. B **59**, 2267 (1999).
19. A. Rochefort, F. Lesage, D.R. Salahub and P. Avouris, http://xxx.lanl.gov/abs/cond-mat/9904083, preprint (1999).
20. M. Buongiorno Nardelli, Phys. Rev. B, in press (1999).
21. J.-L. Fattebert and J. Bernholc, to be published (1999); M. Buongiorno Nardelli, J.-L. Fattebert and J. Bernholc, to be published (1999).
22. J.C. Charlier, Ph. Lambin and T.W. Ebbesen, Phys. Rev. B **54**, R8377 (1996).
23. M. Buongiorno Nardelli and J. Bernholc, to be published.

ELECTROCHEMICAL STORAGE OF HYDROGEN IN CARBON SINGLE WALL NANOTUBES

Christoph Nützenadel, Andreas Züttel, Christophe Emmenegger, Patrick Sudan, and Louis Schlapbach

Université de Fribourg, Institute de Physique, Pérolles,
CH-1700 Fribourg
e-mail: christoph.nuetzenadel@unifr.ch

ABSTRACT

One still not satisfactory solved problem for the use of hydrogen as clean fuel is the safe and efficient storage of hydrogen. Currently cryo tanks, gas cylinders or metal hydrides are used. Important parameters are weight and volume density, cost and safety. Recent publications claimed that large amounts of hydrogen could be stored reversibly in carbon nanotubes from the gas phase. Similar to a gas phase experiment where the storage material absorbs hydrogen by increasing the pressure one can also do this electrochemically. We report that carbon nanotubes can be charged reversibly with amounts of hydrogen exceeding metal hydrides. Samples with different degrees of purity and different production methods were used and are compared.

EXPERIMENTAL

Different kinds of SWNT samples have been investigated. Most of them were produced by the arc discharge method. Only the sample from tubes@rice was synthesised by laser vaporisation. Samples were obtained from Patrick Bernier, Université de Montpellier (FR), MER Corporation (USA) [4], CarboLex (USA), and Dynamic Enterprise Limited (UK) and tubes@rice (USA). Some of their properties are listed in Table 1.

The samples which were not delivered as powders were grinded. The CP-grade sample from Carbolex is delivered as a paper sheet. The paper was cut in small pieces and grinded. Then the powder was mixed with gold powder (purity 99.95%, diameter <53 μm, Goodfellow UK) as compacting additive. Gold was used because it is noble and does not participate in any

Science and Application of Nanotubes, edited by Tománek and Enbody
Kluwer Academic / Plenum Publishers, New York, 2000

electrochemical reaction in the applied potential range. A pellet was pressed (500 MPa) of 20 mg carbon material and 80 mg gold powder. This pellet was used as negative electrode.

A particular case is the sample from tubes@rice which contains chemically purified nanotubes in a suspension. Two methods were used to produce electrodes:

The suspension was filtered on a paper, rinsed and dried in vacuum. When the nanotube layer is thick enough one can easily peel of a nanotube paper sheet. This paper was cut into small pieces and grinded. It is impossible to make a fine powder out of the paper. There are still small pieces of the sheet remaining. Therefore it is very difficult to produce mechanically stable electrodes. The second method was to filter the suspension through nickel foam. The suspension has a high viscosity and most of the nanotubes remain in the foam. The nanotubes remain stuck on the foam when the electrode is immersed in the electrolyte.

All hydrogen weight densities given in this paper are normalised to the sample weight, which is the weight of the entire carbon material; i.e. no correction due to the purity of the samples was done.

The experiments were performed in a half-cell in 6 M KOH electrolyte. A nickel plate was used as counter electrode. The potentials were referred to a Hg/HgO/OH^- reference electrode.

TABLE 1. Properties and max. discharge capacity of the samples

Producer	SWNT	Remainder	Catalyst	Process	Max discharge Capacity [mAh/g]
Patrick Bernier	70%	amorphous Carbon	N, Y	arc, as produced	550
MER	a few %	traces of C-60, C70, amorphous Carbon	Ni,Fe	arc, as-produced	170
CarboLex	50-70%	amorphous Carbon	Ni	arc, as produced	170
CarboLex	85%	amorphous Carbon	Ni	arc, selected grade	175
CarboLex	>90%	amorphous Carbon	Ni	arc, chemically purified	58
DEL	50%	traces of C-60, C70, amorphous Carbon	Ni, Fe	arc, purified	552
tubes@rice	>90%		Ni,Co	laser vaporisation, chemically purified	55 135 (on Ni-foam)

The electrodes were charged for 16 hours with a current of 25 mAg^{-1}. After completion of the charge process gaseous hydrogen is evolved and escapes from the open cell to the environment.

The normal discharge was performed with 25 mAg^{-1}. The discharge ended when the cut-off potential of 0.0 V vs. Hg/HgO/OH^- is reached. The capacity was calculated by multiplication of the current with the discharge time.

A special cycle was applied to measure the maximal discharge capacity at different currents. After the charge a discharge current of 5000 mAg^{-1} is applied. After reaching the cut-off potential the cell is switched of for 3 minutes for recovery and subsequently a new discharge with half of the current applied before starts. This procedure is repeated until a current of 0.5 mAg^{-1} is reached. This cycle is further called deep discharge.

The potential of the electrode is a function of the hydrogen concentration. This was measured by dividing the charge and the discharge in 100 current pulses of 10 mA/g during 24 minutes. After 3 minutes in the open cell state to equilibrate the potential was measured and subsequently the next pulse was applied.

For the voltage step measurements the potential was set from 0.0 V to 0.9 V vs. $Hg/HgO/OH^-$ for charging and vice versa for discharging.

The hydrogen weight density was calculated from the amount of discharged current: 1 Ah/g corresponds to 3.54 wt% hydrogen in carbon.

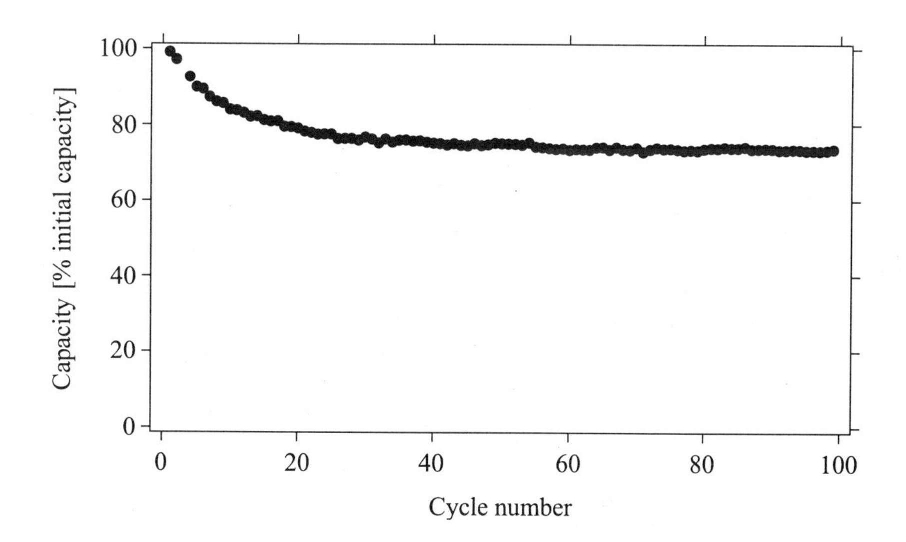

Figure 1. Discharge Capacity as a function of the number of cycles. The initial capacity is 73 $mAhg^{-1}$.The cycles were performed at 25 mAg^{-1} discharge current.

CYCLIC STABILITY

An important requirement for a storage material is the reversibility of the process. A large number of subsequent charge/discharge cycles has been performed on a SWNT sample produced by Patrick Bernier in an arc-discharge process. After 100 charge/discharge cycles the electrode still delivers more than 70% of the initial capacity (Fig. 1).

There are mainly two reasons for the capacity loss. The first is mechanical instability of the entire electrodes. Due to the hydrogen intercalation one would expect a volume increase of the sample. This leads to a decrepitation of the pellet. Black powder was found on the bottom of the cell. An improved electrode design might lead to an even better stability.

The second possibility is a deterioration of the storage material itself. In the case of metal hydrides oxidation of the surface and segregation of elements with low surface energy cause this effect. Both is not expected to appear for the nanotubes.

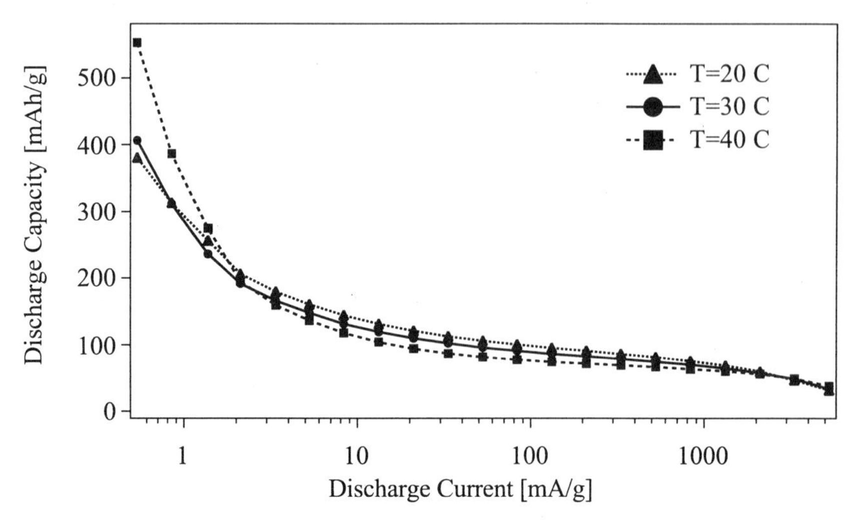

Figure 2. Discharge Capacity as a function of the discharge current

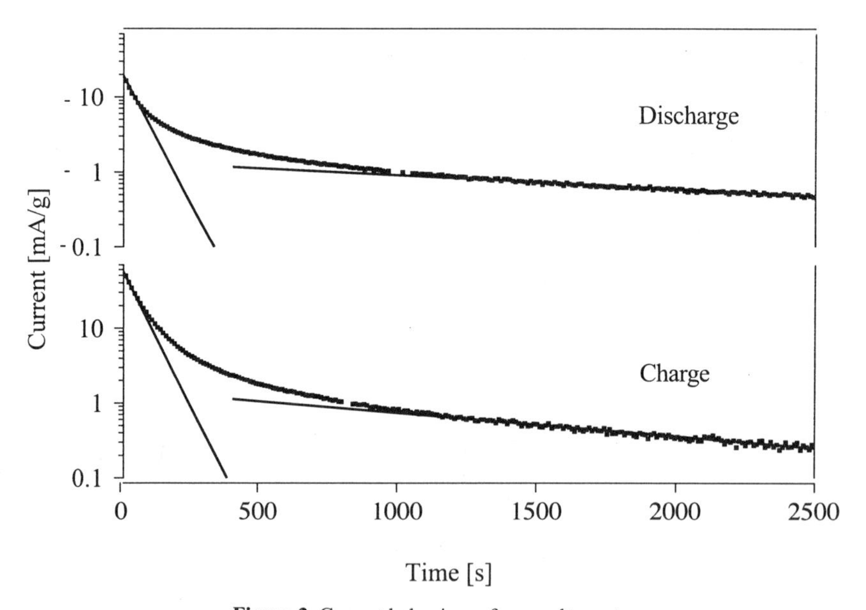

Figure 3. Current behaviour after a voltage step

KINETIC PROPERTIES

Due to a limited reaction rate the over potential increases with increasing discharge current. The cell voltage is reduced and the cut-off potential is reached earlier. Figure 1 shows that behaviour for an electrode made of the DEL sample. The capacity exceeds the capacity of typical metal hydrides for a discharge current lower than $1 mAg^{-1}$ but relatively low capacities are achieved when a high discharge current was used. One can calculate the time in which a discharge can be performed for a certain current. The upper axis is labelled with this unit.

The deep discharge was performed at different temperatures. The influence of the temperature is relatively weak. This indicates a low activation energy. Currently only little is known about the absorption process. It is not clear whether the hydrogen is stored molecular or atomic or whether the hydrogen is in or on the nanotubes. For the hydrogen intercalation a sequence of reactions is necessary: physi-, chemisorption, dissociation, absorption and diffusion. It is still a subject of investigation which part of the listed processes is involved in this particular case. Therefore we can only speculate which reaction is rate limiting for the entire process and which one is responsible for the influence of the temperature.

Figure 3 shows the charge and the discharge current after applying a voltage step. One can distinguish two processes. The first with a time constant of $t_{1/2}$= 38.1 s is due to the double layer capacity of the electrodes. The high surface area of the nanotubes leads to a high double layer capacity. This effect can be used for energy storage in electrochemical double layer capacitors.

The second with a time constant of $t_{1/2}$= 870 s represents the absorption and desorption of the hydrogen into the sample.

CHARGE/DISCHARGE CURVES

The potential of the electrode is a function of the hydrogen concentration. From the section above we concluded that one has to wait at least several times 870 seconds to get close to equilibrium of the electrode. It is not possible to keep the cell at open circuit for this time without adulteration of the result due to parasitary effects like self-discharge, secondary reactions or leakage currents. Figure 4 shows a charge discharge curve where the waiting time was 180 seconds. The curvature of the absorption and of the desorption are different. The shape of the curves is typical for a reaction where the absorption/desorption rate is dependent of the coverage.

CHEMICAL TREATMENT TO IMPROVE THE KINETIC PROPERTIES

From the sections above we concluded the necessity to improve the kinetic properties of the electrodes. Assuming that the dissociation of the hydrogen is the rate limiting step it is advantageous to add catalysts. In most cases the samples contain already nickel nanoparticles. This particles are used to catalyse the growth of the nanotubes. The particles are often placed at the end of the nanotubes. Nickel is known to be a good catalyst for the hydrogen

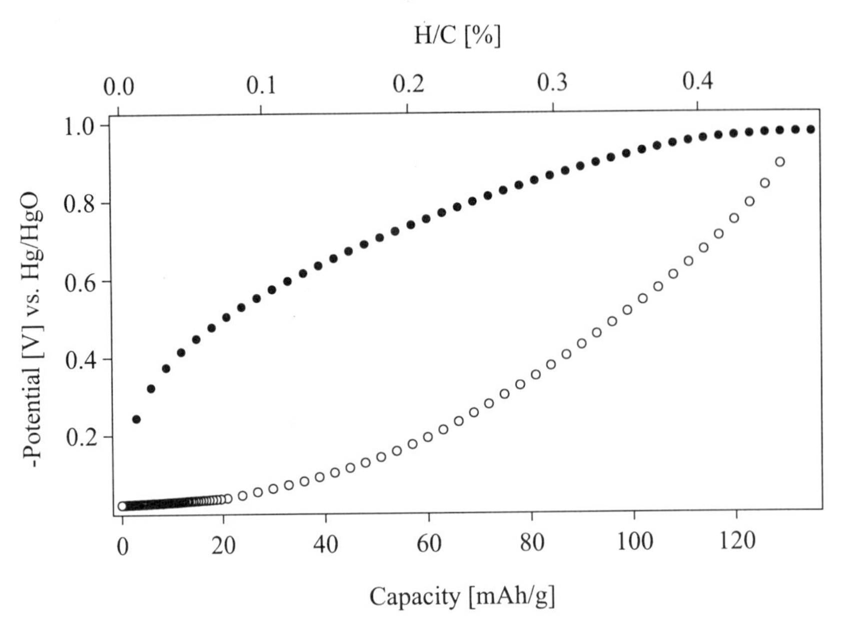

Figure 4. Charge/Discharge Curve: The closed circles show the potential as a function of hydrogen concentration during charge. The open circles show the discharge.

dissociation. If the hydrogen is stored inside the nanotubes and the reaction rate is limited because the hydrogen has to penetrate into the nanotubes opening of the end caps could improve the absorption. Nanotubes can also be cut into shorter tubes by oxidation in a concentrated acid solvent in an ultrasonic bath [5].

The treatments were applied to different samples. Not all of the samples are available in large quantities. Considering that some treatments consume a large percentage of the material some of the samples are not suitable for all treatments. This means that the samples used for chemical treatments are not necessarily the samples with the best performance.

The sample from MER was heated at low oxygen pressure (1mbar) for 30 minutes to 700°C. The comparison with an untreated sample is shown in figure 5. The absolute capacity is low in comparison with other samples because in this case copper was used as compacting additive. To prevent the copper from oxidising the cut-off potential was set to -0.6 V vs. $Hg/HgO/OH^-$. An improvement is visible over the entire range of discharge currents that were applied. It is difficult to say if the treatment resulted only in an opening of the nanotube of if there are secondary effects to the sample. Oxidation of the metallic particles can influence the sorption process as well. The catalytic properties change with the oxidation state.

Purification of as produced samples can be done with acidic solvents. Figure 6 compares the discharge capacity as a function of current for three different samples from Carbolex. All samples are produced in an arc discharge. The first is an untreated as produced sample. The second is a selected grade. The selection is done by Carbolex using Raman spectroscopy. The third sample was purified by washing for 3 days in nitric acid and subsequently filtered. This reduces the content of metallic and amorphous impurities. The difference of the discharge capacity between the as-produced material and the selected grade is negligible while the purified sample shows a distinct reduction in capacity. It is not known whether the final state of the hydrogen in the nanotubes is atomic or molecular. But it is probable the absorption

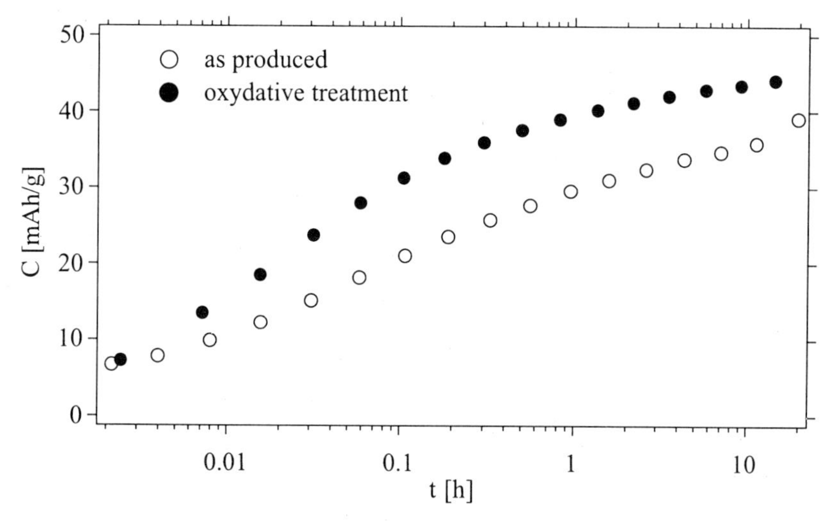

Figure 5. Oxidative treatment to improve the hydrogen storage

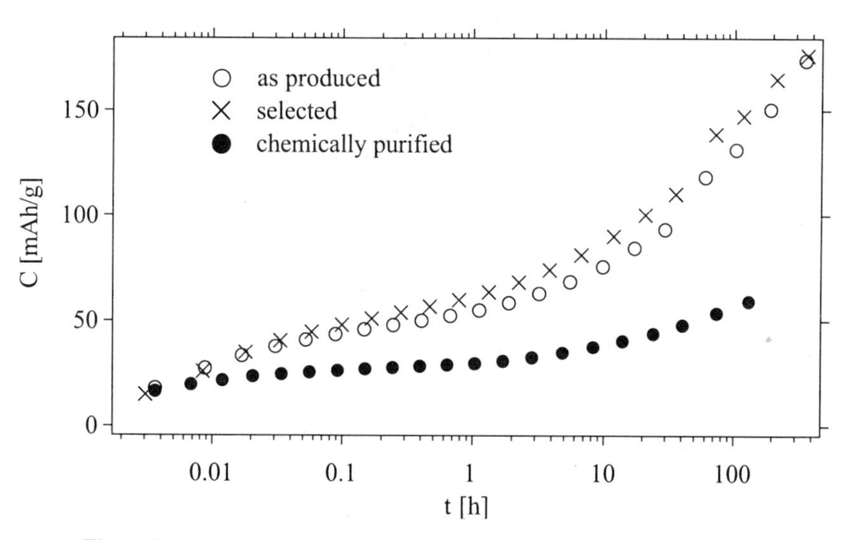

Figure 6. Influence of a chemical purification on the hydrogen storage capacity

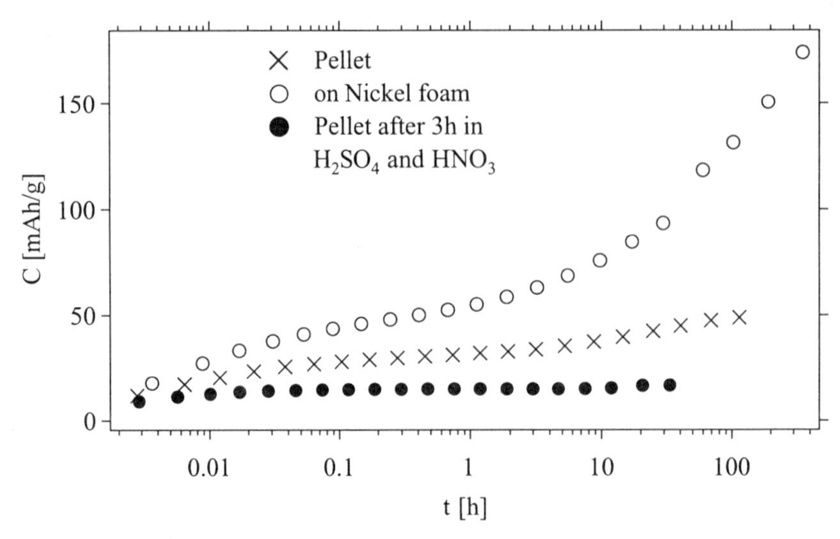

Figure 7. The discharge capacity is increased when nickel is used as substrate. A chemical cutting of the nanotubes does not improve the storage capacity.

process requires dissociation either of hydrogen or water molecules. It is known that metallic impurities, especially nickel, can serve as catalyst and can host H-atoms in an intermediate state. Therefore the capacity loss due to the removal of the metallic impurities indicate that they play an important role in the absorption process.

In the last part of this section we will discuss if it is possible to improve the kinetic properties of a purified nanotube material. SWNTs produced by laser vaporisation and purified in a nitric acid reflux for 2 days were supplied by tubes@rice. The content of metallic particles is below 1%. As tubes end are rarely observed the tubes are either very long or looped. Figure 7 shows that the storage properties of an untreated sample are poor as compared with the unpurified samples. In order to find out if either the absence of metallic particles or the good quality of the sample causes this effect two kind of treatments were applied. The first was to filter the nanotube suspension directly on nickel foam. About 10 mg nanotube are suspended on 4 square centimetres. Subsequently the foam was rinsed with water and ethanol. The electrodes are stable when immersed in the electrolyte. The resulting deep discharge curve is shown in Fig. 7. The capacity improves by a factor of three.

Liu et al. report that nanotubes can be cut in short pieces when treated in a 3:1 mixture of concentrated sulphuric and nitric acid in an ultrasonic bath. We applied this treatment for 3 hours. The suspension was filtered and rinsed with water and with ethanol and dried in vacuum.

The resulting paper was grinded to a powder and pressed to a pellet with gold as compacting additive. The measured capacity for this sample was smaller as compered to the purified material without treatment (Figure 7). This demonstrates again the importance of the metallic impurities for the hydrogen storage. If cutting of the nanotubes improves the kinetics it is not observable because it is hidden by the effect of the removal of the metallic remainder due to the acidic treatment.

CONCLUSIONS

Carbon nanotubes are an interesting material for hydrogen storage. The capacity per weight of some SWNT containing samples exceeds the capacity of typical metal hydrides. We applied different kind of treatments to improve the kinetic properties. Different treatments were applied to shorten or to open the nanotubes. All acidic treatments failed to improve the storage. Only the oxidative heating in the oven increased to hydrogen uptake as compared to an untreated sample. This can be explained because this treatment does not remove the metallic remainder. This model is supported by the observation that the capacity of a highly purified sample increases distinctly when a nickel substrate is used to produce the electrodes.

REFERENCES

[1] Hill, 'Green Cars go farther with graphite', New Scientist, 12/28 December 1996

[2] Dillon, 'Storage of hydrogen in single-walled carbon nanotubes', Nature Vol 386, 27 March 1997, pp 377

[3] Chambers, J. Phys. Chem. B 102 (1998) 4253

[4] Nützenadel, Electrochem. and Sol. State Lett., 2 (1) 30 (1999)

[5] Liu,Science 280 (1998) 1253

DIRECT MEASUREMENT OF BINDING ENERGY VIA ADSORPTION OF METHANE ON SWNT

S. Weber[1], S. Talapatra[1], C. Journet[2] and A. Migone[1]

[1]Department of Physics
Southern Illinois University
Carbondale, IL 62901-4401

[2]Max-Planck-Institut fuer Festkoerperforschung
Heisenbergstraasse 1, D-70569
Stuttgart, GERMANY

ABSTRACT

Low coverage adsorption isotherms for CH_4 on single-walled carbon nanotubes have been measured at various temperatures between 144 and 184 K. The nanotubes used in this study were untreated; hence, they remain capped. Adsorption takes place preferentially in the interstitial channels between the individual SWNT strands which constitute nanotube bundles. The isosteric heat of adsorption for methane on the nanotubes has been obtained from the isotherm results. The isosteric heat is directly related to the binding energy of CH_4 to the nanotube substrate; the binding energy of methane in the interstitial channels of the SWNT's was determined to be 226 meV. This value of the binding energy is 80% larger than the reported value for the binding energy for methane on planar graphite.

INTRODUCTION

Among all the many different substrates studied to date, SWNT's offer a truly unique environment for adsorption. Because of their very large aspect ratios and very small diameters,[1] SWNT's offer the possibility of the realization of effectively one-dimensional systems. Hence, films adsorbed on SWNT's are of interest from a fundamental point of view, because they can provide an experimental testing ground for the study of the properties of 1-D matter.

Additionally, the very large number of carbon atoms near every molecule adsorbed on a SWNT bundle results in binding energies which are much larger than those for planar substrates.[2] This high binding energy makes the study of adsorption on SWNT's relevant from a practical point of view, because it results in the adsorbed species having very large densities

Science and Application of Nanotubes, edited by Tománek and Enbody
Kluwer Academic / Plenum Publishers, New York, 2000

at relatively low pressures. Thus, nanotubes may result in new, and potentially very important, approaches to gas storage technology.[3]

It is this second possibility that has spurred much of the current interest[3-6] in the study of adsorption on SWNT's: Dillon et al. performed thermal programmed desorption measurements for H_2 on SWNT's and indirectly found a value for the binding energy which is far larger than that for H_2 on graphite.[3] The article by Dillon et al. suggested that SWNT's could potentially be used for revolutionizing H_2 gas storage technology and, eventually, could result in H_2-fueled vehicles.[3]

A series of articles[2,7-9] by Cole's group have explored a number of theoretical aspects of the adsorption of H_2, ^{4}He and Ne on SWNT's. These articles probe adsorption at the interior of uncapped nanotubes[2,9], and adsorption in the interstitial spaces of the strands that make up a bundle or rope.[7,8] In particular, these theoretical studies found that the interstitial channels (IC's) between the individual strands which constitute a nanotube rope or bundle, have binding energies for ^{4}He and Ne which are larger than those found for any other known charge-neutral system.[7] Very recently a programmed desorption mass spectrometer study of ^{4}He has largely confirmed these predictions.[6] The binding energy was indirectly determined in this study, as a fitting parameter to desorption data.

In this paper we present a series of results which provide direct approach for the experimental determination of the binding energy of methane molecules on SWNT's. The SWNT's used in this study were not subjected to any post-production treatment, hence their ends are capped. As a result, any enhanced adsorption in the SWNT sample is taking place at the interstitial channels (IC's) of the nanotube ropes.

Experimental

The SWNT samples utilized in this study were prepared using the electric arc discharge method by C. Journet in the laboratory of Prof. Bernier. Details of the nanotube production method have been provided elsewhere.[10] The typical nanotube yields obtained with this method are on the order of 80 % by weight. The bundles which result from this production method have typically about twenty single nanotube strands each.[10]

As previously stated, we did not subject the tubes to any opening or uncapping treatment. The nanotube sample was placed in a copper cell and evacuated to a pressure below 1×10^{-6} Torr for a period of at least 12 hours, at room temperature, prior to the performance of the adsorption measurements.

The apparatus used to measure the adsorption isotherms has been previously discussed in detail.[11] Three electro-pneumatic valves were employed to dose a computer-stipulated amount of gas from a reservoir to the gas-handling system, and from the gas-handling system to the experimental cell. MKS Baratron capacitance pressure gauges and an IBM PC compatible computer were used to measure and record the pressures. Equilibrium conditions were controlled and monitored by the computer.

The working cryogenic temperatures were achieved by using a helium-closed-cycle refrigerator. Stable temperatures were obtained using a two-stage temperature control arrangement.

For the isosteric heat determinations, four isotherms were measured at temperatures between 144 and 184 K. All the pressures involved in these measurements were below 1 Torr, so that it was necessary to apply thermomolecular corrections to all the data.[12] It was necessary to work at these relatively high temperatures because the equilibrium pressures at low coverages fall below the operating range of the gauge at temperatures not much below those utilized here.

Results

Prior to performing the low-coverage measurements which are used in the determination of the binding energy, we performed two complete adsorption isotherms at two temperatures, one below (82 K) and one above (92 K) the bulk triple point of methane (90.7 K). These two isotherms are displayed, respectively, in figures 1 and 2. The low-pressure portions of both adsorption isotherms represent the amount of gas being adsorbed in the interstitial channels in the SWNT sample. In this case, an amount adsorbed of approximately 6000 cc-Torr corresponds to filing all the space in the IC's of the SWNT ropes in the sample. This amount was determined from the "knee" in the adsorption isotherm.

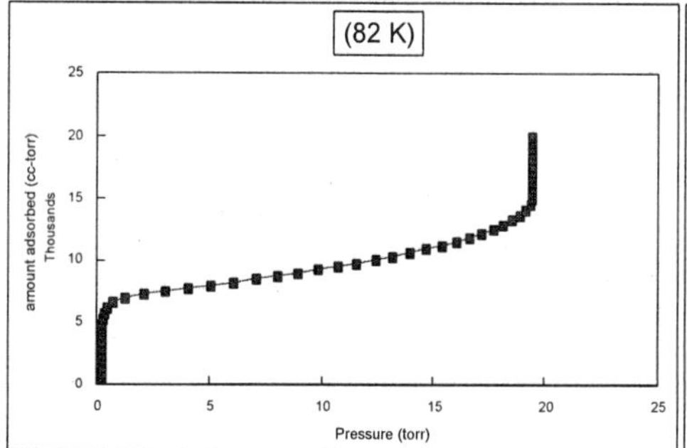

Figure 1: Full adsorption at 82 K. The amount adsorbed at low pressure corresponds to methane being adsorbed in the IC channels of the SWNT bundles. Complete filling of this pace corresponds to roughly 6000 cc-Torr. The amount adsorbed is given in cc-Torr (Y axes) and the pressure is in Torr (X axes).

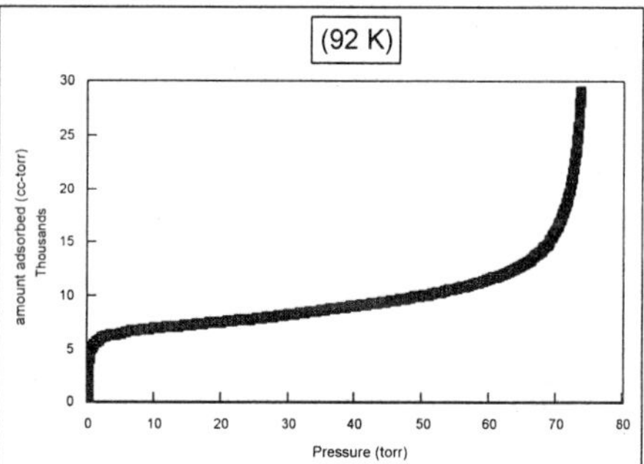

Figure 2: Full absorption isotherm at 92 K. The low pressure portion of the data is the same for Figure 1. The different behavior seen near saturation is due to the fact that this isotherm is above the bulk triple point and that of Fig. 1 is below it.

The majority of our measurements were centered in the determination of a particular temperature derivative of the adsorption data, which is known as the isosteric heat of adsorption, q_{st} . This quantity is defined as:[13]

$$q_{st} = kT^2(\partial \ln(P)/\partial T)_N \tag{1}$$

In this expression N is the 1-D density of the adsorbed gas in the interstitial channel, P is the pressure of the coexisting gas which is present in the vapor phase inside the cell, and k is Boltzmann's constant.

The thermodynamic path specified by the definition of q_{st}, i.e. constant one dimensional density, is not one which corresponds to the way the experiments are performed. The amount of gas adsorbed is measured as a function of pressure at constant temperature in each isotherm. The information needed to determine q_{st} can then be obtained from a set of adsorption isotherms at different temperatures by looking at data for the same amount of gas adsorbed at all the different temperatures.

While the isosteric heat of adsorption can be defined for any value of the one-dimensional density N, for the purpose of determining the binding energy, it is necessary to find the value of the isosteric heat at low coverages. The reason for this is that at low coverages one can assume that the adsorbate molecules on the substrate are not interacting much at all with one another, and hence that the value measured for the isosteric heat reflects only the interaction between the adsorbate and the substrate.[13] Figures 3, 4, 5, and 6 display the low coverage adsorption isotherm data measured at, respectively, 143.91, 164.82, 174.82 and 184.80 K. Note that in all the cases displayed here the amount adsorbed is under 700 cc-Torr. This amount is less than 15 % of the amount corresponding to the IC filling (which corresponds to approximately 6000 cc-Torr in figures 1 and 2).

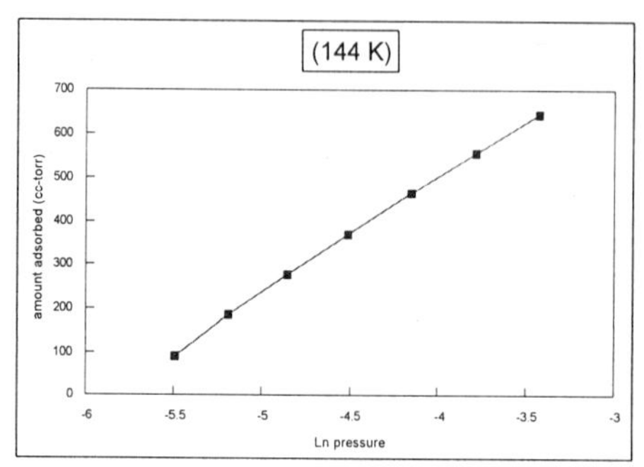

Figure 3: Adsorption measured at 143.91 K. The amount adsorbed, in cc-Torr, is plotted as a function of the natural logarithm of the pressure in Torr.

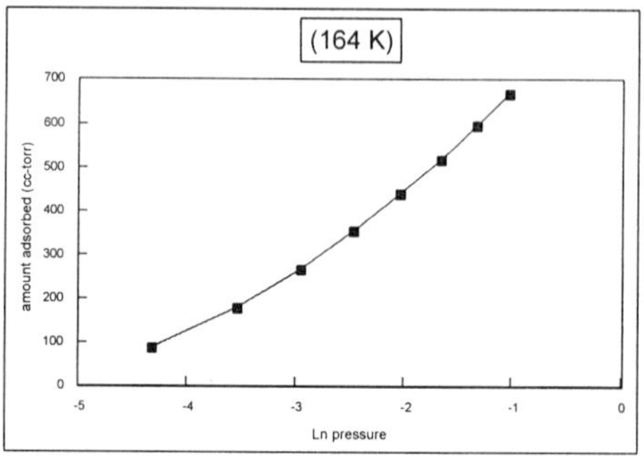

Figure 4: Adsorption isotherm measured at 164.82K. The axes are the same as those for Figure 3. In Figures 3, 4, 5 and 6, all the values of the pressures Have been corrected for thermal transpiration effects.

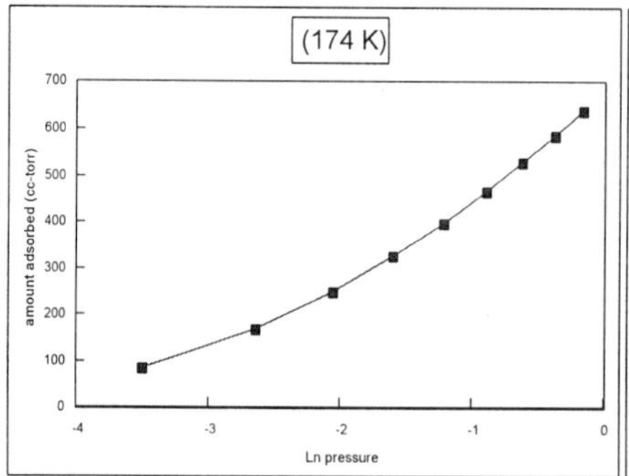

Figure 5: Adsorption isotherm measured at 174.82 K. The axes are the same as those for Figure 4. Note that in this figure, as well as in Figures 3, 4, and 6, the coverage range is much smaller (approximately 1/40) of those used in either Figure 1 or 2.

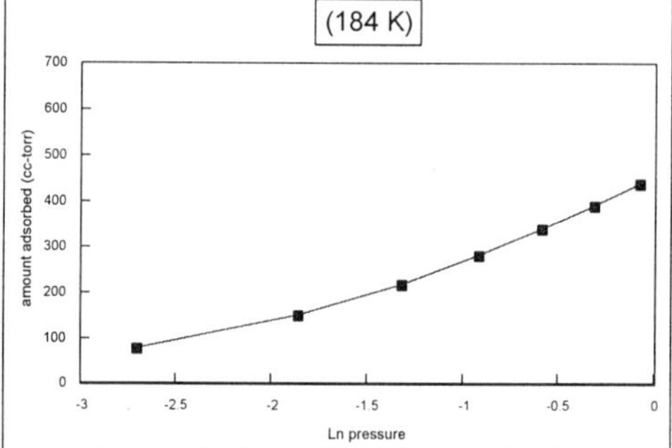

Figure 6: Adsorption isotherm measured at 184.80 K at low coverages. The axes are the same as those for Figure 5. All data in Figures 3, 4, 5, and 6 were measured using a 1 Torr capacitance pressure gauge.

In Table 1 we provide the values of the isosteric heat of adsorption at a series of different coverages calculated by using different pairs of temperatures. Isosteric heats calculated using the 143.91 K isotherm data show a systematic deviation, so they were not included in the calculations. We believe that the pressures measured at 143.91 K were too close to the lower operating limit of the gauge and were significantly affected by any slight change in the zero.

Table 1. Values of isosteric heat of adsorption (in meV) at different coverages for a given pair of temperatures.

Temperatures (K)	Coverage (cc-torr)								Average q_{st} (meV)
	100	150	200	250	300	350	400	450	
164 / 174	243.9	227.9	261.1	261.1	263.5	258.1	254.1	252.3	247.289
174 / 184	255.5	266.4	255.5	255.5	251.8	263.5	258.6	255.5	265.459
164 / 184	250.0	246.7	259.1	264.6	258.6	261.3	256.8	254.5	256.484

The data isosteric heat data calculated between the other three pairs of temperatures show considerable independence from the coverage in the range measured. This is what we expect for this quantity in the low coverage regime. Here the isosteric heat measured is dictated by the adsorbate-substrate interaction. The data shown in Table 1 are also quite substantially independent of the pair of temperatures chosen for the calculation of the isosteric heat, as should be the case for temperatures which are all relatively close to one another.

The isosteric heat of adsorption is related to the binding energy of the adsorbate to the substrate.[13] The connection between these two quantities can be most easily obtained by starting from the equality which exists at equilibrium between the chemical potentials of the one dimensional gas adsorbed in the nanotube and the three dimensional vapor present in the cell, at fixed temperature. What is obtained, after a short algebraic manipulation of the equilibrium condition is:

$$q_{st} = -\varepsilon - 2kT \tag{2}$$

Here ε is the binding energy. As in eqn. (1), k is Boltzmann's constant and T is the temperature. (The factor of 2 which appears in this expression is different from the 3/2 that one obtains when calculating the standard two-dimensional isosteric heat of adsorption.)

Equation (2) shows that the thermodynamic measurements described here provide a direct approach to measuring the binding energy. In Table 2 we display the average values of the isosteric heats of adsorption measured for the different pairs of temperatures used, as well as the corresponding values for the binding energy ε. The average value obtained from the three pairs of temperatures for the binding energy of methane on the interstitial channels between the single strands in carbon nanotube ropes is 225.8 meV. As a point of comparison, the reported value of the binding energy for methane on planar graphite is 126 meV.[14] Thus the value of the binding energy measured on the SWNT's is 80 % larger than on planar graphite.

One of the consequences of the larger binding energy for methane on the IC's of SWNT bundles is rather dramatically illustrated by comparing the estimates of the values of the coexisting 3D pressure of the vapor needed to obtain a given value for the inter-particle separation, at a fixed temperature, on the nanotubes and on planar graphite. The

Table 2. Values of binding energy calculated from the average isosteric heats of adsorption.

Temperature (K)	Average q_{st} (meV)	Binding Energy (meV)
164 / 174	247.289	217.089
174 / 184	265.459	235.259
164 / 184	256.484	226.284

expression for the 1-D density, N, in terms of the 3-D vapor density, the temperature, and the binding energy, ε, is shown in equation (3) below:[7]

$$N = n_v \lambda^2 e^{\varepsilon/kT} \quad (3)$$

In this expression, n_v is the density of the three dimensional vapor (the 3-vapor is assumed to be an ideal gas, so that $n_v = P/kT$) and λ is the de Broglie thermal wavelength. The inter-particle separation is obtained from N. One can write entirely analogous expressions for the two-dimensional density of a gas adsorbed on planar graphite.

Using equation (3), and the corresponding 2-D equation for the 2-D density[13] with, respectively $\varepsilon =$ 226 meV for methane adsorbed on the nanotubes, and $\varepsilon = 126$ meV for methane adsorbed on planar graphite[14], we find that in order to get on both substrates an interparticle separation on the order of 0.9 nm at a temperature of 165 K, it would be necessary for the pressure of the 3-D vapor to be approximately 756 Torr over the planar graphite while over the nanotubes a pressure of 2.3 Torr would suffice. That is, in order to obtain the same interparticle distance in the adsorbed phase, the pressure over the planar substrate needs to be larger by a factor of more than 300.

The larger value of the binding energy which we have measured for methane on the IC's on the SWNT bundles (when compared to the same quantity on planar graphite), and, the estimated very large increase in the pressure required to obtain the same interparticle distance in the adsorbed phase on planar graphite and on the IC's, are both in good agreement with what we expect qualitatively based on the results of the calculations done by Cole's group for ^{4}He and Ne on nanotube substrates.[7]

Conclusions

We have measured the isosteric heat of adsorption for CH_4 on the interstitial channels of SWNT bundles at low coverages for temperatures between 143.91 and 184.80 K. From these results we have determined that the binding energy of CH_4 on the IC's has a value of 226 meV. This value is 80 % larger than that found for the same adsorbate on planar graphite. We have also determined that there is, for the same value of the temperature and of the sure of the coexisting vapor, a considerable enhancement of the amount adsorbed on the nanotubes relative to the value found on planar graphite. Our results are in qualitative agreement with the results of the calculations of Cole for ^{4}He and Ne on SWNT substrates.

Acknowledgments

This research was supported by an award from Research Corporation. We have benefitted from a number of illuminating conversations with Prof. Milton W. Cole.

BIBLIOGRAPHY

1. S. Ijima, *Nature* **354**, 56 (1991).
2. G. Stan and M. W. Cole, *J. Low Temp. Phys.* **110** 539 (1998).
3. A. C. Dillon, K. M. Jones, T. A. Bekkedahl, C. H. Kiang, D. S. Bethune and M. J. Heben, *Nature* **386**, 377 (1997).
4. S. Inoue, N. Ichikuni, T. Suzuki, T. Uematsu, and K. Kaneko, *J. Phys. Chem. B (Letters)* **102**, 4690 (1998).
5. C. Nutzenadel, A. Zuttel, D. Chartouni and L. Schlapbach, *Electrochem. Sol. State Lett.* **2**, 30 (1999).
6. W. Teizer, R. B. Hallock E. Dujardin and T. W. Ebbesen, *Bull. Am. Phys. Soc.* **44**, 519 (1999); W. Teizer, R. B. Hallock, E. Dujardin and T. W. Ebbesen, *Phys. Rev. Lett.* **82**, 5305 (1999).
7. G. Stan, V. H. Crespi, M. W. Cole and M. Boninsegni, *J. Low Temp. Phys.* **113** (1998)
8. G. Stan, S. Gatica, M. Boninsegni, S. Curtarolo and M. W. Cole (submitted to *Am. J. of Phys.*).
9. G. Stan and M. W. Cole, *Surf. Sci.* **395**, 280 (1998).
10. C. Journet, W. X. Maser, P. Bernier, M. Lamy de la Chapelle, S. Lefrants, P. Deniards, R. Lee and J. E. Fischer, *Nature* **388**, 756 (1997).
11. P. Shrestha, M.T. Alkhafaji, M. M. Lukowitz, G. Yang and A. D. Migone, *Langmuir* **10**, 3244 (1994); R. A. Wolfson, L. M. Arnold, P. Shrestha and A. D. Migone, *Langmuir* **12**, 2868 (1996).
12. T. Takaishi and Y. Sensui, *Trans. Faraday Soc* **53**, 2503 (1963).
13. J. G. Dash *"Films on Solid Surfaces"* Academic Press, New York, 1975.
14. G. Vidali, G. Ihm, H-Y Kim, and M. W. Cole, *Surf. Sci Reps.* **12**, 133 (1991).

ELECTRICAL PROPERTIES OF CARBON NANOTUBES: SPECTROSCOPY, LOCALIZATION AND ELECTRICAL BREAKDOWN

Phaedon Avouris,[1] Richard Martel,[1] Hiroya Ikeda[1], Mark Hersam,[1] Herbert R. Shea,[1] and Alain Rochefort[2]

[1]IBM Research Division, T.J. Watson Research Center
Yorktown Heights, New York 10598, USA
[2]Centre de Recherche en Calcul Appliqué (CERCA)
Montréal, Québec, H3C 3J7, Canada

INTRODUCTION

Perfect carbon nanotubes (NTs) are predicted to have rather unique electrical properties. This expectation is one of the main reasons for the interest in these systems. The resistance of a perfect NT should be quantized, and carrier transport can be ballistic with little energy dissipation along the NT. Of course nanotubes, like any other material, are rarely perfect. They can have structural defects, they can be bent or twisted, or contaminated by impurities. In order to evaluate the potential of nanotubes for applications in nanoelectronics, one has to examine the properties of real materials and determine the nature of the deviations from ideal behavior. Here we summarize some of our work on the electronic structure and electrical properties of both single-walled (SWNTs) and multi-walled (MWNTs) nanotubes. Specifically, in order to obtain insight into the transport mechanism in SWNTs we prepare and perform magneto-resistance (MR) measurements on nanotube rings composed of coils of SWNTs. This configuration allows us to determine the coherence length of the carriers in the SWNTs as a function of temperature. The negative MR observed and the temperature dependence of the resistance indicate that the SWNT rings are in a state of weak localization over a wide temperature range. The dominant dephasing mechanism at low temperatures is found to involve electron-electron interactions. Strong electron-electron interactions can lead to the development of a Fermi level singularity. Indeed, we find such a singularity at low temperatures. At ~1 K, the state of the SWNTs changes from weak to strong localization characterized by activated conduction. Below ~0.7 K, a weak positive MR is observed at low magnetic fields turning into negative MR at high fields. Scattering mechanisms that may be responsible for the "anti-localization" behavior are discussed. We then explore briefly the transport and electrical breakdown characteristics of MWNTs over a large range of applied voltage. Scanning tunneling spectroscopy (STS) and transport studies are used to obtain information on the character of the carriers in semiconducting tubes. Evidence is found of not only charge

Science and Application of Nanotubes, edited by Tománek and Enbody
Kluwer Academic / Plenum Publishers, New York, 2000

transfer doping, but also for other forms of doping. Band-bending is also observed in STS for the first time. Finally, we discuss briefly the effect of structural distortions, specifically bending and twisting of the NTs, on their electronic structure and transport properties. The operation of a field effect transistor based on a distorted MWNT is demonstrated.

ELECTRICAL PROPERTIES OF SINGLE-WALLED NANOTUBES: NANOTUBE RINGS

Large diameter MWNTs are essentially two-dimensional (2D) conductors. Transport mechanisms in these systems have been examined using the technique of magneto-resistance (MR). In this technique a perpendicular magnetic field acts on multiply-connected electron (hole) paths. The conclusion of these studies has been that MWNTs are in a state of weak localization (WL).[1,2] SWNTs are very close to ideal one-dimensional (1D) systems and have attracted even more attention that MWNTs. However, in a nearly 1D conductor any multiply-connected paths produced by scattering processes will have a very small area, too small to enclose a significant magnetic flux. An artificially produced multiply connected geometry such as a nanotube ring is needed to make possible MR measurements on such 1D objects. Another obstacle in transport studies of SWNTs has often been their high contact resistance which leads to charging and Coulomb blockade phenomena at low temperatures.[3,4]

Recently, we discovered conditions under which straight SWNTs with an average diameter of 1.4 nm can be induced to coil up to form nanotube rings stabilized by van der Waals attraction between the NTs.[5] As shown in Fig. 1, our technique can produce stable rings with a narrow distribution of diameters (600-800 nm) and yields as high as 50 %. The rings are formed by treating a sample of raw SWNTs with acids and hydrogen peroxide while irradiating them with ultrasound. After filtration the NTs are dispersed in a dichloroethane solution in which they are stable.

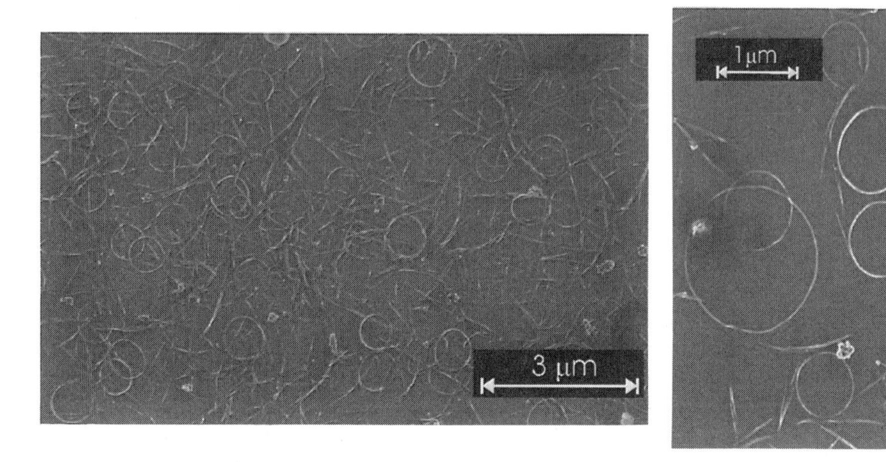

Figure 1. Scanning electron microscope (SEM) images of a ring-containing SWNT samples dispersed on hydrogen-passivated silicon substrates.

The insert in figure 2 shows an atomic force microscope image of a SWNT ring dispersed on gold electrodes so that transport measurement can be performed. Typically, the two-probe resistance of the rings ranges from 20 to 50 kΩ at 300 K. The wall thickness of the rings is usually in the range of 10 to 20 nm, consisting of a mixture of semiconducting and metallic SWNTs.

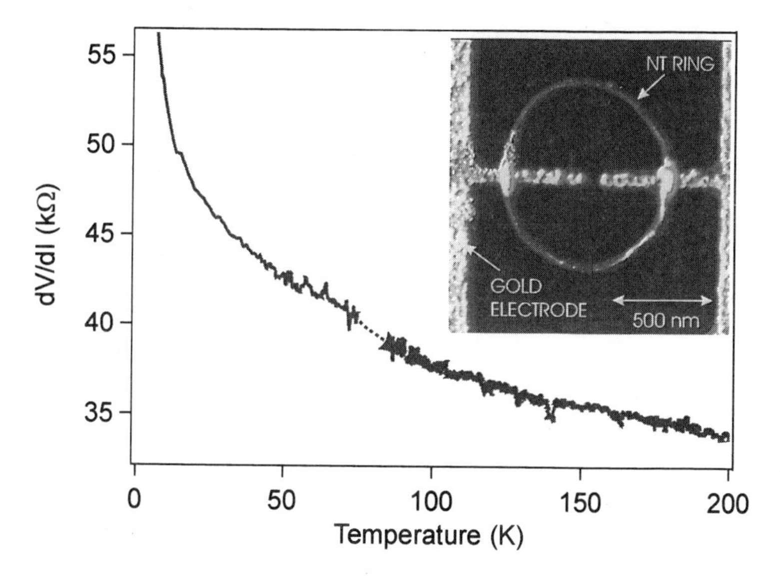

Figure 2. Temperature dependence of the resistance of a single-walled nanotube ring. The insert shows an atomic force microscope image of a ring positioned on gold electrodes.

The temperature dependence of the resistance R of a 0.82 μm diameter, 20 nm thick ring is shown in Fig. 2. The resistance is seen to increase with decreasing temperature. This behavior is suggestive of electrical transport dominated by semiconducting nanotubes. However, this possibility can be ruled out for two reasons. First, there is no effect of the back gate bias on the ring conductance, while a strong modulation is expected for semiconducting NTs.[6,7] Second, I-V curves (not shown) taken at low temperature are essentially linear and do not show a band-gap characteristic of semiconductors. At very low temperatures the conduction should be dominated by metallic nanotubes. In addition, the poor tube-tube conduction compared to conduction along the tube tends to confine electrical transport to only metallic nanotubes that are in direct contact with both gold electrodes. We can think of the ring as if it was composed of a few independent metallic SWNT coiled in parallel. Furthermore, since the ends of the nanotubes forming the ring are not covalently bonded, an electrically continuous ring is formed through tube-metal-tube coupling of the tube ends. Thus, while all metallic tubes in contact with electrodes contribute to conductance, only those, which have their overlapping ends over one of the electrodes, are responsible for the magneto-resistance (see below).

The temperature dependence of the resistance of the ring shown in Figure 2 is characteristic of a 1D metallic conductor in a state of weak localization.[8] This is confirmed by the MR results shown in Figure 3. The resistance of the ring is seen to decrease with increasing magnetic field strength. This negative MR is characteristic of conductors in the weak localization regime, and arises from the quenching of quantum interference effects. Electron waves entering the nanotube ring can move around it in a clockwise or anti-clockwise direction. When the two counter-propagating waves meet again at the origin there would interfere. Due to time-reversal invariance the two conjugate waves acquire the same phase change travelling around the ring. If the scattering processes that the carriers have undergone are all elastic then the backscattering probability would double, leading to a quantum mechanical contribution to the resistance. If the carriers undergo some inelastic collisions but the transversal time of the ring is shorter than the phase relaxation time, then the interference will be still present but weaker. An external magnetic field perpendicular to the ring removes the time reversal invariance, the two conjugate paths acquire opposite phase changes, and the constructive interference is destroyed. We can fit (solid lines) the MR curves of Fig. 3 to the predictions of the theory of 1D weak localization,[9,10] and from the fit extract the coherence length (L_φ) of the carriers. The fit is very good with some deviations appearing at high fields and the lowest temperatures. The L_φ obtained range from 300 nm at 6K to 520 nm at 3K, i.e. they are roughly 4 times smaller than the ring circumference.

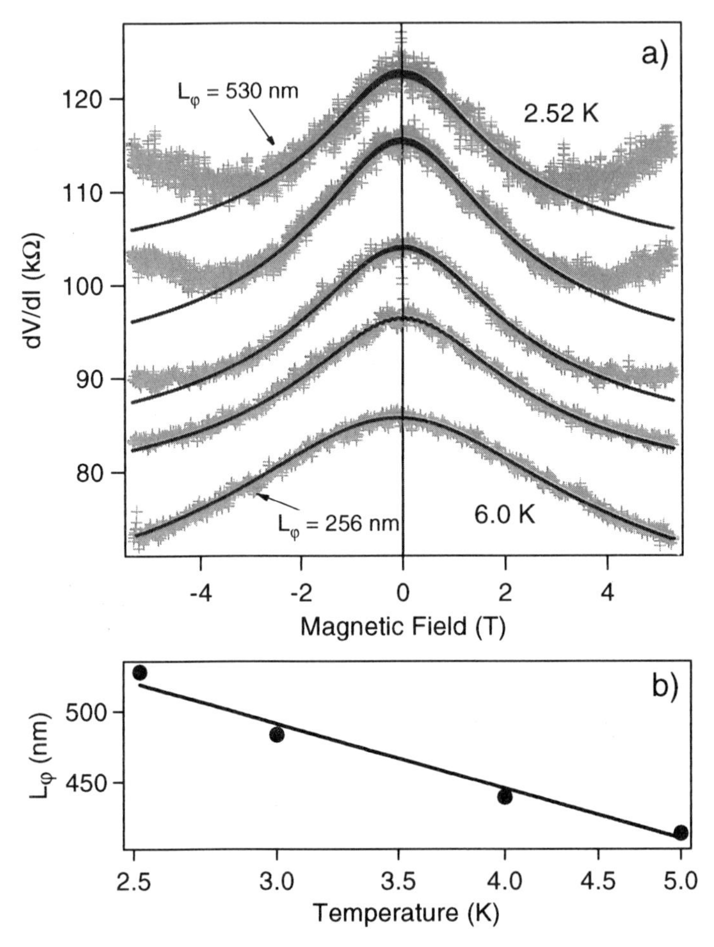

Figure 3. (a) Differential resistance, dV/dI, of a nanotube ring as a function of the strength of a magnetic field applied perpendicular to the plane of the ring. The probe current was 10 pA. The solid lines are fits to 1D weak localization theory. (b) Temperature dependence of the coherence length. The line is a fit to $L_\varphi \propto T^{-1/3}$.

These coherence lengths are larger than those obtained in MWNTs,[2,11] but clearly conduction in the SWNT ring is not ballistic, but involves weakly localized states. The coherence length varies with temperature (at the low temperature limit) as $L_\varphi \propto T^{1/3}$. This dependence is characteristic of dephasing through electron-electron interactions, specifically interactions involving a Nyquist type of electromagnetic field coupling.[12] This type of dephasing predicts a temperature dependence of the resistance of the form: $1/R=\sigma_o-C_oT^{-1/3}$, where σ_o and C_o are sample specific constants. The data of Fig. 1 in the range of 6-60 K indeed show this behavior.

As is shown in Fig. 4a, if we continue cooling the ring to T<6 K we observe a very interesting behavior. At about 1 K there is an exponential increase in resistance by almost an order of magnitude within one degree. The temperature dependence of the resistance can now be expressed as $R \propto \exp(T_o/T)$, with T=0.8 K. It is clear that the ring has undergone a transition from a weakly localized to a strongly localized state brought about by the increased L_φ. In the strong localization regime the width of the individual coherent states $\hbar/\tau_\varphi$ is smaller than the energy separation of the states Δ, so that thermal activation is necessary for transport.[13] The MR in the strong localization regime (Fig. 4b) is strong and shows fine structure due to universal conductance fluctuations (UCF). The UCF nature of this structure is verified by annealing the ring, which results in modifications of these side bands. The increased strength of UCF at low T is due to the longer L_φ.

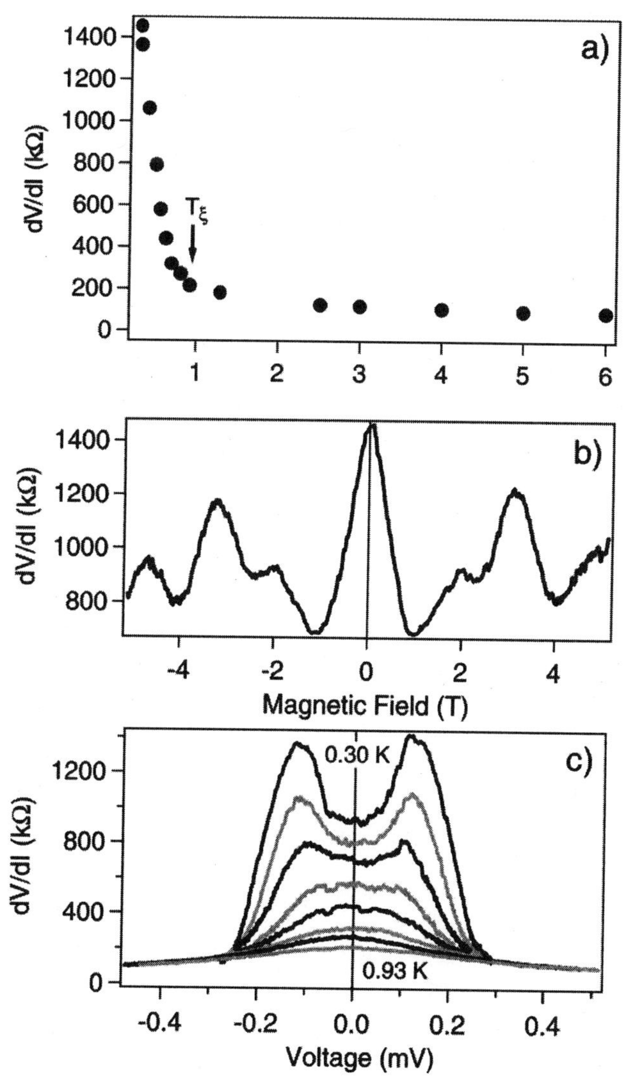

Figure 4. (a) Temperature dependence of the resistance of a SWNT ring. (b) Magneto-resistance of the ring at 0.40K. (c) Differential resistance of the ring vs. bias V, taken from top to bottom, at 0.30K, 0.40K, 0.50K, 0.56K, 0.63K, 0.70K, 0.81K, and 0.93K. The curves are not offset.

As we saw above electron-electron interactions are strong in carbon nanotubes and dominate the scattering at low temperatures. It is known that such interactions are further enhanced in the presence of disorder, which is expected to be present in real nanotubes. As Altshuler pointed out,[14] the scattered electron waves experience the same random potential and they become correlated in space. This spatial correlation then enhances the amplitude of scattering of carriers of the same energy. This can lead to the appearance of a Fermi level singularity.[14] Indeed by looking at the I-V curves of the ring we find such a singularity. Fig. 4c shows such a dV/dI vs. V plot for the ring at temperatures between 0.9 K and 0.3 K. A clear resistance peak is observed. When the temperature is decreased below about 0.7 K, in addition to an overall increase in the magnitude of the resistance peak, a local resistance minimum (or maximum in dI/dV) is observed around zero bias: a zero-bias anomaly (ZBA). Strong electron correlation is expected in carbon nanotubes as a result of disorder-enhanced correlation, or formation of a Luttinger liquid.[15] Evidence for the formation of a Luttinger phase has been produced in resent studies.[16] One way to distinguish between the two possibilities is the use of tunneling contacts, which we have tried to avoid so that we do not induce charging.

The ZBA is likely to involve scattering processes that can compete with electron-electron scattering only at temperatures < 0.7 K. The new scattering process(es) leads to a decrease in resistance. One such process is spin-orbit scattering.[17] The strength of this scattering depends strongly on the atomic number of the atoms of the material involved. Carbon is a light element so that it is unlikely that spin-orbit scattering will be strong enough. However, the heavy gold atoms of the electrodes may be responsible for the observed behavior. If the spin of the electron moving around the ring in one direction is rotated from σ_+ to $\sigma_+'=Q\sigma_+$, then the rotation in the conjugate path will be $\sigma_-'= Q^{-1}\sigma_-$. If the relative rotation is 2π then the electron waves will interfere destructively at the origin leading to a decrease of the backscattering below its statistical value, i.e. an anti-localization. Behavior very similar to our observations was reported in MR studies where a small amount of gold was deposited on thin films of light metals such as Mg.[17,8] It is interesting to note that theoretical work by Ando et al. has suggested the absence of backscattering in carbon nanotubes.[18] This work has predicted a very strong positive MR. Here we see a weak positive MR only at very low temperatures. Moreover, as we show in Fig. 5, rather weak magnetic fields, which have little effect on the resistance peak, tend to quench the positive MR.

Another process that can lead to a ZBA of the type observed here involves third-order scattering via a Kondo term in the Hamiltonian.[19,20] The anomaly results when an electron is back-scattered by the exchange interaction with a localized magnetic moment near the contact, and the reflected and transmitted waves interfere. Localized magnetic moments may be present due to the incomplete removal of the magnetic metals (Co, Fe, Ni) used as catalysts in the growth of the nanotubes. The quenching of the ZBA by a weak magnetic field supports the magnetic (spin) origin of the ZBA.

In conclusion, we have found that in SWNT rings transport is not ballistic but that weak 1D localization prevails over a wide temperature range. The maximum coherence length measured is about 0.5 μm at 3 K, longer that the coherence lengths measured in MWNTs.[1,2,11] The dominant dephasing mechanism at low temperatures was found to involve inelastic electron-electron collisions. Strong electron-electron interactions are also invoked to explain a Fermi level singularity. Below 1 K we observed a transition to a strongly localized state with thermally activated conductance. Finally, a zero bias anomaly was observed and ascribed to spin-orbit or Kondo scattering.

ELECTRICAL TRANSPORT AND BREAKDOWN IN MULTI-WALLED TUBES

An important property of carbon nanotubes is their ability to carry very high current densities. This is particularly true of multi-walled nanotubes (MWNTs). In our work we have been able to pass currents as high as 2.3 mA through an individual MWNT. Because the coupling between different shells of an MWNT is weak it is believed that essentially all of the current is carried by the outermost layer of the tube. This implies that the nanotube can withstand current densities up to 10^{13} A/m^2. This capability is sometimes considered as an indication that transport in pure MWNTs is ballistic with all power dissipation being confined to the NT-electrode contact areas.[21] However, it should be pointed out that NTs do not have grain boundaries that are critical in the electromigration-induced breakdown of

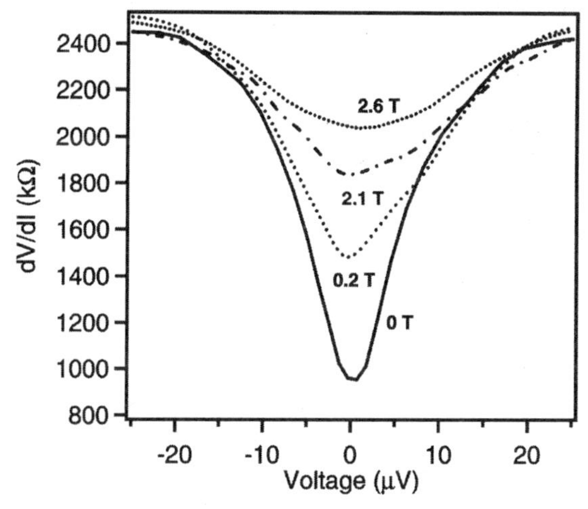

Figure 5. (a) Differential resistance vs. applied bias at different values of a magnetic field perpendicular to the ring.

metal wires, which becomes important at 10^{10} A/m^2.[22] Furthermore, NTs have good thermal conductivity and very high thermal stability.[23] To shed some light into the conduction properties and breakdown processes of MWNTs we have performed voltage stress studies on MWNTs. Here we report some preliminary findings.

In Fig. 6b we show I-V data over a wide voltage range for a 18 nm diameter MWNT deposited on top of gold electrodes. At low bias an ohmic behavior is observed, but already at 0.5 V the I-V curve becomes non-linear (super-linear). Above ~2 V, the dependence turns sub-linear indicating the onset of current saturation. This behavior is qualitatively similar to that observed in conventional conductors. It can be understood qualitatively by describing transport as involving carrier injection from one contact, transport through the tube, and finally exit of the carriers at the other contact. The tube itself has an intrinsic

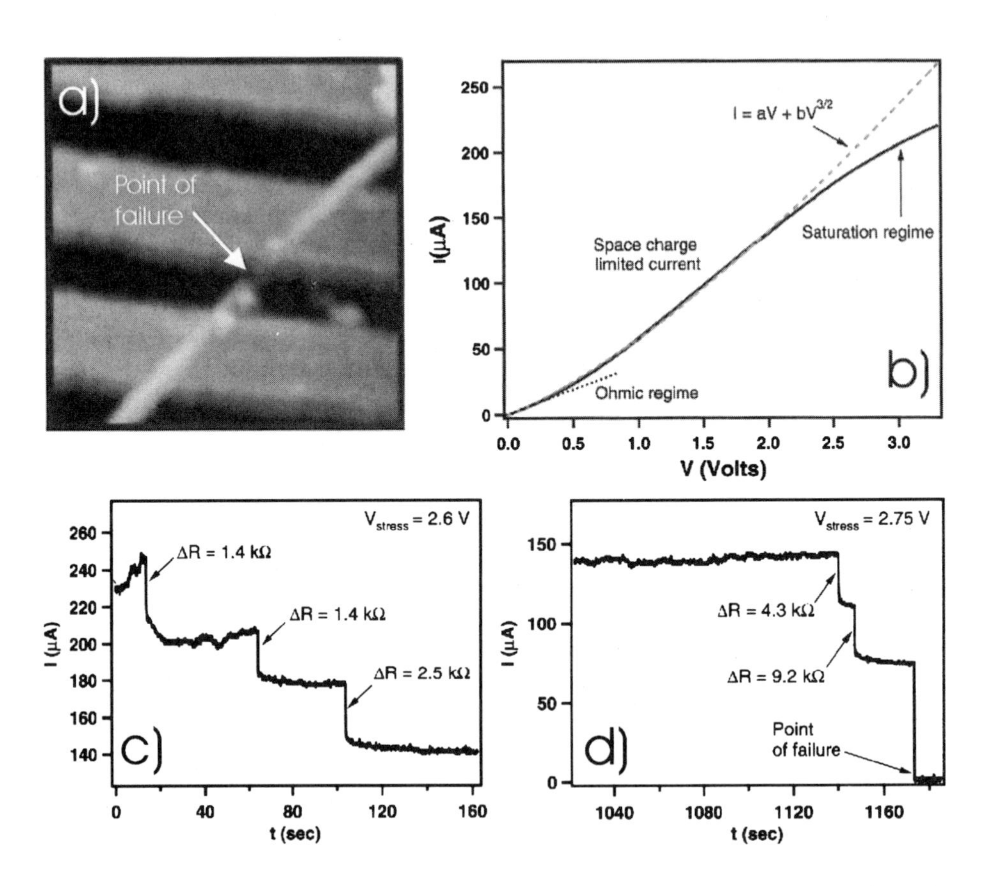

Figure 6. (a) AFM image of a MWNT taken after current-induced breakdown. (b) Current vs. applied voltage curves for a multi-walled carbon nanotube bridging two gold electrodes. (c) and (d) Time evolution of the resistance of a MWNT during current stress and breakdown.

carrier density p_o, while the contacts inject a carrier density p_i. These carriers traverse the tube in a time t_t, while the dielectric relaxation time in the tube is τ_r. A low bias an ohmic regime exists when $p_i<p_o$ and $t_t>\tau_r$. Under these conditions, the carriers injected from one contact have time to redistribute themselves and maintain charge neutrality before exiting the tube. In other words, the injected carriers replace exactly the carriers flowing out the other contact so the carrier density remains constant across the length of the tube. As the bias is increased, the transit time is decreased to the point where $t_t<\tau_r$, while, at the same time, the injection rate increases. Under these conditions, the injected carriers can dominate the intrinsic carriers, and their short transit time does not allow their relaxation by the thermally produced intrinsic carriers. As a result, a space charge develops in the tube that limits current flow. This space charge leads to non-ohmic (super-linear) behavior with the exact dependence being determined primarily by the geometry of the contact. For point-like injection contacts it has been shown that the dependence is $J\propto V^{3/2}$.[24] The dashed line in Fig. 6b is a fit of the data to the expression $J=aV+ bV^{3/2}$. As can be seen from Fig. 6b, a good fit is obtained supporting the notion that the observed non-linearity is due to space charge-limited transport. By increasing the applied voltage, the carrier energy is increased and so are the scattering processes (primarily electron/hole-phonon scattering) that control carrier mobility. The resulting decrease in mobility can be responsible for the onset of the saturation regime seen in Fig. 6b above ~2 V. While the situation may be more complex that the above description implies, nevertheless we believe that the above simple model provides a valid zeroth order description of the two-probe transport in the system. Although four probe measurements are needed before a definite claim can be made, it appears that classical effects such as charge-injection and space charge formation dominate any quantum effects[21] present in MWNTs.

Eventually, the nanotubes break down when stressed at high bias for a sufficiently long time (see 6d). Depending on the particular tube and the quality of the contacts, breakdown occurs when the current is in the space charge limited or saturated regimes. As shown in Figs. 6c,d, the breakdown process usually involves stepwise increases in the resistance. Quite often the resistance steps are of the same magnitude, suggesting that multiple defects of the same kind are generated sequentially. In some cases, steps corresponding to a decrease in resistance are observed, suggesting either the annealing of defects (or contacts), or an enhanced contribution from sub-surface carbon shells.

ELECTRICAL PROPERTIES OF SEMICONDUCTING NANOTUBES

In our discussion up to this point we concentrated on the properties of metallic nanotubes. We will now consider some aspects of the electrical properties of semiconducting SWNTs. An important question involves the nature of the carriers, that is whether the tubes are intrinsic or externally doped. In the work of Tans et al.[6] and Martel et al.[7] in which field-effect transistors (FETs) were constructed using semiconducting SWNTs as their channel, it was found that the FETs behaved as p-type devices. A simple explanation for the p-character of the NTs involves electron transfer from the nanotube (graphite work function ~4.5 eV) to the Au (work function ~5.3 eV) or Pt (work function ~5.6 eV) electrodes leaving the NTs hole rich. Further support for this picture came from scanning tunneling spectra (STS) of nanotubes on Au which showed that the Fermi level is located at the top of the valence band.[25] "Charge-transfer doping" was also observed in recent first-principles calculations involving carbon atom chains instead of NTs.[26] We have also examined NTs on non-metallic surfaces. Specifically we have studied the STS of individual SWNTs on silicon (100) surfaces passivated by hydrogen, i.e. Si(100)-(2x1):H. The latter were prepared by exposing Si(100)-(2x1) to atomic H produced by thermal dissociation of H_2 by a hot W filament. The nanotubes were deposited from solution on this surface; the sample was immediately transferred to a UHV environment where the spectra were obtained. As shown in Fig. 7a, NTs were found to be only lightly p-doped on this passivated Si surface, which has a workfunction slightly higher than graphite.

While such observations tend to suggest that the NTs themselves are intrinsic and become p-doped as a result of their contact with the metal electrodes, we have also observed many deviations from this behavior. First, by depleting the carriers of p-type NTs with a gate in a MOSFET configuration we found carrier densities of ~1 hole per 250 C-atoms; a value that is too high to be accounted for by only charge transfer doping.[7] In some cases, the STS spectra of NTs showed an intrinsic behavior (see Fig. 7b) despite the fact that they were deposited on Au surfaces. Even more surprising was finding n-type NTs on an Au surface as shown in Fig. 8a. The STM image indicates that this NT is a chiral tube, and tunneling spectra (Fig. 8b) were recorded at the points marked by X across its length, from a "bulk" position (spectrum A) to near the nanotube cap (spectrum F). Spectrum A clearly shows that the tube is n-doped. However, interestingly no contamination or adsorbates are

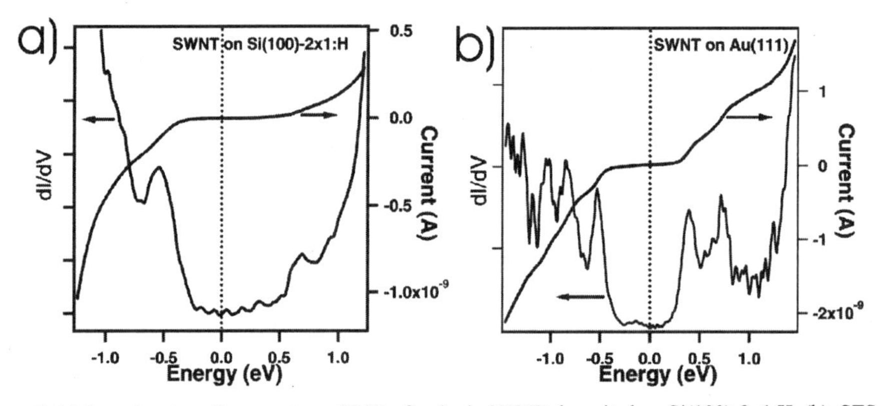

Figure 7. (a) Scanning tunneling spectrum (STS) of a single SWNT deposited on Si(100)-2x1:H. (b) STS data of an individual SWNT on a Au(111) surface.

observed on the surface of the tube in the STM image. This suggests that either substitutional impurities or species trapped inside the tube are responsible for the doping. Most importantly, Fig. 8b shows that the Fermi level position shifts gradually towards the center of the gap as the location of the spectrum moves closer to the cap; a clear band-bending behavior. The shifted bands are plotted in Fig. 8c, and their curvature indicates that the cap of the tube is negatively charged. The curved geometry of the cap requires the presence of pentagon rings. According to Euler's theorem, six pentagons suffice to form a closed cap of an NT. These pentagons introduce localized donor and acceptor states. Their energies depend on the particular topology of the cap;[27] the STM image of a cap with different geometry is shown in Fig. 8d. The cap in Fig. 8a appears similar to the model cap (#1d) whose spectra were calculated by de Vita et al..[27] This structure was found to have a deep acceptor state (~2.7 eV below E_F) which will be filled by electrons from the valence band of the NT leading to local charging, and possibly to the observed band-bending. We have attempted to obtain STS of the apex of the cap for tubes on Au, but our efforts were

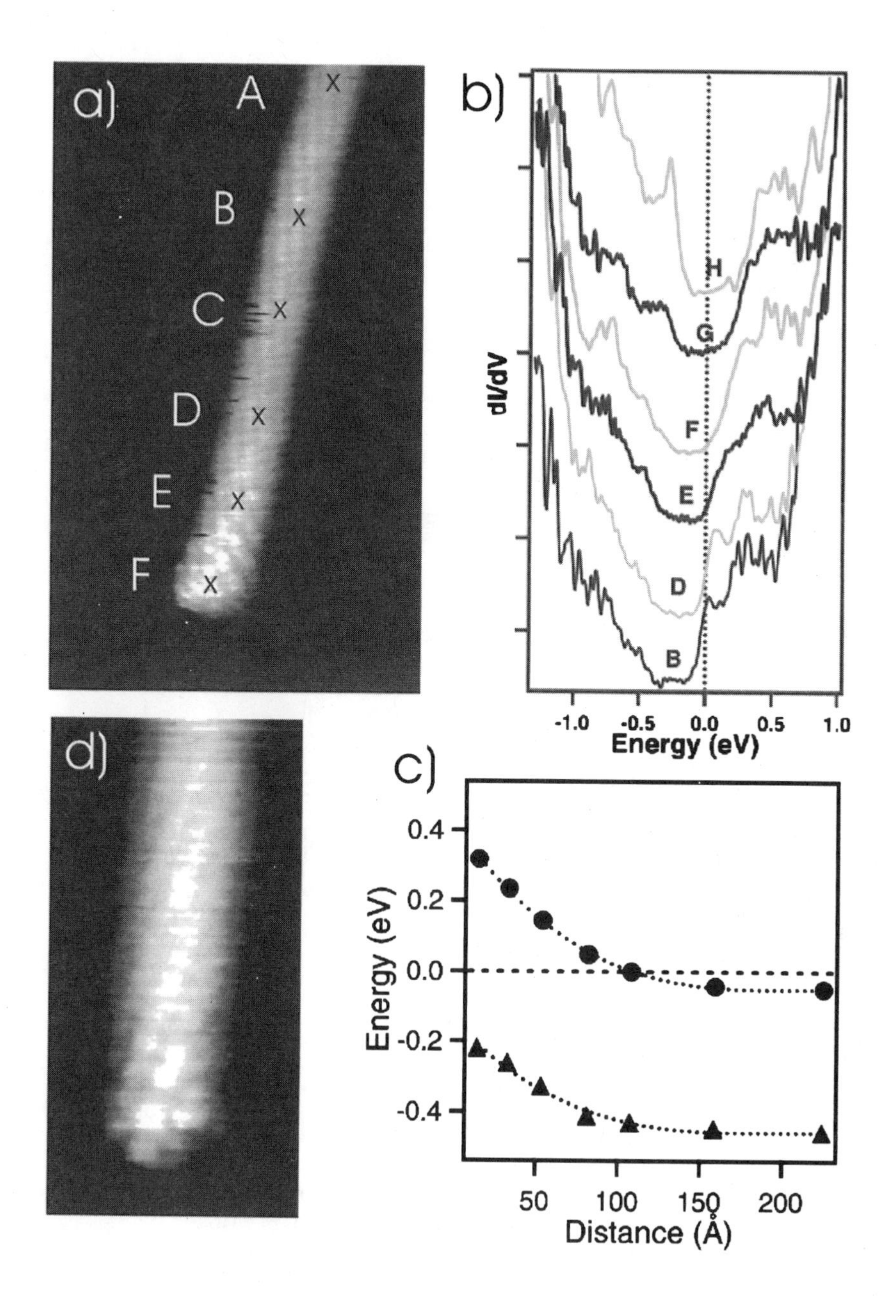

Figure 8. Band-bending in semiconducting nanotubes. (a) Scanning tunneling microscope (STM) image of a single chiral nanotube on Au(111). (b) Scanning tunneling spectra (STS) obtained at the locations indicated on the image in (a). (c) STM image of a chiral tube with a different structure cap. (d) Plot of the bending of the valence and conduction bands of the nanotube based on the ST spectra shown in (a).

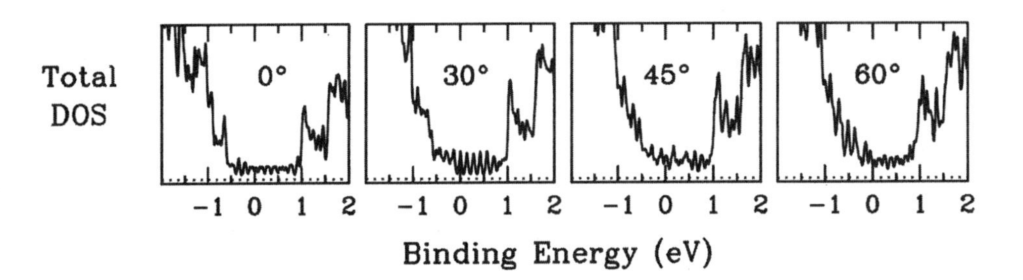

Figure 9. Variation of the total density of states (DOS) near the Fermi level for a (6,6) armchair nanotube as a function of the bending angle.

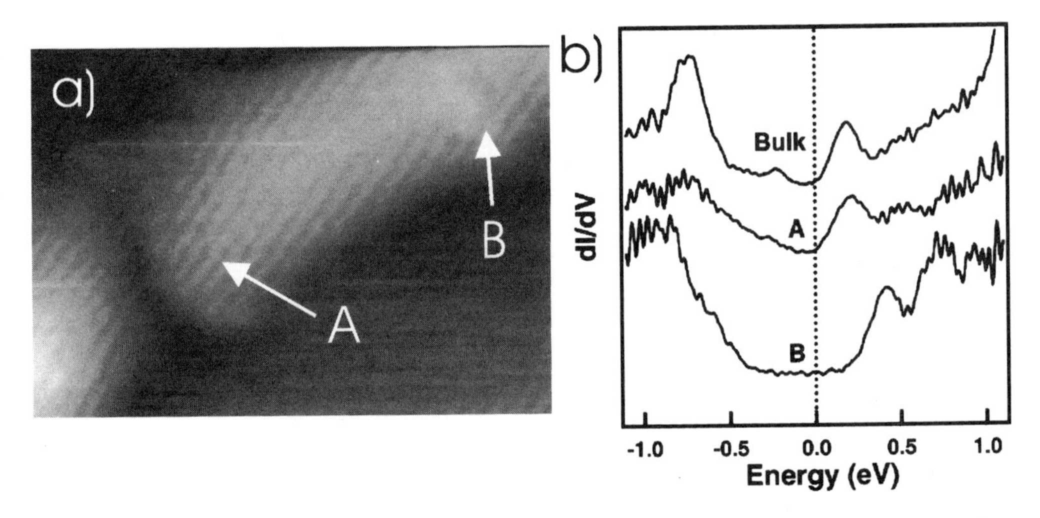

Figure 10. (a) STM image of a kink and a defect site B on a SWNT deposited on Au(111). (b) Scanning Tunneling Spectra (STS) obtained at the corresponding locations on the STM image.

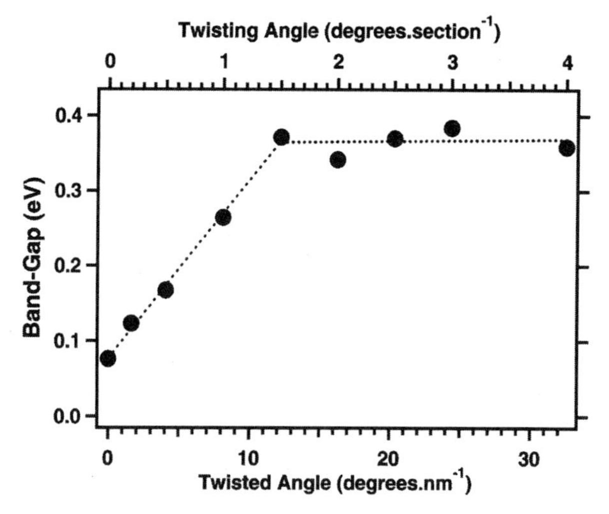

Figure 11. Band-gap opening upon twisting an armchair (6,6) metallic nanotube.

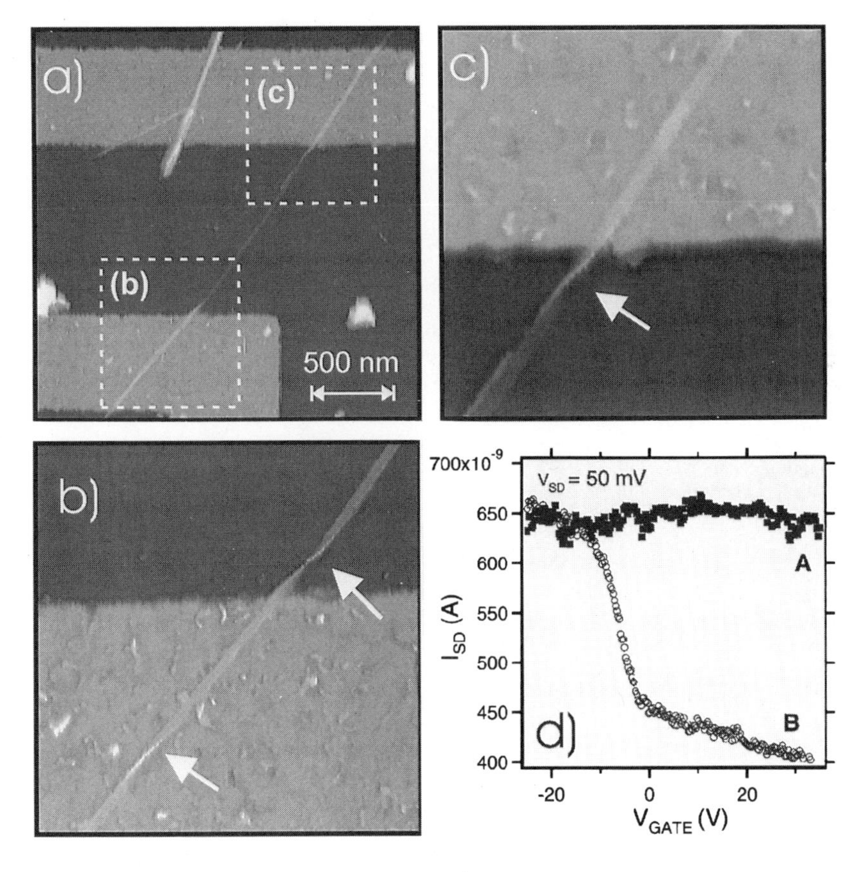

Figure 12. Field effect transistor based on a collapsed and twisted MWNT. (a) and (b) Non-contact AFM image of the MWNT bridging two gold electrodes. The MWNT is twisted in 3 locations as shown by the arrows in (b) and (c). (d) Source-drain current (I_{SD}) vs. gate bias (V_{GATE}) curves of a typical undistorted MWNT (curve A) in comparison with that of the collapsed MWNT of similar diameter (curve B).

frustrated because of strong tunneling from the substrate. On the hydrogen passivated Si surface, however, symmetric-looking caps gave a peak at ~0.25 eV above E_F.

We will now explore briefly the electrical properties of distorted NTs. In past theoretical work we have explored two types of distortions of the structure of an NT: bending[28,29] and twisting.[29] We found that bending the NT leads to the mixing of σ and π states as a result of the increased curvature. The resulting re-hybridized σ–π states have reduced conductivity. The effect of bending is weak at small bending angles, but becomes increasingly more important for bending angles above 45°, when kinks start appearing in the structure of the tube. As is evident from the total density of states plot of a bent (6,6) tube shown in Fig. 9, this mixing leads to an increased DOS on the valence band side near E_F. The effect is even more pronounced in plots of the local density of states.[28] In Fig. 10 we show STS spectra of a kinked semiconductor NT. Spectra in the increased curvature region (A) near the kink show an increased DOS similar to that predicted by the calculations on armchair tubes.[29] A wider gap is observed at the site of a structural defect (B).

Twisting a nanotube introduces drastic changes in its electronic structure.[29,30] In Fig. 11 we display the results of extended-Hückel calculations on the effect of twisting on the electronic structure of a metallic (6,6) armchair tube.[29] We find that upon twisting a band-gap opens up whose magnitude scales linearly with the twist angle for small angles, thus turning the metallic NT to a semiconducting one.[29] Twisting above a certain critical angle (Θ_T=10° nm^{-1} for a (6,6) tube) leads to a collapsed NT structure similar to that of a twisted ribbon.[29] This twisted ribbon structure is also a semiconductor. We found possible experimental evidence for this behavior in our studies of MWNTs. Because of the fact that the band-gap of an NT scales inversely with its diameter,[31] large diameter (>10 nm) MWNTs behave as metals at room temperature. Indeed, as is shown in Fig. 12b using MWNTs as channels in a MOSFET arrangement we found that the gate has little effect on the source-drain current. However, when the collapsed and twisted MWNT shown in Fig. 12a, b, c was used as the FET channel, it behaved as a semiconductor and the gate was able to modulate the source-drain current[7] (Fig. 12d).

REFERENCES

1. L. Langer, V. Bayot, E. Grivei, and J.-P. Issi, Phys. Rev. Lett. 76:479 (1996).
2. A. Bachtold, C. Strunk, J.-P.Salvetat, L. Forró, T. Nussbaumer, and C. Schönenberger, Nature(London) 397:673 (1999).
3. S.J. Tans et al., Nature(London) 386:474 (1997).
4. M. Bockrath et al., Science 275:1922 (1997).
5. R. Martel, H. R. Shea, and Ph. Avouris, Nature(London) 398:299 (1999).
6. S.J. Tans, A.R.M. Vershueren, and C. Dekker, Nature(London) 393:49 (1998).
7. R. Martel, T. Schmidt, H.R. Shea, T. Hertel, and Ph. Avouris, Appl. Phys. Lett. 73:2447 (1998).
8. For reviews see G. Bergmann, Phys. Rep., 107:1 (1984); P. A. Lee, and T.V. Ramakrishnan, Rev. Mod. Phys. 57:287 (1985).
9. A.G. Aronov, Y.V. Sharvin, Rev. Mod. Phys. 59:755 (1987).
10. B.L. Al'tshuler, A.G. Aronov, and B.Z. Spivak, JETP Lett. 33:94 (1981).
11. A. Bachtold, C. Strunk, C. Shönenberger, J.-P. Salvetat, and L. Forró, Proceedings of the 12th IWEPNM AIP, New-York (1998).
12. B.L. Altshuler, A.G. Aronov, and D.E. Khmelnitsky, J. Phys. C 12:7367 (1982).
13. D. J. Thouless, Phys. Rev. Lett. 39, 1167 (1977).
14. B.L. Al'tshuler and A.G. Aronov, Sov. Phys. JETP 50:968 (1979).
15. J. Voit, Rep. Prog. Phys. 57:977 (1994).
16. M. Bockrath et al., Nature(London) 397:598 (1999).
17. G. Bergmann, Phys. Rev. Lett. 17:95 (1982).
18. T. Ando, T. Nakanishi, and R. Saito, J. Phys. Soc. Jpn. 67:2857 (1998).
19. J.A. Appelbaum, Phys. Rev. Lett. 17:91 (1966).
20. P.W. Anderson, Phys. Rev. Lett. 17:95 (1966).
21. S. Frank, P. Poucharal, Z.L. Wang, and W.A. de Heer, Science 280:1744 (1998).

22. For a review see, A. Christou, *Electromigration and Electronic Device Degradation*, Wiley, New-York (1994).
23. J. Hone et al., Phys. Rev. Lett. 80:1042 (1998).
24. M.A. Lampert, A. Many, and P. Mark, Phys. Rev. 135:A1444 (1964).
25. J.W.G. Wildöer, L.C. Venema, A.G. Rinzler, R.E. Smalley, and C. Dekker, Nature(London) 391:59 (1998).
26. N.D. Lang and Ph. Avouris, submitted to Phys. Rev. Letters.
27. A. De Vita, J.-C. Charlier, X. Blasé, and R. Car, Appl. Phys. A 68:283 (1999).
28. A. Rochefort, D.R. Salahub, and Ph. Avouris, Chem. Phys. Lett. 297:45 (1998).
29. A. Rochefort, Ph. Avouris, F. Lesage, and D.R. Salahub, to appear in Phys. Rev. B (1999).
30. C.L. Kane and E.J. Mele, Phys. Rev. Lett. 78:1932 (1997).
31. J.W. Mintmire and C.T. White, Carbon 33:893 (1995).

FIELD EMISSION OF CARBON NANOTUBES FROM VARIOUS TIP STRUCTURES

Jisoon Ihm and Seungwu Han

Department of Physics and Center for Theoretical Physics
Seoul National University
Seoul 151-742, Korea
E-mail: nsjisoon@phya.snu.ac.kr

INTRODUCTION

From the beginning of its discovery, the carbon nanotube (CNT) has been regarded as an ideal material to make field emitters because of its unusually high aspect ratio as well as mechanical and chemical stability. Many experiments with singlewall[1,2] and multiwall[3–7] carbon nanotubes have demonstrated a relatively low threshold voltage of the electron emission with little sample degradation. Recently, flat panel displays fabricated with the CNT's as emitters are demonstrated[8]. Furthermore, the advancements in fabrication technology make it possible to generate a self-aligned or a patterned CNT on a glass[9] or a silicon substrate[10], implying that a commercial production of the CNT-based field emission display (FED) may be possible in near future.

In spite of the accumulating experimental data, realistic quantum mechanical calculations on the field emission of the CNT are still scarce. The simplistic view of the CNT as a jellium metallic emitter based on the conventional Fowler-Nordheim theory could not explain the entire range of the I-V data[1,7] and a critical analysis of the electronic structure on a more fundamental level is required. In this study, we perform *ab initio* pseudopotential electronic structure calculations[11,12] with or without applied fields. We focus on the end geometries of the CNT where the electron emission actually occurs and try to find the optimized tip structure for field emission. Some of the results presented here have been submitted for publication elsewhere[13].

MODEL SYSTEM AND COMPUTATIONAL METHOD

We consider various end geometries of the CNT. The side views of the upper part of the (5,5) CNT's are shown in Fig. 1. The bottom (not shown) of the tube body is terminated with hydrogen atoms for computational convenience. We call (a)-(d) as H(hydrogen-attached)-, FC(flat-cut)-, CAP(capped)-, and SC(slant-cut)-CNT, respectively. The H-CNT represents a situation where the dangling bonds are chemically stabilized by ambient hydrogen. In the CAP-CNT model, the tube end is capped with a patch (hemisphere) of C_{60} having six pentagons. In the FC-CNT and SC-CNT, on the other hand, dangling bonds remain unsaturated. The model systems are periodi-

Science and Application of Nanotubes, edited by Tománek and Enbody
Kluwer Academic / Plenum Publishers, New York, 2000

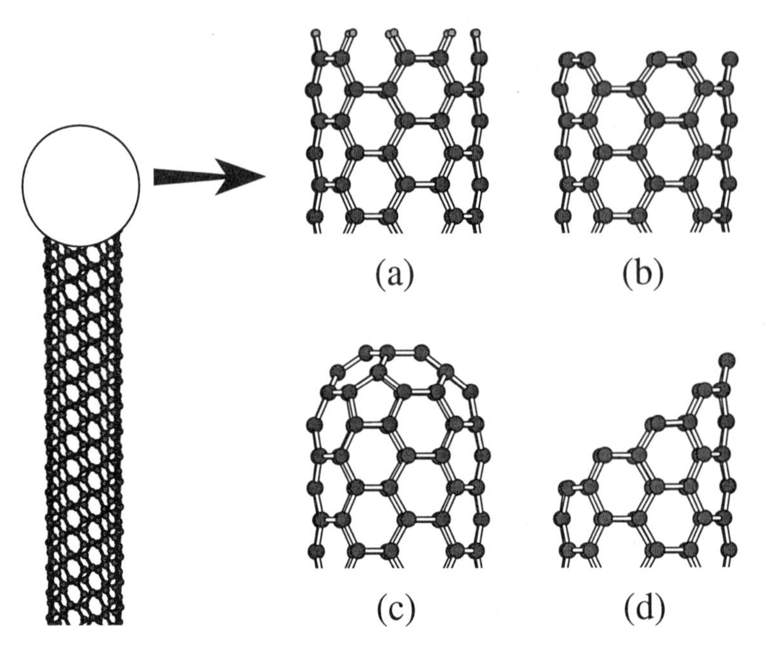

Figure 1: Edge structures studied in this work. (a) H-CNT, (b) FC-CNT, (c) CAP-CNT, and (d) SC-CNT (see text for abbreviations).

cally repeated in a tetragonal supercell with the tube axis lying in the z-direction. In order to make the simulation as realistic as possible, various computational schemes are adopted in this study. The plane-wave based calculations have been done mostly for a $\sim 20\mathring{A} \times 20\mathring{A} \times 30\mathring{A}$ unit supercell. To check the convergence of our results for larger systems, we have also used localized basis functions[14] in the *ab initio* pseudopotential calculations and have been able to enlarge the tube size up to four times. The base grid of 60 Ry. is used for the projection of the orbital and wave function. Relaxed geometries are obtained from computationally less demanding tight-binding calculations[15].

ELECTRONIC STRUCTURE OF MODEL SYSTEMS

In Fig. 2, the total density of states (TDOS) around the Fermi level is plotted in solid lines. All systems except the H-CNT have localized states near the nanotube edge. To investigate these localized states more closely, the local density of states (LDOS) around the end region is calculated and shown in dashed lines in Fig. 2(b) and (c). Energy levels of the localized states induced by the topological defects (pentagons) in the CAP-CNT are known to depend on the relative positions between the defects[16–18] and the pentagons are assumed to be separated from each other here as found in Ref. 17. In Fig. 2(d), the LDOS of the SC-CNT at the two topmost atoms (the sharp edge) and that at the eight side atoms on the slant-cut cross section are shown in dotted and dashed lines, respectively. Descriptions about different characters of these two will be given in the following section. In the SC-CNT, localized π states predicted for the graphite ribbons[19] are also found, but they are less prominent than the σ-bond-derived states and unimportant to field emission.

In Fig. 3(a), we show one of the localized states in the FC-CNT whose energy level is indicated by an arrow in Fig. 2(b). Interestingly, two adjacent dangling bonds

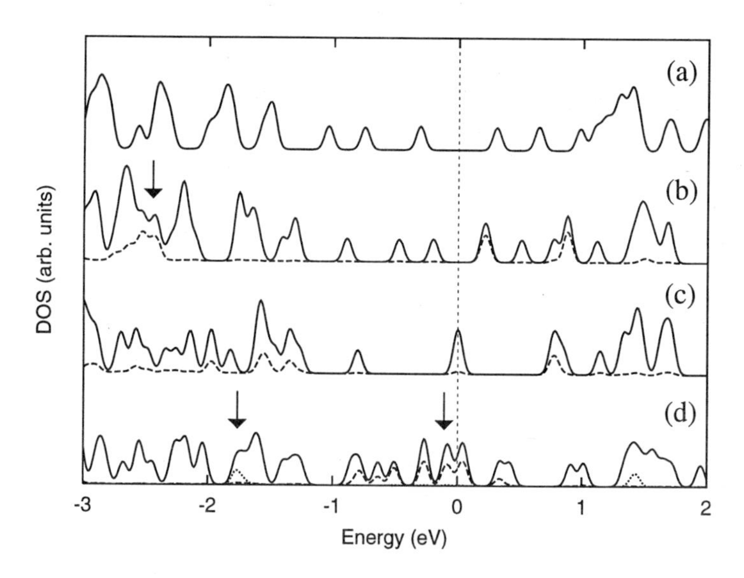

Figure 2: TDOS (solid line) and LDOS (dashed or dotted lines) corresponding to each structure [(a)-(d)] in Fig. 1. The Fermi level is set to zero. The LDOS represents the states localized at the top of the nanotubes in Fig. 1. In (d), the dashed and dotted lines are differentiated as explained in the text.

join at the backbond position of the original sp^2 configuration. The character of the s-orbital is slightly lost in the sp^2 hybridization. Because of these additional bonds, the C-C bond length in the top layer is 1.27$\AA$ compared with the ideal C-C distance in a perfect nanotube (1.41$\AA$), corresponding to a strong double bond. Two types of localized states are identified in the SC-CNT model. In Fig. 3(b), the topmost atoms on the armchair-like edge (states at -1.8 eV indicated by an arrow in Fig. 2(d)) show a similar backbonding as in the FC-CNT, while the individual dangling bonds on the side edge (indicated by an arrow at -0.1 eV in Fig. 2(d)) remain unpaired as shown in Fig. 3(c) since they are far apart from each other along the zigzag edge. Since the interaction between them is weak, the energy splitting is small, resulting in a large LDOS at the Fermi level as indicated in dashed line in Fig. 2(d). On the other hand, the energy splitting between the bonding and antibonding states is rather significant ($\sim$3 eV) for the states localized at the armchair edge as shown in two separate dotted curves.

EFFECT OF EXTERNAL FIELDS

A sawtooth-type potential is applied along the CNT axis to simulate a uniform external electric field in the supercell geometry. In the experimental situation, the applied field strength is usually a few V/μm and the length of the CNT is a few μm, resulting in $\sim$ 10 V difference in the applied potential between the tube ends. It has been well established that the field emission typically occurs when the actual field at the tip is $\sim$ 0.5 V/$\AA$, corresponding to a field enhancement factor of $\sim$ 1000 in the present experimental situation. Since the system size is severely limited in the *ab*

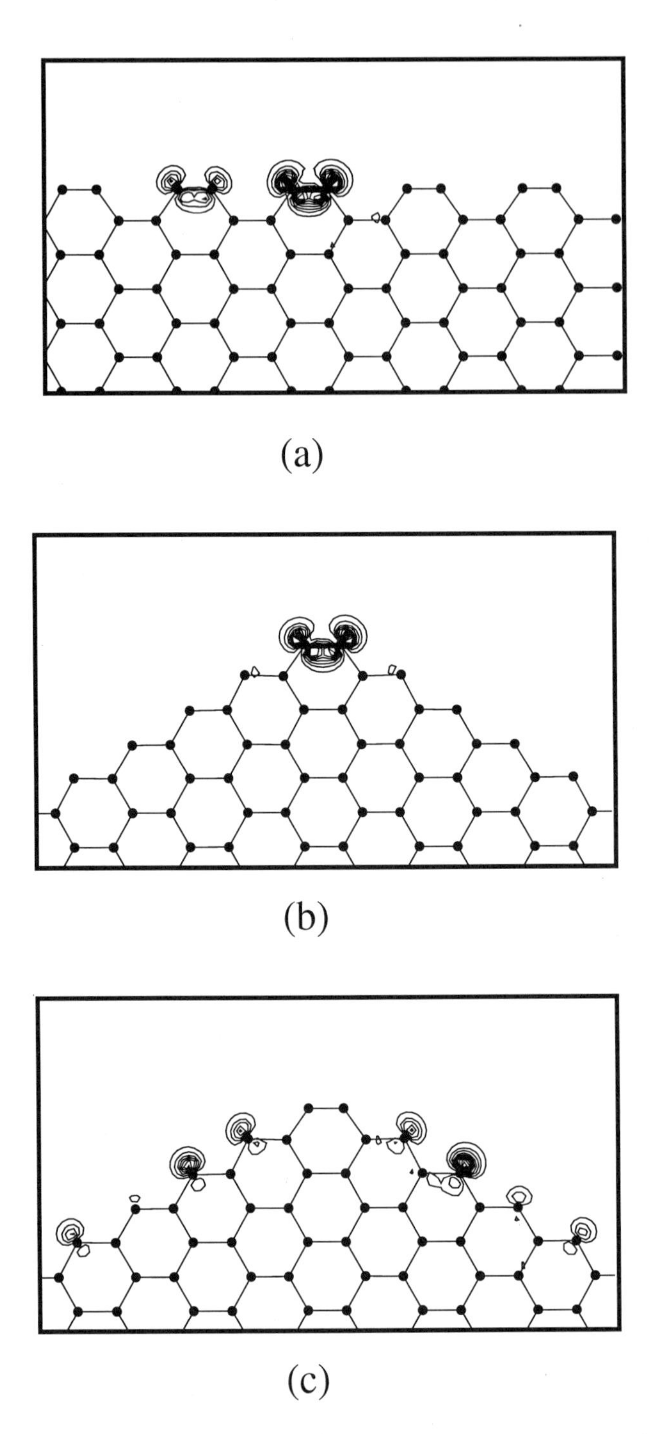

Figure 3: Electron density of the localized states whose energy levels are indicated by arrows in Fig. 2(b) and (d). (a) is the localized state of the FC-CNT at −2.6 eV and (b) and (c) are the localized states of the SC-CNT at −1.8 eV and −0.1 eV, respectively. The electron density is evaluated on the nanotube surfaces and the contour interval is 0.08 $e/\mathring{A}^3$.

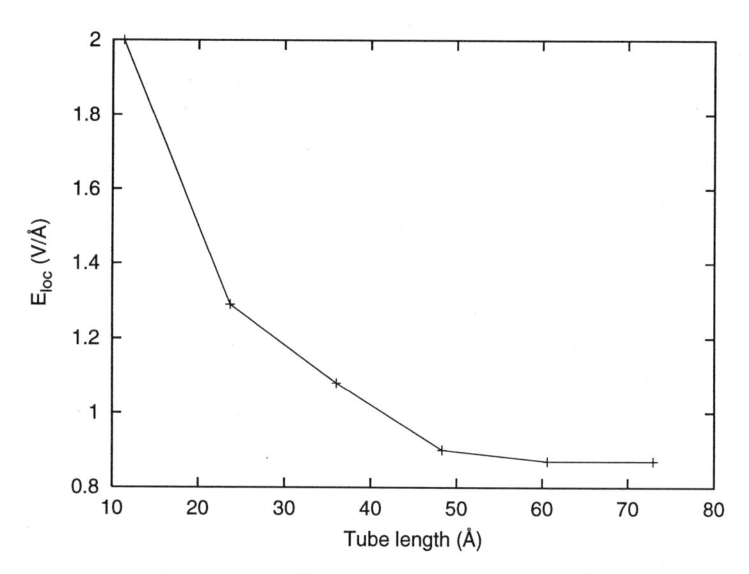

Figure 4: Averaged local field within 3 $\mathring{A}$ outside the tip as a function of the tube length for a CAP-CNT. The applied field is varied so that its product with the tube length is fixed at 10 V.

initio calculation, it is a serious issue whether one can mimic the realistic situation and extract meaningful information on the field emission through the *ab initio* calculation for small-size tubes. After many different calculations, we find the following trends as shown in Fig. 4. We apply a uniform field across the supercell such that the potential difference across the tube (i.e., between the two tube ends) is fixed at 10 V for each case. For instance, for 23.7, 48.3, and 72.9 $\mathring{A}$ tubes, the applied field strengths are 0.42, 0.21, and 0.14 V/$\mathring{A}$, respectively. The resulting total (applied + induced) field at (to be precise, just outside) the tip of the tube is presented in the figure, showing the saturation behavior of the total field when the tube length is beyond $\sim 60\mathring{A}$. We also find that the total field inside the tube is practically zero (see Fig. 5), indicating that the screening of the applied field inside the tube is very effective. Since the discontinuity in the electric field is equal to the surface charge density, the above behavior implies the convergence (saturation) of the induced electronic charge at the tip for the tube length $\gtrsim 60\mathring{A}$. (Even for a tube as short as 36 $\mathring{A}$, the deviation from the converged value is within $\sim$ 25 %.) This behavior is useful for extrapolating to the realistic long-tube-small-field case which is hardly accessible with *ab initio* methods. Another test we have made is that, for a given tube geometry (35 $\mathring{A}$-long FC-CNT), we increase the applied field and obtain the corresponding local field just outside the tip. A linear relation is found as long as the applied field is $\gtrsim$ 0.1 V/$\mathring{A}$. Figure 6 shows such a relation and it may be fit with $\mathrm{E}_{tot} = 4.53\ \mathrm{E}_{appl} - 0.19$ (V/$\mathring{A}$). The shift (-0.19) is due to the fact that the screening by localized states is not initiated for a small field and the linear increase in the induced field sets off only when the localized states begin to fill. Such a linear relation is again useful for the extrapolation or interpolation purpose. Combining these two results, we can deduce a scaling rule as follows. Since the shift (-0.19 in the above case) is generally small, we can approximately express E_{tot} as

$$\mathrm{E}_{tot}(l, \mathrm{E}_{appl}) = a(l) \cdot \mathrm{E}_{appl}, \tag{1}$$

where l is the tube length. On the other hand, we already know that, for a given condition $\mathrm{E}_{appl} \cdot l = c_1$(constant), $\mathrm{E}_{tot}(l, \mathrm{E}_{appl}) \rightarrow c_2$(constant) as l gets large. Therefore,

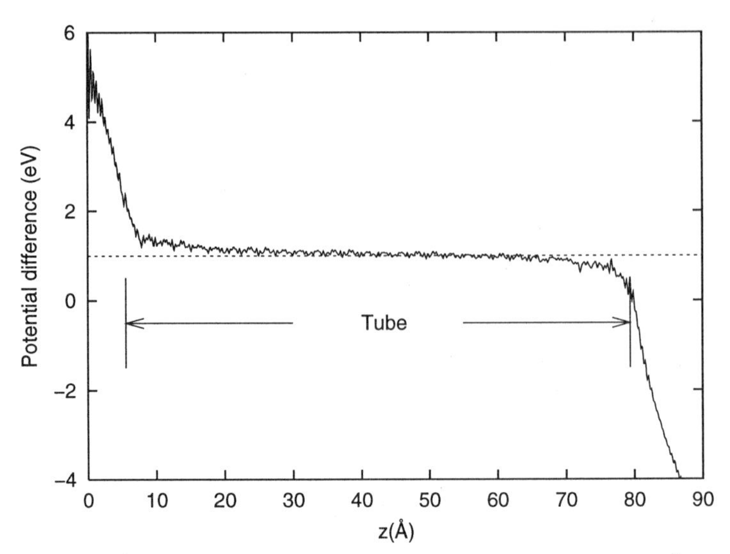

Figure 5: Difference in the total potential when the external field is applied to a 70Å-long CAP-CNT. The horizontal dashed line is a guide to an almost constant potential inside the tube. The origin of the xy-plane is set to the position of a carbon atom in the top pentagon.

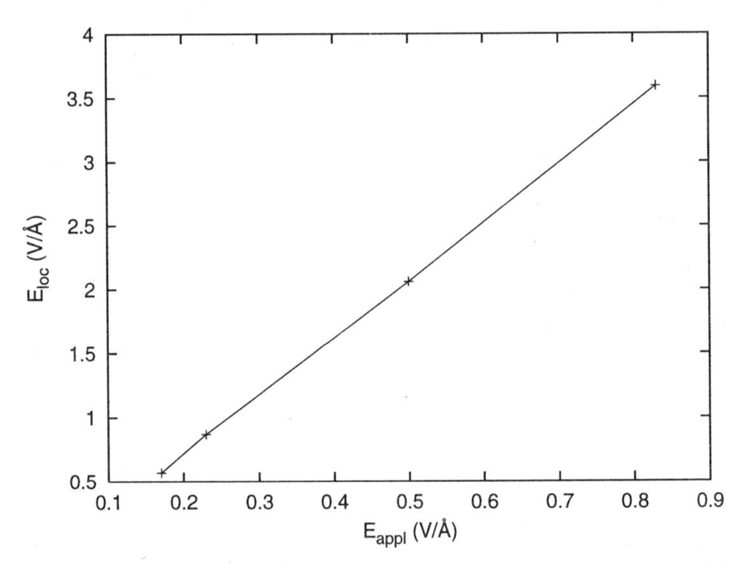

Figure 6: Averaged local field within 3 Å outside the tip as a function of the applied field for a 35 Å-long FC-CNT. The data are well fit with a straight line.

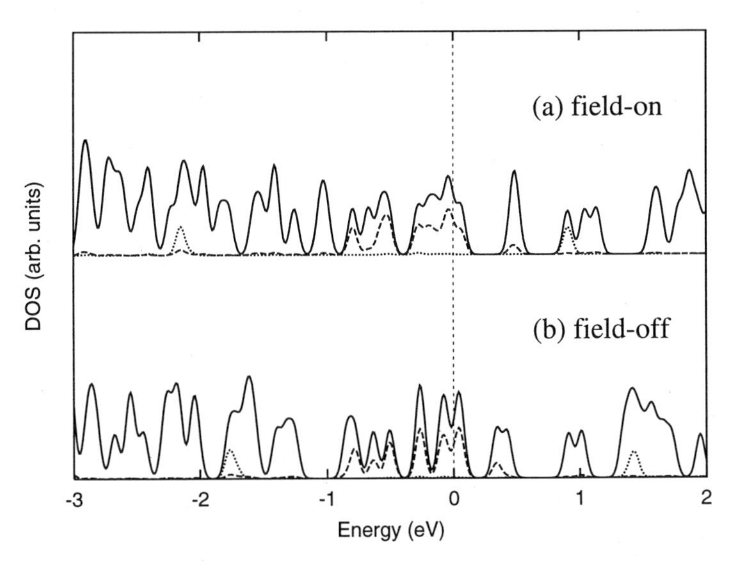

Figure 7: Change in the DOS of the SC-CNT under the applied field ~0.4 V/Å. Solid lines are the total DOS. Dashed and dotted lines are the local DOS and they are differentiated as in Fig. 2(d).

with $E_{appl} = \frac{c_1}{l}$,

$$E_{tot}(l, \frac{c_1}{l}) = a(l) \cdot \frac{c_1}{l} = c_2, \tag{2}$$

for large l. Then we have $a(l) = \frac{c_2}{c_1} l$, namely, $a(l)$ is proportional to l, $E_{tot} = \alpha l \cdot E_{appl}$. Using results of independent calculations as well, we obtain that, for a given applied field, the field enhancement factor $\eta(= E_{tot}/E_{appl})$ at the edge scales linearly with l,

$$\eta \cong \alpha l \quad . \tag{3}$$

For instance, α of a SC-CNT with the diameter of 7Å is approximately 0.17 (with l given in Å) as will be described below. Equation (1) is distinguished from the result for a conventional metallic prolate spheroid, $\eta \sim m^2/\ln(2m) - 1$ for large m, where m is the aspect ratio, which can be deduced from the depolarization factor[20] of a perfect conductor.

Now we examine the change in the electronic structure by the field in detail. In Fig. 7(a) shown above, the DOS for the SC-CNT under the applied field (~ 0.4 V/Å) is plotted. Comparing it with Fig. 7(b) (the same as Fig. 2(d)) for the zero field case, we notice that the localized levels at the armchair edge (dotted curves) undergo a downward shift. The states localized on the zigzag chain sides of the edge are partially occupied for the zero field in Fig. 7(b) (dashed curves) and their occupation increases as the field is turned on. Such a situation is commonly observed in the field emission of metallic tips with atomic adsorbates[21]. Since these states are located at the Fermi level under the applied field, they are the first to be emitted across the barrier to the anode. The increase in the occupation of the dangling bond states is approximately $0.1e$ per edge atom. The downward shift of the localized states by the field occurs in the FC-CNT and CAP-CNT as well. (See Fig. 8 for the case of the CAP-CNT.) Shifts

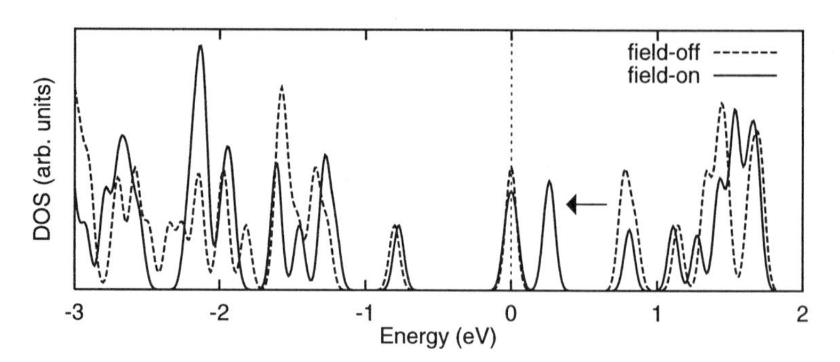

Figure 8: Shift of the localized levels in CAP-CNT. Note that other π states are not much changed.

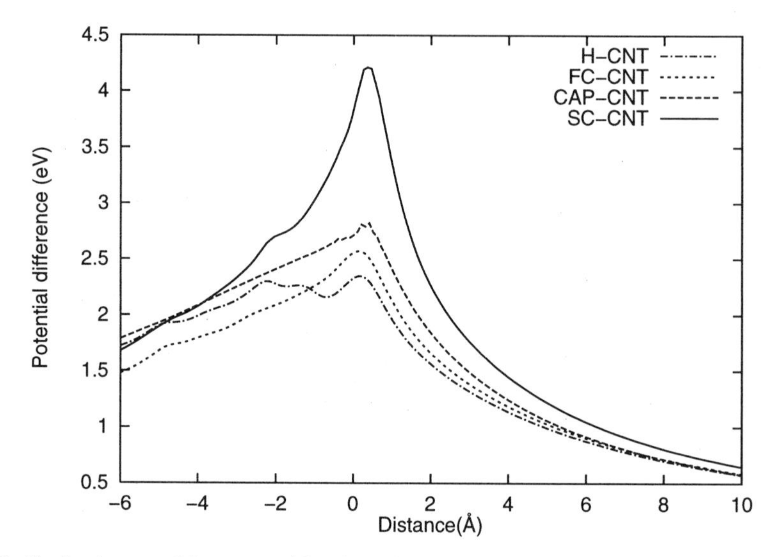

Figure 9: Coulomb potential generated by the induced charge and plotted in the tube axis direction when the external field of 0.17 V/$\AA$ is applied for $\sim$ 35 $\AA$-long tubes. The distance is measured from the edge of the tube.

of the localized states are pinned as they begin to be occupied, hence these localized states stay at the Fermi level for a wide range of the field strength. The occupation of the localized states has an effect of repulsing the π-electrons and suppresses the increase in the amplitude of the π states at the edge. Actually, the sum of the accumulated charge does not vary too much from structure to structure.

LOCAL FIELD ENHANCEMENT

The accumulated charge at the end of the CNT as described above enhances the local electric field, which is the driving force of the field emission. For a quantitative analysis, we calculate the *change* in the charge density under the applied field of 0.17 V/$\AA$ using the *ab initio* method with localized orbitals and plot the resulting Coulomb potential for an isolated tube along the tube axis in Fig. 9. The down slope on the left of the peak is essentially the negative of the applied field which screens out the

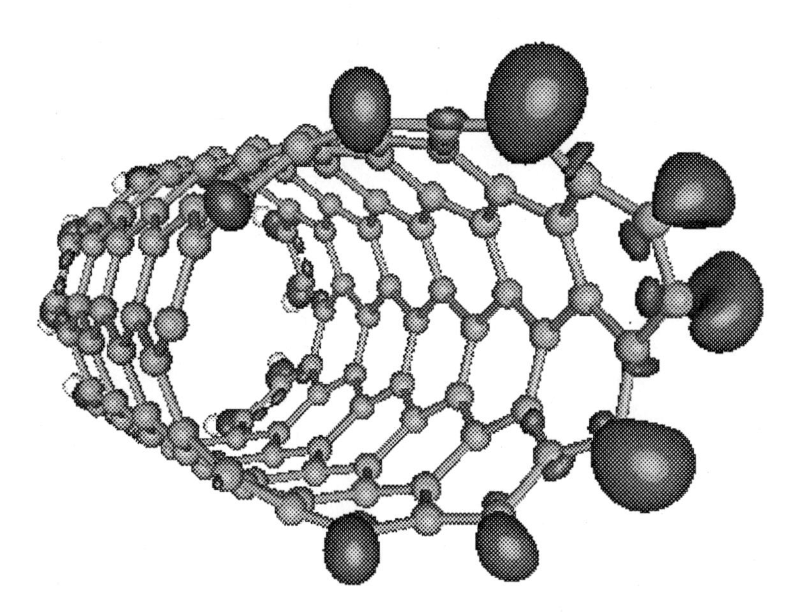

Figure 10: Perspective view of the constant density (0.01 $e/\AA^3$) surface showing the density *change* in the SC-CNT under the external field of 0.4 V/Å. The σ-bond-derived character is clearly visible.

applied field inside the tube. The down slope on the right of the peak, on the other hand, is the induced field that reduces the potential barrier when electrons are emitted across the vacuum barrier to the anode side. The difference between the H-CNT and other systems can be considered as contributions from the localized states inherent to the tip structures in Fig. 1(b)-(d).

A striking feature in Fig. 9 is the prominent potential peak and the corresponding high field enhancement of the SC-CNT among the model systems. The average induced fields within 3 Å from the peak position to the right (vacuum region) are 0.53, 0.58, 0.65, and 1.03 V/Å for (a)-(d) geometries, respectively. The aforementioned extrapolations to a realistic 2 μm-long tube lead to the field enhancement factors of 1700, 1900, 2100, and 3300, respectively, which compare favorably with the experimental values of 3600[2] or 1300[4]. A greater enhancement factor of the SC-CNT than the CAP-CNT is consistent with experiment[5] considering that the multiwall open CNT used in experiment should have some unsaturated dangling bonds like the SC-CNT studied here. One might expect that the highest induced field in the SC-CNT should originate from its sharpest armchair edge (hence the highest field enhancement factor according to classical electrostatics) among the studied structures. However, when we probe the induced electric field very closely all over the space, we find that the region around the zigzag side edge has a field at least as large as that of the topmost atoms. To identify the origin of the high field unambiguously, we have performed the same calculation for a flat-cut (12,0) zigzag-type nanotube. We find a sharp Coulomb potential peak identical to the SC-CNT. We conclude that the unpaired dangling-bond states at the Fermi level under the applied field, which exist both in the SC-(n,n) CNT and the FC-(3n,0) CNT, are mainly responsible for the highest potential peak, i.e., the largest induced field toward the anode side. Therefore, the most favorable tip geometry for the field emission in our study is the open tube with the zigzag-type edge where unsaturated dangling bond states can exist. In Fig. 10, we show an isodensity surface of the induced charge accumulated at the end of the SC-CNT in the presence of the applied field of 0.4 V/Å. The plot exhibits dangling bond states as expected from the above analysis. We also emphasize that the lobes of these dangling bond wave functions are directed toward the anode side and their overlap integral with the tail of the free-electron-like state from the anode side of the barrier is relatively large (meaning a large emission probability), while the π-electron states on the tube body lie perpendicular to the emission direction and the corresponding overlap integral is much smaller.

It is interesting that the CAP-CNT has the next highest peak in Fig. 9, higher than that of the FC-CNT. The occupation of the π states in the hemispherical region of the CAP-CNT easily increases as the external field is turned on and contributes to the induced field efficiently because they are at the Fermi level. These π-electrons are the first to be emitted under the applied field for the CAP-CNT. Since the π-electron at the top of the hemisphere points to the anode side, the overlap integral mentioned above is reasonably large in this case. We also note that, although the H-CNT has a relatively small induced Coulomb potential, its magnitude is still appreciable in Fig. 9. The accumulated charge here is the usual metallic π states of the tube whose amplitude has increased at the edge to screen the applied field as in typical metallic samples. Our results show that both the localized states and the metallic π states can contribute to the field enhancement. To see the contributions of the two separately in the SC-CNT, we attach H atoms to the dangling bonds in the SC-CNT (not shown). We find that the passivation of the dangling bonds quenches the field enhancement to the level of the H-CNT.

We have also studied the effects of the image potential for the H-CNT and CAP-CNT by artificially putting an electron represented by $1/r$ potential to a point in the hypothetical tunneling trajectory. The image potential is calculated from the change in the self-consistent potential at that point. The result is shown in Fig. 11 and the data set obtained in this way are fitted with the classical form of the image potential for an infinite metal plane, $-\alpha\frac{e}{4(z-z_0)}$. The fitted α and z_0 are shown in Table 1. Since the geometry of the CNT is cylindrical rather than flat, the distance between the image charge generated on the CNT and the displaced electron would be effectively greater than those in the flat metal surface. The deeper potential for the CAP-CNT comes from the existence of the cap perpendicular to the z-axis where large amount of positive charge could gather close to the displaced electron. Another physical quantity crucial to the current profile in the field emission is the work function of the nanotube. We find that the calculated work function of the isolated CNT ($\gtrsim$ 4 eV, which is not fully converged in our finite unit cell calculation, but still reasonably close to experimental values of 4.5 - 5.3 eV[2,23,24]) is insensitive to the tip geometry, giving small variations within 0.1 eV.

Field Emitter Array

Another interesting issue we pursue here is the dependence of the field enhancement factor on the separation and packing of the nanotubes. We have already presented the results for an isolated nanotube and now show other cases. We have performed the same electronic structure calculations with the rope (infinite hexagonal array of nanotubes in the xy plane separated by 3.4 Å from each other with a finite length in z direction). The self-consistent electronic structures with and without the applied elec-

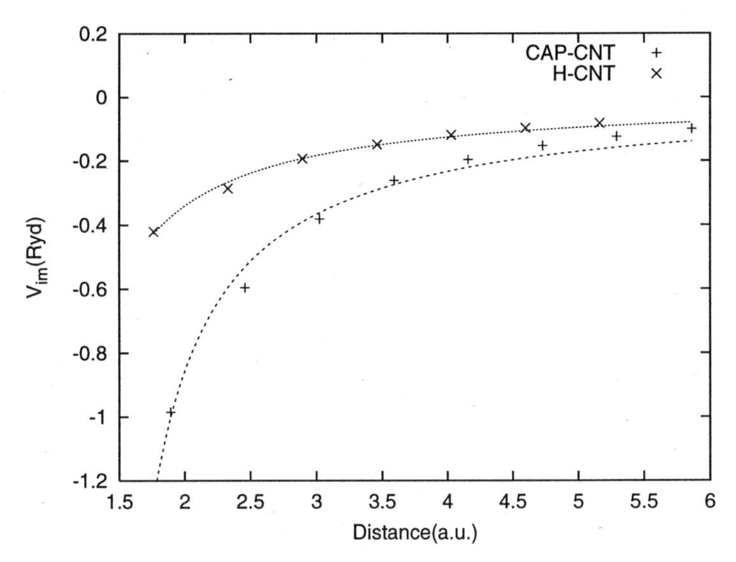

Figure 11: Image potential for the H-CNT and CAP-CNT. The origin of the z-axis is set to the position of the outermost atom for each system. The curves are the best fit of the function $-\alpha e/4(z - z_0)$ to the data points.

Table 1: The fitted parameters of the curves in Fig. 11.

	α	z_0
H-CNT	0.64	1.25
CAP-CNT	0.4	0.82
metal plane	1	0.0

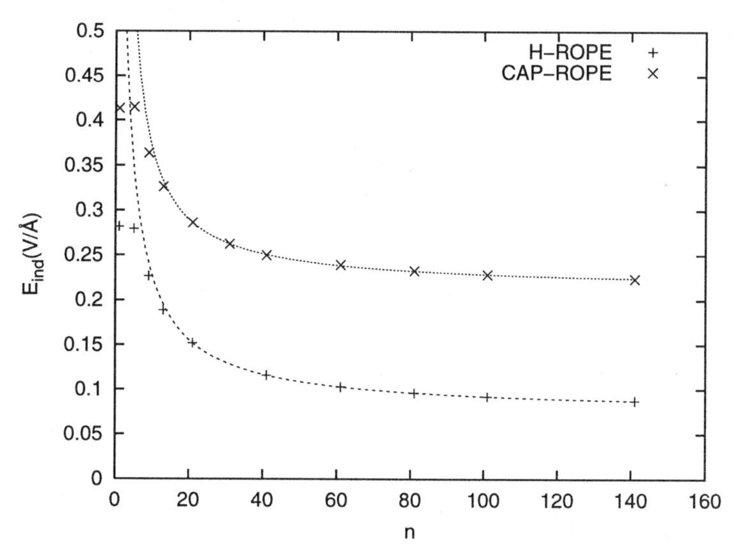

Figure 12: The induced field at the edge of the central tube in a rope. The abscissa is the number of close-packed tubes along a side of square that contribute to the field of the central tube. The asymptotic behavior is nicely fitted with the curve $a/(n - b) + c$.

tric field are calculated and the induced charge is obtained. Then we calculate the field at the tube edge produced by an isolated tube with this induced charge, that produced by a (3×3) array of tubes (at the edge of the central tube), that by a (5×5) array of tubes, etc. Figure 12 shows the field obtained in this way and this plot implies two crucial effects. First, as the number of close-packed nanotubes increases, the field decreases. This is expected because the larger the packed area is, the more similar to the planar geometry (which does not enhance the field) is the configuration. The effective curvature of the tube at the edge is substantially reduced when the tubes form a rope. The converged value of E_{ind} in the large n limit here is only about a quarter of the field produced by a tube in the isolated configuration studied above. Secondly, the induced charge per tube in the rope configuration is also significantly diminished compared with that in the isolated case. The field for $n = 1$ in Fig. 12 is only about half of the corresponding field produced by the isolated tube under the some applied field. Therefore it is desirable to deposit (or grow) nanotubes far apart from each other for efficient field emission. The enhancement is effective if the separation between tubes is beyond a few nanometers. Another desirable, yet difficult thing to achieve is to have singlewall nanotubes rather than multiwall tubes. It is also found that the ratio of E_{ind} between CAP-CNT and H-CNT is significantly enlarged from that in the isolated situation, which implies that the effects of localized states and the shape of the end structure become more important in the close-packed configuration.

DISCUSSION

In summary, we have studied the electronic structure of various edge shapes of a singlewall nanotube with or without external fields. A linear scaling behavior of the field enhancement factor as a function of the tube length has been obtained. Open tubes with dangling bonds at the zigzag-type edge are predicted to be the most favorable for the field emission among the structures considered here. The capped nanotubes with π-bonding states localized at the cap region are the next favorable structure. In a real situation, it is likely that hydrogen atoms or other ambient atoms and molecules cap the reactive dangling bonds of the open nanotubes. However, the attached atoms or molecules may be desorbed by an initial clean-up procedure before emission or by a strong local field during the emission process. Fortunately, the strong sp^2 bond of the tube body is not dissociated at all under the typical field ($\lesssim 1$ V/Å) in experiment. For multiwall CNT cases, the saturated (3-fold) coordination of all edge atoms by bridging carbon atoms cannot occur in general because of the the incommensurability between neighboring coaxial tubes, and there should remain some unsaturated dangling bonds contributing actively to the field emission[22]. Existing experimental data do not yet give information on the dependence of the emission efficiency on the tip structure. Our study proposes a high-efficiency edge structure which can significantly improve the field emission performance. We also note that, for the practical display application, it is enough for only a small fraction of nanotubes to actually emit electrons because of huge redundancy in the total number of nanotube emitters.

Acknowledgment. - The study on the carbon nanotube field emitter array was supported by the Samsung Electronics. This work was supported by the BK21 Project of the MOE, the SRC Program of the KOSEF, and the BSRI Program of the KRF.

REFERENCES

1. Y. Saito, K. Hamaguchi, T. Nishino, K. Hata, K. Tohji, A. Kasuya, and Y. Nishina, Jpn. J. Appl. Phys. **36**, L1340 (1997).
2. J.-M. Bonard, J.-P. Salvetat, T. Stockli, W. A. de Heer, Appl. Phys. Lett. **73**, 918 (1998).
3. A. G. Rinzler, J. H. Hafner, P. Nikolaev, L. Lou, S. G. Kim, D. Tomanek, P. Nordlander, D. T. Colbert, and R. E. Smalley, Science **269**, 1550 (1995)
4. W. A. de Heer, A. Chatelain, and D. Ugarte, Science **270**, 1179 (1995).
5. Y. Saito, K. Hamaguchi, S. Uemura, K. Uchida, Y. Tasaka, F. Ikazaki, M. Yumura, A. Kasuya, and Y. Nishina, Appl. Phys. A **67**, 95 (1998).
6. Q. H. Wang, T. D. Corrigan, J. Y. Dai, and R. P. H. Chang, Appl. Phys. Lett. **70** 3308 (1997).
7. P. G. Collins and A. Zettl, Appl. Phys. Lett. **69**, 1969 (1996); P. G. Collins and A. Zettl, Phys. Rev. B **55**, 9391 (1997).
8. Q. H. Wang, A. A. Setlur, J. M. Lauerhaas, J. Y. Dai, E. W. Seelig, and R. P. H. Chang, Appl. Phys. Lett. **72**, 2912 (1998); Y. Saito, S. Uemura, and K. Hamaguchi, Jpn. J. Appl. Phys. **37**, L346 (1998); W. B. Choi, D. S. Chung, S. H. Park, and J. M. Kim, in Proceedings of the MRS Spring Meeting, San Francisco (1999) (unpublished).
9. Z. F. Ren, Z. P. Huang, J. W. Xu, J. H. Wang, P. Bush, M. P. Siegal, and P. N. Provencio, Science **282**, 1105 (1998); S. Fan, M. G. Chapline, N. R. Franklin, T. W. Tombler, A. M. Cassell, and H. Dai, *ibid.* **283**, 512 (1999).
10. S. Fan, M. G. Chapline, N. R. Franklin, T. W. Tombler, A. M. Cassell, and H. Dai, Science **283**, 512 (1999).
11. J. Ihm, A. Zunger and M. L. Cohen, J. Phys. C:Solid State Phys., **12**, 4409 (1979).
12. N. Troullier and J. L. Martins, Phys. Rev. B **43**, 1993 (1991).
13. Seungwu Han and Jisoon Ihm (to be published).
14. O. F. Sankey and D. J. Niklewski, Phys. Rev. B **40**, 3979 (1989); The r_C is set to 4.0 for C and H atoms.
15. C. H. Xu, C. Z. Wang, C. T. Chan, and K. M. Ho, J. Phys.: Condens. Matter **4**, 6047 (1992).
16. D. L. Carroll, P. Redlich, P. M. Ajayan, J. C. Charlier, X. Blase, A. De Vita, and R. Car, Phys. Rev. Lett. **78**, 2811 (1997).
17. P. Kim, T. W. Odom, J.-L. Huang, and C. M. Lieber, Phys. Rev. Lett. **82**, 1225 (1999); R. Tamura and M. Tsukada, Phys. Rev. B **52**, 6015 (1995).
18. A. De Vita, J. -Ch. Charlier, X. Blase, and R. Car, Appl. Phys. A **68**, 283 (1999).
19. M. Fujita, K. Wakabayashi, K. Nakada, and K. Kusakabe, J. Phys. Soc. Japan. **65**, 1920 (1996).
20. J. A. Osborn, Phys. Rev. **67**, 351 (1945).
21. V. T. Binh, S. T. Purcell, N. Garcia, and J. Doglioni, Phys. Rev. Lett. **69**, 2527 (1992); M. L. Yu, N. D. Lang, B. W. Hussey, T. H. P. Chang, and W. A. Mackie, Phys. Rev. Lett. **77**, 1636 (1996).
22. J. Charlier, A. De Vita, X. Blase and R. Car, Science **275**, 646 (1997).
23. O. M. Kuttel, O. Groening, C. Emmenegger, and L. Schlapbach, Appl. Phys. Lett. **73**, 2113 (1998).
24. S. J. Tans, A. R. M. Verschueren, and C. Dekker, Nature **393**, 49 (1998).

FIRST AND SECOND-ORDER RESONANT RAMAN SPECTRA OF SINGLE-WALLED CARBON NANOTUBES

M.S. Dresselhaus,[1,2] M.A. Pimenta,[3] K. Kneipp,[4,5] S.D.M. Brown,[1] P. Corio,[1] A. Marucci,[1] and G. Dresselhaus[6]

[1] *Department of Physics, Massachusetts Institute of Technology, Cambridge, MA 02139, USA*
[2] *Department of Electrical Engineering and Computer Science, Massachusetts Institute of Technology, Cambridge, MA 02139, USA*
[3] *Departamento de Fisica, Universidade Federal de Minas Gerais, Belo Horizonte, 30123-970 Brazil*
[4] *George R. Harrison Spectroscopy Laboratory Massachusetts Institute of Technology, Cambridge, MA 02139, USA*
[5] *Department of Physics, Technical University of Berlin, Berlin, Germany*
[6] *Francis Bitter Magnet Laboratory, Massachusetts Institute of Technology, Cambridge, MA 02139, USA*

INTRODUCTION

The first-order and second-order Raman spectra of single-wall carbon nanotubes (SWNTs) are very special relative to other crystalline systems in exhibiting features associated with resonant enhancement phenomena arising from their unique one-dimensional (1D) density of electronic states. Both the first-order and second-order spectra are selective of specific carbon nanotubes.[1–3] In this paper we review the unique behavior of the first-order Raman spectra and show asymmetric behavior in a comparison between the Stokes and anti-Stokes Raman processes which also arise from the unique 1D density of electronic states. We also relate the observed overtones and combination modes in the second-order Raman spectra to the dominant features in the first-order spectra.

Figure 1 shows the radial breathing mode (around 150–160 cm^{-1}) and the tangential mode (1500–1600 cm^{-1}) forming the two dominant features of the first-order Raman spectrum of single wall carbon nanotubes. A strong feature in the 2610–2680 cm^{-1} range in the second-order Raman spectra for single-wall nanotubes is also observed, in addition to weaker overtones and combination modes for these three laser excitation energies. The relative intensities of both the first- and second-order features are dependent on the laser excitation energy E_{laser} and resonant Raman effects are important in the analysis of all the spectra.

Science and Application of Nanotubes, edited by Tománek and Enbody
Kluwer Academic / Plenum Publishers, New York, 2000

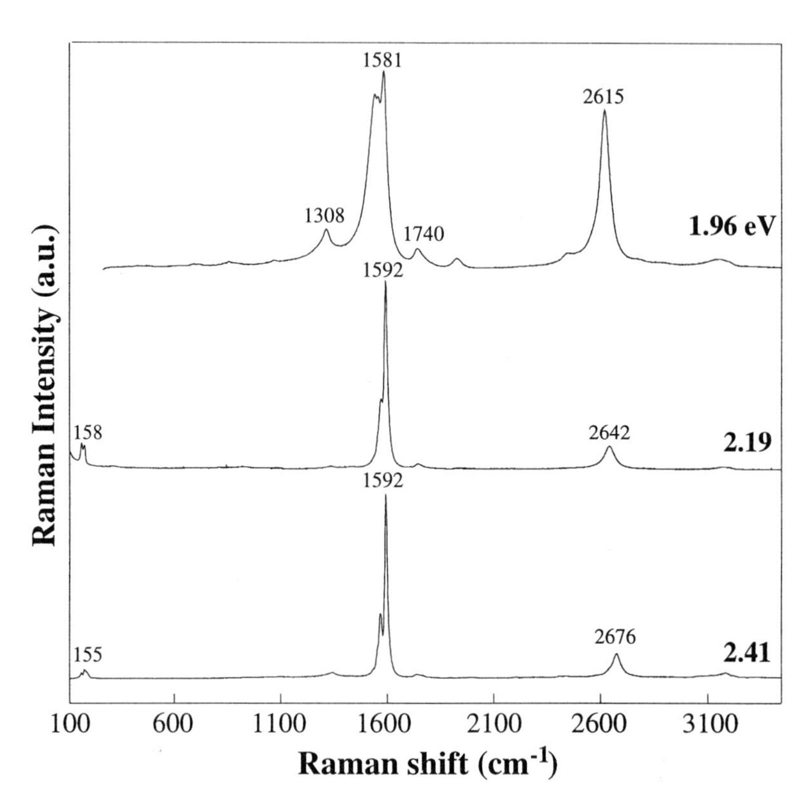

Figure 1: First and second-order Raman spectra for single wall carbon nanotubes over the broad phonon frequency range 100–3700 cm^{-1} for $E_{\mathrm{laser}} = 1.96, 2.19$, and 2.41 eV.

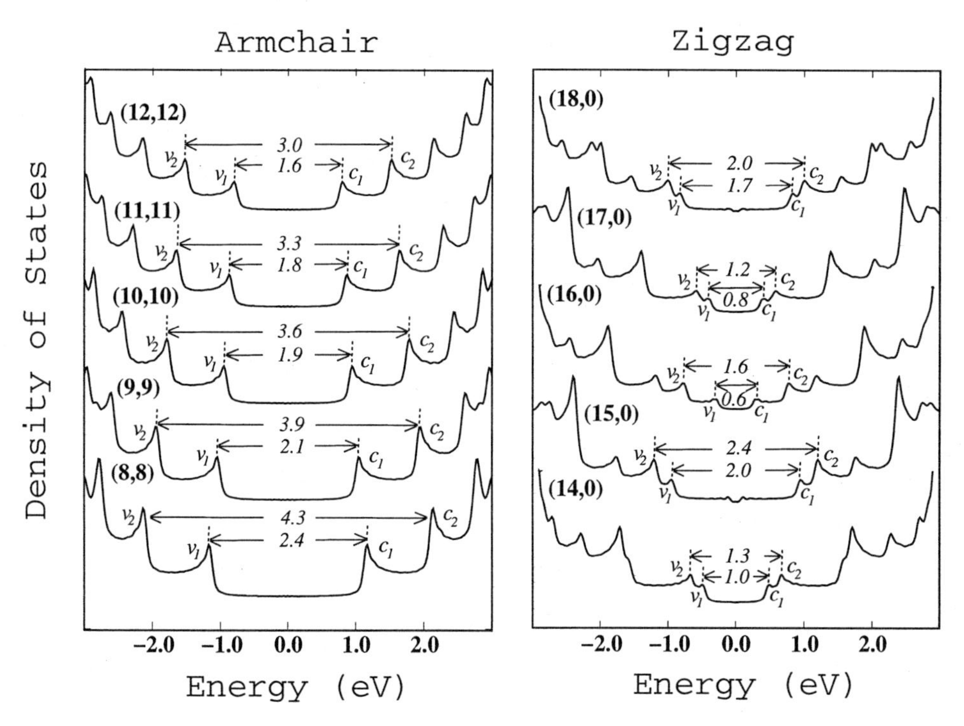

Figure 2: Electronic 1D density of states (DOS) calculated with a tight binding model for (8,8), (9,9), (10,10), (11,11), and (12,12) armchair nanotubes and for (14,0), (15,0), (16,0), (17,0), and (18,0) zigzag nanotubes.[4] Wavevector-conserving optical transitions can occur between mirror image singularities in the 1D density of electronic states of the valence and conduction bands, i.e., $v_1 \to c_1$ and $v_2 \to c_2$, etc. These optical transitions, which are given in the figure in units of eV and are denoted in the text by E_{11}, E_{22}, etc., are responsible for the resonant Raman effect. The 1D density of electronic states for chiral nanotubes (n, m) tends to be similar to the results shown here for the zigzag nanotubes. Metallic nanotubes require that $2n + m = 3q$, where q is an integer.

1D DENSITY OF ELECTRONIC STATES

The origin of the unique behavior of the first- and second-order resonant Raman processes is directly related to the 1D density of electronic states shown in Fig. 2.[1,4] The radial breathing mode and the tangential stretching mode in the first-order Raman spectrum and their overtones and combination modes are resonantly enhanced when the laser excitation energy (E_{laser}) is equal to the energy separation between a singularity in the 1D electron density of states in the valence and conduction bands, such as the energy separations $E_{ii} = E_{11}, E_{22}, \ldots$ shown in Fig. 2 by the horizontal arrows. Resonant enhancement occurs not only for the incident photon but also for the scattered photon, so that the conditions for resonant Raman scattering include $E_{\text{laser}} = E_{ii}$ and $E_{\text{laser}} = E_{ii} + \hbar\omega_{\text{phonon}}$ for the Stokes process which creates a phonon, while for the anti-Stokes process which absorbs a phonon, resonant Raman scattering occurs for $E_{\text{laser}} = E_{ii}$ and for $E_{\text{laser}} = E_{ii} - \hbar\omega_{\text{phonon}}$. Since the energy separations E_{ii} in Fig. 2 are strongly dependent on the diameter of the nanotubes, a change in E_{laser} results in the resonant Raman excitation of *different* nanotubes. Specifically, the radial breathing mode frequency ω_{RBM} depends on the nanotube diameter d_t in accordance with $\omega_{\text{RBM}} \propto 1/d_t$, independent of the chirality of the nanotube. Therefore, a variation of E_{laser} results in a variation of the observed ω_{RBM}, as has been studied extensively by many authors,[1,3,5–11] and has been used as a characterization tool for the diameter distribution of the single wall nanotubes in a given nanotube sample because of the diameter-specific resonant Raman process for SWNTs. Since the radial breathing mode is observed in the second-order Raman spectrum as both a harmonic (overtone) and as a combination mode,[12] the special resonant features of the radial breathing mode in the first-order spectrum affect the second-order spectrum in important ways, as discussed below. Since $\hbar\omega_{\text{phonon}}$ can be as large as 0.4 eV for second-order scattering processes, the specific nanotubes in a given sample that are excited in the first-order Raman scattering process may be different from the nanotubes that participate in the second-order scattering process, and using the same argument, different nanotubes are resonantly enhanced for the Stokes and anti-Stokes processes for a given incident laser energy E_{laser} (see Fig. 2). Several examples of phenomena directly related to the 1D electronic density of states are presented in the present work.

Kataura et al.[14] have published a very useful plot, an adaption of which is shown in Fig. 3 for the energy separations $E_{ii}(d_t)$ between the i^{th}singularity in the conduction and valence bands, such as shown in Fig. 2, as a function of nanotube diameter d_t. Selection rules indicate that the intensity for optical transitions $E_{ii}(d_t)$ are much stronger than for $E_{ij}(d_t)$ for $i \neq j$. The plot in Fig. 3 includes all (n, m) nanotubes with diameters d_t up to 3 nm, and the calculations are based on a tight binding calculation using a value of the nearest neighbor C-C overlap energy $\gamma_0 = 2.90$ eV. The plot includes both metallic nanotubes ($2n - m = 3q$) and semiconducting nanotubes ($2n - m \neq 3q$), where q is an integer.

As stated above, for both the Stokes and anti-Stokes processes, resonance with the incident photons occurs for $E_{\text{laser}} = E_{ii}(d_t)$, whereas for the scattered light the resonance occurs at $E_{\text{laser}} = E_{ii}(d_t) + E_{\text{phonon}}$ for the Stokes process and at $E_{\text{laser}} = E_{ii}(d_t) - E_{\text{phonon}}$ for the anti-Stokes process. This implies that, in general, different nanotubes will be in resonance with the scattered light beam in the Stokes and anti-Stokes processes, as demonstrated below.

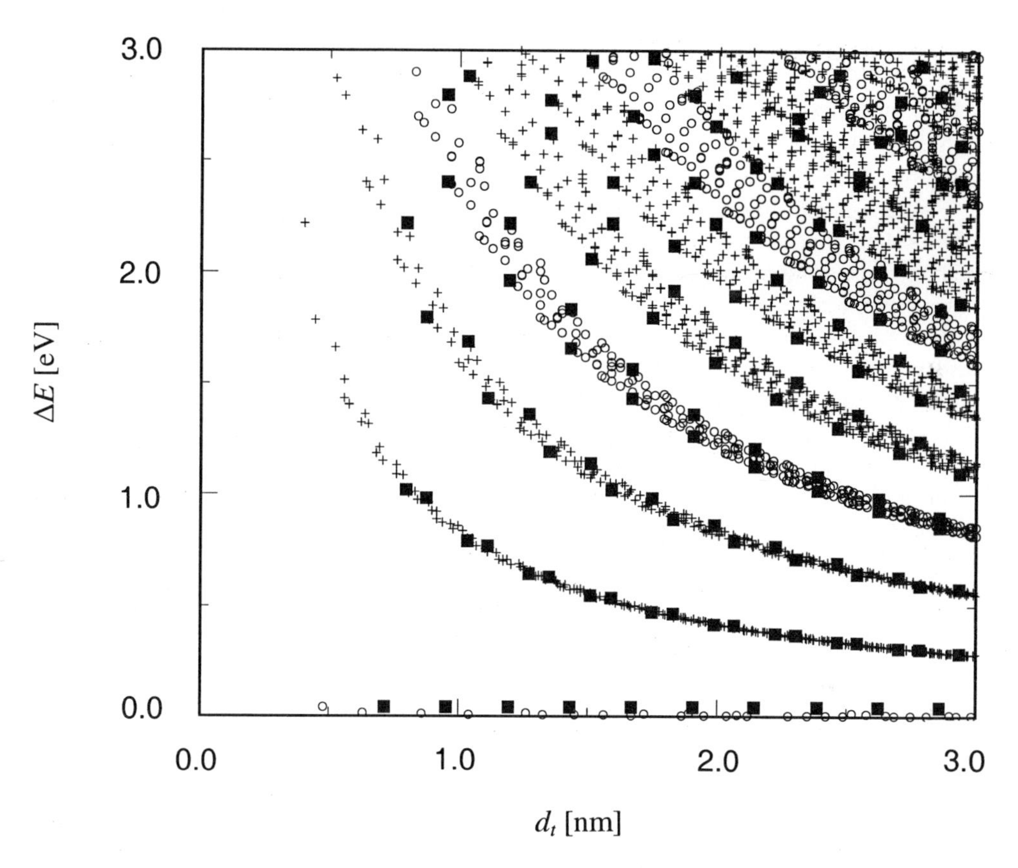

Figure 3: Calculation of the energy separations $E_{ii}(d_t)$ or gap energies for all (n, m) nanotubes as a function of nanotube diameter up to 3 nm based on a tight binding model calculation with $\gamma_0 = 2.90\,\text{eV}$.[13] The armchair (n, n) nanotubes are indicated by the circled dots. Note the points in the figure for zero gap metallic nanotubes along the abscissa.

FIRST-ORDER STOKES SPECTRA

The characteristic features of the tangential modes of the first-order spectrum are due to resonant Raman effects associated with singularities in the 1D electron density of states, as shown in Fig. 4 for a single wall carbon nanotube sample with an average diameter d_t of 1.49 ± 0.20 nm.[12] Since the tangential modes of single-wall carbon nanotubes have only a very weak dependence on nanotube diameter,[3,15] the characteristic features of the Raman spectra in Fig. 4 for the tangential band over a wide range of E_{laser} ($2.19 < E_{\text{laser}} < 2.71$ eV) are essentially independent of E_{laser}.[2] However, in the range of $1.58 < E_{\text{laser}} < 1.96$ eV, additional peaks in the first-order resonant Raman spectra for the tangential band are observed, and, in addition, the spectra for this range of E_{laser} show extensive broadening. We attribute these additional peaks and the line broadening effects to laser-induced electronic resonances involving predominantly metallic nanotubes and to the coupling between phonons and conduction electrons in metallic nanotubes.[2]

A Lorentzian lineshape analysis of the spectra in Fig. 4 has been carried out. The results show that the same dominant Lorentzian components (at 1567 cm^{-1}, 1592 cm^{-1}, 1599 cm^{-1}) are found in the spectra taken at 2.19, 2.41, 2.54, and 2.71 eV, where the resonant contributions from semiconducting nanotubes dominates all spectra, showing similar peak frequencies, linewidths and relative intensities. Also, three additional, intense broad Lorentzian components are found in the Stokes spectra for $E_{\text{laser}} = 1.58, 1.84, 1.92$, and 1.96 eV. The broad features in the Raman spectra at 1540 and 1515 cm^{-1} and the narrower feature at 1581 cm^{-1} [see Fig. 5(a)] are identified with metallic nanotubes.[2] For this sample, the intensity of the most intense feature at 1540 cm^{-1} associated with metallic nanotubes is dominant relative to the intensity of the most intense 1592 cm^{-1} feature associated with semiconducting nanotubes for $E_{\text{laser}} = 1.8$ eV.

In Fig. 5 a lineshape analysis for the tangential band for two Stokes spectra is presented showing the relative intensities and linewidths of the three components (with peak intensities at 1515, 1540, and 1581 cm^{-1}) associated with metallic nanotubes and the three components at 1567, 1591, and 1601 cm^{-1} associated with semiconducting nanotubes, which all can be found in the Stokes spectrum at 1.92 eV. For comparison, we also show in Fig. 5, the Stokes spectrum for the tangential band at 1.49 eV where only semiconducting nanotubes are in resonance. From Fig. 3 we can see that the laser beam is in resonance with the E_{22}^S transition for semiconducting nanotubes for $E_{\text{laser}} = 1.49$ eV, while for $E_{\text{laser}} = 1.96$ eV, most of the nanotubes in the sample ($d_t = 1.49 \pm 0.20$ nm) are in resonance with the E_{11}^M transition for metallic nanotubes, with some nanotubes on the high end of the diameter distribution being in resonance with the $E_{33}^S(d_t)$ transition for semiconducting nanotubes. The low frequency tail of the Stokes spectrum at 1.49 eV may have some contributions from metallic nanotubes of large diameter in resonance with the E_{11}^M transition.

When we now look at the anti-Stokes spectra (see Fig. 6) at the same two laser energies as in Fig. 5, a very different picture emerges, as seen by comparison of the anti-Stokes spectra for the tangential band (Fig. 6) with the corresponding Stokes spectra taken at the same two values of E_{laser} (Fig. 5). For the anti-Stokes spectra at $E_{\text{laser}} = 1.49$ eV we see a spectrum due to metallic nanotubes arising from resonance of the $E_{11}^M(d_t)$ transition for metallic nanotubes with the incident beam and of $E_{11}^M(d_t) + E_{\text{phonon}}$ with the scattered beam. Comparison of the tangential band for the metallic nanotubes in the anti-Stokes process with the corresponding spectrum for the Stokes process shows that the anti-Stokes band has no contributions from semicon-

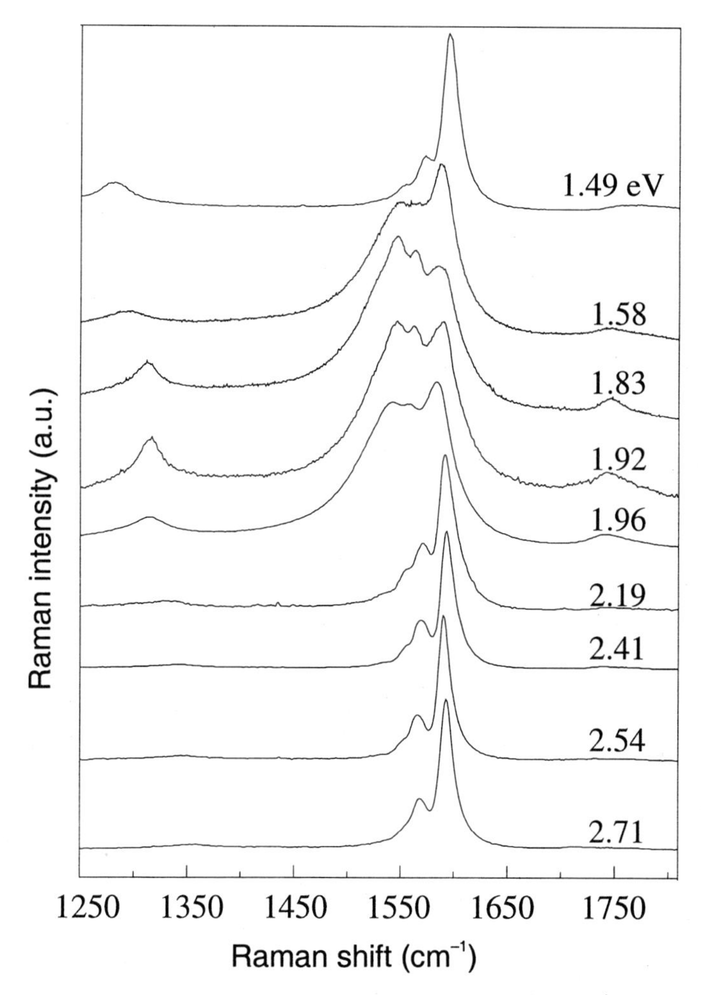

Figure 4: First-order Raman spectra for the tangential band (1500–1650 cm^{-1}) taken for laser excitation energies between 1.58 eV and 2.71 eV on a single wall nanotube sample with nanotubes of diameter $d_t = 1.49 \pm 0.20$ nm. At lower frequencies (1300–1350 cm^{-1}) the 'D-band' feature is observed, and at higher frequencies near 1750 cm^{-1} a feature tentatively identified with a combination mode is seen for some values of $E_{\rm laser}$ in the red.

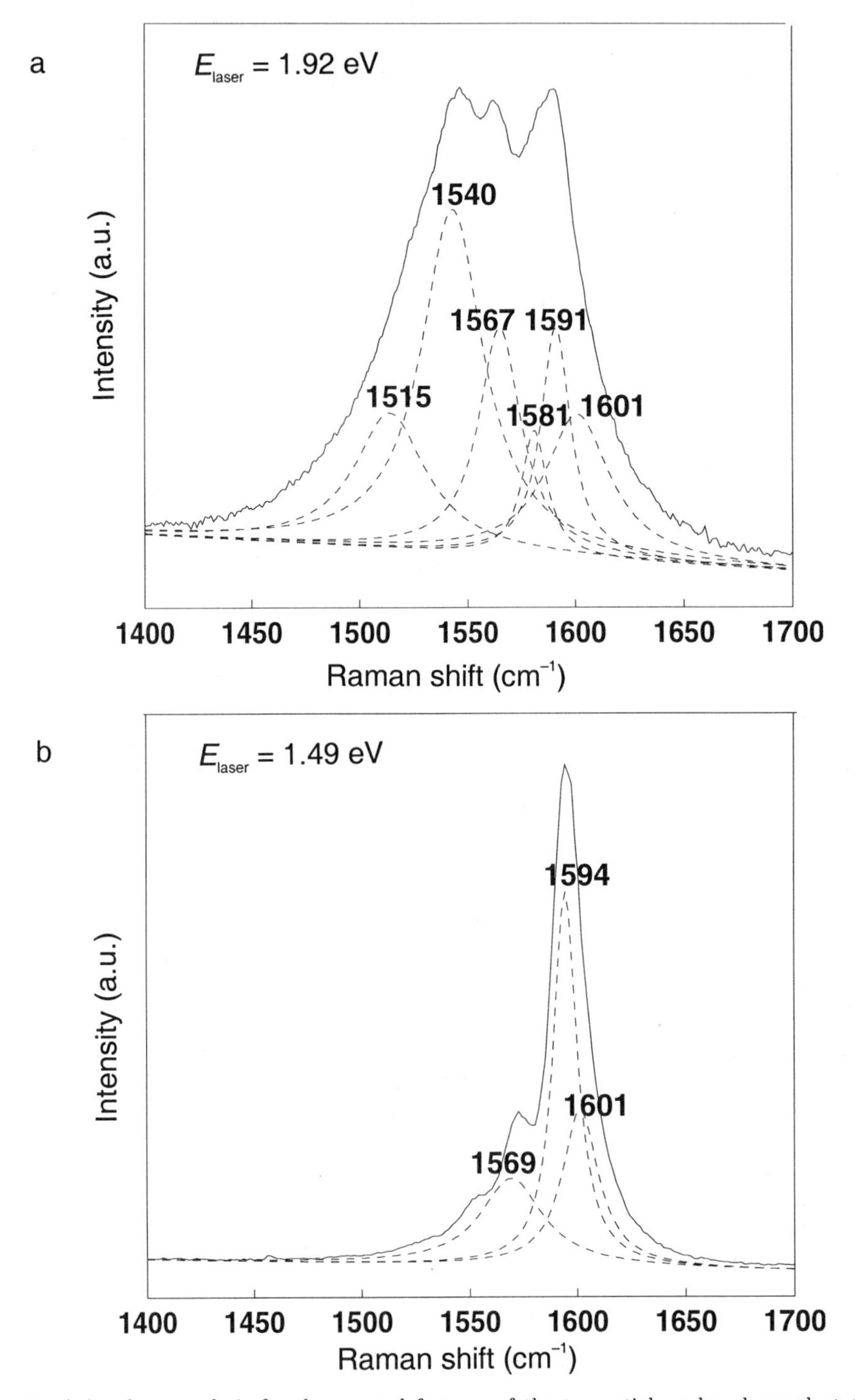

Figure 5: A lineshape analysis for the spectral features of the tangential modes observed at (a) $E_{laser} = 1.92\,eV$ and (b) $E_{laser} = 1.49\,eV$ for the Stokes process. We note that the spectrum at 1.92 eV shows Raman intensity near $1591\,cm^{-1}$ where the most intense component for the semiconducting nanotubes occurs, as shown in (b) for the spectrum taken at 1.49 eV.

ducting nanotubes. This absence of the semiconducting contributions can be understood by examination of Fig. 3, showing that the nanotube sample does not contain nanotubes of small enough diameters to be resonant with the $E_{22}^{S}(d_t)$ semiconducting nanotube transition or large enough in diameter to be in resonance with the $E_{33}^{S}(d_t)$ electronic transition for semiconducting nanotubes. For the anti-Stokes tangential band at $E_{\text{laser}} = 1.92\,\text{eV}$, both the incident and scattered beams are resonant with the $E_{33}^{S}(d_t)$ and perhaps also $E_{44}^{S}(d_t)$ transitions for semiconducting nanotubes.

The importance of the scattered beam in the Stokes and anti-Stokes tangential bands can be appreciated by looking at the results for the E_{laser} dependence of the normalized intensity for the $1540\,\text{cm}^{-1}$ Lorentzian component $\tilde{I}_{1540}$ which is the component with the maximum intensity for the metallic nanotubes. Expressions for the E_{laser} dependence of the normalized intensity for the $1540\,\text{cm}^{-1}$ metallic component in the Stokes process $\tilde{I}_{1540}^{S}(d_0, E_{\text{laser}})$ are given by

$$\begin{aligned}\tilde{I}_{1540}^{S}(d_0, E_{\text{laser}}) &= \textstyle\sum_{d_t} A \exp\left[\frac{-(d_t-d_0)^2}{\Delta d_t^2/4}\right] \\ &\times \left[(E_{11}^{M}(d_t) - E_{\text{laser}})^2 + \gamma_e^2/4\right]^{-1} \\ &\times \left[(E_{11}^{M}(d_t) - E_{\text{laser}} + E_{\text{phonon}})^2 + \gamma_e^2/4\right]^{-1},\end{aligned} \tag{1}$$

where d_0 is the average nanotube diameter of the sample, and for the anti-Stokes process $\tilde{I}_{1540}^{AS}(d_0, E_{\text{laser}})$ is given by

$$\begin{aligned}\tilde{I}_{1540}^{AS}(d_0, E_{\text{laser}}) &= \textstyle\sum_{d_t} A \exp\left[\frac{-(d_t-d_0)^2}{\Delta d_t^2/4}\right] \\ &\times \left[(E_{11}^{M}(d_t) - E_{\text{laser}})^2 + \gamma_e^2/4\right]^{-1} \\ &\times \left[(E_{11}^{M}(d_t) - E_{\text{laser}} - E_{\text{phonon}})^2 + \gamma_e^2/4\right]^{-1},\end{aligned} \tag{2}$$

where the sum in Eqs. (1) and (2) is taken over the tube diameters, Δd_t is the FWHM width of the distribution and γ_e is a damping energy. A plot showing the predictions of this model for a sample with $d_t = 1.49 \pm 0.20\,\text{nm}$ is given in Fig. 7, which clearly shows the range of E_{laser} where resonance with metallic nanotubes can be observed for both the Stokes and anti-Stokes processes. It is interesting to note that for the two values of E_{laser} in Figs. 5 and 6, the Stokes and anti-Stokes spectra at the same E_{laser} values are very different. For almost all materials, the Stokes and anti-Stokes spectra have similar lineshapes. The asymmetry between the Stokes and anti-Stokes spectra is unusual and is due to the sensitivity of these Raman bands to resonances of the scattered photons with *different* nanotubes in the sample for the Stokes and anti-Stokes processes.

THE *D*-BAND AND *G′* BAND FEATURES

Figure 4 also shows the presence of a Raman band near $1330\,\text{cm}^{-1}$ in the first-order spectrum of single wall carbon nanotubes which is analogous to the '*D*-band' observed in disordered sp^2 carbons or in sp^2 carbons of small crystallite size. This '*D*-band' in sp^2 carbons and in carbon nanotubes has a large dispersion in peak frequency as a function of E_{laser}, as well as a strong dependence of its intensity on E_{laser}.

All sp^2 carbons (including both 2D and 3D structures) exhibit a resonant Raman band near $2660\,\text{cm}^{-1}$ (see Fig. 1) called the G'-band, with a frequency and intensity that are strongly dependent on E_{laser}, and a structure that dominates the second-order spectrum.[18] The physical origin of this resonant Raman band has recently been explained in sp^2 carbons in terms of a strong coupling between electrons and phonons

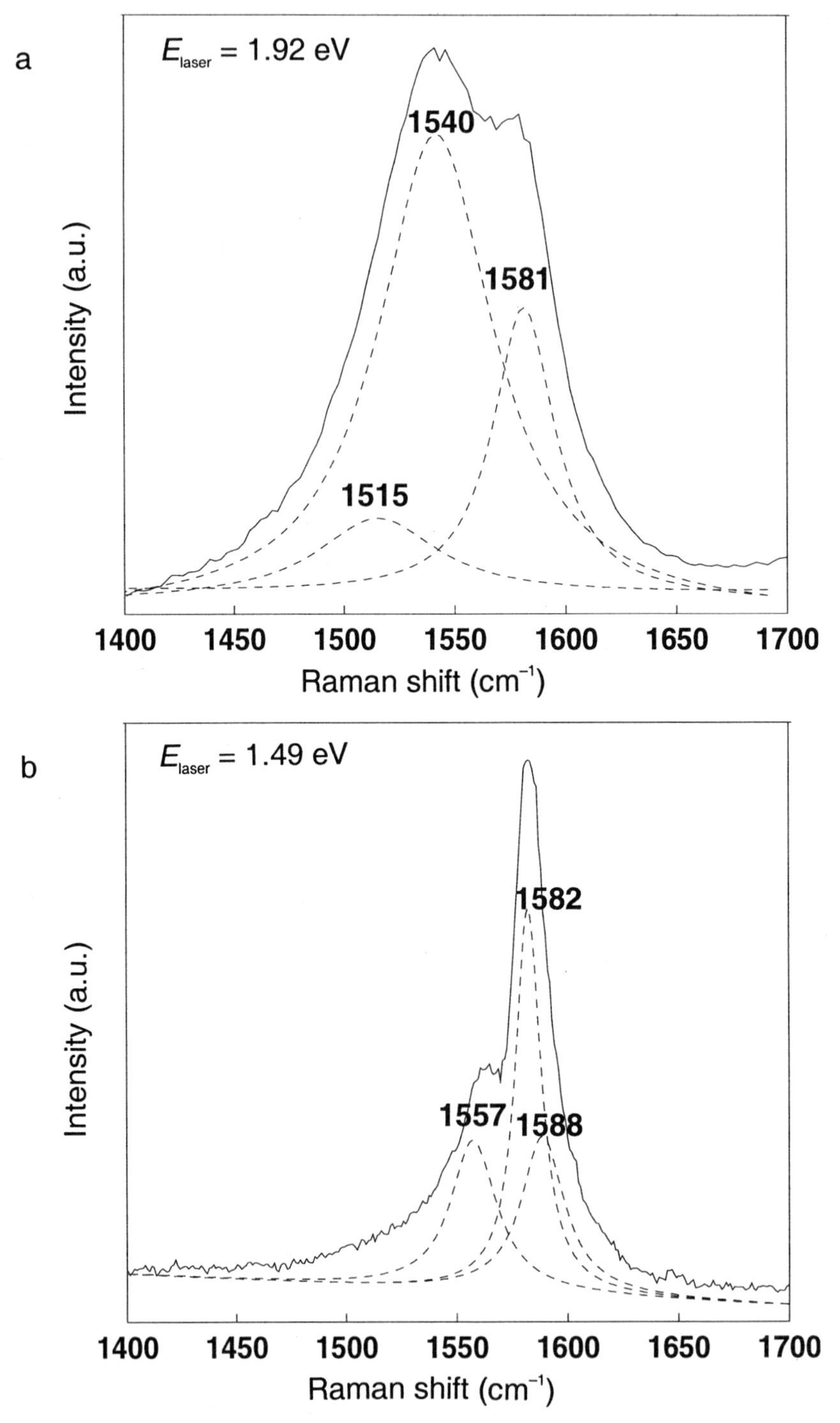

Figure 6: A lineshape analysis for the spectral features of the tangential modes observed at (a) $E_{\text{laser}} = 1.49\,\text{eV}$ and (b) $E_{\text{laser}} = 1.92\,\text{eV}$ for the anti-Stokes process. The metallic nanotubes dominate the spectra at 1.49 eV, while the semiconducting nanotubes dominate the spectra at 1.92 eV. No semiconducting nanotube scattering is seen at $E_{\text{laser}} = 1.49\,\text{eV}$ and almost no metallic nanotube scattering is seen at $E_{\text{laser}} = 1.92\,\text{eV}$.

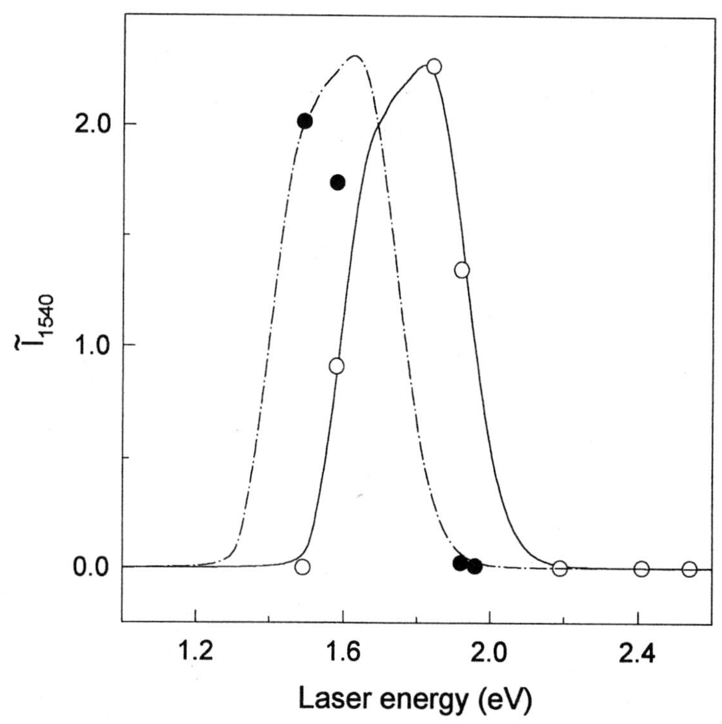

Figure 7: Stokes and anti-Stokes normalized intensity $\tilde{I}_{1540}$ for the feature at 1540 cm^{-1} for metallic nanotubes as a function of laser energy for the same SWNT sample ($d_0 = 1.49$ nm). The experimental Stokes data are shown as open circles and the solid curve is the fit of Eq. (1) to the experimental data. The experimental normalization is done in terms of I_{1540}/I_{1593} for the Stokes data, where I_{1540} and I_{1593}, respectively, denote the maximum intensity components for metallic and semiconducting nanotubes. The predicted plot of $\tilde{I}^{AS}_{1540}$ vs. E_{laser} for the anti-Stokes process uses the same parameters as for the Stokes process, except that Eq. (2) is used for the functional form for $\tilde{I}^{AS}_{1540}$. Preliminary experimental anti-Stokes data are given by the solid points and here the normalization is made with respect to the FWHM linewidth of the tangential band (see Fig. 6) which includes the 1540 cm^{-1} component associated with metallic nanotubes.

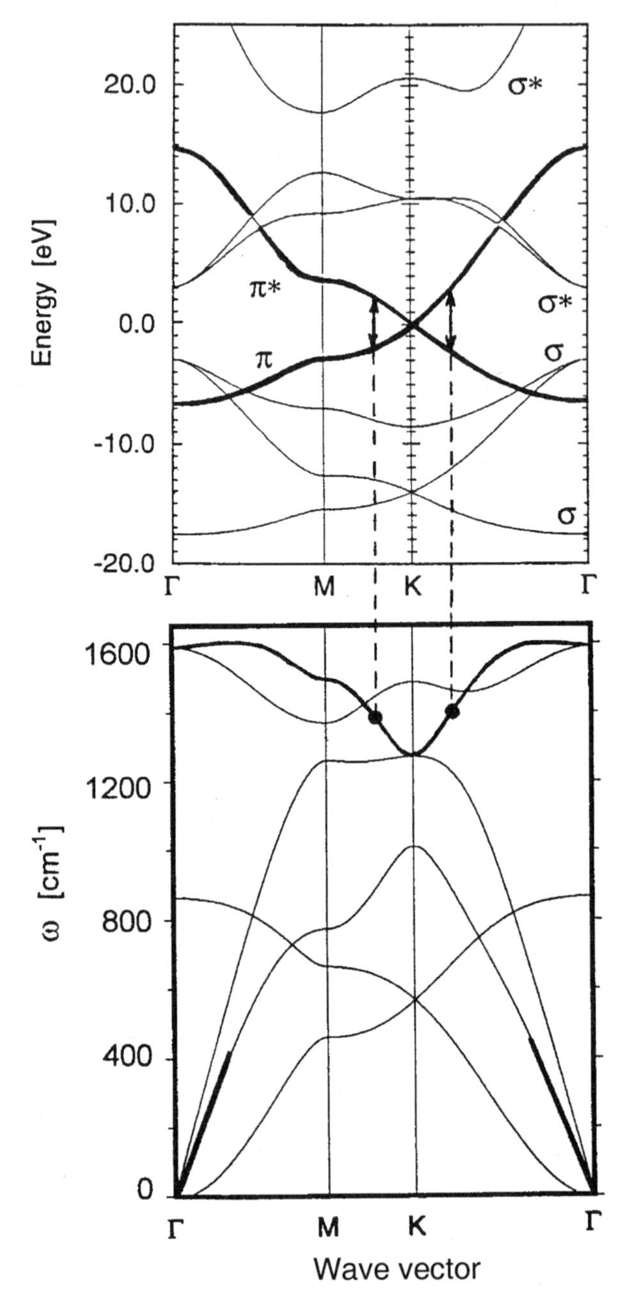

Figure 8: Electronic energy bands of 2D graphite (top).[16,17] Phonon dispersion curves of 2D graphite (bottom).[16,17] Both the phonon branch that is strongly coupled to electronic bands in the optical laser excitation, and the electronic bands near the Fermi level ($E = 0$) that are linear in wave vector k near the K point in the Brillouin zone are indicated by heavy lines. The slope for the TA phonon branch is also indicated by heavy lines.[18] The condition $\Delta q = \Delta k$ is indicated by the vertical dashed lines.

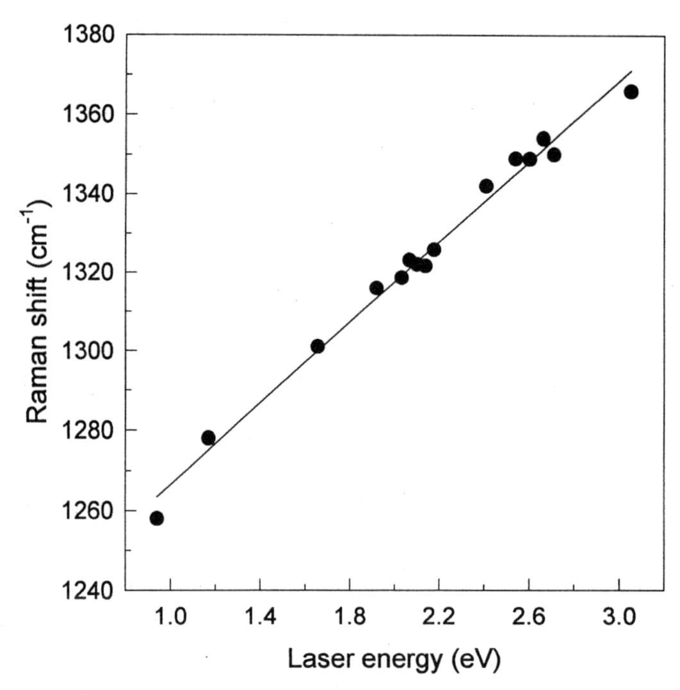

Figure 9: Frequency of the 'D-band' for single wall carbon nanotubes as a function of laser excitation energy. The line is a least squares fit to the data points for one SWNT sample. The fit to the data points yields a slope of $51.2\,\text{cm}^{-1}/\text{eV}$ and an intercept of $1215\,\text{cm}^{-1}$ at $E_{\text{laser}} = 0$.

with the same wave vector in the vicinity of the K-point of the 2D Brillouin zone (see Fig. 8), where the electronic energy bands exhibit unique linear k-dependent phonon and electron dispersion relations.[18] Figure 8 shows the dispersion for the electrons and the phonons near the K-point in the Brillouin zone. The conditions where Δq for the phonon wave vector as measured from the K-point is equal to the corresponding electron wave vector Δk is indicated. In this figure it is seen that the slope of the TA acoustic branch (in bold face) at ΓK is equal to the slope of the optical branch at $K\Gamma$. This leads to a relation between the dispersion of the phonon frequencies of the optical phonon near the K-point in the Brillouin zone $\partial\omega/\partial E_{\text{laser}}$ (that is measured experimentally) and the velocity of sound. This change in optical phonon frequency with wave vector near the K-point is the same as $\partial\omega/\partial\Delta q$ for the TA phonon at the Γ point. Measurement of $\partial\omega/\partial E_{\text{laser}}$ for carbon nanotubes is shown in Fig. 9 and an analysis such as is given in Ref. [18] yields a value of $51.2\,\text{cm}^{-1}/\text{eV}$ and an intercept at $E_{\text{laser}} = 0$ of $1215\,\text{cm}^{-1}$. Measurements on a number of SWNT samples give an average value of $52 \pm 1\,\text{cm}^{-1}/\text{eV}$ which is to be compared with an average value for graphite of $48 \pm 3\,\text{cm}^{-1}/\text{eV}$.[18] Although the frequency of the G' band occurs at twice the frequency of the D band in disordered or nano-size sp^2 carbons, the G' band is also observed with approximately equal intensity for highly crystalline graphite, which has no first-order 'D-band' intensity.

Analogous resonant Raman phenomena occur for carbon nanotubes, for which the most intense feature in the second-order Raman spectra (see Fig. 1) is associated with the 'G'-band'. All the nanotubes in the sample contribute resonantly to the G' band feature, thereby accounting for its high intensity. The laser excitation energy ($E_{\text{laser}} = 1.92\,\text{eV}$) where the intensity of the 'G'-band' is the largest is also the E_{laser} value where the 'D-band' intensity is the largest. A comparison between the 'G'-

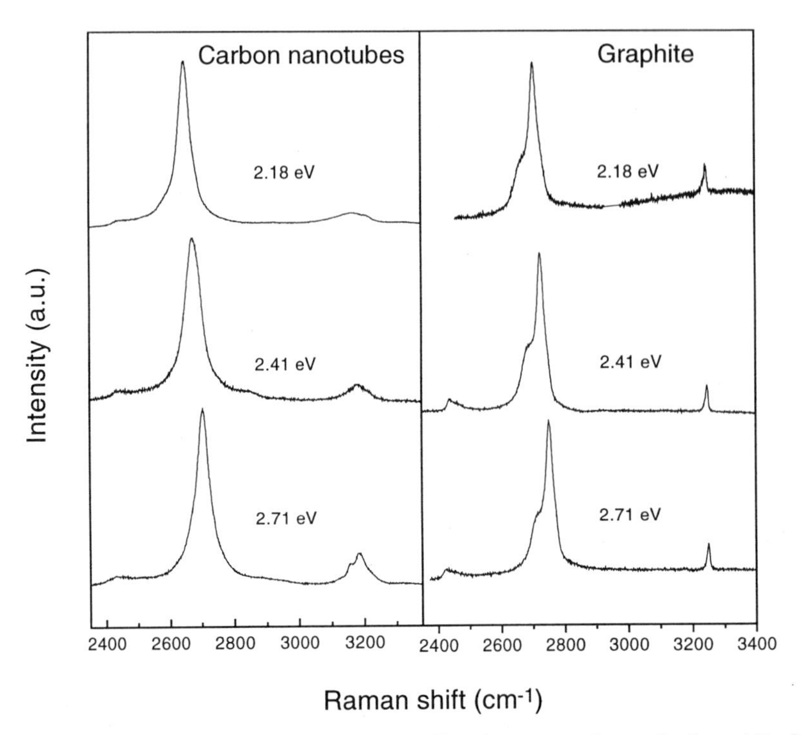

Figure 10: Second-order Raman spectra of single wall carbon nanotubes and of graphite for three different laser excitation energies.

band' in graphite and in carbon nanotubes is shown in Fig. 10 where it is seen that for both graphite and carbon nanotubes, the 'G'-band' is the dominant feature in the second-order spectrum. However, the lineshape for the 'G'-band' in the nanotubes is Lorentzian, whereas the graphite G' lineshape shows a well-defined shoulder at the low frequency side of the spectrum, which is attributed to the two crystallographically distinct layers in the graphite crystal structure. These inequivalent layers result in a splitting of the phonon bands in the vicinity of the K-point in the Brillouin zone. Also seen in Fig. 10 are the second-order features at ~2430 cm^{-1} and ~3200 cm^{-1} which are discussed below.

Measurements of the dispersion of the frequency of the G' band (denoted by $\omega_{G'}$) with laser excitation energy E_{laser} have been carried out. Here a value of $\partial\omega_{G'}/\partial E_{\text{laser}} = 106\,\text{cm}^{-1}/\text{eV}$ has been obtained.[19] This shows that the slope for the 'G'-band' for carbon nanotubes is very close to twice that for the 'D-band', consistent with results obtained for other sp^2 carbons.[18]

OVERTONES AND COMBINATION MODES

We now discuss the various features in the second-order spectrum associated with the harmonics (overtones) and combination modes of the two dominant features of the first-order spectrum, the radial breathing mode at low frequencies and the tangential mode at high frequencies. The second harmonic (overtone) of the radial breathing mode $2\omega_{\text{RBM}}$ at two different laser excitation energies is shown in Fig. 11. For example, at $E_{\text{laser}} = 1.58\,\text{eV}$ (785 nm), the first-order spectrum gives resonantly-enhanced radial breathing modes at 150 and $162\,\text{cm}^{-1}$, with linewidths (FWHM) of $14\,\text{cm}^{-1}$ and $11\,\text{cm}^{-1}$, respectively, to be compared with the second-order lines which occur at $301\,\text{cm}^{-1}$ and $330\,\text{cm}^{-1}$ and have relatively narrow linewidths ($21\,\text{cm}^{-1}$).[12] The first-order radial breathing mode spectrum and its overtone spectrum are both inhomogeneously broadened because of the different nanotubes that are simultaneously in resonance with the laser excitation. A change in E_{laser} results in the excitation of *different* nanotubes, so that for $E_{\text{laser}} = 2.54\,\text{eV}$ (488 nm), the first-order Raman spectrum shows a radial breathing mode at $159\,\text{cm}^{-1}$ and a small shoulder at $176\,\text{cm}^{-1}$. The second harmonic of this first-order band shows a well resolved second-order feature at $320\,\text{cm}^{-1}$ (see Fig. 11). Similar trends are observed at other values of E_{laser}.

In contrast, the second harmonic of the tangential band $2\omega_{\text{tang}}$ which occurs in the range 3100–$3250\,\text{cm}^{-1}$ (see Fig. 12) has very different characteristics from the second harmonic of the radial breathing mode $2\omega_{\text{RBM}}$ shown in Fig. 11. We see in Fig. 12 the evolution with E_{laser} of the second harmonic of the tangential band for five laser energies in the range 1.96–2.71 eV. The central frequency and linewidth of this second-order band is relatively weakly dependent on E_{laser} for $2.19 < E_{\text{laser}} < 2.71\,\text{eV}$, where the semiconducting nanotubes dominate the first-order spectra for this sample. However, for $E_{\text{laser}} = 1.96\,\text{eV}$, where the dominant contribution to the first-order spectrum in Fig. 4 comes from metallic nanotubes, the second-order spectrum is downshifted and much broader than for the higher E_{laser} values, consistent with the behavior of the first-order tangential band shown in Fig. 4.

The feature in the second-order Raman spectrum for graphite near $3240\,\text{cm}^{-1}$ (see Fig. 10) is strongly affected by the mode frequency dispersion of the phonon dispersion curves.[20–22] This mode frequency dispersion in graphite gives rise to a peak in the phonon density of states near $1620\,\text{cm}^{-1}$, associated with non-zone center phonons.[20] This peak in the phonon density of states is responsible for the feature in the second-order spectrum of graphite near $3240\,\text{cm}^{-1}$. This frequency is upshifted by $76\,\text{cm}^{-1}$ from twice the zone-center phonon mode in graphite at $1582\,\text{cm}^{-1}$. Since there is no corresponding dispersion effect near $1620\,\text{cm}^{-1}$ in carbon nanotubes, we expect the mode frequency of the second-order tangential band to be close to twice that for the first-order tangential band.

A Lorentzian lineshape analysis of the second-order spectrum associated with the second harmonic of the tangential band has been carried out.[12] At $E_{\text{laser}} = 1.96\,\text{eV}$, a very broad Lorentzian component is found at $3082\,\text{cm}^{-1}$ which is in good agreement with twice the dominant line at $1540\,\text{cm}^{-1}$ in the first-order spectrum for metallic nanotubes. Also the linewidth of the second-order feature is roughly twice that for the corresponding first-order feature. Likewise at $E_{\text{laser}} = 2.71\,\text{eV}$, the dominant peak in the second-order spectrum is at $3181\,\text{cm}^{-1}$ which corresponds well to twice the dominant peak frequency at $1592\,\text{cm}^{-1}$ in the first-order spectrum for the semiconducting nanotubes.

As E_{laser} decreases from 2.71 eV, the peak frequency of the entire second-order band (see Fig. 12) downshifts, especially for the lowest values of E_{laser}, where new tangential peaks associated with metallic nanotubes are resonantly enhanced. For example, the dominant Lorentzian components in the second-order spectra at 2.19 eV and 2.41 eV are at $3171\,\text{cm}^{-1}$ and $3166\,\text{cm}^{-1}$, both downshifted relative to $2(1592) = 3184\,\text{cm}^{-1}$. Furthermore, Fig. 12 shows an increase in linewidth (FWHM) of the entire second-order band with decreasing E_{laser}. The large linewidth below $\sim 2.0\,\text{eV}$ in the first-order spectra in Fig. 4 is associated with a large contribution to the spectral intensity from metallic nanotubes.[2] The broadening of the second-order features extends to much

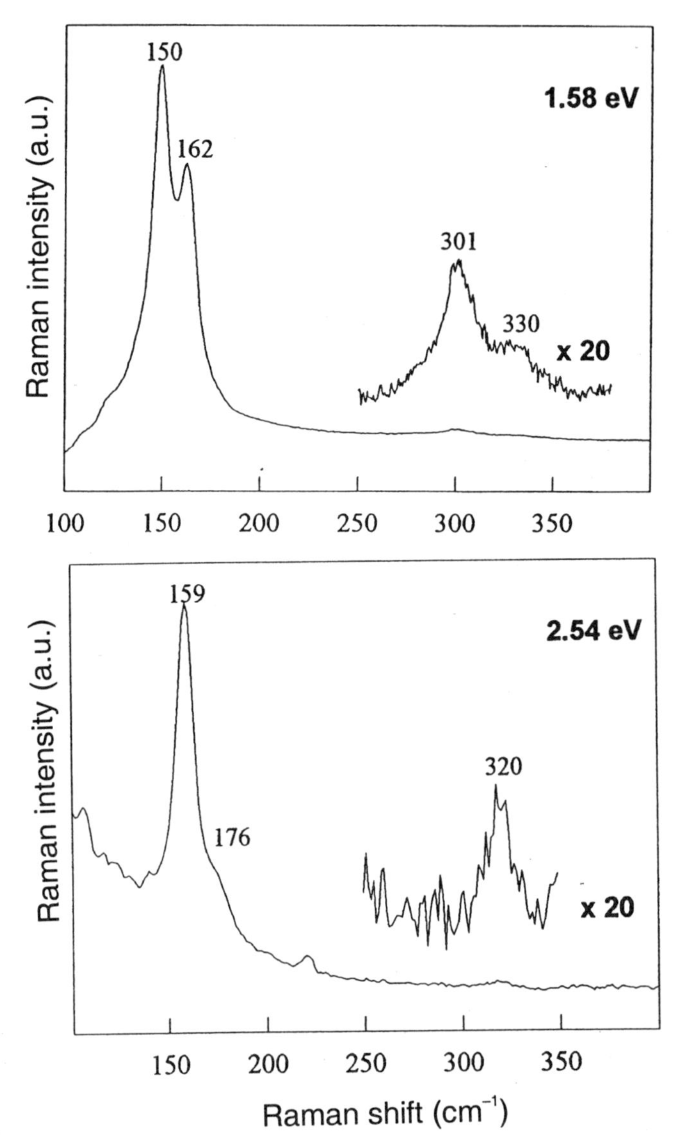

Figure 11: The Raman spectra for the radial breathing mode band and its second-harmonic at two laser excitation energies 1.58 eV (785 nm) and 2.54 eV (488 nm).[12]

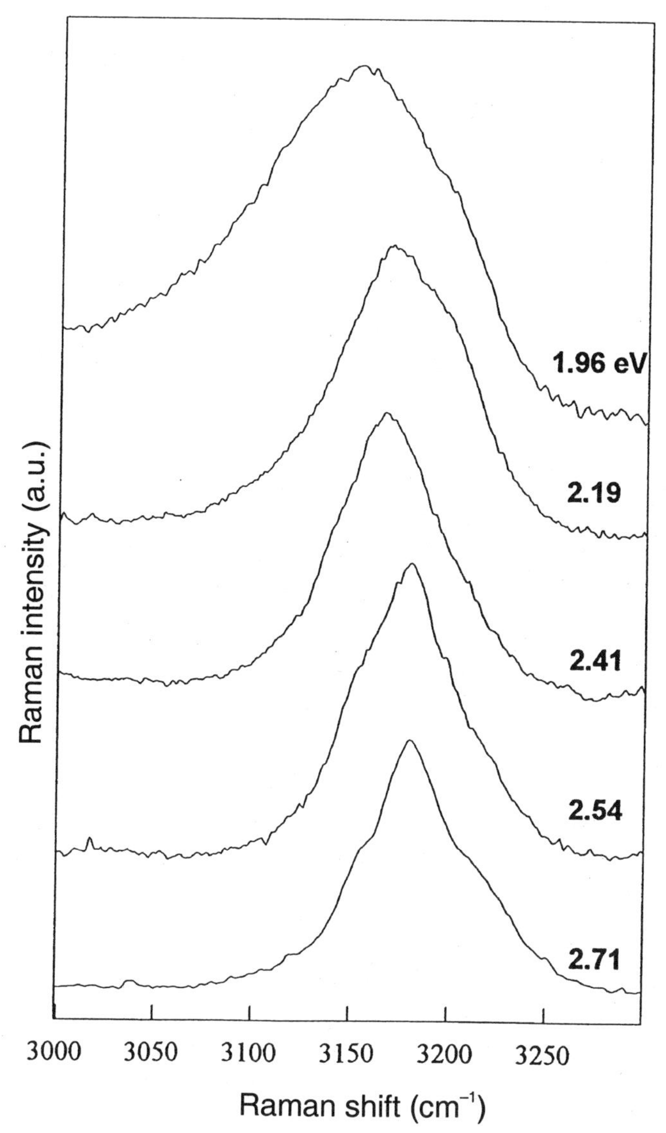

Figure 12: The Raman spectra for the second harmonic of the tangential mode, collected at five laser excitation energies.[12] The corresponding first-order spectra are displayed in Fig. 4.

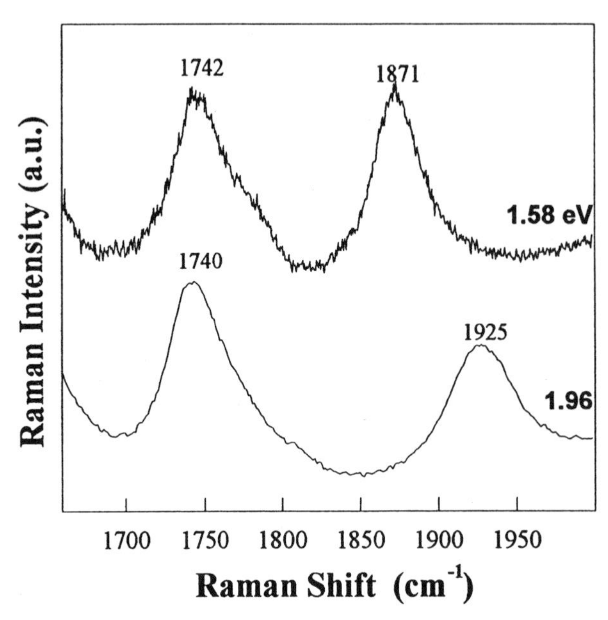

Figure 13: Combination bands associated with $\omega_{\text{tang}}+\omega_{\text{RBM}}$ near $1740\,\text{cm}^{-1}$ and with $\omega_{\text{tang}}+2\omega_{\text{RBM}}$ at $1871\,\text{cm}^{-1}$ and $1925\,\text{cm}^{-1}$ in the second-order Raman spectra of carbon nanotubes at $E_{\text{laser}} = 1.58\,\text{eV}$ (785 nm) and 1.96 eV (632.8 nm).[12]

higher values of E_{laser} when compared to the first-order spectra, due to the high phonon energy ($\sim$0.4 eV). Figure 10 shows the second-order overtone for the tangential mode for another single wall nanotube sample. For this sample, the lineshape of the second-order band at $E_{\text{laser}} = 2.71\,\text{eV}$ can be nicely seen in the spectrum.

The resonant enhancement of the phonon modes for metallic nanotubes occurs when the energy of either the incident or the scattered photon is in resonance with the first electronic transition E_{11} for the metallic nanotubes in the sample (see Figs. 2 and 3). When $1.92\,\text{eV} \leq E_{\text{laser}} \leq 1.96\,\text{eV}$, both the incident and scattered photons for the Stokes process are in the resonance window for metallic nanotubes, so that the overtone band is expected to broaden considerably, and this is observed experimentally (see Fig. 12). For $2.19\,\text{eV} \leq E_{\text{laser}} \leq 2.41\,\text{eV}$, even though the incident photon is higher in energy than the resonance energy window E_{laser} for metallic nanotubes in the first-order spectrum (see Fig. 4), the energy of the scattered photon falls within the interval of resonant enhancement for metallic nanotubes (see Fig. 3), because the phonon frequency of $3200\,\text{cm}^{-1}$ in the Stokes process corresponds to a large energy upshift ($\sim$0.4 eV). We thus attribute the onset of line broadening in the first-order spectrum for $E_{\text{laser}} < 2.2\,\text{eV}$ to the contribution from metallic nanotubes. Because of the large value of ω_{phonon}, the second-order spectrum shows line broadening below about 2.4 eV. Thus at $E_{\text{laser}} =$ 2.19 eV and 2.41 eV, the specific nanotubes contributing to the first-order and second-order spectra are likely to be different nanotubes.

Resonant Raman effects associated with the 1D electron density of states singularities also give rise to resonant effects in the *combination* modes. One example of a combination mode occurs at the sum frequency between a tangential and a radial breathing mode $\omega_{\text{tang}} + \omega_{\text{RBM}}$, as shown in the spectra in Fig. 13 taken at $E_{\text{laser}} = 1.58\,\text{eV}$ and 1.96 eV. Since the radial breathing mode ω_{RBM} spectra at $E_{\text{laser}} = 1.58\,\text{eV}$ consist of two lines at $150\,\text{cm}^{-1}$ and $162\,\text{cm}^{-1}$ (see Fig. 11), then using the most intense Lorentzian component at 1591 cm^{-1} for the tangential band ω_{tang} yields a sum of 1741 cm^{-1}, in good agreement with the dominant observed peak at 1742 cm^{-1} in Fig. 13.

The relative intensities of this combination mode at various values of E_{laser} shows that this feature at 1742 cm^{-1} is most prominent near ~1.8 eV where the metallic nanotube contribution is dominant. Since the tangential mode frequency ω_{tang} is expected to be almost independent of E_{laser}, while the radial breathing mode frequency varies as $\omega_{\text{RBM}} \propto 1/d_t$, the shifts in the peak frequencies of the 1740 cm^{-1} band for various E_{laser} lines are expected to reflect the variation in ω_{RBM} as E_{laser} is varied, since different nanotubes are resonantly excited at each value of E_{laser}. Since $\hbar\omega_{\text{phonon}} \sim 0.2$ eV for this combination mode, it is also possible that, at a given value of E_{laser}, different nanotubes within the sample are resonantly enhanced in the first-order spectrum as compared with the second-order spectrum.[12] The features at 1871 cm^{-1} and 1925 cm^{-1} in Fig. 13 are tentatively identified with the $(\omega_{\text{tang}} + 2\omega_{\text{RBM}})$ combination mode, corresponding in the $E_{\text{laser}} = 1.58$ eV trace to ω_{tang} for metallic nanotubes and in the $E_{\text{laser}} = 1.96$ eV trace to ω_{tang} for semiconducting nanotubes.

As mentioned above, the most intense feature in the second-order spectrum of carbon nanotubes (see Fig. 1) is the peak located at ~2680 cm^{-1}, and this feature has an especially strong intensity near $E_{\text{laser}} = 1.96$ eV (Fig. 1). This feature, which is closely related to the G' feature in sp^2 carbons, shows an upshift in frequency as E_{laser} increases (see Fig. 10). The lineshape of the G' band is fit by a single Lorentzian component, with a linewidth that has a very weak dependence on E_{laser}. Assuming that the frequency of the G' band depends linearly on E_{laser}, consistent with the behavior of the G' band in other sp^2 carbons, the experimental G' band frequencies for the nanotube sample extrapolate to 2421 cm^{-1} at $E_{\text{laser}} = 0$, and this phonon frequency is close to twice the K-point phonon frequency in the 2D Brillouin zone of sp^2 carbons. From Fig. 9 we find a value of 1215 cm^{-1} for the extrapolation of the 'D-band' data to $E_{\text{laser}} = 0$. The extrapolation of the data for the G' band to $E_{\text{laser}} = 0$ agrees quite well with the direct measurement of this same phonon frequency, shown in Fig. 14.

The feature that appears at 2440 cm^{-1} in the second-order Raman spectrum of 2D graphite (see Fig. 10) has been attributed to the sum of two K-point phonons, where each has a frequency of ~1220 cm^{-1} based on analysis of the 2D resonant Raman effect for the 'D-band' in sp^2 carbons.[18] The phonon density of states in graphite shows a weak peak associated with these K-point phonons where an optical and an acoustic branch of the phonon spectrum for 2D graphite are degenerate at the zone edge K-point (see Fig. 8). The K-point feature which appears in the phonon spectrum of graphite and disordered carbons is non-resonant, because the valence and conduction bands in the electronic structure are degenerate at the K-point in the Brillouin zone and the photon that would be necessary to be in resonance with this phonon has $E_{\text{laser}} \simeq 0$.

For $E_{\text{laser}} = 1.84, 1.92$, and 1.96 eV, the results in Fig. 14 show that the peak frequency and linewidth of the second harmonic (overtone) of the K-point phonon band are independent of E_{laser}, though at higher values of $E_{\text{laser}} = 2.41, 2.54$ and 2.71 eV, the peak downshifts, broadens and becomes more asymmetric. We currently have no explanation for this frequency downshift or for the line broadening of this spectral feature for E_{laser} above 2.4 eV.

SUMMARY

In summary, overtones and combination modes have been identified in the second-order spectra for the two dominant features in the first-order spectra (the radial breathing mode and the tangential mode) that are associated with the resonant Raman enhance-

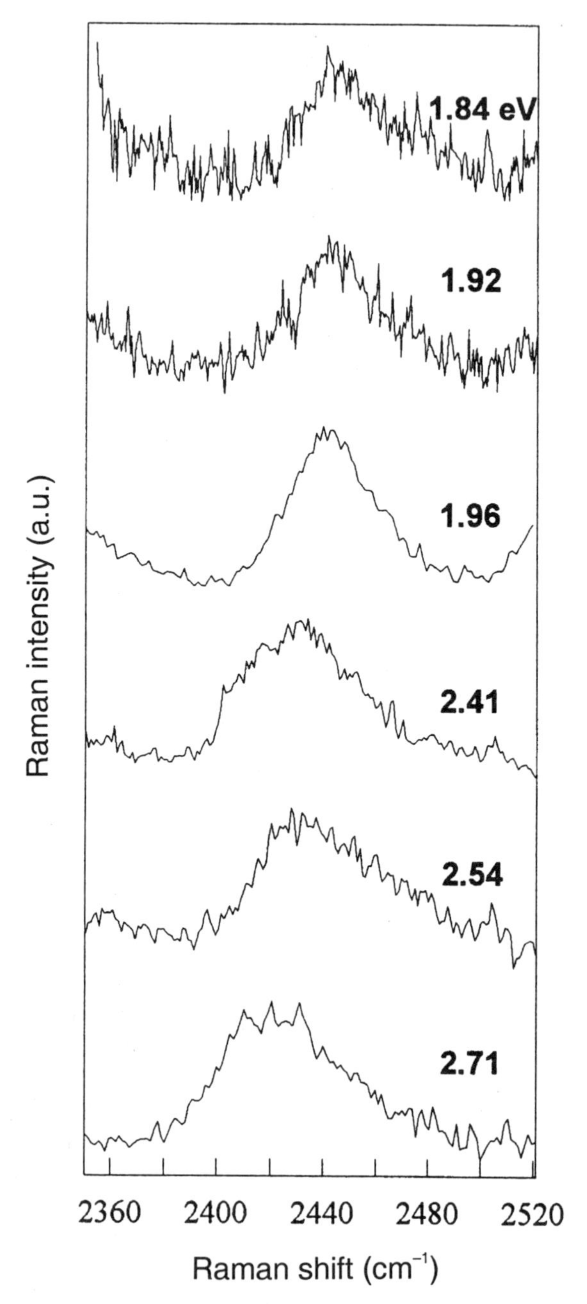

Figure 14: The weak non-resonant feature associated with the second harmonic (overtone) of the K-point phonon in the 2D Brillouin zone for six values of E_{laser}.

ment process arising from the 1D electronic density of states. Just as for the case of the first-order spectra, the resonant contributions to the second-order spectra also involve a different set of (n, m) nanotubes at each laser excitation energy E_{laser}. A second-order analog is observed for the broad spectral band identified with contributions from metallic nanotubes to the first-order tangential mode spectra. The unique feature of the second-order tangential overtone band shows a larger E_{laser} range over which the metallic nanotubes contribute, and this effect is attributed to the large ($\hbar\omega_{\text{phonon}} \sim 0.4\,\text{eV}$) energy of these phonons. Combination modes associated with $(\omega_{\text{tang}} + \omega_{\text{RBM}})$ and $(\omega_{\text{tang}} + 2\omega_{\text{RBM}})$ have been identified. These combination modes show behaviors as a function of E_{laser} that are consistent with the behavior of their first-order constituents, namely that different nanotubes contribute to the spectra at each value of E_{laser}. The behavior of the 'D-band' and G'-band features show a very large phonon frequency dependence on E_{laser}, and show a resonant 2D behavior when the electron and phonon wave vectors coincide, as also occurs in other sp^2 carbons.

Future Raman studies are likely to explore the relation between the Stokes and the anti-Stokes spectra as a function of E_{laser}. Surface-enhanced Raman scattering (SERS) is likely to be explored as a method for achieving much higher sensitivity, allowing exploration of the vibrational spectra of a small number of nanotubes and perhaps even eventually the Raman spectrum of a single nanotube.

ACKNOWLEDGMENTS

The authors are thankful to Dr. Gene Hanlon for his help with the measurements. We gratefully acknowledge valuable discussions with Drs. P.C. Eklund, H. Kataura and C. Dekker. The authors gratefully acknowledge support for this work under NSF grant DMR 98–04734. M.A.P acknowledges support by the Brazilian agencies FAPEMIG, CAPES and FINEP and P.C. acknowledges support by the Brazilian agency FAPESP. The measurements performed at the George R. Harrison Spectroscopy Laboratory at MIT were supported by the NIH grant P41-RR02594 and NSF grant CHE97-08265.

REFERENCES

1. A. M. Rao, E. Richter, S. Bandow, B. Chase, P. C. Eklund, K. W. Williams, M. Menon, K. R. Subbaswamy, A. Thess, R. E. Smalley, G. Dresselhaus, and M. S. Dresselhaus, Science **275**, 187–191 (1997).
2. M. A. Pimenta, A. Marucci, S. Empedocles, M. Bawendi, E. B. Hanlon, A. M. Rao, P. C. Eklund, R. E. Smalley, G. Dresselhaus, and M. S. Dresselhaus, Phys. Rev. B **58**, R16012–R16015 (1998).
3. M. S. Dresselhaus, G. Dresselhaus, M. A. Pimenta, and P. C. Eklund. pages 367–434, Blackwell Science Ltd., Oxford, UK, 1999. Analytical Applications of Raman Spectroscopy.
4. Jean-Christophe Charlier. Private communication.
5. C. Journet, W. K. Maser, P. Bernier, A. Loiseau, M. Lamy de la Chapelle, S. Lefrant, P. Deniard, R. Lee, and J. E. Fischer, Nature (London) **388**, 756–758 (1997).
6. Y. Saito, Y. Tani, N. Miyagawa, K. Mitsushima, A. Kasuya, and Y. Nishina, Chem. Phys. Lett. **294**, 593–598 (1998).
7. M. Sugano, A. Kasuya, K. Tohji, Y. Saito, and Y. Nishina, Chem. Phys. Lett. **292**, 575–579 (1998).

8. H. Kuzmany, B. Burger, M. Hulman, J. Kurti, A. G. Rinzler, and R. E. Smalley, Europhys. Lett. **44**, 518–524 (1998).
9. A. Kasuya, M. Sugano, Y. Sasaki, T. Maeda, Y. Saito, K. Tohji, H. Takahashi, Y. Sasaki, M. Fukushima, Y. Nishina, and C. Horie, Phys. Rev. B **57**, 4999 (1998).
10. E. Anlaret, N. Bendiab, T. Guillard, C. Journet, G. Flamant, D. Laplaze, P. Bernier, and J-L Sauvajol, Carbon **36**, 1815–1820 (1998).
11. H. Kuzmany, B. Burger, A. Thess, and R. E. Smalley, Carbon **36**, 709–712 (1998).
12. S. D. M. Brown, P. Corio, A. Marucci, M. A. Pimenta, M. S. Dresselhaus, and G. Dresselhaus, (unpublished).
13. G. Dresselhaus, M. A. Pimenta, R. Saito, J.C. Charlier, S. D. M. Brown, P. Corio, A. Marucci, and M. S. Dresselhaus. In *Science and Applications of Nanotubes*, edited by D. Tománek and R. J. Enbody, Kluwer Academic, New York, 1999. Proceedings of the International Workshop on the Science and Applications of Nanotubes, Michigan State University, East Lansing, MI, USA, July 24-27, 1999.
14. H. Kataura, Y. Kumazawa, N. Kojima, Y. Maniwa, I. Umezu, S. Masubuchi, S. Kazama, X. Zhao, Y. Ando, Y. Ohtsuka, S. Suzuki, and Y. Achiba. In *Proc. of the Int. Winter School on Electronic Properties of Novel Materials (IWEPNM'99)*, edited by H. Kuzmany, M. Mehring, and J. Fink, page unpublished (in press), American Institute of Physics, Woodbury, N.Y., 1999. AIP conference proceeding.
15. M. S. Dresselhaus, G. Dresselhaus, and P. C. Eklund, *Science of Fullerenes and Carbon Nanotubes* (Academic Press, New York, NY, 1996).
16. R. Saito, G. Dresselhaus, and M. S. Dresselhaus, *Physical Properties of Carbon Nanotubes* (Imperial College Press, London, 1998).
17. R. A. Jishi, L. Venkataraman, M. S. Dresselhaus, and G. Dresselhaus, Chem. Phys. Lett. **209**, 77–82 (1993).
18. M. J. Matthews, M. A. Pimenta, G. Dresselhaus, M. S. Dresselhaus, and M. Endo, Phys. Rev. B **59**, R6585 (1999).
19. M. A. Pimenta, E. B. Hanlon, A. Marucci, P. Corio, S. D. M. Brown, S. Empedocles, M. Bawendi, G. Dresselhaus, and M. S. Dresselhaus, (unpublished).
20. M. S. Dresselhaus and G. Dresselhaus, Light Scattering in Solids III **51**, 3 (1982). edited by M. Cardona and G. Güntherodt, Springer-Verlag Berlin, Topics in Applied Physics.
21. M. S. Dresselhaus, P. C. Eklund, and M. A. Pimenta. In *Raman Scattering in Materials Science*, edited by W. Weber and R. Merlin, Springer-Verlag, Berlin, 1999. in press.
22. P. C. Eklund, J. M. Holden, and R. A. Jishi, Carbon **33**, 959 (1995).

ON THE $\pi - \pi$ OVERLAP ENERGY IN CARBON NANOTUBES

G. Dresselhaus,[1] M.A. Pimenta,[2,3] R. Saito,[4] J.C. Charlier,[5] S.D.M. Brown,[2] P. Corio,[2] A. Marucci,[2] and M. S. Dresselhaus[2,6]

[1] *Francis Bitter Magnet Laboratory, Massachusetts Institute of Technology, Cambridge, MA 02139, USA*
[2] *Department of Physics, Massachusetts Institute of Technology, Cambridge, MA 02139, USA*
[3] *Departamento de Fisica, Universidade Federal de Minas Gerais, Belo Horizonte, 30123-970 Brazil*
[4] *Department of Electronics-Engineering, University of Electro-Communications, Tokyo, 182-8585 Japan*
[5] *Unite de Physico-Chimie et de Physique des Materiaux, Universite Catholique de Louvain, Louvain la Neuve, Belgium*
[6] *Department of Electrical Engineering and Computer Science, Massachusetts Institute of Technology, Cambridge, MA 02139, USA*

INTRODUCTION

The π–π overlap energy γ_0 or the closely related transfer integral t enters many discussions of the physical properties of single wall carbon nanotubes,[1,2] including their electronic structure, lattice vibrations, optical properties, and elastic properties. Several experiments have recently been reported which provide determinations of this interaction energy. In this paper, we present some theoretical background, bring together presently available experimental data, and try to form a consistent picture relevant to the π–π overlap energy, including resonant Raman scattering data for the tangential modes, Raman measurements relevant to the 'D-band', scanning tunneling spectroscopy data,[3–6] and optical data.[7,8]

THEORETICAL ISSUES

The physical origin of the overlap energy γ_0 of the Slonczewski–Weiss–McClure band model for graphite[9–11] can be understood in terms of the tight binding approximation for 2D graphite[1] which has two inequivalent lattice sites A and B in the honeycomb network, leading to a (2×2) Hamiltonian $\mathcal{H}$ and a (2×2) overlap integral matrix $\mathcal{S}$ given by

$$\mathcal{H} = \begin{pmatrix} \epsilon_{2p} & tf(k) \\ tf(k)^* & \epsilon_{2p} \end{pmatrix} \quad \text{and} \quad \mathcal{S} = \begin{pmatrix} 1 & sf(k) \\ sf(k)^* & 1 \end{pmatrix} \tag{1}$$

Science and Application of Nanotubes, edited by Tománek and Enbody
Kluwer Academic / Plenum Publishers, New York, 2000

where t and s denote the transfer integral and overlap integral, respectively, and

$$f(k) = e^{ik_x a/\sqrt{3}} + 2e^{-ik_x a/2\sqrt{3}} \cos \frac{k_y a}{2}. \tag{2}$$

Solution of the secular equation

$$\det(\mathcal{H} - E\mathcal{S}) = 0 \tag{3}$$

implied by Eq. (1) leads to the eigenvalues

$$E^{\pm}_{g2D}(\vec{k}) = \frac{\epsilon_{2p} \pm t w(\vec{k})}{1 \pm s w(\vec{k})} \tag{4}$$

for $t < 0$, where the $+$ signs in the numerator and denominator go together, to give the dispersion relation for the bonding π energy band, and likewise for the $-$ sign, which gives the dispersion relation for the antibonding π^* band. If $t > 0$ in Eq. (4), then the $\pm$ in the numerator becomes $\mp$. The function $w(\vec{k})$ in Eq. (4) is given by

$$\begin{aligned} w(\vec{k}) &= \sqrt{|f(\vec{k})|^2} \\ &= \sqrt{1 + 4\cos\frac{\sqrt{3}k_x a}{2}\cos\frac{k_y a}{2} + 4\cos^2\frac{k_y a}{2}}. \end{aligned} \tag{5}$$

Near the K-point at the corner of the hexagonal Brillouin zone of graphite, $w(\vec{k})$ has a linear dependence on $k \equiv |\vec{k}|$ as

$$w(\vec{k}) = \frac{\sqrt{3}}{2}ka + \ldots, \qquad \text{for } ka \ll 1. \tag{6}$$

Thus the expansion of Eq. (4) for small k yields

$$E^{\pm}_{g2D}(\vec{k}) = \epsilon_{2p} \pm (|t| - s\epsilon_{2p}) w(\vec{k}) + \ldots \tag{7}$$

so that the antibonding and bonding bands are symmetric near the K point, independent of the value of s. When we adopt $\epsilon_{2p} = 0$ and $s = 0$ for Eq. (4) and assume a linear approximation for $w(k)$, then $\gamma_0 = -t$ and we get the linear dispersion relation for graphite given by McClure[10],

$$E(k) = \pm\frac{\sqrt{3}}{2}\gamma_0 ka = \pm\frac{3}{2}\gamma_0 k a_{\mathrm{C-C}}, \tag{8}$$

where $a_{\mathrm{C-C}}$ is the nearest neighbor carbon-carbon distance.

The asymmetry in the valence and conduction bands in Eq. (4) arises from the quadratic terms in $w(\vec{k})$ and this band asymmetry becomes important for large k values. For a 2D graphene sheet, the values of the tight binding parameters $\epsilon_{2p} = 0$, $t = -3.033$ eV and $s = 0.129$ fit both the first principles calculation of the electronic energy bands of 2D turbostratic graphite[12,13] and experimental data.[1,11] The non-zero value of s leads to an overall asymmetry between the bonding and anti-bonding states for large k as measured with respect to the K point. This symmetry is shown in Fig. 1, where the symmetry-imposed degeneracy between the valence and conduction bands at the K-point in the Brillouin zone is seen. At the M point and Γ point, which are far from the K point, the energies are $E^{\pm}_{g2D}(M) = \pm t/(1 \pm s)$ and $E^{\pm}_{g2D}(\Gamma) = \pm 3t/(1 \pm 3s)$, so that the energy differences $E^{-}_{g2D}(M) - E^{+}_{g2D}(M) = -2t/(1 - s^2)$ and $E^{-}_{g2D}(\Gamma) - E^{+}_{g2D}(\Gamma) =$

$-6t/(1-9s^2)$ become larger when $s \neq 0$. Thus if s is assumed to be non-zero, the γ_0 value estimated from measurements relevant to the Γ or M points in the Brillouin zone should be smaller than the γ_0 obtained from analysis of experiments relevant to the K point. This means that if $s > 0$, different values can be obtained for γ_0 when different physical phenomena are measured, since different physical phenomena will in general be sensitive to $E(\mathbf{k})$ at different $\mathbf{k}$ values. On the other hand, the different physical phenomena can be used to determine an experimental value for s, which would provide a more complete description of the dispersion relations for carbon nanotubes.

In tight binding calculations made for $E(k)$ for carbon nanotubes, zone folding techniques are employed to match the wave functions where the seamless cylindrical joint of the nanotube is made. Although it would be expected that $s \neq 0$ in general for carbon nanotubes, the various tight binding calculations that have been performed thus far for carbon nanotubes have taken $s = 0$. At present, no experiments have been interpreted to yield a non-zero value of s for SWNTs. However, the asymmetry of the energy dispersion relations of the graphite π bands as shown in Fig. 1 should be experimentally observable in carbon nanotubes using angle resolved, photo-electron and inverse photo-electron spectroscopy, similar to work carried out for the valence and conduction bands of single crystal graphite. It would be expected that the values of s and t would approach those for graphite in the limit of large nanotube diameter, while for nanotube diameters of $\approx 1.0\,\text{nm}$ (corresponding to typical experimental values for SWNTs) the values of s and t could depend on nanotube diameter and chirality and could be somewhat different in the circumferential direction relative to the direction along the nanotube axis.

Models for the electronic band structure of 3D graphite near the Fermi surface include 7 band parameters, whose values are well established.[11,14–17] The largest and most important of the band parameters for 3D graphite is $\gamma_0 = 3.16\,\text{eV}$, which is one order of magnitude larger than any of the other graphite band parameters. All of the other band parameters for 3D graphite involve overlap and transfer integrals for carbon atoms on *different* layer planes. The Slonczewski–Weiss–McClure band model[16] is constructed using symmetry and tight-binding concepts to obtain the form of the $E(\mathbf{k})$) relations. Experiments are used to evaluate the band parameters, and the wave functions in the model are all assumed to be orthonormal, whereas in the tight-binding model, overlaps of the wave functions are considered explicitly. Thus by setting $s = 0$ in Eq. (1), we can equate the overlap integral γ_0 for 2D graphite to the transfer integral t by $\gamma_0 = -t$. The work of Charlier et al.[17] illustrates the problem of using density functional theory in the local density approximation (DFT-LDA) for extracting accurate values for band structures close to the Fermi energy. For example, the value for γ_0 deduced by Charlier et al.[17] for 3D graphite was 2.6 eV, while the experimental value is 3.16 eV.[11]

Lambin[18] derived an effective π-electron Hamiltonian taking the nanotube curvature into account. Due to the curvature, he showed that the effective hopping interaction between two nearest-neighbor carbon atoms depends of the direction of the bond with respect to the nanotube axis. The arithmetic average of the three bonds per carbon atom in armchair and zig-zag nanotubes leads to an effective hopping interaction given by

$$\gamma_0(d_t) = \gamma_0(\infty)[1 - \frac{1}{2}(a_{\text{C-C}}/d_t)^2] \tag{9}$$

where $a_{\text{C-C}}$ is the carbon-carbon nearest neighbor distance and d_t is the nanotube diameter. Thus a nanotube with a C_{60} fullerene diameter would have only about a 1% decrease in the value for γ_0 due to nanotube curvature. He showed that this small correction is sufficient to account for the strain energy of $0.08\,\text{eV nm}^2/d_t^2$ found in

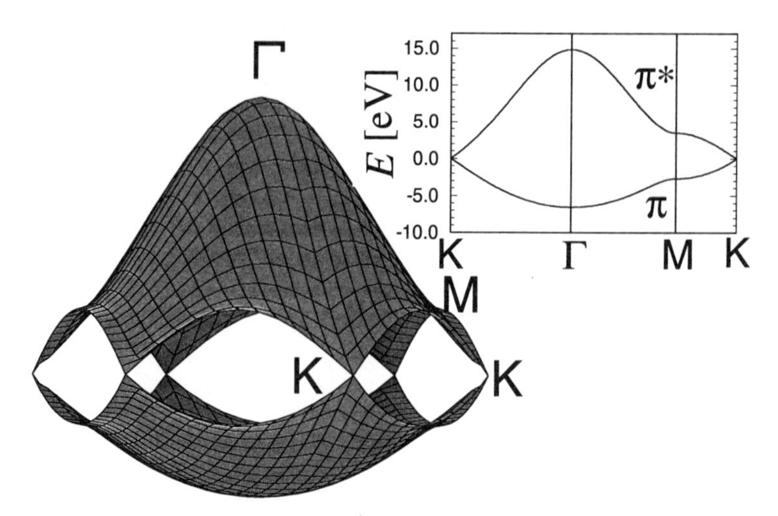

Figure 1: The energy dispersion relations $E(k)$ for the π and π^* bands in 2D graphite are shown throughout the whole region of the 2D Brillouin zone. The inset shows $E(k)$ along the high symmetry directions of the 2D Brillouin zone.

carbon nanotubes due to the curvature of the graphene sheet.

Conceptually there are two principles involved in the various experimental determinations that have been made of γ_0 thus far. The first involves measurement of the energy separation $E_{ii}(d_t)$ between singularities in the 1D density of states in the valence and conduction bands for a nanotube with diameter d_t, as shown in Fig. 2. The second principle involves the strong coupling between electrons and phonons with the same wavevector separation relative to the K-point in the Brillouin zone. This condition leads to resonant Raman transitions occurring when the phonon vector, as measured from the K-point in the Brillouin zone Δq, is equal to the corresponding electron wave vector Δk, i.e., $\Delta \mathbf{q} = \Delta \mathbf{k}$. Below we discuss each of these conceptual approaches.

THE 1D ELECTRONIC DENSITY OF STATES

The 1D electronic density of states for armchair and zigzag nanotubes of various (n, m) values is shown in Fig. 2. The corresponding 1D density of states curves for chiral nanotubes are similar to those shown in this figure, depending of whether the K-point degeneracy corresponds to $k = 2\pi/3T$ or to $k = 0$, where T is the magnitude of the translation vector along the nanotube axis. Examination of the energy difference $E_{11}(d_t)$ between the lowest-lying conduction band singularity E_{c_1} and the highest-lying valence band singularity E_{v_1} in the 1D density of states shows that $E_{11}^M(d_t)$ is much larger for the metallic nanotubes (for which $n - m = 3q$, where q is an integer), and much smaller for the semiconducting nanotubes $E_{11}^S(d_t)$. In Fig. 2 it is also seen that as the tube diameter d_t increases, the energy difference $E_{11}(d_t)$ decreases.

In the small k approximation where the linear dispersion relation of Eq. (8) is valid, we can then write simple analytic expressions for metallic $[E_{11}^M(d_t)]$ and semiconducting $[E_{11}^S(d_t)]$ nanotubes that are given by

$$E_{11}^M(d_t) = 6\gamma_0 a_{\mathrm{C-C}}/d_t \tag{10}$$

and

$$E_{11}^S(d_t) = 2\gamma_0 a_{\mathrm{C-C}}/d_t \tag{11}$$

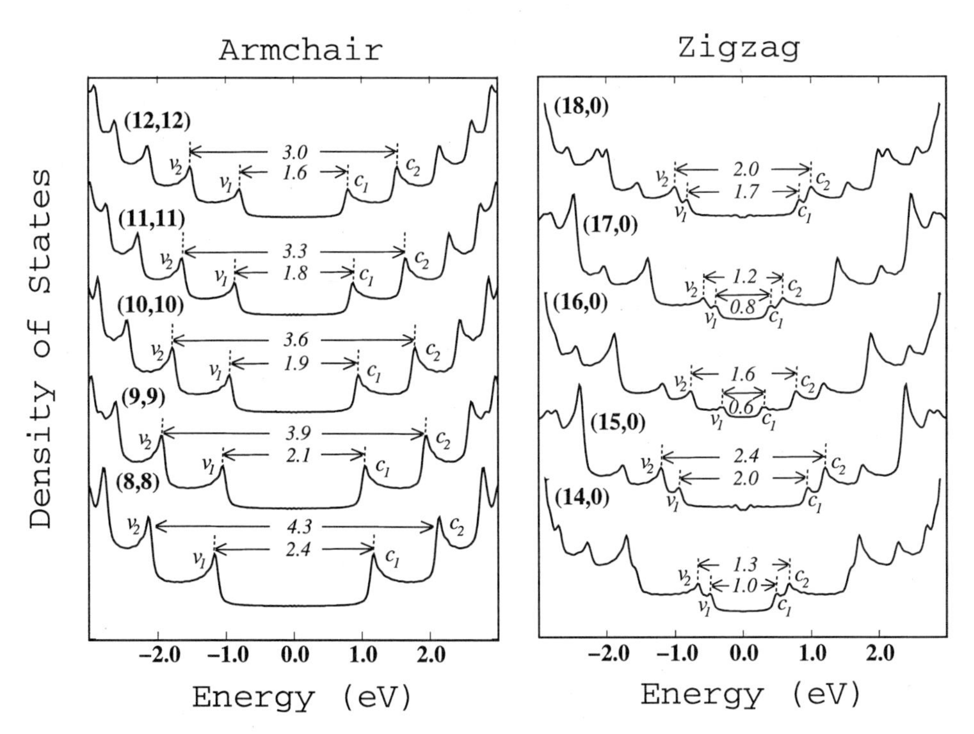

Figure 2: Electronic 1D density of states (DOS) calculated with a tight binding model for (8,8), (9,9), (10,10), (11,11), and (12,12) armchair nanotubes and for (14,0), (15,0), (16,0), (17,0), and (18,0) zigzag nanotubes.[19] Wavevector conserving optical transitions can occur between singularities in the 1D density of electronic states of the valence and conduction bands, i.e., $v_1 \rightarrow c_1$ and $v_2 \rightarrow c_2$, etc. These optical transitions, which are given in the figure in units of eV and are denoted in the text by E_{11}, E_{22}, etc., where $E_{ii} = E_{c_i} - E_{v_i}$ are responsible for the resonant Raman effect. The 1D density of electronic states for chiral nanotubes (n, m) tends to be similar to the results shown here for armchair and zigzag nanotubes.

where the nanotube diameter d_t is related to the integers (n, m) that specify the nanotube by

$$d_t = \sqrt{3}a_{\mathrm{C-C}}(m^2 + mn + n^2)^{1/2}/\pi. \tag{12}$$

If $E_{11}^M(d_t)$ and $E_{11}^S(d_t)$ can be determined experimentally and d_t is known, then the overlap energy parameter γ_0 can be determined from Eqs. (10) and (11). A value of $a_{\mathrm{C-C}} = 0.144$ nm is taken for the nearest neighbor carbon-carbon distance in carbon nanotubes in Eqs. (10), (11) and (12).

Equations (10) and (11) are derived as follows. In Fig. 3 we show the allowed wave vectors k as bold lines and we show the reciprocal lattice wave vectors K_1 and K_2 for a one-dimensional carbon nanotube in the hexagonal Brillouin zone of two-dimensional graphite for (a) metallic and (b) semiconducting carbon nanotubes. The periodic boundary condition for a carbon nanotube (n, m) gives N discrete k values in the circumferential direction, where N is the number of hexagons in the nanotube unit cell given by

$$N = \frac{2(n^2 + m^2 + nm)}{d_R} \tag{13}$$

and d_R is the greatest common divisor of $(2n + m)$ and $(2m + n)$.[1] For example, $N =$ 20 for (10,10) armchair nanotubes. The direction of the discrete k vectors and the separation between two adjacent sets of k vectors are both given by the K_1 vector shown in Fig. 3. Here we show one K point and only a few K_1 vectors for our discussion. In the direction of the carbon nanotube axis which is expressed by the K_2 vector, we can define a set of continuous k vectors in the one-dimensional Brillouin zone for each K_1 vector.

The one-dimensional van Hove singularities near the Fermi energy come from the energy dispersion $E(k)$ for carbon nanotubes along the bold lines in Fig. 3 near the K point of the Brillouin zone of 2D graphite. For metallic carbon nanotubes, one bold line in Fig. 3 goes right through a K point, and this intersection gives rise to a zero band gap at the K point, while in the semiconductor nanotubes, the K point always exists in a position one-third of the distance between two adjacent K_1 lines.[20] The energy minimum of each subband near the K point corresponds to a peak in the 1D electronic density of states that is a van Hove singularity.

Using the linear approximation for the energy dispersion of Eq. (8), the energy differences between the van Hove singularities are expressed by substituting for k the values of K_1 for metallic nanotubes and of $K_1/3$ and $2K_1/3$ for semiconducting nanotubes, respectively. After a simple calculation using the general formula for K_1 for carbon nanotubes,[1] we get the important relation $|K_1| = 2/d_t$. Substituting the value of $|K_1| = 2/d_t$ into Eq. (8), we get the formulae for E_{11} given by Eqs. (10) and (11) in the linear k approximation. For isolated single wall nanotubes, use of the linear k approximation in Eq. (8), allows us to write[21,22] the relation $E_{11}^M(d_t) = 3E_{11}^S(d_t)$ at the same value of d_t. With these approximations, the resonant energy differences for a semiconducting nanotube of given d_t, occur at $E_{11}^S(d_t)$, $2E_{11}^S(d_t)$, $4E_{11}^S(d_t)$, $5E_{11}^S(d_t)$, $7E_{11}^S(d_t)$, ... and for metallic nanotubes resonant energy differences occur at $E_{11}^M(d_t)$, $2E_{11}^M(d_t)$, ... using these approximations.

When the value of $|K_1| = 2/d_t$ is large, which implies smaller values of d_t, the linear dispersion approximation is no longer correct. When we then plot equi-energy lines near the K point (see Fig. 4), we can get circular contours for small k values near the K and K' points of the Brillouin zone, but for large k values the equi-energy contour eventually becomes a triangle which connects the three M points nearest to the K-point (see Figs. 3 and 4). This distortion of the equi-energy lines away from a circular contour in materials with a 3-fold symmetry axis is known as the trigonal

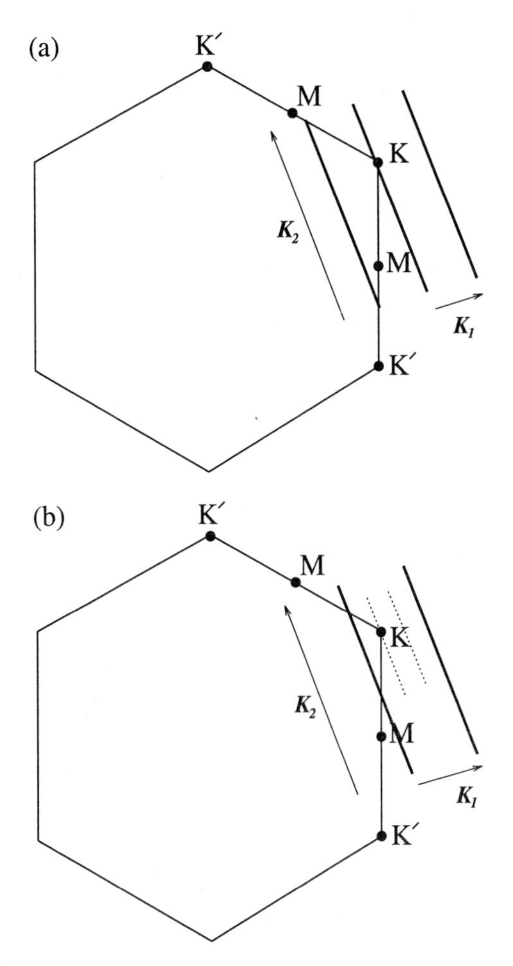

Figure 3: The wave vector k for one-dimensional carbon nanotubes is shown in the two-dimensional Brillouin zone of graphite (hexagon) as bold lines for (a) metallic and (b) semiconducting carbon nanotubes. In the direction of K_1, discrete k values are obtained by applying periodic boundary conditions for the circumferential direction of the carbon nanotubes, while in the direction of the K_2 vector, continuous k vectors are shown in the one-dimensional Brillouin zone. For metallic nanotubes (a), the bold line intersects a K point (corner of the hexagon) at the Fermi energy of graphite. For the semiconducting nanotubes (b), the K point always appears one-third of the distance between two bold lines. It is noted that only a few of all the possible bold lines are shown near the indicated K point. For each K_1 vector, there is an energy minimum in the valence and conduction energy subbands, giving rise to the energy differences E_{ii} shown in Fig. 2.

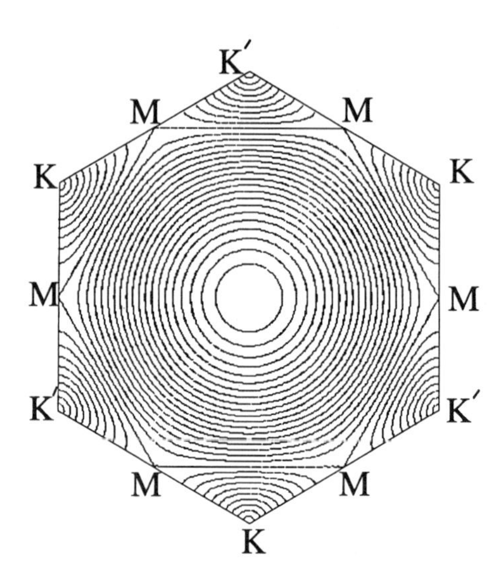

Figure 4: A plot of the equi-energy contours of 2D graphite. The equi-energy lines are circles near the K point and near the center of the Brillouin zone, but are straight lines which connect the nearest M points.

warping effect, and is briefly discussed below for the case of carbon nanotubes. Thus the energy minima positions are not always at the closest positions to the bold lines around the K point (see Fig. 3), so that the energy minima positions (see Fig. 4) now depend on the direction of the K_2 vectors, that is, on the chirality of the carbon nanotubes, as shown in Fig. 5.

Trigonal warping effects generally split the singular peaks of the 1D density of states which come from k-vectors in different directions from the K point. It is only the armchair nanotubes that do not show such a splitting. The splitting becomes a maximum for zigzag nanotubes, as shown in Fig. 6, where we see that the trigonal warping effect is important for single wall zigzag nanotubes with diameters $d_t < 2\,\text{nm}$, and the figure furthermore shows that this trigonal warping effect is especially important for metallic zigzag nanotubes. In Fig. 5 the energy of the van Hove singularity is plotted for metallic (open circles) and semiconducting (closed circles) zigzag nanotubes along $K \to M$ and along $K \to \Gamma$ for different diameter zigzag nanotubes, showing the trigonal warping effect to be largest for the smallest diameter (9,0) nanotube ($d_t \sim 0.7\,\text{nm}$).

When all k values within the Brillouin zone are included in the calculation of the energies where the singularities in the 1D density of states occur, it is seen, as for example in Fig. 2, that for the armchair nanotubes the energy separation $E_{22}^{M}(d_t)$ is somewhat less than $2E_{11}^{M}(d_t)$, and that for the zigzag nanotubes, $E_{22}^{S}(d_t)$ is also not equal to $2E_{11}^{S}(d_t)$. Thus it is important to make a plot for $\Delta E = E_{11}(d_t)$ vs. d_t which includes all nanotubes, includes trigonal warping effects, and allows contributions to $E_{ii}^{M}(d_t)$ to be calculated for all k values.

Kataura et al.[8] were the first to publish such a plot of the energy separations $E_{ii}(d_t)$ as a function of d_t. In Fig. 7 we have re-plotted the work of Kataura et al.[8] for $\gamma_0 = 2.9\,\text{eV}$ and we present $E_{ii}(d_t)$ values for a range of nanotube diameters d_t from 0.7 nm to 3.0 nm. We will make use of Fig. 7 in discussing various experiments that are sensitive to $E_{ii}(d_t)$. Both Fig. 7 and the Kataura et al.[8] plot includes all (n, m) values. The original Kataura plot includes nanotube diameters d_t from 0.7 nm to 1.8 nm and the calculations are based on a value of $\gamma_0 = 2.75\,\text{eV}$. Both plots include both metallic nanotubes ($n - m = 3q$) and semiconducting nanotubes ($n - m \neq 3q$). Both plots also

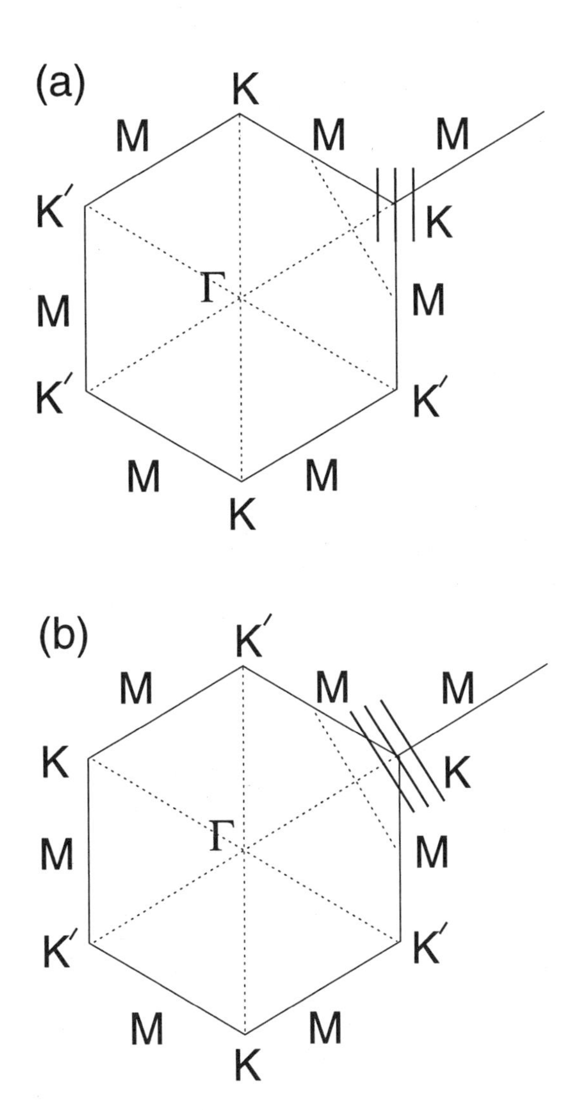

Figure 5: The effect of chirality on the van Hove singularities. The three bold lines near the K point are possible k vectors in the hexagonal Brillouin zone of graphite for metallic (a) armchair, and for (b) zigzag carbon nanotubes. All chiral nanotubes with chiral angles $|\theta| \leq \pi/6$ have lines for their k vectors with the directions making a chiral angle θ measured from the bold lines for the zigzag nanotubes shown in (b). Although there are N inequivalent lines which give $2N$ energy subbands for (n, m) carbon nanotubes, only the line which goes through the K point and neighboring lines are shown for simplicity in the figure. The minimum energy along two neighboring lines gives the energy position of the van Hove singularity. It is clear that the armchair nanotube has the same energy minimum for two neighboring lines, but that the zigzag carbon nanotubes have different energy minimum positions for the two neighboring lines. Similarly, for semiconducting nanotubes, we consider the minima for the energy subbands by changing the distance between the allowed k vector lines to 1/3 of that between the two lines near the K point in (b).

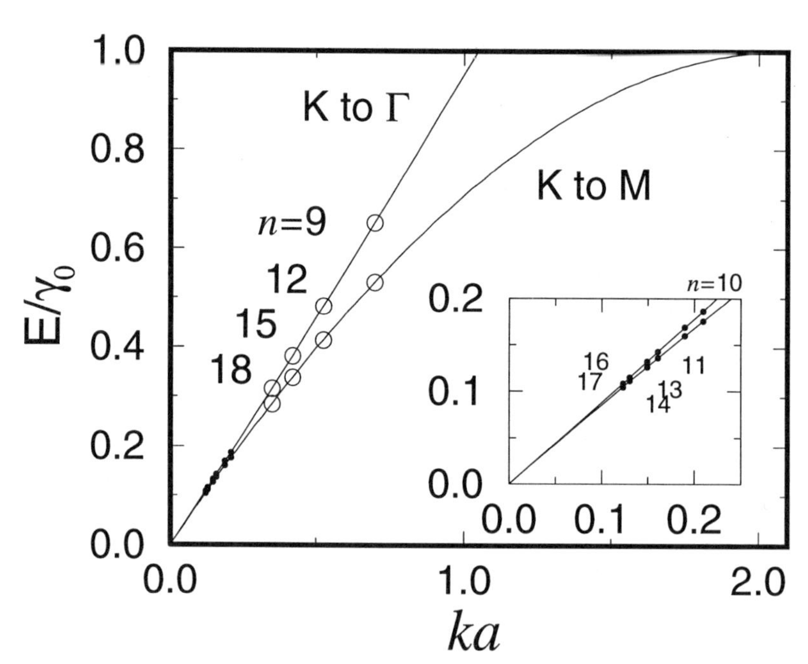

Figure 6: Trigonal warping effect for zigzag nanotubes. Two energy minimum positions in the conduction band for zigzag nanotubes, $(n, 0)$, are measured from the energy at the K point and the energy is normalized to γ_0. Open circles denote metallic carbon nanotubes for $k = |K_1|$ vectors away from the K point along the $K \to M$ and $K \to \Gamma$ lines. Figure 5(b) shows that $K \to \Gamma$ is the direction of the energy minimum. The closed circles denote semiconductor carbon nanotubes for $k = |K_1|/3$ vectors. (The inset shows an expanded view of the figure at small E/γ_0 and small ka for semiconducting nanotubes.) Note that the maximum of the horizontal axis corresponds to the M point, $ka = 2\pi/3$ which is measured from the K point (where $ka = 0$). A nanotube diameter of 1 nm corresponds to a (13,0) carbon nanotube.

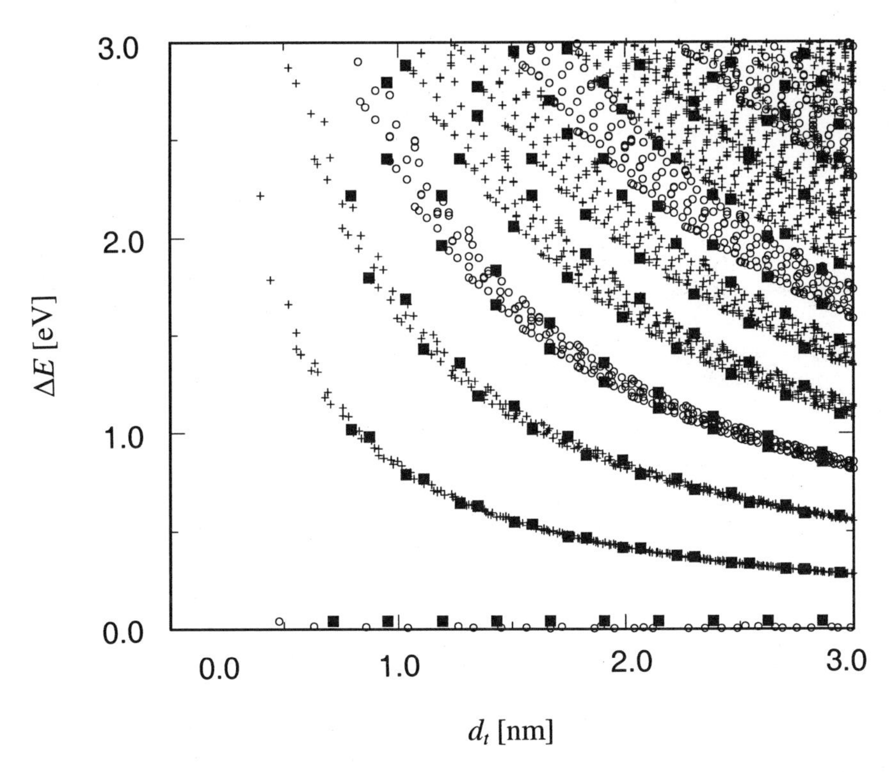

Figure 7: Calculation of the energy separations $\Delta E = E_{ii}(d_t)$ for all (n, m) values as a function of nanotube diameter between $0.7 < d_t < 3.0\,\mathrm{nm}$ (based on the work of Kataura et al.[8]). The results are based on the tight binding model of Eqs. (4) and (5), with $\gamma_0 = |t| = 2.90\,\mathrm{eV}$ and $s = 0$. The semiconducting and metallic nanotubes are indicated by crosses and open circles, respectively. The filled squares denote the zigzag nanotubes.

show that, for a given d_t value, the points for $E_{11}^S(d_t)$ are not precisely at $2a_{\mathrm{C-C}}\gamma_0/d_t$, but show some deviations from this value. The width of the $E_{ii}(d_t)$ curves in Fig. 7 is seen to increase with increasing energy and with increasing i. In Fig. 7 it is seen that the width for a given $E_{ii}(d_t)$ increases as d_t decreases. For a given carbon nanotube diameter, the resonant width for metallic carbon nanotubes $E_{11}^M(d_t)$ is larger than $E_{11}^S(d_t)$ and $E_{22}^S(d_t)$ for the semiconducting nanotubes. In Fig. 7, we also plot the singular peaks for the zigzag nanotubes, $(n, 0)$, denoted by filled squares and we note that the energy splitting associated with these points comes from trigonal warping effects shown in Fig. 6. It is clear from Fig. 7 that the widths of not only $E_{ii}^M(d_t)$ but also of $E_{ii}^S(d_t)$ are determined by the zigzag nanotubes. It is noted here that the $E_{ii}(d_t)$ position for the armchair nanotube is almost at the center of the width of the points for all (n, m) nanotubes of diameter d_t. It should thus be possible to make an experimental study of the dependence of the splitting of the van Hove singularities on nanotube diameter and chirality as predicted in Fig. 7 using STM/STS measurements.

RELEVANT EXPERIMENTAL RESULTS

Four different types of experiments can be used to determine the energy separations $E_{11}(d_t)$, as discussed in this section. One of these utilizes the STS/STM (scanning tunneling spectroscopy and microscopy) techniques, which can at least in principle determine the geometric structure (n, m) by STM and the energy separation $E_{11}(d_t)$ by STS on the same nanotube. The other methods depend on optical measurements, and resonance Raman studies of the Stokes and anti-Stokes processes.

STS/STM Measurements

Of all techniques used to date, the STS/STM technique is the most attractive because it is matched to making measurements on single isolated carbon nanotubes. The STS/STM results of Wildöer et al.[3] yield $\gamma_0 = 2.7 \pm 0.1\,\text{eV}$ for their semiconducting nanotubes on the basis of Eq. (11) and they obtain $\gamma_0 = 2.9 \pm 0.1\,\text{eV}$ for their metallic nanotubes on the basis of Eq. (10). This deviation in γ_0 should be considered to be within the width of $E_{11}^M(d_t)$ shown in Fig. 7. Odom et al.[4] reported results for semiconducting nanotubes, yielding a value of $\gamma_0 = 2.5\,\text{eV}$, significantly lower than the values reported by Wildöer et al.[3] and by other techniques discussed below. From an experimental standpoint, the STS/STM experiments can be made more accurate by determining $E_{ii}(d_t)$ for the specific nanotube for which (n, m) has also been measured.[23] A determination of $E_{11}^M(d_t) = 1.57\,\text{eV}$ along these lines was recently provided[6] for a (13,7) nanotube, yielding a low value for $\gamma_0 = 2.54\,\text{eV}$. The STM/STS method for determining γ_0 is direct, but no corrections have been made for the perturbation of the nanoprobe electric field on the 1D electron density of states. Accurate measurements of d_t by STM is also subject to a number of experimental problems.

Furthermore, although the measurements can be made on isolated carbon nanotubes rather than on nanotubes immersed in nanotube bundles, the nanotubes are normally placed on substrates during the measurement process and the effect of the charge transfer between the nanotube and the substrate is generally not taken into account. Another possible source of error in interpreting STS data for $E_{11}(d_t)$ is associated with the trigonal warping effect and asymmetry of the density of states singularities which should be taken into account when broadening effects in the actual STS data occur. Also greater accuracy would be obtained in the γ_0 determination, if the linear k approximation were relaxed, as is done in Fig. 7, in calculating the $E_{ii}(d_t)$ resonant energies. A possible problem in the evaluation of γ_0 relates to the deformation of the nanotube cross-section by the probe. This distortion leads to a perturbation of the local DOS and thus influences the measurement of γ_0.

Optical Measurements

A second method for determining $E_{11}(d_t)$ comes from optical spectra, where the measurements are made on ropes of single wall carbon nanotubes, so that appropriate corrections should be made for inter-tube interactions in interpreting the experimental data. Optical transmission spectra were taken for single wall nanotubes synthesized using four different catalysts,[7,8] namely NiY (1.24–1.58 nm), NiCo (1.06–1.45 nm), Ni (1.06–1.45 nm), and RhPd (0.68–1.00 nm). For the NiY catalyst, three large absorption peaks were observed at 0.68, 1.2 and 1.7 eV, yielding a value of $\gamma_0 = 3.0 \pm 0.2\,\text{eV}$, using Eqs. (10) and (11). Optical spectra were also reported for nanotubes produced with the NiCo, Ni and RhPd catalysts, but the peak values for the absorption bands were not explicitly quoted.[7,8,24] In interpreting the optical transmission data, corrections for the diameter distribution of the nanotubes, for trigonal warping effects, and for the non-linear k dependence of $E(k)$ away from the K-point need to be considered. The calculations given in Fig. 7 provide a firm basis for a detailed analysis of the optical data.

Resonant Raman Scattering Experiments

The third determination for single-wall carbon nanotubes (SWNTs) discussed here relates to the analysis of the tangential phonon modes,[25–30] which can be sensitively probed by resonant Raman spectroscopy. This technique provides two independent determinations of γ_0, one involving the Stokes spectra (phonon emission), the other involving anti-Stokes spectra (phonon absorption) for the tangential phonon bands. Thus far, almost all of the resonant Raman experiments have been done on the Stokes spectra.

Recently,[25,26] an analysis was made of the dependence on the laser excitation energy E_{laser} of the intensity of the $1540\,\text{cm}^{-1}$ Stokes Raman component (denoted by $\tilde{I}^S_{1540}$) of the special tangential phonon band associated with metallic nanotubes that are resonantly enhanced over a narrow range of laser energies. Referring to Fig. 7, we see that, for nanotubes with a diameter (1.38 nm) equal to that of the (10,10) nanotube, resonant enhancement of metallic nanotubes through the $E^M_{11}(d_t)$ van Hove singularity in the electronic density of states occurs over the range $1.9 < E_{\text{laser}} < 2.2\,\text{eV}$ considering only the incident laser excitation energy $E_{\text{laser}} = E^M_{11}(d_t)$. For the Stokes process, the normalized intensity $\tilde{I}^S_{1540}$ of the dominant Stokes component for metallic nanotubes at $1540\,\text{cm}^{-1}$ has a dependence on E_{laser} that includes both the incident and the scattered photon beams, and $\tilde{I}^S_{1540}(d_0, E_{\text{laser}})$ is given by

$$\begin{aligned}\tilde{I}^S_{1540}(d_0, E_{\text{laser}}) &= \textstyle\sum_{d_t} A \exp\left[\frac{-(d_t-d_0)^2}{\Delta d_t^2/4}\right] \\ &\times \left[(E^M_{11}(d_t) - E_{\text{laser}})^2 + \gamma_e^2/4\right]^{-1} \\ &\times \left[(E^M_{11}(d_t) - E_{\text{laser}} + E_{\text{phonon}})^2 + \gamma_e^2/4\right]^{-1},\end{aligned} \tag{14}$$

where d_0 and Δd_t are, respectively, the mean diameter and the width of the Gaussian distribution of nanotube diameters within the SWNT sample, E_{phonon} is the average energy (0.197 eV) of the tangential phonons, γ_e is a damping factor that is introduced to avoid a divergence of the resonant denominator, and the sum in Eq. (14) is carried out over the diameter distribution. Equation (14) indicates that the normalized intensity $\tilde{I}^S_{1540}(d_0, E_{\text{laser}})$ for the Stokes process is large when either the incident laser energy is equal to $E^M_{11}(d_t)$ or when the scattered laser energy is equal to $E^M_{11}(d_t)$. The dependence of the normalized intensity $\tilde{I}^S_{1540}(d_0, E_{\text{laser}})$ for the actual SWNT sample on E_{laser} is primarily sensitive[25,26] to the energy difference $E^M_{11}(d_t)$ for the various d_t values in the sample, and the resulting normalized intensity $\tilde{I}^S_{1540}(d_0, E_{\text{laser}})$ is expressed here in terms of the average nanotube diameter in the sample. The determination of the diameter distribution of each SWNT sample d_t is most accurately from made TEM measurements,[25,26] though measurement of the radial breathing mode for a number of E_{laser} values provides a useful secondary method for determining d_t.[1]

Curve fitting of experimentally measured $\tilde{I}^S_{1540}(d_0, E_{\text{laser}})$ for the Stokes spectra to Eq. (10) can then be used to determine γ_0 through Eq. (10),[21,31] where the nearest neighbor carbon–carbon distance $a_{\text{C-C}}$ is taken to be 1.44 Å for SWNTs. Thus far $\tilde{I}^S_{1540}(d_0, E_{\text{laser}})$ has been found through the normalization of the intensity of the dominant component I_{1540} for the metallic nanotubes with respect to the intensity I_{1593} for the dominant component for the semiconducting nanotubes.[25,26] Figure 7 shows that for the range of photon energies over which the metallic nanotubes are observed for specific nanotube samples (such as $d_t = 1.37 \pm 0.18\,\text{nm}$), there are semiconducting nanotubes in the sample that are resonant either with respect to the $E^S_{22}(d_t)$ transition or

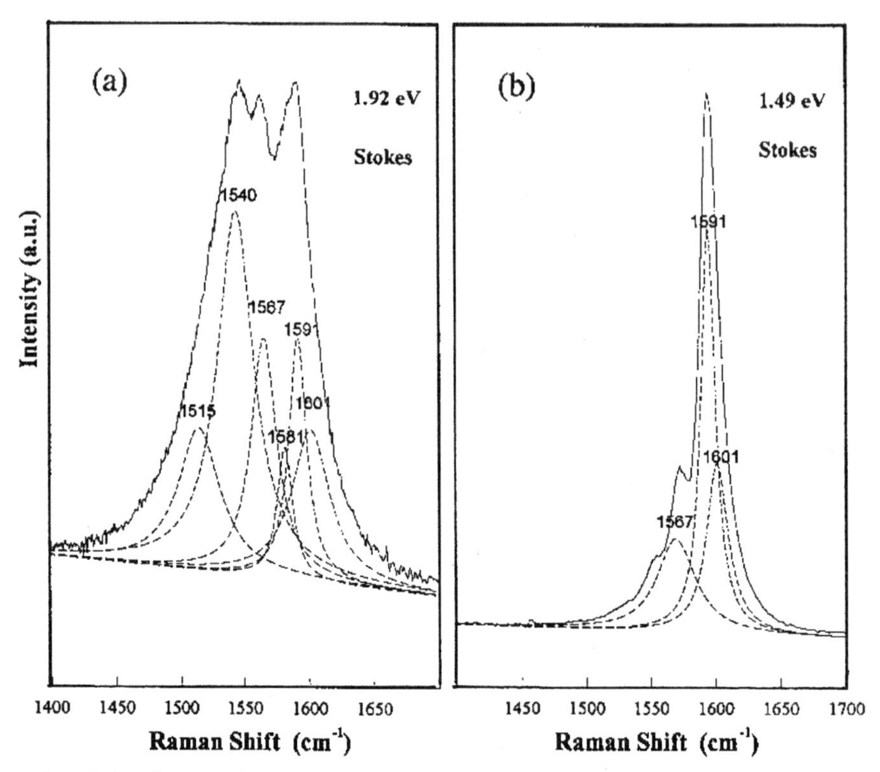

Figure 8: A lineshape analysis for the spectral features of the tangential modes observed at (a) $E_{\text{laser}} = 1.92\,\text{eV}$ and (b) $E_{\text{laser}} = 1.49\,\text{eV}$ for the Stokes process. We note that the spectrum at 1.92 eV shows Raman intensity near $1591\,\text{cm}^{-1}$ where the most intense component for the semiconducting nanotubes occurs, as shown in (b) for the spectrum taken at 1.49 eV.

to the $E_{33}^S(d_t)$ transition. Thus the normalized experimental $\tilde{I}_{1540}^S(d_0, E_{\text{laser}})$ was approximated by the ratio (I_{1540}/I_{1593}) in an early effort[25] to fit the experimental Stokes data to Eqs. (10) and (14).

We show in Fig. 8 the line shape analysis of the tangential Raman band for the Stokes process at $E_{\text{laser}} = 1.92\,\text{eV}$ for which the metallic nanotubes are dominant [Fig. 8(a)] and for $E_{\text{laser}} = 1.49\,\text{eV}$ where the semiconducting nanotubes are dominant [Fig. 8(b)]. The data for Fig. 8 are for a SWNT sample with a diameter distribution of $d_t = 1.49 \pm 0.20\,\text{nm}$. Figure 8(a) suggests that for nanotubes of $d_0 = 1.49\,\text{nm}$ average diameter, resonant Raman scattering at $E_{\text{laser}} = 1.92\,\text{eV}$ occurs predominantly for metallic nanotubes within the sample (see Fig. 7), although there are always semiconducting nanotubes within the sample that are in resonance with either the $E_{22}^S(d_t)$ or the $E_{33}^S(d_t)$ resonant transitions. The contributions of semiconducting nanotubes in Fig. 8(a) is seen by the presence of Raman intensity at 1591, 1601, and $1567\,\text{cm}^{-1}$, the dominant components associated with semiconducting nanotubes, as shown in Fig. 8(b).

However for $E_{\text{laser}} = 1.49\,\text{eV}$, the contributions to the resonant Raman process are almost entirely due to semiconducting nanotubes (see Fig. 7) and there is essentially no Raman intensity at the dominant phonon frequency for the metallic nanotubes $1540\,\text{cm}^{-1}$, as seen in Fig. 8(b). In Refs. [[25, 26]] it is shown that for a sample with nanotube diameters $d_t = 1.37 \pm 0.18\,\text{nm}$, the range of energies where the Stokes spectra for metallic nanotubes are resonantly enhanced, is found experimentally to lie between 1.7 and 2.2 eV. The Stokes resonant Raman spectra for semiconducting nanotubes were

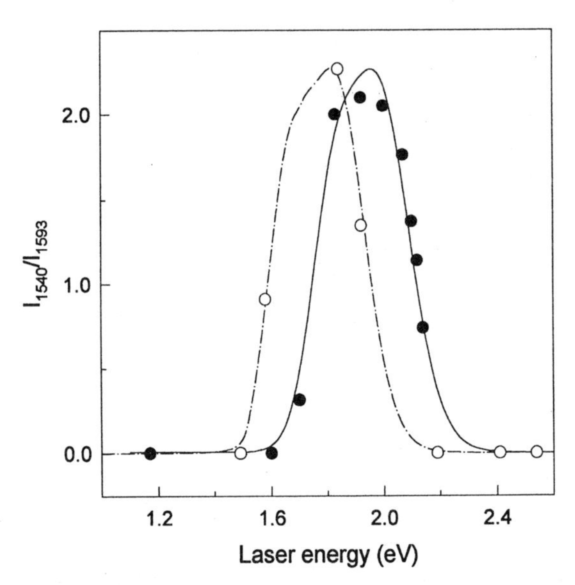

Figure 9: The intensity ratio of the Raman peaks at 1540 and 1593 cm^{-1} as a function of laser excitation energy. The experimental measurements for the $d_0 = 1.37$ nm and $d_0 = 1.49$ nm diameter SWNT samples are, respectively, shown as closed and open circles, and the solid and dot-dashed curves represent the corresponding fits to the experimental data using Eqs. (10) and (14).

found to occur for $E_{\text{laser}} < 1.7$ eV and for $2.2 < E_{\text{laser}} < 3.0$ eV for the same SWNT sample.[25] The experimental results for the two SWNT samples on the dependence of the normalized intensity I_{1540}/I_{1593} on E_{laser} are shown in Fig. 9 and the fit to the experimental points was made using Eqs. (10) and (14).

The best fit to the experimental data for I_{1540}/I_{1593} in the Stokes spectra was obtained for $\gamma_0 = 2.95 \pm 0.05$ eV for the $d_0 = 1.37$ nm sample,[25] and by $\gamma_0 = 2.93 \pm 0.05$ eV for the $d_0 = 1.49$ nm sample (see Fig. 9). The damping parameter $\gamma_e = 0.04$ eV is much less sensitive to the fit of the data to Eq. (14) and may be largely determined experimentally through contributions to a particular $\Delta E_{11}^M(d_t)$ from nanotubes with different chiralities (see Fig. 7). The average value of $E_{11}^M(d_t)$ for the $d_0 = 1.37$ nm SWNT sample was found to be $\langle E_{11}^M \rangle = 1.84 \pm 0.24$ eV from Eq. (10) and for the $d_0 = 1.49$ nm sample, $\langle E_{11}^M \rangle = 1.69 \pm 0.23$ eV was obtained. The results for the larger diameter tubes show a downshift in E_{laser} for the resonance window for metallic nanotubes. This downshift in E_{laser} can be expressed in terms of the E_{laser} values defining the FWHM (full width at half maximum intensity) for the laser energy range of the metallic window (discussed below) which occurs between $E_{\text{laser}} \simeq 1.60$ and 2.00 eV for the $d_0 = 1.49$ nm sample, and between 1.76 and 2.11 eV for the smaller diameter sample with $d_0 = 1.37$ nm.

Anti-Stokes Resonant Raman Spectra

We now discuss the corresponding behavior for the anti-Stokes (phonon absorption) process, which, in principle, can be used independently to determine γ_0. In fact, the best determination of γ_0 is obtained by the measurements of both the Stokes and anti-Stokes spectra for the same sample, as discussed below. The normalized intensity for the dominant phonon mode for the metallic nanotubes at 1540 cm^{-1} as a function of E_{laser} for the anti-Stokes process, denoted by $\tilde{I}_{1540}^{AS}(d_0, E_{\text{laser}})$, depends on terms similar

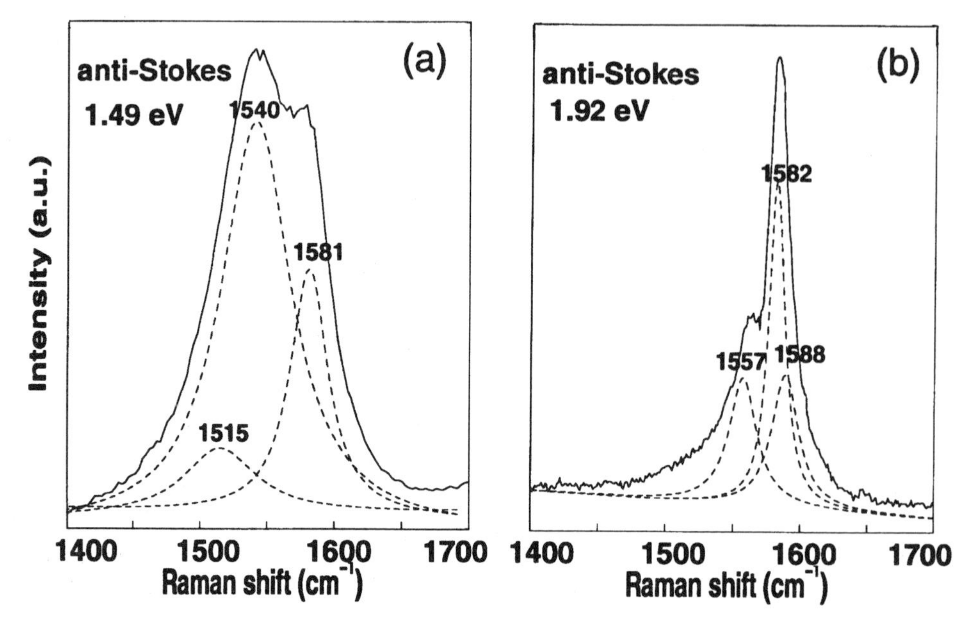

Figure 10: A lineshape analysis for the spectral features of the tangential modes observed at (a) $E_{\text{laser}} = 1.49\,\text{eV}$ and (b) $E_{\text{laser}} = 1.92\,\text{eV}$ for the anti-Stokes process. The metallic nanotubes dominate the spectra at 1.49 eV, while the semiconducting nanotubes dominate the spectra at 1.92 eV. No semiconducting nanotube scattering is seen at $E_{\text{laser}} = 1.49\,\text{eV}$ and very little metallic nanotube scattering is seen at $E_{\text{laser}} = 1.92\,\text{eV}$.

to those appearing in Eq. (14), except for the change in the sign of the term E_{phonon}

$$\begin{aligned} \tilde{I}^{AS}_{1540}(d_0, E_{\text{laser}}) &= \textstyle\sum_{d_t} A \exp\left[\frac{-(d_t-d_0)^2}{\Delta d_t^2/4}\right] \\ &\times [(E^M_{11}(d_t) - E_{\text{laser}})^2 + \gamma_e^2/4]^{-1} \\ &\times [(E^M_{11}(d_t) - E_{\text{laser}} - E_{\text{phonon}})^2 + \gamma_e^2/4]^{-1}. \end{aligned} \tag{15}$$

For the anti-Stokes process we note that $\tilde{I}^{AS}_{1540}(d_0, E_{\text{laser}})$ is large when $E^M_{11}(d_t) = E_{\text{laser}}$ for the incident beam or $E^M_{11}(d_t) = E_{\text{laser}} + E_{\text{phonon}}$ for the scattered beam. Therefore the anti-Stokes spectra for metallic nanotubes appears at *lower* E_{laser} values than for the case of the Stokes spectra. Because of the great similarity between Eqs. (14) and (15), it would appear that the anti-Stokes band for laser energies which excite metallic nanotubes (see Fig. 10) should be similar to the corresponding Stokes band (see Fig. 8) for the same nanotube sample. A comparison between Figs. 8 and 10 for the Stokes and anti-Stokes bands, respectively, show that this is not the case. The anti-Stokes band taken for $E_{\text{laser}} = 1.49\,\text{eV}$ in Fig. 10 shows the strongest Lorentzian component to be at $1540\,\text{cm}^{-1}$, just as for the Stokes spectrum at $E_{\text{laser}} = 1.92\,\text{eV}$ [Fig. 8(a)] where the metallic nanotubes are dominant. However, Fig. 10(a) shows that there is no Raman intensity at $1593\,\text{cm}^{-1}$, where the Raman band for the semiconducting nanotubes is dominant. The absence of the $1593\,\text{cm}^{-1}$ mode from the anti-Stokes spectrum at $E_{\text{laser}} = 1.49\,\text{eV}$ can be understood by referring to Fig. 7, where it is seen that for carbon nanotubes with diameters of $d_t = 1.49 \pm 0.20\,\text{nm}$, no semiconducting nanotubes in the sample are resonant for the range of E_{laser} where the metallic nanotubes are resonant for the anti-Stokes process, in agreement with the experimental observations.

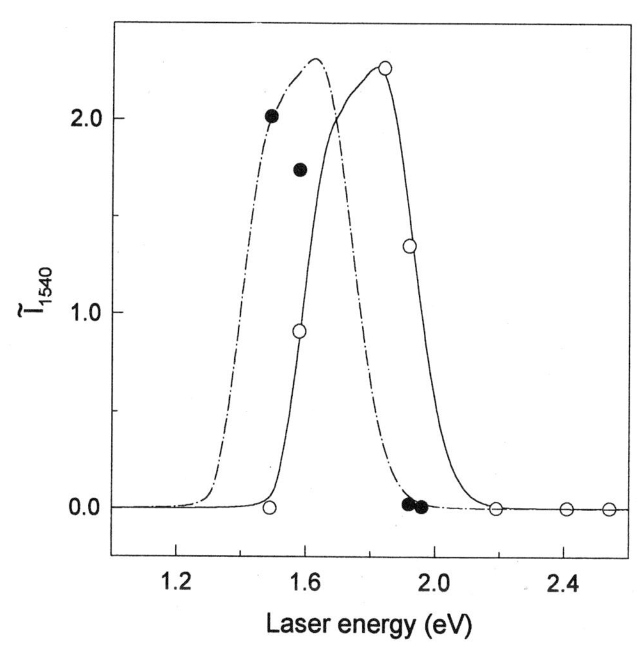

Figure 11: Stokes and anti-Stokes normalized intensity for the $\tilde{I}_{1540}$ metallic feature as a function of laser energy for the same SWNT sample ($d_0 = 1.49\,\text{nm}$). The experimental Stokes data are shown as open circles and the solid curve is the fit of Eq. (14) to the experimental data. The experimental normalization is done in terms of I_{1540}/I_{1593} for the Stokes data. The predicted plot of $\tilde{I}^{AS}_{1540}$ vs. E_{laser} for the anti-Stokes process uses the same parameters as for the Stokes process, except that Eq. (15) is used for the functional form for $\tilde{I}^{AS}_{1540}$. Preliminary experimental anti-Stokes data are given by the solid points and here the normalization is made with respect to the FWHM linewidth of the tangential mode band (see Fig. 10) which includes the $1540\,\text{cm}^{-1}$ component associated with metallic nanotubes.

Therefore, a normalization different from I_{1540}/I_{1593} is needed for $\tilde{I}^{AS}_{1540}$ in the anti-Stokes tangential mode spectrum as a function of E_{laser}.

Since properly normalized experimental data are not yet available for $\tilde{I}^{AS}_{1540}$ for the anti-Stokes process, we show in Fig. 11 a theoretical plot of the expected dependence of the intensity of the normalized metallic anti-Stokes Raman component $\tilde{I}^{AS}_{1540}$ on E_{laser}. Also plotted in Fig. 11 is the E_{laser} dependence of $\tilde{I}^{S}_{1540}$ for the Stokes process, showing both the experimental data and the theoretical fit using Eqs. (10) and (14). From Fig. 11 we see the importance of the resonant denominator term for the scattered beam in producing the profile shown in this figure. The metallic window which is defined as the range of E_{laser} where the metal nanotubes contribute to both the Stokes and the anti-Stokes processes is found from this definition using Fig. 11. It is the resonance with the *scattered* beam that is responsible for the downshift of the anti-Stokes profile relative to that for the Stokes profile that is seen in Fig. 11. The point at $E_{\text{laser}} = 1.69\,\text{eV}$ where the intensities of the Stokes and anti-Stokes $\tilde{I}_{1540}$ are equal represents the center of the metallic window for the nanotube sample and therefore provides the most accurate determination of γ_0 from the resonant Raman measurements. The experimental points at 1.92 eV for the anti-Stokes Raman spectra show very little intensity at $1540\,\text{cm}^{-1}$ [see Fig. 10(b)]. The anti-Stokes spectrum at 1.92 eV is in sharp contrast to the Stokes spectra at 1.92 eV where the $1540\,\text{cm}^{-1}$ feature associated with the metallic nanotubes is very strong, as seen in Fig. 8.

The anti-Stokes Raman spectra taken at $E_{\text{laser}} = 1.49\,\text{eV}$ and $1.58\,\text{eV}$ show strong Raman intensity at $1540\,\text{cm}^{-1}$, but the absence of the $1593\,\text{cm}^{-1}$ contribution to the spectrum prevents use of the normalization that was employed in Fig. 9. Instead we used the FWHM linewidth of the spectra observed within the metallic window to provide a qualitative normalization for the relative contribution of the metallic nanotubes. The experimental points (solid circles) thus obtained are shown in Fig. 11. A more detailed experimental study of the anti-Stokes Raman spectra for the tangential band in the metallic window is presently in progress.[32]

It is noted that the asymmetry between the Stokes and anti-Stokes spectra (as shown for example at $E_{\text{laser}} = 1.49\,\text{eV}$ and $1.93\,\text{eV}$) is unique for single wall carbon nanotubes, reflecting the one-dimensional nature of the electronic structure. In fact, for graphite there is no difference between the Stokes and anti-Stokes lineshapes.[33]

THE '*D*-BAND' DISPERSION EFFECTS

Another approach to the determination of γ_0 comes from a study of the dispersion of the '*D*-band' for carbon nanotubes, which bears a close relationship to the *D*-band observed in the Raman spectra in disordered sp^2 carbons. The *D*-band in these Raman spectra is associated either with structural disorder relative to the ideal graphite structure, or with finite size effects associated with the small crystallite size in some sp^2 carbon samples. It is likely that some of the '*D*-band' contribution in presently available samples, nominally labelled single wall carbon nanotubes, is due to disordered carbon constituents in the samples, because of the observed spot to spot variation in the '*D*-band' intensity. It is, however, believed that some of the signal is due to intrinsic scattering associated with the nanotubes themselves, because of the different dependence on E_{laser} of the intensity of the '*D*-band' for nanotubes relative to other sp^2 carbons.[34]

It has recently been shown that the dispersion of the *D*-band mode frequency is related to a strong coupling between electrons and phonons with the same wave vector ($\Delta k = \Delta q$) near the K-point in the 2D Brillouin zone.[35,36] Study of the dispersion of this feature allows a determination of γ_0 to be made through the relation[35]

$$\frac{\partial \omega}{\partial \Delta q} = \sqrt{3} a_0 \gamma_0 \frac{\partial \omega}{\partial E_{\text{laser}}}. \tag{16}$$

The quantity $\partial\omega/\partial E_{\text{laser}}$ is determined experimentally, while $\partial\omega/\partial\Delta q$ has been determined from the phonon dispersion curves for graphite. With these assumptions, γ_0 has been determined for sp^2 carbons using Eq. (16) where $a_0 = \sqrt{3}a_{\text{C-C}}$, assuming that $\partial\omega/\partial\Delta q$ for phonons in sp^2 carbons has the same value as for graphite, based on measurements of the sound velocity for the TA phonon branch.[35] While it is not obvious that $\partial\omega/\partial\Delta q$ has the same value for single wall carbon nanotubes as it does for graphite, we nevertheless make the same approximation for $\partial\omega/\partial\Delta q$ that was made for sp^2 carbons in general.[35]

The experimental Raman spectra for single wall carbon nanotubes show a feature at $\sim 1350\,\text{cm}^{-1}$ for $E_{\text{laser}} \sim 2.4\,\text{eV}$ that is highly dispersive,[27] and as stated above, is sensitive to the location of the laser beam on the SWNT sample. This feature has been studied as a function of E_{laser}. From the arguments given above, measurement of the dispersion of this '*D*-band' feature $\partial\omega/\partial E_{\text{laser}}$ permits determination of γ_0 through Eq. (16). Values of $\partial\omega/\partial E_{\text{laser}}$ have been found by several groups using SWNT samples with different distributions of diameters. Results for $\partial\omega/\partial E_{\text{laser}}$ for a particular SWNT

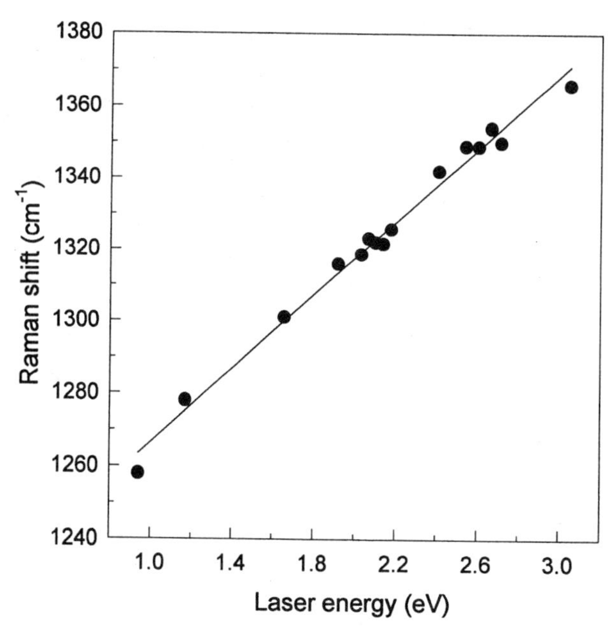

Figure 12: Frequency of the 'D-band' for single wall carbon nanotubes as a function of laser excitation energy. The line is a least squares fit to the data points for one SWNT sample. The fit to the data points yields a slope of $51.2\,\mathrm{cm}^{-1}/\mathrm{eV}$ and an intercept of $1215\,\mathrm{cm}^{-1}$ at $E_{\mathrm{laser}} = 0$.

sample can be obtained from the slope of the ω vs E_{laser} plot shown in Fig. 12, yielding a value of $51\,\mathrm{cm}^{-1}/\mathrm{eV}$. Raman scattering spectra taken over a range of E_{laser} values show this feature to have a dispersion of $52\pm1\,\mathrm{cm}^{-1}/\mathrm{eV}$ close to that of graphite for which the dispersion is $48 \pm 3\,\mathrm{cm}^{-1}/\mathrm{eV}$.[35,37] Using an average value of $\partial\omega/\partial E_{\mathrm{laser}} = 52\,\mathrm{cm}^{-1}/\mathrm{eV}$ for single wall carbon nanotubes and $a_0 = 2.46\,\mathrm{Å}$ yields a value of $\gamma_0 = 2.9 \pm 0.2\,\mathrm{eV}$. A similar analysis based on the dispersion of the G' band[35,37] which is observed at approximately twice the 'D-band' phonon frequency[35] yields a value of $\gamma_0 = 2.8\pm0.2\,\mathrm{eV}$ based on average values of the slope $\partial\omega/\partial E_{\mathrm{laser}}$ of 96.5 and $106\,\mathrm{cm}^{-1}/\mathrm{eV}$ for the G'-band for graphite and carbon nanotubes, respectively.

SUMMARY AND CONCLUSIONS

From comparison of resonant Raman tangential band spectra, optical transmission data and STM/STS measurements we conclude that $\gamma_0 = 2.9 \pm 0.2\,\mathrm{eV}$ for SWNTs of diameter $d_t \sim 1.4\,\mathrm{nm}$, which is approximately 10% lower than the value of γ_0 for graphite. The discussion in this paper shows that most of the experiments reported to date, which yield information relevant to the value of γ_0, yield values consistent with $\gamma_0 = 2.9\pm0.2\,\mathrm{eV}$. It is expected that additional experimental determinations with reduced uncertainties will be available in the near future, as well as determinations of γ_0 using new experimental techniques, such as the resonant Raman effect for the anti-Stokes spectra. Perhaps measurement of the magnetic susceptibility, which has been important for the determination of γ_0 for 3D graphite,[16,38] could also provide interesting results regarding γ_0 for SWNTs and the dependence of γ_0 on d_t. Furthermore, we can anticipate future experiments on SWNTs which could illuminate phenomena showing differences in the $E(k)$ relations for the conduction and valence bands of SWNTs. Such information would be of particular interest for the experimental determination of the overlap integral s as a function of nanotube diameter.

The discussion presented in this paper for the determination of γ_0 depends on assuming $s = 0$, in order to make direct contact with the tight-binding calculations. If $s \neq 0$, then the determination of γ_0 would depend on the physical experiment that is used because different experiments emphasize different k points in the Brillouin zone. Theoretical tight binding calculations for nanotubes should also be refined. Higher order (more distant neighbor) interactions should yield corrections to the lowest order theory discussed here.

The authors gratefully acknowledge valuable discussions with Drs. P.C. Eklund, K. Kneipp, H. Kataura, C.M. Lieber and C. Dekker. One of us (M.A.P.) is thankful to the Brazilian agencies FAPEMIG, CAPES, and FINEP for financial support during his visit to MIT., P.C. acknowledges support by the Brazilian agency FAPESP, and R.S. acknowledges a grant from the Japanese Ministry of Education (No. 11165216) and the Japan Society for the Promotion of Science for his visit to MIT. The MIT authors acknowledge support for this work under NSF grant DMR 98-04734 and INT 98-15744. The resonant Raman measurements performed at the George R. Harrison Spectroscopy Laboratory at MIT were supported by the NIH grant P41-RR02594 and NSF grant CHE 97-08265.

REFERENCES

1. R. Saito, G. Dresselhaus, and M. S. Dresselhaus, *Physical Properties of Carbon Nanotubes* (Imperial College Press, London, 1998).
2. M. S. Dresselhaus, G. Dresselhaus, and P. C. Eklund, *Science of Fullerenes and Carbon Nanotubes* (Academic Press, New York, NY, 1996).
3. J. W. G. Wildöer, L. C. Venema, A. G. Rinzler, R. E. Smalley, and C. Dekker, Nature (London) **391**, 59–62 (1998).
4. T. W. Odom, J. L. Huang, P. Kim, and C. M. Lieber, Nature (London) **391**, 62–64 (1998).
5. T. W. Odom, J. L. Huang, P. Kim, M. Ouyang, and C. M. Lieber, J. Mater. Res. **13**, 2380–2388 (1998).
6. T. W. Odom. Private communication.
7. H. Kataura, Y. Kumazawa, Y. Maniwa, I. Umezu, S. Suzuki, Y. Ohtsuka, and Y. Achiba, Synthetic Metals **103**, 2555–2558 (1999).
8. H. Kataura, Y. Kumazawa, N. Kojima, Y. Maniwa, I. Umezu, S. Masubuchi, S. Kazama, X. Zhao, Y. Ando, Y. Ohtsuka, S. Suzuki, and Y. Achiba. In *Proc. of the Int. Winter School on Electronic Properties of Novel Materials (IWEPNM'99)*, edited by H. Kuzmany, M. Mehring, and J. Fink, page unpublished (in press), American Institute of Physics, Woodbury, N.Y., 1999. AIP conference proceeding.
9. P. R. Wallace, Phys. Rev. **71**, 622 (1947).
10. J. W. McClure, Phys. Rev. **104**, 666 (1956).
11. M. S. Dresselhaus, G. Dresselhaus, K. Sugihara, I. L. Spain, and H. A. Goldberg, *Graphite Fibers and Filaments* (Springer-Verlag, Berlin, 1988), Vol. 5 of *Springer Series in Materials Science*.
12. R. Saito, M. Fujita, G. Dresselhaus, and M. S. Dresselhaus, Phys. Rev. B **46**, 1804–1811 (1992).

13. G. S. Painter and D. E. Ellis, Phys. Rev. B **1**, 4747 (1970).
14. J. C. Slonczewski and P. R. Weiss, Phys. Rev. **99**, 636 (1955).
15. J. C. Slonczewski and P. R. Weiss, Phys. Rev. **109**, 272 (1958).
16. J. W. McClure, Phys. Rev. **108**, 612 (1957).
17. J. C. Charlier, X. Gonze, and J. P. Michenaud, Phys. Rev. B **43**, 4579 (1991).
18. Ph. Lambin and V. Meunier. In *Proceedings of the Winter School on Electronic Properties Novel Materials*, edited by H. Kuzmany, J. Fink, M. Mehring, and S. Roth, pages 504–508, 1998. Kirchberg Winter School, (AIP Conference proceedings 442, Woodbury, 1998).
19. J. C. Charlier. unpublished.
20. R. A. Jishi, D. Inomata, K. Nakao, M. S. Dresselhaus, and G. Dresselhaus, J. Phys. Soc. Jpn. **63**, 2252–2260 (1994).
21. C. T. White and T. N. Todorov, Nature (London) **393**, 240 (1998).
22. J. W. Mintmire and C. T. White, Phys. Rev. Lett. **81**, 2506–2509 (1998).
23. C. H. Olk and J. P. Heremans, J. Mater. Res. **9**, 259–262 (1994).
24. H. Kataura. Private communication.
25. M. A. Pimenta, A. Marucci, S. Empedocles, M. Bawendi, E. B. Hanlon, A. M. Rao, P. C. Eklund, R. E. Smalley, G. Dresselhaus, and M. S. Dresselhaus, Phys. Rev. B **58**, R16012–R16015 (1998).
26. M. A. Pimenta, A. Marucci, S. D. M. Brown, M. J. Matthews, A. M. Rao, P. C. Eklund, R. E. Smalley, G. Dresselhaus, and M. S. Dresselhaus, J. Mater. Research **13**, 2396–2404 (1998).
27. A. M. Rao, E. Richter, S. Bandow, B. Chase, P. C. Eklund, K. W. Williams, M. Menon, K. R. Subbaswamy, A. Thess, R. E. Smalley, G. Dresselhaus, and M. S. Dresselhaus, Science **275**, 187–191 (1997).
28. A. Kasuya, M. Sugano, Y. Sasaki, T. Maeda, Y. Saito, K. Tohji, H. Takahashi, Y. Sasaki, M. Fukushima, Y. Nishina, and C. Horie, Phys. Rev. B **57**, 4999 (1998).
29. M. Sugano, A. Kasuya, K. Tohji, Y. Saito, and Y. Nishina, Chem. Phys. Lett. **292**, 575–579 (1998).
30. M. J. Matthews, M. A. Pimenta, S. D. M. Brown, A. Marucci, M. S. Dresselhaus, M. Endo, and C. Kim. In *Extended Abstract for the International Symposium on Carbon Science and Technology for New Carbons*, edited by M. Endo, 1998. Tokyo, November 8-12, 1998.
31. J.-C. Charlier and Ph. Lambin, Phys. Rev. B **57**, R15037 (1998).
32. K. Kneipp, H. Kneipp, P. Corio, S. D. M. Brown, K. Shafer, J. Motz, L. T. Perelman, E. B. Hanlon, A. Marucci, G. Dresselhaus, and M. S. Dresselhaus, (unpublished).
33. P.-H. Tan, Y.-M. Deng, and Q. Zhao, Phys. Rev. B **58**, 5435–5439 (1998).
34. M. A. Pimenta, E. B. Hanlon, A. Marucci, P. Corio, S. D. M. Brown, S. Empedocles, M. Bawendi, G. Dresselhaus, and M. S. Dresselhaus, (unpublished).
35. M. J. Matthews, M. A. Pimenta, G. Dresselhaus, M. S. Dresselhaus, and M. Endo, Phys. Rev. B **59**, R6585 (1999).
36. I. Pocsik, M. Hundhausen, M. Koos, and L. Ley, J. Non-Cryst. Solids **227-230 B**, 1083 (1998).
37. Huiming Cheng. Private communication.
38. K. S. Krishnan, Nature (London) **133**, 174 (1934).

ELECTRONIC AND MECHANICAL PROPERTIES OF CARBON NANOTUBES

L. Forró[1], J.-P. Salvetat[1], J.-M. Bonard[1], R. Bacsa[1], N.H. Thomson[1], S. Garaj[1], L. Thien-Nga[1], R. Gaál[1], A. Kulik[1], B. Ruzicka[2], L. Degiorgi[2], A. Bachtold[3], C. Schönenberger[3], S. Pekker[4], K. Hernadi[5]

[1]Département de Physique, EPF-Lausanne, CH-1015 Lausanne, Switzerland
[2]Laboratorium für Festkörperphysik, ETH-Zürich, CH-8093 Zürich, Switzerland
[3]Institut für Physik, Universität Basel, Klingelbergstr. 82, CH-4056 Basel, Switzerland
[4]Research Institute for Solid State Physics and Optics, H-1525 Budapest, Hungary
[5]Applied and Environmental Chemistry Department, JATE, H-6720 Szeged, Hungary

INTRODUCTION

Interest in carbon nanotubes has grown at a very rapid rate because of their many exceptional properties, which span the spectrum from mechanical and chemical robustness to novel electronic transport properties. Their physics, chemistry and perspectives for applications are very challenging. Below we highlight the main results of the Lausanne group and their collaborators on transport, electron spin resonance, elastic and field emission properties of single wall (SWNT) and multi-wall (MWNT) carbon nanotubes.

SAMPLES

We use MWNTs produced by arc discharge or by thermal decomposition of hydrocarbons, and SWNTs either prepared by the arc discharge method in the presence of catalysts or commercially available (Carbolex, Rice University, MER, DEL). The first step in the study of CNTs is technological: their purification. This is especially true for SWNTs, which are severely contaminated with magnetic catalyst particles. The purity of the arc-discharge fabricated MWNTs is much better, since magnetic materials are not used in their production. Nevertheless they have to be separated from graphitic flakes, polyhedral particles and amorphous carbon present in the raw soot. For MWNTs, we have developed a

Science and Application of Nanotubes, edited by Tománek and Enbody
Kluwer Academic / Plenum Publishers, New York, 2000

soft purification method, which uses the properties of colloidal suspensions[1]. We started the purification with a suspension prepared from 500 ml of distilled water, 2.5 g of SDS (sodium dodecyl sulfate; a common surfactant) and 50 mg of MWNT arc powder sonicated for 15 minutes. Sedimentation and centrifugation (at 5000 rpm for 10 minutes) removed all graphitic particles larger than 500 nm from the solution, as confirmed by low magnification SEM observations (upper part of Fig. 1). We then added surfactant to the solution to reach 12 CMC (critical micelle concentration). At these surfactant concentrations, micelles form and induce flocculation, i.e. the formation of aggregates. These aggregates mostly contain large objects, while smaller objects remain dispersed, and sediment after a certain time, typically a few days. After decanting the suspension one week later, we repeat the procedure once or twice. Fig. 1 (lower part) shows scanning electron microscopy images of a MWNT deposit after the separation procedure. The untreated material contains a large proportion of nanoparticles (typically 70 % in number and 40 % in weight). After the purification, the material remaining in suspension consisted nearly exclusively of nanoparticles, while the sediment contained nanotubes with a content of over 80 % in weight.

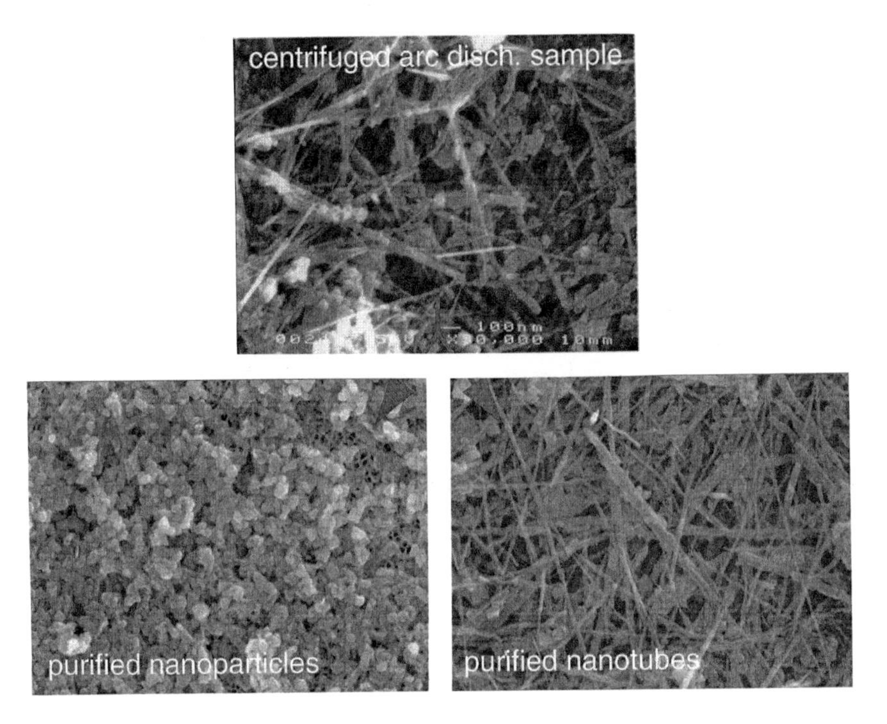

Figure 1. Scanning electron (SEM) micrographs of a MWNT deposit (top) and MWNTs after purification and as a "side-product ", nanoparticles of carbon.

For SWNTs, the purification of the raw soot has been carried out by oxidative dissolution of the carbon encapsulated metal particles with concentrated acid, which ensures maximum efficacy of the process of metal elimination. A weighed amount of the raw soot was sonicated in an ultrasonic bath at 25 •C with concentrated nitric acid for a few minutes and subsequently refluxed for 4-6 h. Thick brown fumes containing oxides of nitrogen were seen, indicating the rapid oxidation of carbon to carbon dioxide. After cooling, water was added so as to leave the samples in 6M HNO_3 for the next 8-12h, after which, it was centrifuged several times and the supernatant rejected until the pH of the solution was around 6.5. High resolution transmission electron microscopy of the material from this solution showed the presence of long ropes of bundled nanotubes accompanied by small amounts of carbon-coated metal particles. Parallel examination of the unpurified soot indicated that more than 80% of the metal had been dissolved. The SWNTs in suspension were stabilized by using a surfactant such as SDS and left undisturbed for 3-5 days until the slow aggregation of the nanotubes allowed their separation from the nanoparticles in solution. The nanotube suspension was filtered through a polycarbonate membrane (1 μm pore size) in order to eliminate most of the particles. On drying, the sediment on the filter paper peeled away to form a self supporting sheet of carbon nanotubes. Scanning electron micrographs of such a sediment, like the one in Fig. 2 shows a network of SWNTs.

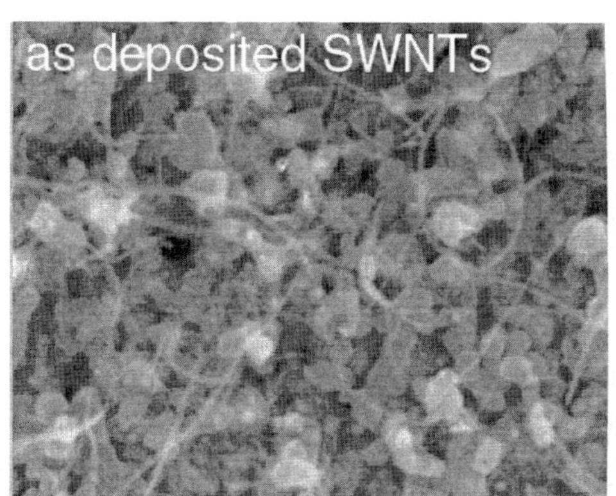

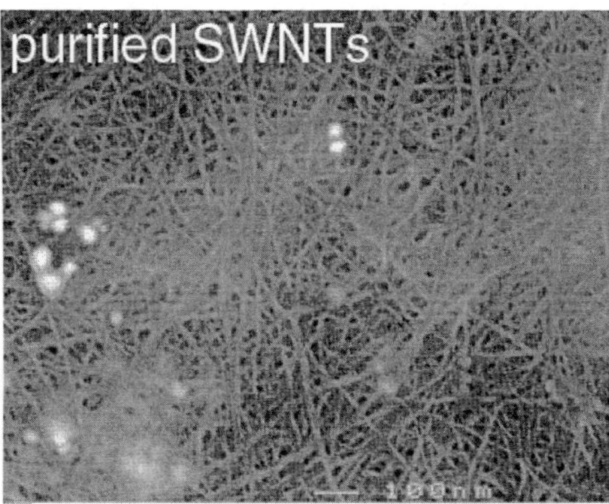

Figure 2. Scanning electron (SEM) micrographs of (a) a SWNT deposit and (b) a mat of SWNTs after purification. A few nanoparticles and embedded catalyst particles are still present.

MWNTs are also prepared by catalytic decomposition of acetylene (or other carbon-containing materials) over supported transition metal catalysts in a temperature range of 700-800°C. This reaction can be carried out under relatively mild conditions in a fixed bed flow reactor at atmospheric pressure. After optimization, the catalytic method can be suitable for the production of either single and multiwall or spiral carbon nanotubes.

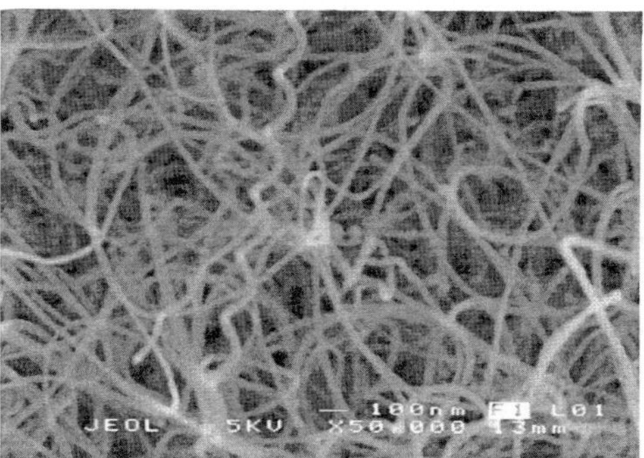
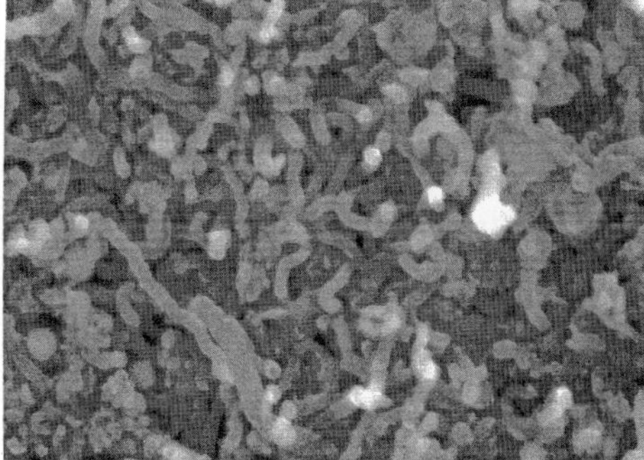

Figure 3. Scanning electron (SEM) micrographs of MWNTs prepared by catalytic decomposition of acetylene.

A further advantage of this method is that it enables the deposition of carbon nanotubes on pre-designed lithographic structures[2], producing ordered arrays which can be used in applications such as thin-screen technology, electron guns etc. The feasibility of the deposition of carbon nanotubes on a ceramic membrane and its field emission properties was demonstrated.

MECHANICAL PROPERTIES

It is becoming clear from recent experiments[3-9] that carbon nanotubes (CNTs) are fulfilling their promise to be the ultimate high strength fibres for use in materials applications. There are many outstanding problems to be overcome before composite materials, which reflect the exceptional mechanical properties of the individual nanotubes, can be fabricated. Arc-discharge methods are unlikely to produce sufficient quantities of nanotubes for such applications. Therefore, catalytically grown tubes are preferred, but these generally contain more disorder in the graphene walls and consequently they have lower moduli than the arc-grown ones. Catalytic nanotubes, however, have the advantage that the amount of disorder (and therefore their material properties) can be controlled through the catalysis conditions, as mentioned before. As well as optimizing the material properties of the individual tubes for any given application, the tubes must be bonded to a surrounding matrix in an efficient way to enable load transfer from the matrix to the tubes. In addition, efficient load bearing within the tubes themselves needs to be accomplished, since, for multi-walled nanotubes (MWNTs), experiments have indicated that only the outer graphitic shell can support stress when the tubes are dispersed in an epoxy matrix[10], and for single wall nanotube (SWNT) bundles (also known as ropes), it has been demonstrated that shearing effects due to the weak intertube cohesion gives significantly reduced moduli compared to individual SWNTs[6]. The reduced bending modulus of these SWNT bundles is a function of their diameter. An individual tube has an elastic modulus of about 1 TPa, but this falls to around 100 GPa for bundles 15 to 20 nm in diameter. In summary, there are two main challenges to address: to enable strong bonding between the CNTs and the surrounding matrix; to create crosslinks between the shells of MWNTs and also between the individual SWNTs in SWNT bundles, so that loads can be homogeneously distributed throughout the CNTs. Ideally, both these goals should be achieved without compromising the mechanical properties of the CNTs too drastically. Efforts within this group have begun to address these problems using post production modification of CNTs via chemical means and controlled irradiation.

High resolution transmission electron microscopy (HRTEM) can be used to give invaluable information about the structure of CNTs, in particular, the amount of order/disorder within the walls of MWNTs. Atomic force microscopy (AFM) can be used to measure the mechanical properties of individual CNTs[4-7]. Use of both techniques has allowed us to make a correlation between the strength of MWNTs, grown in different ways, with the amount of disorder within the graphene walls (see below).

The AFM technique developed in our laboratories has already enabled characterisation of the moduli of SWNT bundles[6] and MWNTs, both arc-grown and catalytically grown[7]. The method has been described in detail previously[6,7]. Briefly, it involves depositing CNTs from a suspension in liquid onto well-polished alumina ultrafiltration membranes with a pore size of about 200 nm (Whatman anodisc). By chance, CNTs occasionally span the pores and these can be subjected to mechanical testing on the nanometer length scale. Contact mode AFM (M5 Park Scientific Instruments) under ambient conditions is used to collect images of the suspended CNTs at various loading forces. Fig. 4 shows an AFM image of a SWNT bundle suspended across a pore and a schematic representation of the mechanical test. The maximum deflection of the CNT into the pore as a function of the loading force can be used to ascertain whether the behavior is elastic. If the expected linear behavior is observed, the Young's modulus can be extracted using a continuum mechanics model for a clamped beam configuration[6,7]. The suspended length of the CNT, its deflection as a function of load and its diameter can all be determined from the images, enabling the modulus to be deduced. The diameter is taken as the height of the tube above the membrane surface at the clamped ends. Although tip convolution can be a problem in measuring lateral dimensions using AFM, the height is a reliable measure because the CNTs are essentially incompressible at these loads (nominal loading forces are in the range 1 to 5 nN). The suspended length can be determined from line profiles taken either side of the tube. A minimum suspended length is measured because of the convolution of the tip with the edge of the pore. This means that all the determined moduli quoted here are minimum values.

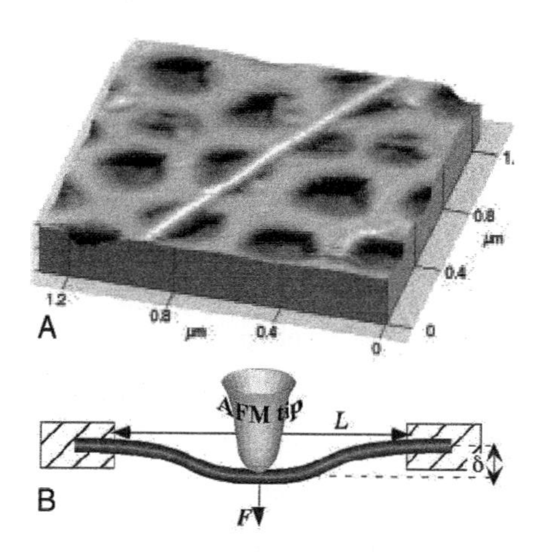

Figure 4. (a) 3D rendering of an AFM image of a SWNT bundle adhered to the alumina ultrafiltration membrane, leading to a clamped beam configuration for mechanical testing. (b) Schematic representation of the measurement technique. The AFM applies a load, *F*, to the portion of nanotube with a suspended length of *L* and the maximum deflection *d* at the center of the beam is directly measured from the topographic image, along with *L* and the diameter of the tube (measured as the height of the tube above the membrane).

The powerful advantage of the AFM technique employed in our laboratory is its simplicity. There is no need for complex lithographic techniques for suspending and clamping tubes. The surface forces between the CNTs and the alumina membrane are sufficiently high to maintain the clamped beam condition in the majority of cases. In addition, the nanotubes are never exposed to electron radiation during measurement, which is the case for TEM studies. Radiation will induce defects, if the energy of the electrons is high enough, and thereby alter the material properties. This is one effect that we are currently utilising in a positive way in an effort to modulate CNTs' mechanical properties. The relative ease of sample preparation in our AFM method enables a high measurement throughput allowing us to measure a variety of CNTs synthesized in different ways and to compare the results. Described below are new data on catalytic MWNTs, and arc-grown SWNT bundles that have been hydrogenated and exposed to a low-level of radiation. Catalytic MWNTs were produced through decomposition of acetylene over a cobalt/silica catalyst. The previously measured catalytic MWNTs were produced at a temperature of 900°C [11], whereas the new data presented here were obtained on MWNTs fabricated at

720°C. The microstructure of the catalytic MWNTs has a strong dependence on the synthesis temperature which can be readily seen via HRTEM.

To produce crosslinks between the shells of MWNTs and between the SWNTs of SWNT bundles, the sp^2 carbon bonding must be disrupted to sp^3 bonding so that dangling bonds are available for crosslinking. Since the sp^2 bonding is the essence of the CNTs strength, this must not be disrupted to such a degree that the properties of the individual shells in MWNTs or individual SWNTs in bundles are degraded. Hydrogenating CNTs is a first step towards producing them with internal crosslinking. The MWNTs and SWNT bundles were hydrogenated using a modified Birch reduction using Lithium and methanol in liquid ammonia[12]. The SWNT bundles were subsequently exposed to 2.5 MeV electrons with a total radiation dose of 11 C/cm^2. A theoretical estimation of the number of displacements that this dose produces suggests that it will create about 1 defect per 360 carbon atoms[13].

AFM measurements on the hydrogenated, irradiated SWNT bundles are shown in Fig. 5, along with the previous measurements of untreated SWNT bundles[6]. As before, the bundles were dispersed in ethanol using an ultrasonic probe and deposited on the alumina membranes for measurement by AFM in air. Within the errors of our measurement technique a strengthening of the bundles was not observed: the Young's modulus still decreases in a similar trend to the as-grown bundles. However, the treatment does not appear to have compromised the strength of the individual SWNTs either, since the lower diameter bundles have comparable moduli. In addition, it was noticed that the treated bundles were more difficult to disperse in ethanol and the morphology of the sample in the AFM showed that the bundles exhibited a higher degree of aggregation. Taken together these data suggest that the radiation treatment produced bonding between the tubes but was not sufficient to produce enough crosslinks within the bundles to reduce shearing effects and produce bundles with higher Young's moduli. Future efforts will concentrate on optimising the chemistry and irradiation doses to improve their mechanical properties.

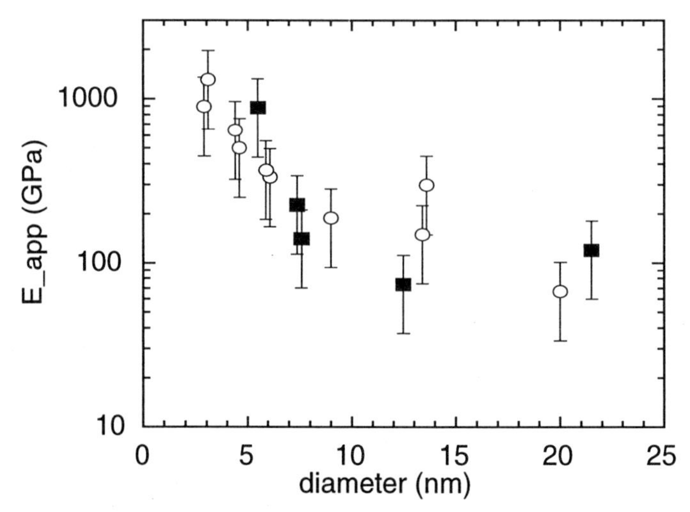

Figure 5. Dependence of the apparent Young's modulus (E_app) on the diameter of SWNT bundles meaured using AFM. The untreated bundles are represented by the open circles and the hydrogenated and irradiated bundles by the filled squares.

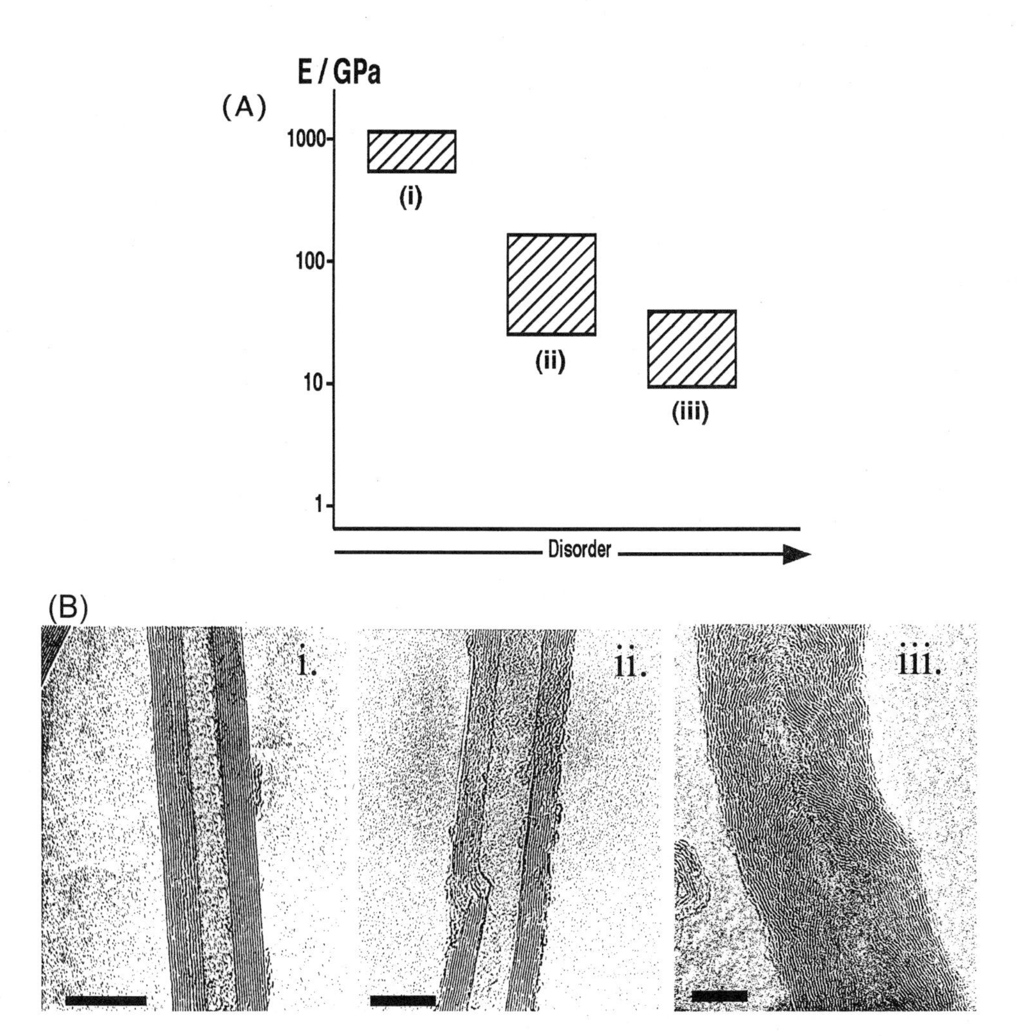

Figure 6. Correlation of the measured Young's modulus of MWNTs with the amount of disorder present with the graphitic walls. (A) Ranges of measured moduli for three different types of MWNT against an arbitrary scale of increasing disorder. (B) The amount of disorder seen in HRTEM data can be qualitatively ranked to make the correlation. MWNTs were produced via, i) arc-discharge, and decomposition of acetylene using a Co / silica catalyst, ii) at 720°C and iii) at 900°C. All scale bars are 10 nm.

Previous AFM measurements of the mechanical properties of MWNTs, arc-grown and catalytically grown, are compared with catalytic MWNTs grown at different temperatures. These data are summarised in Fig. 6, along with HRTEM data showing qualitatively the amount of order/disorder within the walls. As one might expect, the Young's modulus of the MWNTs decreases as the disorder within the walls increases. Arc-grown MWNTs, which contain very few defects, have a modulus comparable with that measured for an individual SWNT[6,7]. The moduli of catalytic MWNTs can vary depending on the structure. Those measured previously[7] were very disordered and had a kind of stacked coffee cup structure (see Fig. 6b, part iii). The other catalytic MWNTs, grown at a lower temperature, showed a higher degree of order within the tubes and consequently had slighty higher moduli. The dispersion of the measured values in this case was large. This could be due to greater uncertainties in the measurement technique, since these MWNTs were usually curved, making the continuum beam approximation less valid. Previous structural characterisation by TEM has shown that they can have a helical morphology, like the cord on a telephone[14].

However, HRTEM images also indicate that the defect morphology varies along the tubes. This would explain why the measured Young's modulus can vary considerably (remembering that our technique measures the modulus over a section of the CNT 200 nm long).

There is some contention about whether the elastic modulus of MWNTs varies as a function of their diameter. Poncharal et al.[8] have recently suggested a rippling mode on the surface of bent MWNTs with diameters greater than about 15 nm, leading to a reduction in the measured modulus. However, a strong dependence of the measured modulus on the diameter was not observed in our previous AFM measurements[7]. It is conceivable that the measurement of the modulus is force dependent and the transition to the rippling mode is not reached with the loading forces used in the AFM experiments. TEM data show rippling on the compressed side of statically bent MWNTs, but these have rather high curvatures[8,15] (much higher than the AFM experiments). Interestingly, we have observed rippling on the compressed sides of the catalytic MWNTs. Fig. 7 shows a high resolution AFM image of a catalytic MWNT (grown at 720°C) lying across a pore, the edge of which can be seen on the left of the image. It clearly shows rippling only on the right-hand side of the tube, the direction in which the CNT is bent, which has a period of roughly 16 nm. These ripples are not perpendicular to the tube axis but are inclined at approximately 30°, making the CNT left-handed. This rippling could be an inherent structure of the catalytic MWNTs or it could arise from the surface forces, which constrain the CNTs on the membrane. However, rippling can also arise when the tubes are loaded in the AFM. Rippling on the upper, compressed side of these MWNTs has also been observed as the imaging force is increased. Non-linear behaviour in the loading/unloading characteristics of the catalytic tubes was frequently noticed. This also contributes to the uncertainties in measuring moduli on these kinds of MWNTs. It is conceivable that the onset of rippling will occur at lower curvatures, i.e. lower forces, in tubes with a higher amount of disorder.

To conclude this section on the measurement of mechanical properties, we have demonstrated that the Young's modulus of MWNTs correlates to the amount of disorder in the graphene walls. Initial attempts at modulating their structure and consequently their mechanical properties have not produced the desired effect, but we believe that chemical modification of CNTs and radiation treatment will be the way forward for improving and/or customising the properties of CNTs for use in composite materials. The radiation doses used to date on the SWNT bundles also did not eliminate the shear behaviour of the SWNTs within the bundles, but observed differences in their propensity to dispersion and greater aggregation of the bundles observed in the AFM, indicates that the surface chemistry of the bundles has been modified, which gives hope that these methods will soon become fruitful. We expect that there will be an optimal radiation dosage, which introduces enough crosslinks between the SWNTs, to improve the moduli of the bundles, but is not sufficient to introduce enough disorder to seriously reduce the modulus of an individual SWNT.

TRANSPORT PROPERTIES

Carbon nanotubes are molecular wires whose electronic properties are largely determined by extended molecular orbitals. Depending on the specific realization, the nanotube may be a true one-dimensional metal or a semiconductor with a gap. By

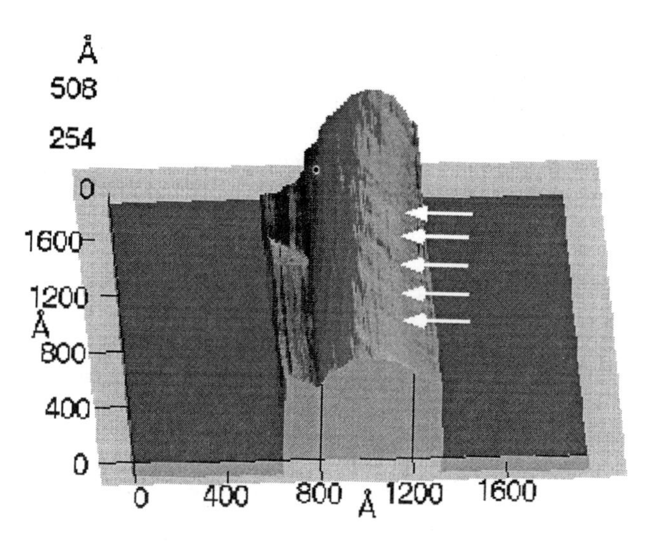

Figure 7. 3D rendering of a high resolution AFM image of a catalytic MWNT, grown at 720°C, suspended on the alumina ultrafiltration membrane. Rippling is observed on the innner side of the natural curvature of the CNT, with a periodicity of about 16 nm, inclined at 30° to the tube axis.

combining metallic and semiconducting tubes the whole span of electronic components ranging from wires, bipolar devices to field-effect transistors[16] may be embodied in nanotubes.

On the fundamental side, a perfect metallic nanotube is supposed to be a ballistic conductor in which only two one-dimensional (1d) subband carry the electric current[17]. Hence, the conductance should be given by $G = 4e^2/h = 6.4\ k\Omega^{-1}$ independent of the NT diameter d.

Carbon nanotubes, especially SWNT are considered as prototypes of 1d conductors. The electronic properties of one-dimensional conductors have generated a lot of interest both experimentally and theoretically. The reason for this excitement lies in the very rich phase diagram of a 1d conductor (expressed in "g-ology") and the prediction that in a 1d system Coulomb interactions should lead to a strongly correlated electron gas, called a Luttinger liquid (LL), instead of the usual quasi-particle picture described by a Fermi-liquid. Experimentally, the systems that were the closest to these expectations were the organic linear chain compounds like TTF-TCNQ, $Qn(TCNQ)_2$, TMTSF and TMTTF salts[18]. There are a lot of interesting transport studies on individual SWNTs, or on ropes of SWNTs (see the contributions of C. Dekker and P. Avouris in this volume). However, there are still unresolved questions concerning their purity (due to their synthesis with magnetic catalysers), the interaction with the substrate, the role of unintentional doping during the purification etc. We have addressed the effect of doping of SWNTs and in order to separate intrinsic and contact effects, we have studied both d.c. and optical conductivities.

Because MWNTs consists of several concentrically arranged SWNTs, one would expect that MWNTs do not qualify as 1d conductors. If adjacent carbon shells interact as in graphite, electrons may not be confined to one shell only, so that much more than just 2 subbands (as for a perfect SWNT) could carry the current. However, we have found very strong evidence that the current in MWNTs, contacted by metallic electrodes from the "outside", is to a large extend confined to the outermost SWNT. MWNTs have certain specific advantages: their larger diameter favours low-ohmic contacts, they do not contain magnetic impurities, they have very well ordered structure, high conductivity and their mesoscopic size enables the observation of quantum interference effects like the Aharonov-Bohm effect. The latter magnetotransport measurements can all very well be understood in the traditional Fermi liquid framework and assuming 2d-diffusive transport. However, recent measurements of the tunneling DOS have revealed anomalies quite similar to the features observed in SWNTs which were assigned to LL behavior[19]. Because LL-like 1d-features seem to coexists with Fermi-liquid 2d-features, more work is needed to pinpoint the origin of these anomalies.

Optical conductivity and d.c. transport of single walled nanotubes

It seems that with the synthesis of single wall carbon nanotubes one can study a real 1D system. Depending on the choice of the chiral vector, which defines the orientation of the unit cell to the tube axis, the nanotube is either a metal, a narrow gap semiconductor or an insulator. Although the chirality cannot be controlled during the synthesis, STM studies show that 30-40 % of the SWNTs are metallic. The LL behaviour is expected from these metallic SWNTs [19]. In single SWNT measurements, however, the manifestation of the metallic state is not straightforward since twists of the nanotube or interactions with the surface can strongly perturb the electronic structure. It is believed that in mats of SWNTs (called bucky paper) the metallic nature of the nanotubes manifests itself much more readily, due to the absence of the substrate, despite the mixing in of the intertube contact resistances. Moreover, the carrier density in SWNTs can be changed by doping, as in the case of graphite[20]. In order to have a clear picture of the effect on the transport properties of the charge transfer upon doping, it is necessary to understand the interplay between different sources of transport: namely, the transport primarily induced by on-tube doping or governed by the doping of the intertube contacts. In this respect, we have performed dc transport measurements and optical experiments on pristine and potassium doped bucky paper samples.

Fig. 8a shows the d.c. transport of the pristine SWNT and the K-doped sample. Fig.8b. displays the optical conductivity for the same samples. The pristine sample shows a Drude component in the optical conductivity despite the non-metallic behaviour in dc resistivity measurements. We attribute this disagreement to effects relating to the non-metallic intertube contacts, which will affect the d.c. transport. Potassium doping renders

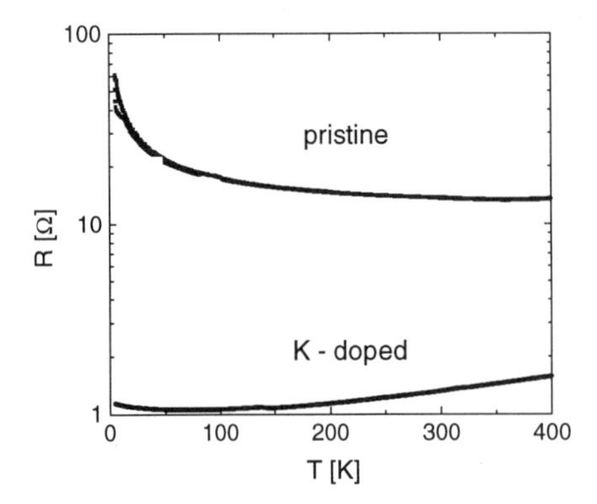

Figure 8a. d.c. resistivity of pristine and potassium doped SWNT thick film. The stoichiometry of the K-doped samples is close to KC_{16}.

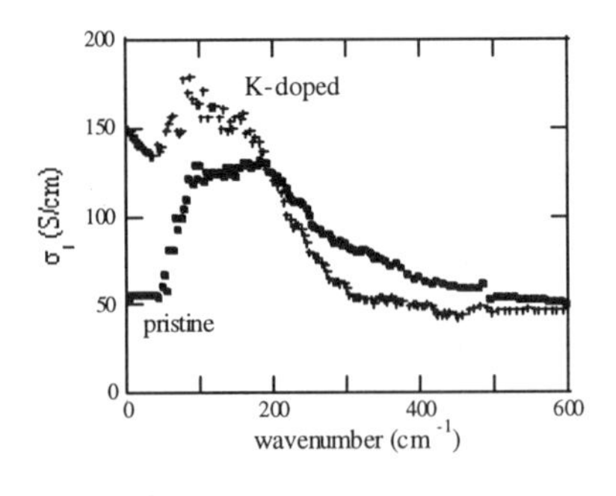

Figure 8b. Optical conductivity of pristine and potassium doped SWNT thick film. The stoichiometry of the K-doped samples is close to KC_{16}.

the nanotubes more metallic, the d.c. resistivity slope is positive in a wide temperature range. This is partly due to improved nanotube rope-rope contacts and to the increased carrier density of the tubes. It also turns out that the potassium doping influences the tube-tube contact regions much more strongly than the intrinsic on-tube transport, so that the average dc-conductivity is higher than the ω=0 optical contribution, averaged over all directions.

Transport of multiwalled nanotubes

Electric transport measurements for single multiwall nanotubes (MWNTs) contacted by four metallic Au fingers from above will be discussed. An example of a device is shown in Fig. 9. Observed interference corrections to the resistance allow to determine the degree of scattering (elastic-scattering length l_e and phase-coherence length l_ϕ). Tunneling spectroscopy, on the other hand, is used to determine how strongly the single-particle density-of states (DOS) is modified by interactions (e.g. Coulomb interaction). Low-ohmic contacts < 10 kΩ are used for measuring the equilibrium (two- or four-terminal) conductance[21] and a high-ohmic contact > 100 kΩ is employed in a tunneling spectroscopy experiment in which the differential conductance dI/dV is measured as a function of V and temperature[22].

Four-terminal electrical resistances R_{4t} have been measured as a function of temperature, T, in a He-3 system down to T≈0.3 K. The resistance always increases with decreasing temperature and appears to saturate around 1-4 K. A typical example is shown in Fig. 10. Note that this saturation has recently been identified to be of extrinsic origin. It is caused by electron heating induced by rf-radiation in the measuring leads. The increase of R from room temperature down to Helium temperatures is moderate amounting to a factor of ≤ 2-3. This is taken as evidence for the metallic nature of the MWNTs.

We emphasize that not only is the temperature dependence of R(T) similar for all samples, but the absolute resistance values also fall into a relatively narrow range of R_{4t}≈2-20 kΩ. The increase of resistance at low temperatures is markedly different from (HOPG) graphite. For HOPG the resistivity decreases with decreasing temperature as commonly associated with metallic behavior. In trying to understand the temperature

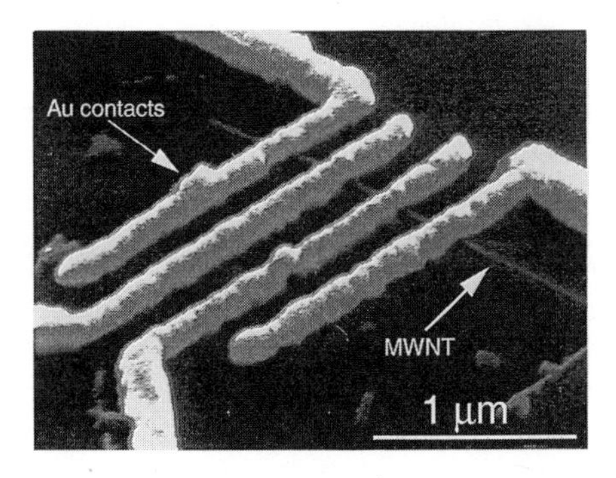

Figure 9. Scanning-electron microscopy image of a single multiwall nanotube (MWNT) electrically contacted by four Au fingers from above. The separation between the contacts is 350 nm center-to-center.

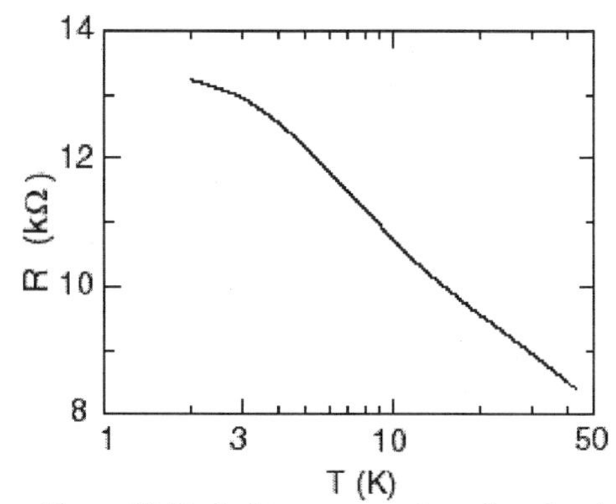

Figure 10. Typical temperature dependent electrical resistance R(T) of a single MWNT measured in a four-probe configuration, i.e. the current is passed through the outer contacts and voltage is measured over the inner ones.

dependence of R, we first consider the simplest possible model. We will compare the absolute measured resistance values with an expression for the classical Drude resistance taking one graphene cylinder and assuming 2d-diffusive transport, i.e. assuming $l_e << \pi d$. We thereby completely disregard the quantization of the wavevector around the tube circumference leading to 1d subbands. The electron states have energy $E = \pm (h/2\pi) v_F|\mathbf{k}|$, where $\mathbf{k}$ is the 2d-wavevector measured with respect to the two independent Brillouin corner points. The Fermi energy is taken to be $E_F=0$ and a reasonable value for the Fermi velocity $v_F=10^6$ m/s is assumed. For the electron density-of-states (DOS) we obtain $n_{2d}(E) = 2E/\pi(hv_F/2\pi)^2$. Using the Einstein equation $\sigma_{2d} = e^2 n_{2d}D$, which relates the conductivity σ_{2d} to the diffusion coefficient $D=v_F l_e/2$ and the electron density, the energy-dependent conductivity is found to be $\propto E$. To obtain the equilibrium sheet conductivity σ_{2d} at finite temperature, the energy has to be replaced by kT, leading to:

$$\sigma_{2d} \approx \left(\frac{2e^2}{h}\right)\frac{kT}{hv_F}l_e$$

Due to the vanishing electron DOS for $E \rightarrow 0$, the resistance of a graphene sheet increases with decreasing temperature following $R(T) \propto T^{-1}$. Although a resistance increase is observed, the increase is not compatible with a T^{-1} dependence. Moreover, putting in numbers results in a unreasonably large mean-free path l_e. Taking from the experiments, $R=10k\Omega$, contact separation L = 350 nm, tube diameter d = 20 nm and T = 4 K, we obtain $l_e \approx 13$ μm. This large mean-free path violates the assumption that diffusion is 2-dimensional. Even more serious, $l_e >> L$. The only way to reconcile this model with the requirement that $l_e \leq L$ is to assume that a large number (30) of graphene cylinders carry the electric current equally. We know from the Aharonov-Bohm experiments that this is not the case (see below)[23]. We therefore conclude that the specific temperature dependence of R cannot be related to the energy-dependent DOS of graphene. Within this simple Drude picture, the discrepancy can, however, be resolved if we take into account the band-structure modifications imposed by the periodic-boundary condition along the circumference of the cylinder leading to 1d-subbands. In contrast to graphene, for which the DOS tends to zero as $E \rightarrow 0$, the DOS is constant in a relatively large energy window centered around the Fermi energy. This energy window is given by the subband separation ΔE_{sb} which should be inserted instead of kT into the previous equation. With $\Delta E_{sb}=100$ meV, typically valid for the outermost cylinders of our MWNTs, one arrives at a mean-free path of $l_e \approx 50$ nm, which is of order of the circumference of the tube. This number is of reasonable magnitude and in agreement with magnetoresistance measurements. This argument suggests that electron transport in MWNTs is not purely 2d-diffusive. More importantly, it demonstrates that the 1d-subbands need to be considered in MWNTs as well.

The classical 1d-Drude resistance due to static-disorder alone predicts a temperature-independent resistance. Temperature dependences can be caused by other scattering mechanisms, like electron-phonon, electron-electron interactions, and quantum interference corrections. Magnetoresistance (MR) measurements indeed show that the intereference and interaction contributions to the resistance are large and display the required temperature-dependence. Particularly attractive for the investigation of interference phenomena in carbon nanotubes is the fact that they are composed of cylindrical graphene sheets. Nanotubes are therefore *hollow* conductors on the molecular scale with diameters ranging from 1 to 20 nm. In such a geometry, with the magnetic field along the nanotube axis, nanotubes provide a unique model system for the investigation of quantum interference in conducting macromolecules by virtue of the Aharonov-Bohm (AB) effect. Fig. 11 displays such magnetoresistance measurements in a parallel field, which show pronounced oscillations (AB effect)[23]. The oscillations are associated with the weak localization, a quantum-mechanical treatment of backscattering of electrons, which contains interference terms adding up constructively in zero field. Backscattering is thereby enhanced, leading to a resistance larger than the classical Drude resistance. In the usual geometry for films/wires, in which the magnetic field is applied perpendicular to the film/wire, these interference terms cancel in a magnetic field of sufficient strength. In contrast, for a hollow conductor in a parallel field, the magnetoresistance is periodic in the magnetic flux penetrating the hollow tube, with period h/2e. This periodic contribution is due to closed counterpropagating electron trajectories that encircle the tube circumference. This result has given compelling evidence that l_φ can exceed the circumference of the tube so that large coherence lengths are possible for MWNTs. The AB-MR agrees with theory *only*, if the current is assumed to flow through one or at most two metallic cylinders with a diameter corresponding to the measured outer diameter of the NT. We note, that it is not possible to relate the peak separation to h/e because a nanotube diameter would result which is larger than the actually measured diameter. In contrast, taking an AB flux of h/2e results in a diameter compatible with the measured one. Because the h/2e period requires backscattering on the scale of the diameter of the NT it is clear that the NTs are *not ballistic*.

In addition to the resistance oscillation, which agrees with the magnetic-field period as derived from the cross-section, unexpected short-period oscillations have also been discovered[23]. It was suggested that the short period oscillations are due to chiral currents in strained nanotubes. Chirality is a well estabilished structural property of carbon nanotubes. In principle, graphitic tubules have isotropic conductivity, however, when they are mechanically stretched due to the interaction with the substrate, or due to thermal contraction, the honeycomb lattice can be distorted resulting in an anisotropic, chiral current. Chirality can select special winding numbers which give the higher order term superimposed on the long period oscillations. In a MWNT the chirality can change from layer to layer, and the period of short oscillations can vary from tube to tube and from sample to sample. This fine structure in the AB oscillations is a completely new phenomenon, which we believe is closely related to the special structure of the carbon nanotubes.

From the width of the zero-field resistance peak of Fig. 11 the phase-coherence length l_φ is estimated to be $l_\varphi \approx 200$ nm. As a test for consistency the WL correction to the conductance ΔG is compared with the measurement. Taking L=350 nm and $l_\varphi = 200$ nm we obtain $\Delta G = 4.4\ 10^5$ S, which is in very good agreement with the measured conductance change of $\Delta G = 4.6\ 10^5$ S. From MR measurements in a perpendicular field similar coherence lengths are deduced. Because $l_\varphi(T)$ has a temperature-dependence which agrees with the expected Nyquist-type dephasing, the comparison with theory allows the elastic-scattering length l_e, to be deduced. We obtain l_e =90-180 nm[22]. Because $l_e \geq \pi d$, transport in our MWNTs should be classified as *quasi-ballistic*.

There are two other independent observations in favor of this result. 1) van Hove type singularities have been seen in a dI/dV tunneling spectrum[22]. Because these features are associated with the formation of 1d-subbands, l_e should be of order of, or larger than, the circumference of the NT. 2) as shown before, the measured resistance values $\approx$10 kΩ can only be explained if the DOS is considered to be 1d. However, the resistance contains a contribution that grows with length. The length-dependence amounts to $\approx$ 6 kΩ/μm at room temperature. Using the Landauer formula one obtains an effective transmission coefficient of T=0.5 per micrometer length. Hence, the mean-free path must be large. The observed van-Hove type dI/dV spectra[22] demonstrates that the peculiar bandstructure effects of SWNTs are also found for MWNTs. We have to emphasize, however, that a spectrum with sharp van-Hove singularities, in close agreement with tight-binding calculations, has only been observed on one sample until now. The prevailing spectra

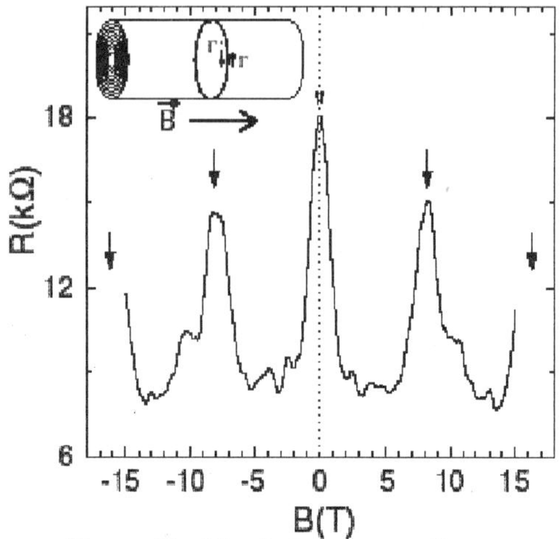

Figure 11. Electrical resistance R as a function of magnetic field B of a MWNT aligned parallel to B. Arrows denote the resistance maxima corresponding to multiples of h/2ein magnetic flux through the nanotube taking the *outer* diameter

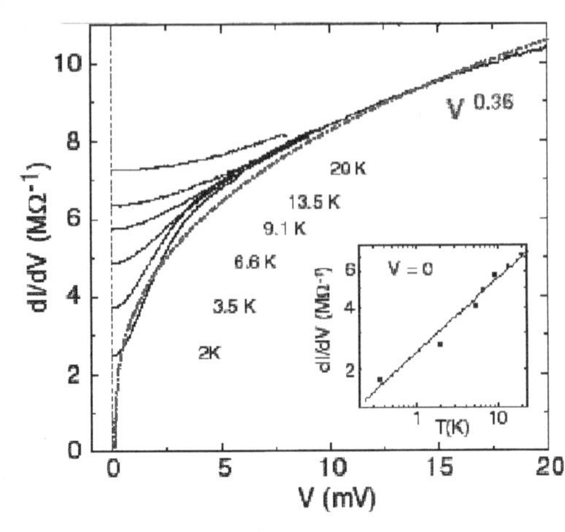

Figure 12. Differential tunneling conductance dI/dV measured on a single MWNT at different temperatures T displaying a pronounced zero-bias anomaly. Inset: log-log representation of dI/dV vs. T for V=0. The dashed-dotted curve displays the power law dI/dV $\propto V^{\alpha}$ with $\alpha = 0.36$ deduced from the inset.

display a pronounced zero-bias anomaly (ZBA) on a smaller energy scale of 1-10 meV. For larger energies, a peak-structure develops in dI/dV on the scale of the subband separation, 0.1eV, which may be associated with (broadened) van-Hove singularities.

A typical ZBA is shown in Fig. 12 for six temperatures ranging from 2-20 K. A suppression of the tunneling DOS is expected for a strongly correlated electron gas [24]. Similar anomalies have recently been observed by Bockrath *et al.* for SWNTs[19]. Their measurement and analysis provide the first demonstration for possible Luttinger liquid (LL) behavior in carbon NTs due to long-range Coulomb interactions. LL theory predicts power laws both for the voltage and temperature dependence with the same exponent α. A power-law with $\alpha \approx 0.36$ is deduced from dI/dV(T,V=0); see the inset of Fig. 12. For comparison with the observed dI/dV-voltage dependence, the dashed-dotted curve $\propto V^{0.36}$ has been plotted. The same exponent $\alpha = 0.36$ was obtained by Bockrath *et al.* This exact agreement has presumably no significance because we use single MWNTs whereas they have used SWNT ropes. On the other hand, the agreement may indicate that the same physics is responsible for the ZBA.

The reported study of electric transport of single MWNTs gives rise to results which appear to be in contradiction. For example, the observation of an Aharanov-Bohm effect with period h/2e suggests diffusive transport on the scale of the circumference of the nanotube, i.e. $l_e \leq \pi d$. On the other hand, we have observed a dI/dV spectrum which agrees with tight binding models assuming the existence of 1d-subbands. This suggests the opposite, i.e. $l_e \geq \pi d$. Our results are therefore consistent only if l_e is of the order of the circumference. There is a second 'contradiction': on the one hand, we have used weak-localization theory which is based on the Fermi liquid hypothesis; on the other hand, the observed suppression of the single-particle of states suggests that NTs may develop a Luttinger liquid (LL) state. If an LL is the correct description for NTs (including MWNTs) we need to know how the observed quantum interference corrections have to be described. Is there something similar as weak localization in the LL picture? What happens in a magnetic field?

ELECTRON SPIN RESONANCE

Since the pioneering work of Wagoner on graphite[25], Electron Spin Resonance (ESR) has been used intensively to study the electronic properties of graphitic or conjugated materials. One of the advantages of the method is to give information both on localised spins and conduction carriers. Three different quantities are provided by ESR: the g-factor, which depends on the chemical environment of the spins via spin-orbit coupling (and hyperfine interaction); the linewidth, which is governed by spin relaxation mechanism; the intensity of the signal, which is proportional to the static susceptibility. In graphite, anisotropy of the band structure induces anisotropy in the g-factor and the linewidth. How is the ESR behaviour modified in carbon nanotubes? Changing the dimensionality from 2D to 1D is expected to significantly modify the main ESR characteristics. In particular, the conduction carrier contribution to the susceptibility should be enhanced in 1D. For metallic tubules, the density of states *per unit length* (independent of the diameter) along the axis is a constant given by: $N(E_f)=8/(\sqrt{3}\pi a_0 \gamma_0)$, where a_0 is the lattice constant and $\gamma_0 \approx 2.5$ eV is the overlap energy. The DOS per carbon atoms is thus a decreasing function of the diameter. It is worth measuring graphitic nanostructures with different diameters to check this prediction.

ESR of multiwalled nanotubes

Conduction carriers and defect density

Pauli susceptibility of MWNTs, measured from X-band ESR, is comparable to that of graphite. Taking into account the rather large distribution of diameters and lengths in the samples, one can estimate the susceptibility in the approximation of independent shells, i.e. supposing that the interlayer interactions do not modify $N(E_f)$ of MWNTs, in contrast to

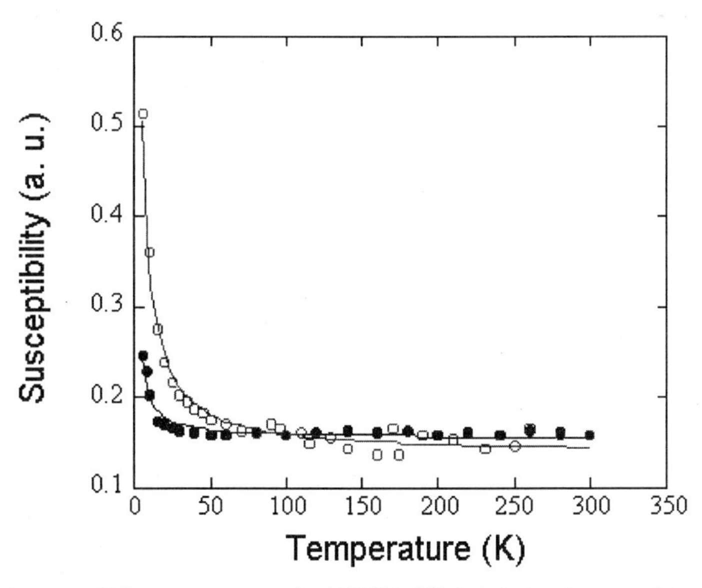

Figure 13. Spin susceptibility vs temperature for MWNTs (filled circles) and nanoparticles (open cirlces). Note that the Curie tail at low temperatureres is smaller for MWNTs than nanoparticles.

graphite. Doing this, supposing that 1/3 of the cylinders are metallic, it is difficult to reach the measured value. Moreover, polyhedral multishelled particles (always present in macroscopic quantities of MWNT) have the same susceptibility as MWNTs within an uncertainty of 20%. We speculate that the interlayer interactions can not be neglected for MWNTs and have an important contribution in $N(E_f)$ in a similar way to graphite, i.e. π/π^* overlap occurs between neighbouring shells.

There are only a few studies of defects in carbon nanotubes. It was suspected that arc-grown MWNTs contain a lot of defects that modify their electronic properties[26]. Among these defects, vacancies and interstitials can generate paramagnetic centres that can be detected by ESR. Pentagon-heptagon pair defects can also be present in the lattice. All these defects, which result from displacement of carbon atoms from their pristine location in the hexagonal lattice, should be produced by irradiation and annealed at high temperature. We carried out energetic electron irradiations[27] and annealing of MWNTs to investigate the influence of defects on the ESR behaviour. We found that arc-grown MWNTs contain a very low density of paramagnetic defects, which manifest themselves as a Curie tail in the susceptibility at low temperature. No particular modifications were observed after annealing at 2800°C, in contrast to ref. 26. The density of paramagnetic defects increases linearly with the electron irradiation dose. We also found that the defects density *per gram* in MWNTs is lower than in polyhedral multishelled nanoparticles, which suggests that imperfections are present in the MWNT tips (Fig. 13).

g-factor, spin relaxation, and contact broadening effect

The conduction electron spin g-factor is determined by spin-orbit splitting of the energy levels in presence of a magnetic field[28]. In the case of degenerate bands (as in graphite at the K point), theory predicts that spin-orbit coupling, which removes the degeneracy, induces a large g-shift which varies inversely with temperature. This is exactly what happens in graphite when the magnetic field is perpendicular to the plane, allowing large orbital currents. This increases the spin-orbit coupling. When the magnetic field is parallel to the planes, orbital currents are suppressed and a small g-shift nearly independent of temperature is observed. When the Fermi level is shifted away from the K point by doping, degeneracy and the g-shift anisotropy disappear. (Despite this understanding of the variation of g-shift a rigorous theory is still missing in graphite due to the complicated band structure near the point of degeneracy.) Using this qualitative theoretical framework, one can speculate about the g-factor of carbon nanotubes. When the magnetic field is parallel to the tube axis, nearly the same value as found in graphite is expected, except at a field for which the cyclotron radius equals that of the nanotubes. Such fields (typically 1T and greater) are higher than those used in X-band spectrometers, but a small increase of the g-factor at the X-band field (0.33 T) is not excluded. When the magnetic field is perpendicular to the tube axis orbital currents can not completely close, as in a plane, and we can expect a smaller g-shift than in graphite. A decrease of this g-shift with the diameter of the tube is then expected. Some of these predictions are indeed realised in MWNTs. First the average g-value is 2.012 compared with 2.018 in graphite and 2.015 in partly graphitised carbon blacks (the difference is small but significant). Second the anisotropy is lower than in graphite.

Spin relaxation in metals and semimetals depends also on the spin-orbit coupling. More precisely it is the modulation of the spin-orbit coupling by lattice vibrations that causes spin relaxation. It can be shown that in the framework of Elliott's theory the spin relaxation time is proportional to the momemtum relaxation time. When it is governed by thermal phonons an increase of the linewidth with increasing temperature is observed. In graphite, the linewidth increases when the temperature decreases. This behaviour is attributed to motional narrowing over the g value distribution. A decrease of the linewidth when temperature increases is also observed in carbon nanotubes, which is probably due to this motional narrowing. An additional feature is a line broadening that appears when the nanotubes are brought into contact with each other. This behaviour is specific to nanotubes (and nanoparticles) and it is contrary to what happens in graphitized carbon blacks. Indeed, with carbon black particles or fine graphite powder, increasing the contact between grains increases the motional averaging of the g-factor anisotropy which decreases the linewidth. We believe that the inverse effect with nanotubes results from their mesoscopic nature. In effect, the intrinsic spin mean free path on a cylindrical graphene shell is much larger than the real nanotube length. The spin mean free path is already large in graphite, and should be even larger in nanotubes due to the reduced dimensionality. Other extrinsic mechanisms participate in the linewidth. Among them the intershell-intertube hopping, and also reflection by the tips. We speculate that intertube hopping is responsible for the broadening of the linewidth when the contact between nanotubes increases.

ESR of singlewall nanotubes

In spite of our efforts, all the SWNT samples we studied are ESR silent, i.e. no line is detected with our conventional X-band spectrometer, except from ferromagnetic particles which are never completely eliminated by the purification treatment. Recent theoretical and experimental works suggest that the electronic properties of SWNTs are best understood in a Luttinger liquid (LL) model. If the ground state of an LL is antiferromagnetic this could explain the absence of the ESR line. Another possibility is the presence in SWNTs of magnetic impurities with strong coupling which shorten the spin relaxation time so much that the line can not be detected. It is worth focusing on the latter assumption since the absence of an ESR signal (from carbon materials) is not only observed on SWNTs but also on catalytic multiwall tubes for which we know that an intrinsic ESR signal exists. Spin relaxation of metals with local magnetic moments is a well-known problem[29]. Such a strong influence of magnetic particles on the spin relaxation means that they could also alter the momentum relaxation that is the transport properties of SWNT.

Concerning the presence of defects, which can also have important consequences on the transport properties of SWNTs, the same remarks stated for MWNTs hold, i.e. if heptagon-pentagons pairs are present, vacancies should be present too, since their

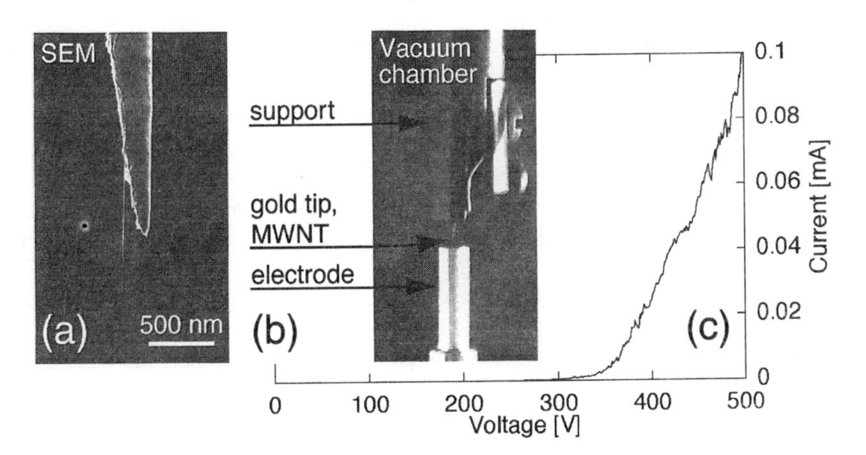

Figure 14. (a) Single MWNT mounted on the tip of an etched gold wire. (b) Optical micrograph of the experimental set-up for field emission : the gold wire is fixed on a support, and placed 1 mm above the cylindrical counter-electrode. (c) I-V characteristics for a single opened MWNT.

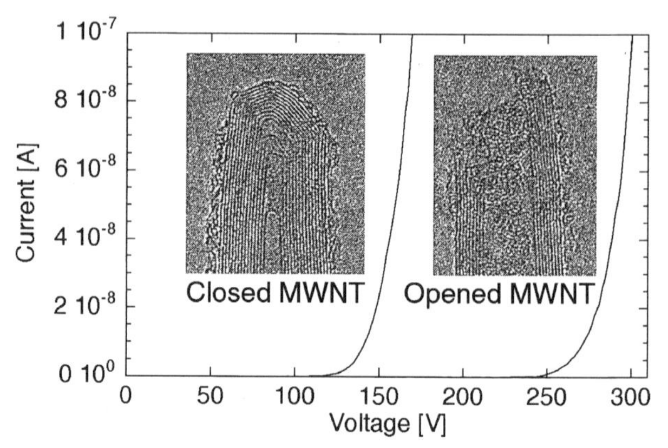

Figure 15. I-V characteristics for a single closed and opened MWNT (the current is given in logarithmic scale), and corresponding TEM micrographs of typical NT tips.

formation energies are similar. Then an ESR signal from defects is expected. To check this, we have irradiated a purified mat of SWNTs with 2 MeV electrons and an ESR line indeed grows with increasing dose. We therefore suggest that ropes contain only a few defects even after the purification treatment. This is an important result since thermal treatment at high temperature destroys the single shell structure so that point defects in SWNTs could not be annealed, contrary to MWNTs.

ELECTRON AND LIGHT EMISSION

Although most of the research conducted on nanotubes since their discovery has been of a fundamental nature, a keen interest is shown for their potential applications. One of these is field emission, since nanotubes rank among the best electron field emitters that are now available[30-35]. Our goal here is to compare the field emission properties of the different types of nanotubes, and to understand why nanotubes are such excellent field emitters.

Emitter fabrication

To realize a field emission source with one nanotube only, we mounted single MWNTs on a supporting gold wire (diameter 20 μm) that was electrolytically etched to a 250 nm radius tip. No adhesive was used, and the tubes were held onto the tip by Van der Waals forces. A typical emitter is shown in Fig. 14(a). We produced nanotube films by a simple and fast preparation method. Nanotubes were dispersed in a solvent (typically water with a surfactant or ethanol) and the resulting suspension was drawn through a 0.2 μm pore silica filter. The film remaining on the surface of the filter was then transferred on a teflon-coated metal surface by applying pressure[30]. The emission surfaces ranged from 0.1 to 25 mm^2 and can be easily scaled up. The main advantages of this method are that it can be used for all types of nanotubes, in contrast to catalytic deposition techniques, and that it is non-destructive, which is not the case for alternative film preparation techniques where the tubes are opened. We studied films made with closed or opened arc-discharge MWNTs, as-produced SWNTs, as well as catalytic MWNTs showing ``coffee-cup" structures. Their morphology is readily comparable to the SEM micrographs of Fig 1-3.

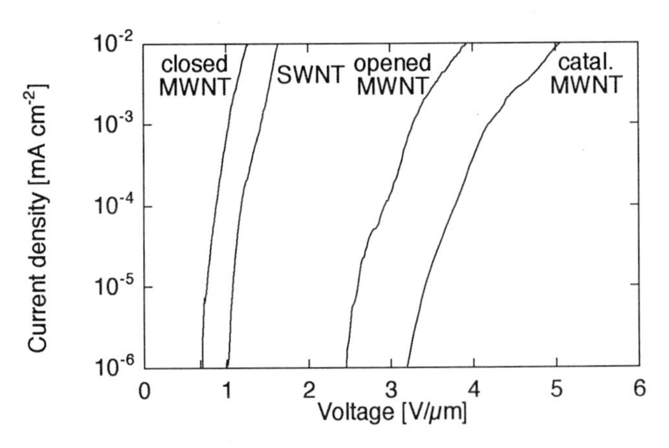

Figure 16. I-V characteristics for different nanotube films.

Emission properties

Single nanotube field emitter

Fig. 14(c) displays a typical I-V curve for a single MWNT (in this case an opened MWNT). Most single MWNT emitters, closed as well as opened, are capable of emitting over an incredibly large current range. The maximal current we succeeded in drawing from one nanotube was 0.2 mA, and MWNTs reach routinely and repeatedly 0.1 mA. This represents a tremendous current density for such a small object, and is actually quite close to the theoretical limit where the tube should be destroyed by resistive heating[32]. This experimental limit is comparable to the one observed by other groups that studied the electronic transport properties of MWNTs[36].

No significant difference was observed between closed and opened MWNTs, except for the most important property: the voltage needed for the emission. In Fig. 15, we compare the I-V performances of a closed and opened MWNT. We noted for all measured samples that opened tubes were far less efficient emitters than as-grown tubes. The voltages needed for a given emission current are typically a factor 2 higher for the opened tubes. In effect, open edges like those depicted in Fig. 15 may have smaller radius of curvature than closed ends, and opened tubes would be expected to emit current at lower applied voltages due to the higher field amplification. Quite surprisingly, the emission characteristics of nanotubes are seriously degraded by opening their ends. Regarding long-term stability and lifetime, stable emission was observed for more than 100 h at 2 μA in high vacuum. Typical behaviour of metallic cold field emitters, i.e., a gradual and reversible decrease due to the formation of absorbed layers, was not observed. Termination of the emission happened on most tips as a catastrophic and irreversible failure. Lifetimes of more than 1400 h have been reported for emission in ultra-high vacuum for one single tube emitting at 0.5 μA[35].

Nanotube film field emitter

We compare in Fig. 16 the field emission performances of four different types of nanotube films. A useful parameter for such a comparison is the electric field (voltage over interelectrode distance V/d) needed to produce a given current density. We found

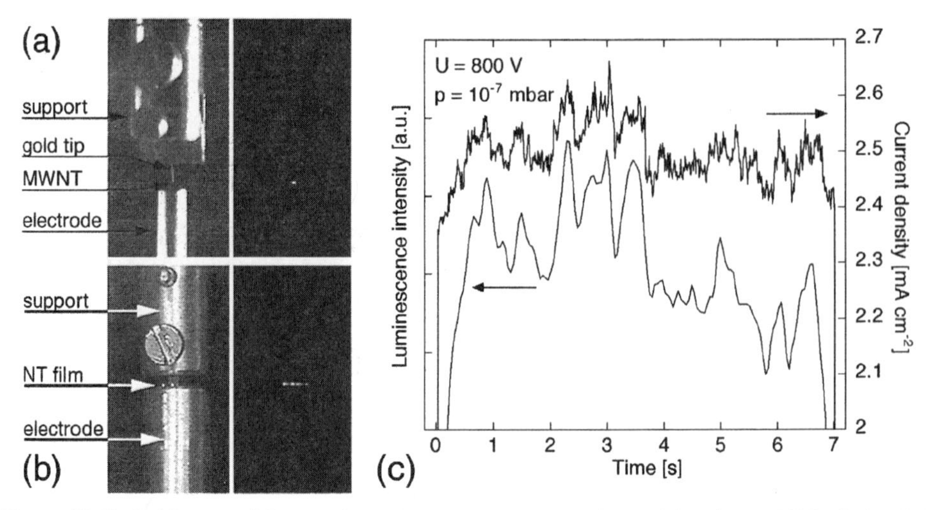

Figure 17. Optical images of the experimental set-up for field emission and the observed light during field emission for a single multiwall nanotube (a) and a multiwall carbon nanotube film (b). Emitted light intensity and current density as a function of time for a MWNT film (c).

systematically that closed MWNT films displayed lower emission voltages, followed by SWNTs, opened MWNTs and finally catalytic "coffee-cup" MWNTs. To understand the observed differences between the different types of nanotubes, it is instructive to estimate the field amplification factor β from I-V characteristics and corresponding F-N plots[32]. Since this field amplification factor depends only on the geometrical shape of the emitter for a given workfunction and interelectrode distance, we can then directly compare the effective emitter shape of the different nanotube films (by assuming that the workfunction is 5 eV for all tubes). We found that the field amplification factors were significantly higher for singlewall (β =3400) than for multiwall nanotube (β =1600) films. This enhancement is most probably due to the smaller tip radius of SWNTs. The tip radius is also responsible for the low field amplification obtained with the catalytic tubes (β =830) with respect to the closed MWNTs, although the disordered structure of the tip and the high defect density may also have an influence. However, the difference between closed (β =1600) and opened MWNTs (β =1100) cannot only arise from geometrical considerations, since the variation of mean diameter and length between opened and closed MWNT is rather small, as can be seen in Figure 11 for example. We speculate that most of this difference is due to changes in the work function that arise from the state of the tip. Finally, the above values show that the high density of nanotubes on the film surface may be a disadvantage. For the same experimental conditions, a single MWNT should produce a field amplification of β =6500, and the average value we observe is lower by a factor 4. It appears that the presence of neighboring tubes near the emitters screen the applied field, which results in a lower field amplification.

In summary, our catalytic tubes showed high emission voltages mainly because of their larger average diameter. The small diameter of SWNT should lead to very low emission voltages, while the SWNT films show "only" comparable performances to closed MWNT films. We suppose that this relative inefficiency arises from the fact that most SWNTs are bundled in ropes, and that these ropes mostly end in catalyst particles. Only few SWNTs tips are detected by TEM, and these protrude only by a few tens of nanometers at most from the sample. This in turn means that the density of free SWNT tips, and thus of potential emission centers, is far lower than for MWNTs. As for the huge difference between closed and opened MWNTs, we noted that our best film emitter with opened tubes didn't even come close in performances to the worst emitter with closed tubes. The observed difference can therefore not be assigned to the quality of the films, and we conclude that it is due in great part to the state of the tip. Interestingly, the only emitters which compete in terms of emission voltage with nanotubes films are diamond-covered Si tips[37]. This indicates that maximal efficiency for film field emitters can be reached only when the emitters are well aligned and placed with their long axis perpendicular to the film substrate, and when the emitters are well separated from one another. We thus infer that carbon nanotube films will have to be realised fulfilling the above-mentioned conditions to reach a maximal emitting efficiency. To our knowledge, the only method to grow aligned and well-separated nanotubes is by catalytic reactions over a patterned substrate, as currently investigated by several groups[34,38,39]. Our results suggest however that care should be taken to obtain well-graphitized, closed MWNTs with small diameters.

Light emission

A rather unusual behaviour linked to the field emission was observed in the form of light emission. This light emission occurred in the visible part of the spectrum, and could sometimes be seen with the naked eye. Typical examples are shown in Fig. 17(a) and (b). This luminescence was induced by the electron field emission since it was not detected without applied potential (and thus emitted current). Furthermore, the emitted light intensity followed closely the variations in emitted current, as can be assessed in Fig. 17(c), where the emitted current and emitted light intensity have been simultaneously measured.

To investigate further this phenomenon, we analysed the spectra of the emitted light for single MWNTs. A typical spectrum is displayed in the inset of Fig. 18. No significant changes in the shape of the spectrum were observed when the current was varied apart from a small shift (<25 meV) of the broad peak. The luminescence intensity I_p as a function of the emitted current I_e, as shown in Figure 14, followed a power law $I_p \sim I_e^{\alpha}$ with α=1.4±0.2 in this case.

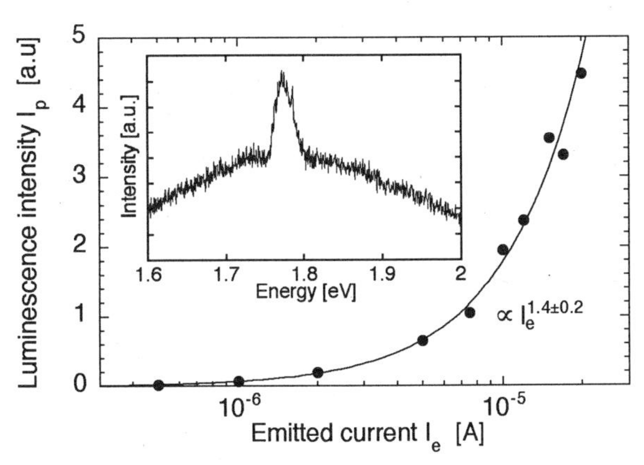

Figure 18. Variation of the total emitted intensity as a function of emitted current. The solid line is a power law fit of the experimental data, and yields an exponent of 1.4±0.2. Inset: spectra of field emission induced luminescence for one MWNT at 20 μA emitted current

There has been one report of observed luminescence on opened nanotubes[31] but it was attributed to an incandescence of carbon chains at the tip of the tube provoked by resistive heating. Our results however strongly suggest that the light emission is directly coupled to the field emission. The narrowness of the luminescence lines and the very small shifts with varying emitted current show that we are not in presence of blackbody radiation or of current-induced heating effects, but that photons are emitted following transitions between well-defined energy levels. Actually, the dependence of I_p versus I_e can be reproduced by a simple two-level model[40], where the density of states at the nanotube tip is simplified to a two level system, with the main emitting level at energy E_1 below or just above the Fermi energy, and a deep level at $E_2 < E_1$. When an electron is emitted from the deep level, it is replaced either by an electron from the tube body, or by an electron from the main level which can provoke the emission of a photon. From the Fowler-Nordheim model, the transition probability D(E) can be evaluated for each level, and in the frame of our model, $I_e \sim D(E_1)$, $I_p \sim D(E_2)$. It appears that I_p varies as a power of I_e with an exponent that depends on the separation of the levels[35], and that amounts to 1.51-1.65 for the energies observed here (typically 1.8 eV). This compares well to the experimental observations. These observations point to the presence of energy levels, and thus of localised states, at the tip. We estimate that one emitted photon corresponds to at least 10^6 field emitted electrons. With localised states at the tip, the greatest part of the emitted current will arise from occupied states with a large local density of states located near the Fermi level. Other, more deeply located electronic levels may also contribute to the field emission. In this case, the emitted electron will be replaced either by an electron from the semi-metallic tube body with an energy comparable to the level energy, or by a tip electron from the main emitting state. Clearly, the second alternative may provoke the emission of a photon. Although the tunneling probability for electrons from deeper state is several orders of magnitude lower than for the main emitting state, it will be readily sufficient to cause the observed light intensities.

Field emission mechanism

The results presented above show that the large field amplification factor, arising from the small radius of curvature of the nanotube tips, is partly responsible for the good emission characteristics. It is however still unclear whether the sharpness of nanotubes is their only advantage over other emitters, or if other properties also influence the emission performances. Most authors conclude that carbon nanotubes are metallic emitters, essentially because the I-V characteristics seem to follow the Fowler-Nordheim law. Our results[32] show, however, systematic deviations from the Fowler-Nordheim model at high emitted currents, which point to the fact that nanotubes cannot be considered as usual metallic emitters. Further observations confirm this conclusion, and strongly suggest that the electrons are not emitted from a metallic continuum as in usual metallic emitters, but rather from well-defined energy levels of ~0.3 eV half-width corresponding to localised states at the tip. First, the energy spread of nanotubes is typically half that of metallic emitters (about 0.2 eV), and the shape of the energy distribution suggests that the electrons are emitted from narrow energy levels[28,30]. Second, the observation of luminescence coupled to the field emission indicates that several of these levels participate to the field emission. Although the greatest part of the emitted current comes from occupied states with a large density of states near the Fermi level, other, deeper levels also contribute to the field emission. In fact, theoretical calculations and STM measurements on SWNTs and MWNTs show that there is a distinctive difference in the electronic properties between the tip and the cylindrical part of the tube. For MWNTs, the tube body is essentially graphitic, whereas SWNTs display a characteristic DOS[41,42] that reflects their one-dimensional character. In contrast, the local density of states at the tip presents sharp localised states that are correlated to the presence of pentagons[43,44]. Interestingly, the FWHM of these states and their separation is readily compatible with our observations. We conclude that the greatest part of the emitted current comes from occupied states just below the Fermi level. The position of these levels with respect to the Fermi level, which depends primly on the tip geometry[44] (i.e., tube chirality and diameter and the eventual presence of defects), would be, together with the tip radius, the major factor that determines the field emission properties of the tube. Indeed, only tubes with a band state just below or just over the Fermi level are good candidates for field emission. Finally, it is worth noting that the presence of such localised states influences greatly the emission behaviour. At and above room temperature, the bodys' of MWNTs behave essentially as graphitic cylinders. This means that the carrier density at the Fermi level is very low, i.e., on the order of $5x10^{18}$ cm^{-3}, which is 3 orders of magnitude less than for a metal. Simulations show that the local density of states at the tip reaches values at least 30 times higher than in the cylindrical part of the tube. The field emission current would be far lower without these localised states for a geometrically identical tip since it depends directly on this carrier density. The crystalline structure also strongly influences the position and intensity of the localised states, which could explain the superiority of closed over opened or disordered MWNTs. Another complementary explanation for this observation is that the coupling of the tip states to the metallic body is probably far better for closed MWNTs, leading to an increased electron supply and thus higher emitted current.

ACKNOWLEDGEMENT

The work is supported by the Swiss National Foundation for Scientific Research.

REFERENCES

1. J.-M. Bonard et al., *Adv. Mater.* **9**, 827 (1997).
2. D. Xu et al., *Appl. Phys. Lett.* **75**, 481 (1999).
3. M.M.J. Treacy, T.W. Ebbesen and J.M. Gibson. *Nature* **381**, 678 (1996).
4. E.W. Wong, P.E. Sheehan and C.M. Lieber. *Science* **277**, 1971 (1997).
5. A. Krishnan,. E. Dujardin, T.W. Ebbesen, P.N. Yianilos and M.M.J. Treacy. *Phys. Rev. B* **58** (20) 14013 (1998).
6. J.-P. Salvetat, G.A.D. Briggs, J.-M. Bonard, R. R. Basca, A.J. Kulik, T. Stöckli, N.A. Burnham and L. Forró. *Phys. Rev. Lett.* **82** (5) 944 (1999).
7. J.-P. Salvetat, A.J. Kulik, J.-M. Bonard, G.A.D. Briggs, T. Stöckli, K. Méténier, S. Bonnamy, F.Béguin, N.A. Burnham and L. Forró. *Adv. Mater.* **11** (2) 161 (1999).
8. P. Poncharal, Z.L. Wang, D. Ugarte and W.A. de Heer. *Science* **283**, 1513 (1999).
9. D.A. Walters, L.M. Ericson, M.J. Casavant, J. Liu, D.T. Colbert, K.A. Smith and R.E. Smalley. *Appl. Phys. Lett.* **74** (25) 3803 (1999).
10. L.S. Schadler, S.C. Giannaris and P.M. Ajayan. *Appl. Phys. Lett.* **73** (26) 3842 (1998).
11. A. Hamwi, H. Alvergnat, S. Bonnamy and F. Béguin. *Carbon* **35**, 723 (1997).
12. S. Pekker, J.-P. Salvetat, E. Jakab, J.-M. Bonnard and L. Forró. *Proc. IWEPNM (Science and Technology of Molecular Nanostructures) - Kirchberg, Austria* - in press (1999).
13. F. Beuneu, C. l'Huillier, J.-P. Salvetat, J.-M. Bonnard and L. Forró. *Phys. Rev. B* **59** (8) 5945 (1999).
14. K. Hernadi, A. Fonseca, J.B. Nagy and D. Bernaerts. Catalytic synthesis of carbon nanotubes p. 81 - 97 in: *Supercarbon : Synthesis, Properties and Applications.* S. Yoshimura and R.P.H. Chang, Eds., Springer-Verlag, Berlin (1998).
15. O.Lourie, D.M. Cox and H.D. Wagner. *Phys. Rev. Lett.* 81 (8) 1638 (1998).
16. S.J. Tans *et al.*, Nature 393, 49 (1998); see also: P.L. McEuen *ibid.*, 15; R. *Martel et al.*, Appl. Phys. Lett. 73, 2447 (1998).
17. R. Saito *et al.,* Appl. Phys. Lett. 60, 2204 (1992); J. W. Mintmire *et al., Phys. Rev. Lett.* 68, 631 (1992).
18. D. Jerome and H. Schulz, *Adv. in Physics* 31, 299 91982).

19. M. Bockrath, D. H. Cobden, J. Lu, A. G. Rinzler, R. Smalley, L. Balents, P. L. McEuen, *Nature* 397, 598 (1999).
20. R. S. Lee, H. J. Kim, J. E. Fischer, A. Thess, R. E. Smalley, *Nature* 388, 255 (1997).
21. Bachtold *et al., Appl. Phys. Lett.* 73, 274 (1998).
22. C. Schönenberger *et al.*, to appear in *Appl. Phys. A*.
23. A. Bachtold, et al., *Nature* 397, 673 (1999).
24. M.P.A. Fisher and L. Glazman, in *Mesoscopic Electron Transport*, L.L.Sohn, L.P. Kouwenhoven, and G. Schön, eds., NATO ASI, Series E: Applied Sciences 345 (Kluwer Academic, Dordrecht 1997); R. Egger and A. O. Gogolin, *Phys. Rev. Lett.* 79, 5082 (1997); C.\ Kane, *et al.*, *ibid.* 5086.
25. G. Wagoner, *Phys. Rev.* 118:647 (1960).
26. M. Kosaka, T. W. Ebbesen, H. Hiura, K. Tanigaki, *Chem. Phys. Lett.* 225:161 (1994).
27. F. Beuneu, C. L'Huillier, J. P. Salvetat, J.-M. Bonard, L. Forró, *Phys. Rev. B* 59:5945 (1999).
28. Y. Yafet, *Solid State Phys.* 14:1 (1963). R. J. Elliott, *Phys. Rev.* 96, 266 (1954).
29. S. Schultz, M. R. Shanabarger, P. M. Platzman, *Phys. Rev. Lett.* 19, 749 (1967).
30. W.A. de Heer, A. Châtelain, D. Ugarte: *Science* 270, 1179 (1995).
31. A.G. Rinzler, J.H. Hafner, P. Nikolaev, L. Lou, S.G. Kim, D. Tomanek, P. Nordlander, D.T. Colbert, R.E. Smalley: *Science* 269, 1550 (1995).
32. J.-M. Bonard, F. Maier, T. Stöckli, A. Chatelain, W.A. de Heer, J.-P. Salvetat, L. Forró: *Ultramicroscopy* 73, 7 (1998).
33. J.-M. Bonard, J.-P. Salvetat, T. Stöckli, L. Forró, A. Chatelain: *Appl. Phys. A*, in press (1999).
34. S. Fan, M.G. Chapline, N.R. Franklin, T.W. Tombler, A.M. Cassell, H. Dai: *Science* 283, 512 (1999).
35. M. Fransen: *Towards high brightness, monochromatic electron sources*, PhD thesis, Technical University Delft (1999).
36. S. Frank, P. Poncharal, Z.L. Wang, W.A. de Heer: *Science* 280, 1744 (1998).
37. V.V. Zhirnov, A.B. Voronin, E.I. Givargizov, A.L. Meshcheryakova, *Proceedings of the IEEE International Vacuum Microelectronics Conference*, IVMC 1995, 340 (1996).
38. W.Z. Li, S.S. Xie, L.X. Qian, B.H. Chang, B.S. Zou, W.Y. Zhou, R.A. Zhao, G. Wang: Science 274, 1701 (1996).
39. H. Kind, J.-M. Bonard, C. Emmenegger, L.-O. Nilsson, K. Hernadi, E. Maillard-Schaller, L. Schlapbach, L. Forró and K. Kern, submitted to Adv. Mater. (1999).
40. J.-M. Bonard, T. Stöckli, W.A. de Heer, A. Chatelain, J.-P. Salvetat, L. Forró: Phys. Rev. Lett. 81, 1441 (1998).
41. J.W.G. Wildoer, L.C. Venema, A.G. Rinzler, R.E. Smalley, C. Dekker: Nature 391, 59 (1998); T.W. Odom, J.-L. Huang, P. Kim, C. M. Lieber: Nature 391, 62 (1998).
42. P. Kim, T.W. Odom, J.-L. Huang, C.M. Lieber: Phys. Rev. Lett. 82, 1225 (1999).
43. A. De Vita, J.-C. Charlier, X. Blase, R. Car: Appl. Phys. A. 68, 283 (1999).
44. D.L. Carroll, P. Redlich, P.M. Ajayan, J.-C. Charlier, X. Blase, A. De Vita, R. Car: Phys. Rev. Lett. 78, 2811 (1997).

LOW ENERGY THEORY FOR STM IMAGING OF CARBON NANOTUBES

C.L. Kane and E.J. Mele

Department of Physics
Laboratory for Research on the Structure of Matter
University of Pennsylvania
Philadelphia PA 19104

ABSTRACT

Scanning tunneling images of carbon nanotubes frequently show electron distributions which break the local sixfold symmetry of the graphene sheet. We present a theory of these images which relates the anisotropies to off diagonal correlations in the single particle density matrix induced by elastic scattering from tube defects. The theory reveals that there are three general effects which one can associate with elastic scattering from defects. Any defect can be characterized by a matrix of reflection coefficients which define its signature in the tunneling image. We provide a theory for tunneling images with broken translational symmetry, "primitive" broken symmetry images with broken rotational symmetry but no broken translational symmetry, and a novel "semiconductor effect" in which the symmetry of the tunneling image switches with the sign of the tunnel bias.

INTRODUCTION

Since the discovery of carbon nanotubes in 1991 there has been interest in the intrinsic electronic properties of these objects. An isolated single wall carbon nanotube is a highly ordered single molecular structure 1 nm wide and more than 1 micron long. The possibility of combining these objects to make nanometer scale circuits, and studying coherent quantum transport effects in them has been particularly intriguing. Recent research in this field has shown that electronic transport in these systems is a very complex physical phenomenon which can be quite sensitive to the degree and type of structural disorder in tubes and tube bundles, the interaction of these objects with substrates and with external electrodes, and the direct electron electron interactions.

Several groups have reported atomic resolution STM imaging of the low energy electronic transport states in SWNTs[1-5]. In this paper we briefly review our theory of the images obtained in these experiments[6,7]. We are particularly interested in understanding the

Science and Application of Nanotubes, edited by Tománek and Enbody
Kluwer Academic / Plenum Publishers, New York, 2000

symmetries of these images, which at low tunnel bias generally do not display the local sixfold symmetry one might expect for a folded graphene sheet[3]. In many cases the tunneling images display a broken translational symmetry with a modulated periodic charge density which is commensurate with (though larger than) the period of the graphite primitive cell[3,5,7]. Very frequently the images contain a striped pattern in which maxima in the charge density are observed in relatively well defined bond chains which coherently spiral around the tube surface[1,5,7]. Initial observation of these spiral patterns[3] actually led to the suggestion that the tubes were under significant mechanical torsion when adsorbed on a substrate or packed in a rope. However, the elastic energy inferred inferred from the observed "torsion" cannot plausibly be accounted for by the relatively weak intertube interactions which are expected to occur in nanotube ropes. In special cases broken rotational symmetries in STM images can be attributed to artifacts arising from aymmetries in the tunnelling tip[8]. However tunneling images which show a periodic superlattice with a period larger than the graphene primitive cell cannot be explained in this way[7]. In addition one occasionally encounters images taken with a single tip in which the symmetry depends on the sign of the tunneling bias; this effect cannot be accounted in any simple way by tip anisotropy[7]. Recently, Venema et al[5]. observed modulations of the charge density when the tunneling images of individual eigenstates on a finite tube were resolved. These authors attributed the modulation to an intrinsic quantum mechanical interference effect in which propagating Bloch waves are reflected from tube ends and interfere with the incident wave to form a standing wave pattern which is then imaged with the STM. We believe that closely related effects are responsible for the tunneling patterns observed in most STM images of carbon nanotubes. Indeed over the last year several theoretical groups have reported theoretical calculations of the charge densities for individual eigenstates on finite tubes and find that they contain a very rich internal spatial structure[6,9,10]. Here we analyze these effects using the appropriate low energy theory for the electronic states[11], which we believe provides the most natural framework for interpreting these images[6]. In particular we show that the symmetry of the STM image can be directly related to off diagonal correlations in the single particle density matrix which inevitably occur when Bloch waves are elastically scattered from defects, tube ends, etc. Because of the simple analytic structure of the unscattered low energy electronic states on a carbon nanotube, the scattering matrix and thus the tunneling density of states near defects can be investigated in a particularly useful and direct way. A short report of our main results using the low energy theory has appeared previously[6].

In this paper we will outline the theory for computing the local tunneling density of states using the low energy effective mass theory. We will illustrate the theory by showing the effects of elastic scattering from defects in the tunneling image. We will discuss the effect of backscattering with a large change in crystal momentum, and the case of backscattering with no change in crystal momentum. This latter effect is peculiar to carbon nanotubes and provides a novel signature in the tunneling image. We will then briefly describe the theory of tunneling images obtained near the band edges for semiconducting tubes. These are predicted to show a novel switching of the symmetry of the tunneling image with the sign of the tunnel bias.

STM IMAGES: GENERAL THEORY

The STM probes the local tunneling density of states resolved in lateral position r and energy E at a height h above a surface. It is therefore convenient to define the STM density of states n(r,h, E):

(1)

$$n(r,h,E) = \langle \Psi^{\dagger}(r,h)\, \delta(E - H)\, \Psi(r,h) \rangle$$

The effective mass theory for the calculating n(r,h,E) on a carbon nanotube is developed in the following way[11]:

(a) The field operator $\Psi(r)$ in equation (1) is expanded using the K and K' point wavefunctions as a basis

(2)

$$\Psi(r) = \sum_{bnq} F_{bn}(r)\, e^{iqx}\, \psi_{bn}(q)$$

Here q defines a crystal momentum with measured with respect to the K point, the sum on n denotes a sum over the two critical (K and K') points and the sum on b denotes a sum over the two branches (bands) at each critical point.

(b) Using equation (2) the STM density of states is rotated into the "ψ-representation" using

(3)

$$\rho(r,E) = \sum_{bn\, b'n'} F^{*}_{bn}(r)\, F_{bn}(r)\, \rho_{bn\, b'n'}(x,E)$$

(c) The density matrix in the "ψ-representation" in equation (3) is obtained

(4)

$$\rho_{bnb'n'} = \langle \psi_{bn}^{\dagger}\; \delta(E - H_{eff})\, \psi_{b'n'} \rangle$$

where H_{eff} is the effective mass Hamiltonian. Equations (1) (3) and (4) allow one to reconstruct the STM tunneling density of states, and thus the tunneling image, directly from the structure of the density matrix which can be studied easily in the low energy theory.

The most general form for the density matrix in this "b-n" representation is(5)

$$\rho = \frac{\rho_o}{4} \begin{pmatrix} 1+\xi_z & \xi^+ & \zeta_a & \zeta_m \\ \xi^- & 1-\xi_z & \zeta_m & \zeta_b \\ \zeta_a^* & \zeta_m^* & 1+\xi_z & \xi^- \\ \zeta_m^* & \zeta_b^* & \xi^+ & 1-\xi_z \end{pmatrix}$$

In equation (5) the convention is that the matrix is broken into 2x2 "blocks" representing forward and backward moving particles; thus for example the upper left hand block represents any coherent superposition of the two "right moving" basis states at the K and K' points. Note that in the absence of scattering, i.e. for a defect free nanotube, all of the ξ and ζ coefficients are zero. However elastic scattering from defects produces a coherent superposition of these fundamental modes which can be represented by nonzero ξ and/or ζ . We show below that these have different physical effects on the tunnelling density of states.

We also note that in a typical tunneling experiment the height variable h is large compared to the "size" of the carbon atom on the graphene sheet. Thus the fields F measured at height h will be smoothly varying as a function of the lateral position r, and it is sufficient to include in our computation only its lowest nonvanishing Fourier components. The calculation can be systematically extended to include higher Fourier components in F, but the image symmetry is correctly obtained using just the lowest components. The relevant amplitudes in the Fourier image of F are obtained in the "three pronged" star $K + G_i$ where $\frac{1}{3} \sum_i G_i = K$, and similarly for the K' point. Pairwise combinations of these fundamental plane waves according to equation (3) yields (in addition to a trivial constant) density waves modulated either at the wavectors (K, K') or in the the first star of reciprocal vectors $\{G_i\}$ A map of the relevant wavevectors is given in the right panel of Figure 1.

Finally, to analyze the tunneling image it is convenient to group the Fourier components of the STM density of states using the "triangular harmonics" which are constructed within each reciprocal space star {k} according to:

$$n_{km} = \sum_{m'} z^{mm'} \, n([C3]^{m'} k) \tag{6}$$

where $z = \exp(2\pi i/3)$ and where the operator $[C3]k$ rotates the wavevector k by $2\pi/3$. Using this representation we find a compact expression which relates the density matrix coefficients to the Fourier amplitudes of the density waves in each star:

For the $\sqrt{3} \times \sqrt{3}$ star {K}:

$$\begin{aligned} n_{00} &= \frac{1}{2} (\zeta_b - \zeta_a^*) \\ n_{0\pm1} &= \frac{1}{2} \, e^{\pm i\theta} \left(\zeta_b + \zeta_a^* - (\pm 2\xi_-) \right) \end{aligned} \tag{7}$$

and for the primitive star $\{G_i\}$:

$$\begin{aligned} n_{10} &= -1 + \sqrt{3} \; i \, \mathrm{Re}[\zeta_m] \\ n_{1\pm1} &= \frac{1}{2} \, e^{\pm i\theta} \left(\xi_z - (\pm 2i \, \mathrm{Im}[\zeta_m] \right) \end{aligned} \tag{8}$$

Finally, it is useful to express these results in terms of the reflection amplitudes describing elastic backscattering of a Bloch wave from a point defect. There are three independent reflection amplitudes describing backscattering from K$\rightarrow$ K' (r_b), backscattering from K' $\rightarrow$ K (r_a) and intravalley backscattering K $\rightarrow$ K or K' $\rightarrow$ K' (r_m). Each of these backscattering processes is illustrated schematically in the left panel of Figure 1. The last of the three is peculiar to the graphene band structure and occurs because pairs of electronic branches cross through the Fermi energy at the K and K' points; thus the electron can reverse its velocity with no change in crystal momentum. In terms of these reflection amplitudes we find:

$$
\begin{aligned}
n_{00} &= \frac{1}{2}\left(r_b\, e^{iqx} - r_a{}^*\, e^{-iqx}\right) \\
n_{0\pm1} &= \frac{1}{2}\, e^{\pm i\theta}\left(r_b\, e^{iqx} + r_a{}^*\, e^{-iqx}\right)
\end{aligned}
\tag{9}
$$

and

$$
\begin{aligned}
n_{10} &= -1 + \sqrt{3}\;\, i\, \mathrm{Re}[r_m\, e^{iqx}] \\
n_{1\pm1} &= -\left(\pm\frac{1}{2}\;\, i\, e^{\pm i\theta}\, \mathrm{Im}[r_m\, e^{iqx}]\right)
\end{aligned}
\tag{10}
$$

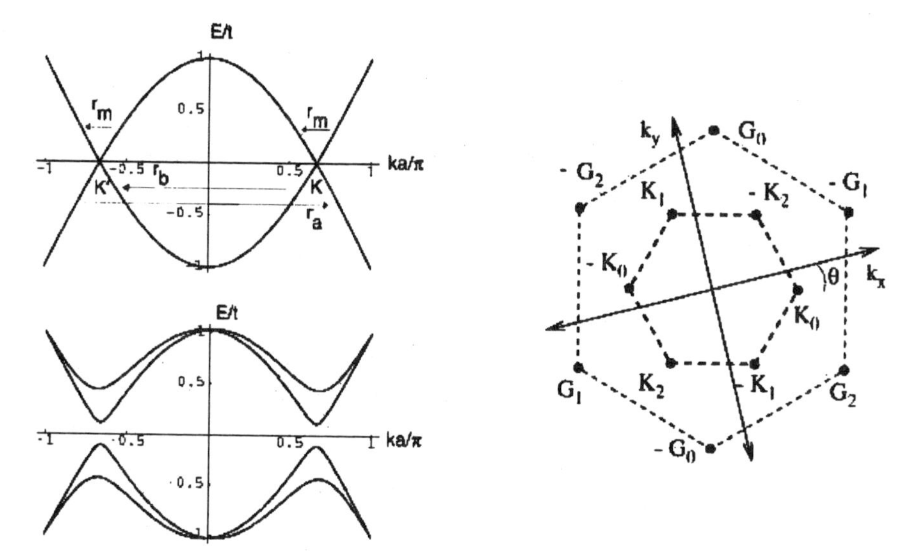

Figure 1. Low energy electronic spectra for a conducting tube (top left panel) and for a semiconducting tube (lower left panel). The right panel shows the momenta in the first two stars in reciprocal space which we use to compute the local tunneling density of states measrued in an STM experiment.

with $q = E/\hbar\, v_F$. Equations (9) and (10) demonstrate that backscattering with no change in momentum generates a "primitive" modulation of the STM density, i.e. one which preserves the translational symmetry of the parent graphene lattice. The large momentum backscattering terms break the translational symmetry, as they would in a conventional metal.

RESULTS FOR STM IMAGES OF CONDUCTING TUBES

In this section we display several STM images which illustrate the theory presented in the previous section.

The simplest scattering effect on a carbon nanotube which produces a modulation of the STM image involves backscattering with a large change in momentum, described by the backscattering amplitudes r_a and r_b. This process requires elastic scattering of a Bloch wave from the K point to the K' point of the graphene Brillouin zone. This scattering wavevector corresponds to a $\sqrt{3} \times \sqrt{3}$ translation vector in direct space. Thus one expects a modulation of the tunneling density of states with a $\sqrt{3} \times \sqrt{3}$ superlattice, as one often observes in atomically resolved images of tubes in tube bundles. This effect is very common and is the tubule analog of the Friedel oscillation found in an ordinary metal.

The magnitudes and phases of the scattering amplitudes r_a and r_b in equation (9) depend on the internal structure of the scattering center. Thus there exists a family of possible but related $\sqrt{3} \times \sqrt{3}$ images. An important case occurs for perfect scattering from a hard wall for which $r_a = r_b = 1$. This produces the tunneling image shown in Figure 2. Here we find a periodic $\sqrt{3} \times \sqrt{3}$ superlattice in which the electronic density is enhanced in an superlattice of dimer bonds on the tube surface. In Figure 2 we also show the tunneling image obtained for the case $r_a = r_b = e^{i\pi/2}$. Here the density wave retains the $\sqrt{3} \times \sqrt{3}$ spatial periodicity, but with its origin "offset" to the center of a sixfold ring so that the tunneling image is maiximized in a series of isolated rings on the tube surface.

A different class of images are produced by backscattering with no change in momentum, described by the amplitudes r_m. This process is peculiar to the parent graphene bandstructure shown in Figure (1) which exhibits two branches of counter propagating electronic states at each of the K and K' points. Intravalley scattering between these branches reflects a propagating electron and produces a standing wave pattern in the tunneling image which preserves the graphene lattice translational symmetry, but which can break its rotational symmetry. The most dramatic such effect occurs when the backscattering amplitude develops a nonzero imaginary part, and is illustrated in Figure 3. Here we obtain an image in which the tunneling density is enhanced in a coherent bond chain which wraps around the tube surface in a spiral pattern: a chiral bond density wave. The real part of the small momentum backscattering amplitude produces a very different effect shown in right hand panel of Figure 3. Here the density wave is enhanced on one of the two graphene sublattices. The sign of the scattering amplitude "selects" one of the two sublattices where the density is enhanced. As for the case of the large momentum backscattering there are a continuous family of patterns which connect these two limiting cases.

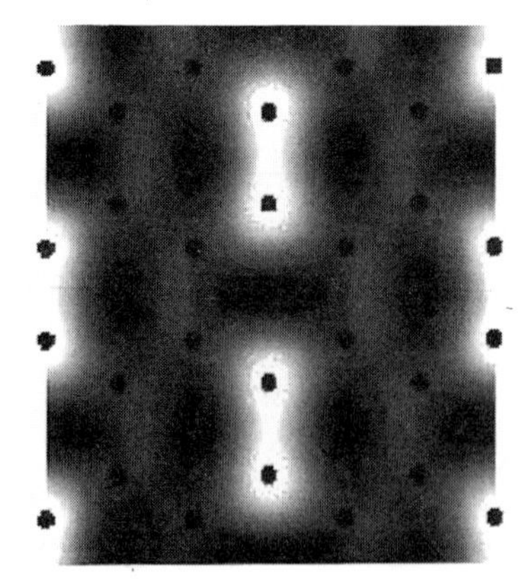

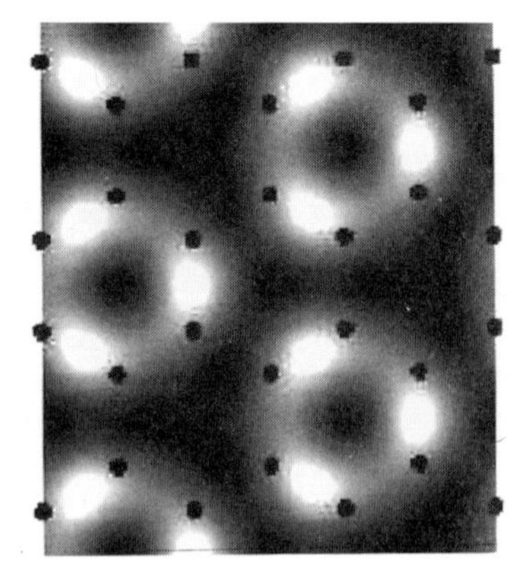

Figure 2: (left) Local tunneling density of states calculated at E=0 for a [10,10] tube in the presence of hard wall scattering $r_a = r_b = 1$. (right) Local TDOS for $r_a = r_b = e^{i\pi/2}$.

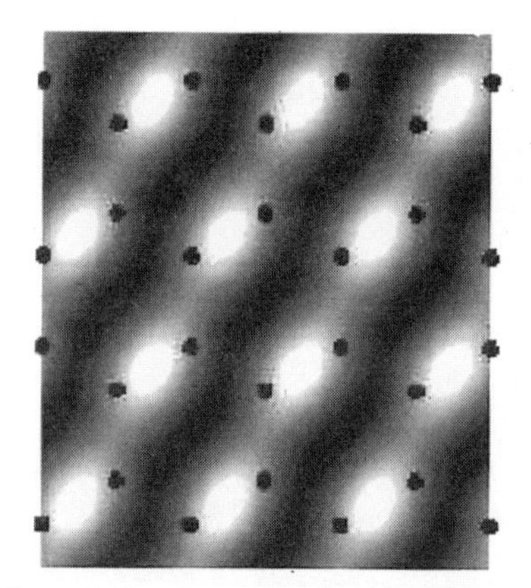

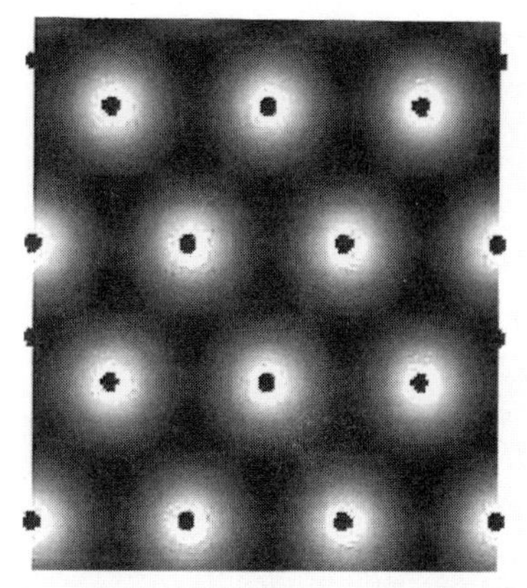

Figure 3: Local tunneling densities of states for a [10,10] tube produced by backscattering with no change in crystal momentum. The left panel shows the effect of a nonvanishing imaginary part of the tunneling amplitude which produces a chiral bond density wave in the tunneling image. The right panel shows the effect of $Re[r_m]$ which modulates the site-centered charge density in the TDOS.

We believe that the chiral bond density wave shown in the left hand panel of Figure 3 provides an explanation for the "striped" pattern observed in STM images of carbon nanotube ropes, and occasionally observed for isolated SWNTs as well. Interestingly an $Im[r_m]$ does not require reflection of a Bloch wave from an exotic scatterering center. Instead we find that the effect is quite common and generally occurs for any scattering center which breaks the two sublattice symmetry of the tube surface. Thus a site centered adsorbate, or even intertube interactions which are not sublattice symmetricwill produce a long range modulation in the tunneling image with this symmetry.

The tunneling images have an interesting dependence on the tunnel bias. If one "measures" the local tunneling density of states precisely at energy E measured with respect to the band center, one finds that the image symmetry "precesses" between the limits displayed in Figures 3 and 4 with a wavelength $L = 2\pi\hbar v_F/E$. The wavelength for this precession diverges if $E \rightarrow 0$, i.e. in the limit of very low tunnel bias. If instead one integrates over a finite energy window determined by the tunnel bias, one obtains a superposition of modes over a finite (though possibly small) distribution in q space, so that the predicted modulation of the electron density produced by a scattering center will be suppressed for $x >> L$ from the scattering center.

IMAGE SWITCHING FOR SEMICONDUCTING TUBES

Standing wave patterns in the tunneling image can also be produced as a intrinsic effect in semiconducting tubes. For example it is well known that [M,N] tubes with $mod(M-N,3) \neq 0$ have a gap in the low energy electronic spectrum, as shown in the lower left panel of Figure 1. The electronic states at the band edges are perfect standing waves, containing an equal admixture of "foward" and "backward" propagating waves. However, for a semiconducting tube the backscattering is not a extrinsic effect produced by interaction with an isolated point defect, but arises instead from the mass operator in the low energy Hamiltonian which explicitly breaks the local rotational symmetry of the tube. Nonetheless, the density matrix formulation presented above can be applied to analyze the tunneling images obtained for the low energy eigenstates at energy E in this case as well. For a gap parameter Δ, we find that the density matrix now takes the form:

$$\rho(E) = \frac{\rho_o}{4}\begin{pmatrix} 1 & 0 & 0 & -i\frac{\Delta}{E} \\ 0 & 1 & -i\frac{\Delta}{E} & 0 \\ 0 & i\frac{\Delta}{E} & 1 & 0 \\ i\frac{\Delta}{E} & 0 & 0 & 1 \end{pmatrix} \tag{11}$$

Thus $\zeta_m = -i\frac{\Delta}{E}$ and the tunneling image can be represented in terms of its triangular harmonic amplitudes

$$n_{10} = -N(E)$$
$$n_{1\pm1} = \pm\ i\, s\, N(E)\, e^{\pm i\theta}\frac{\Delta}{E} \tag{12}$$

where s is the chiral index $(-1)^{mod(m-n,3)}$ adn N(E) is the spatially averaged density of states. Note that only $\zeta_m \neq 0$, so that the only nonzero reflection amplitude is r_m and therefore, as expected, the image retains the translational symmetry of the graphene sheet. Note also that the sense of the broken symmetry depends on the sign of E from equation (12). Thus, and very interestingly, the symmetry of the positive and negative symmetry solutions must be inequivalent. However they are complementary; superposition of $n_{1\pm1}$ (E) and $n_{1\pm1}$ (-E) gives a vanishing symmetry breaking amplitude $n_{1\pm1}$, and thus superposition of positive and negative energy images should recover the sixfold graphene symmetry.

We demonstrate these effects in Figure (4) and Figure (5) which show the tunneling densities of states calculated for the band edge states of a [12,7] and for a [17,0] tube. The left hand panels in each figure give the images for tunneling out of an occupied valence band state, and the right hand panels give the image for tunneling into an unoccupied conduction

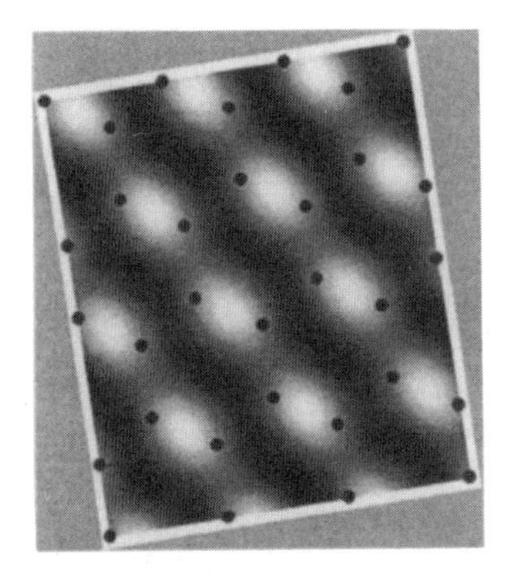

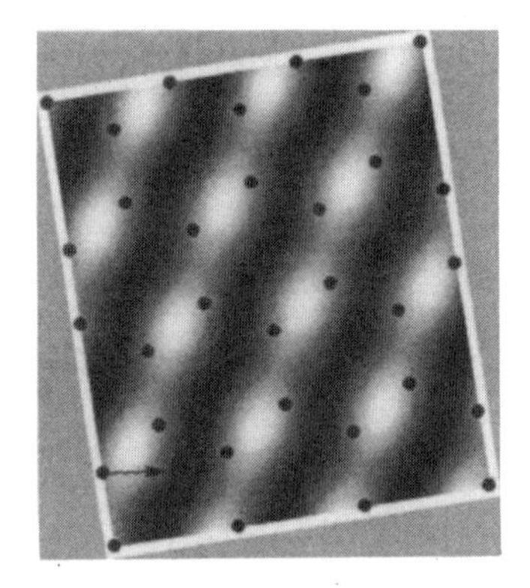

Figure 4: Complementary tunneling images obtained at the valence band edge (left) and conduction band edge (right) of a [12,7] nanotube. A superposition of the two images gives a pattern which recovers the sixfold symmetry of the parent graphene sheet.

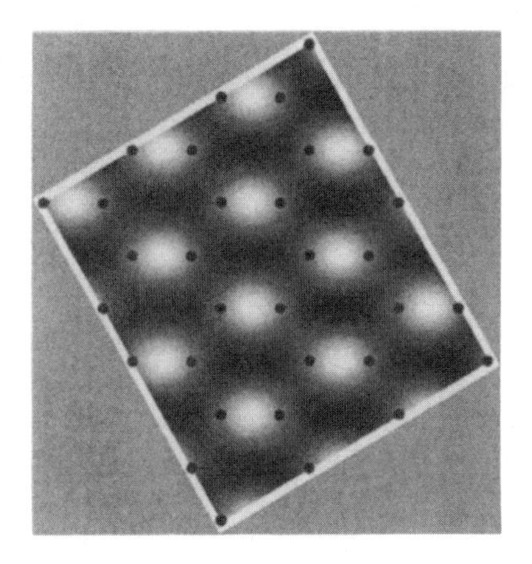

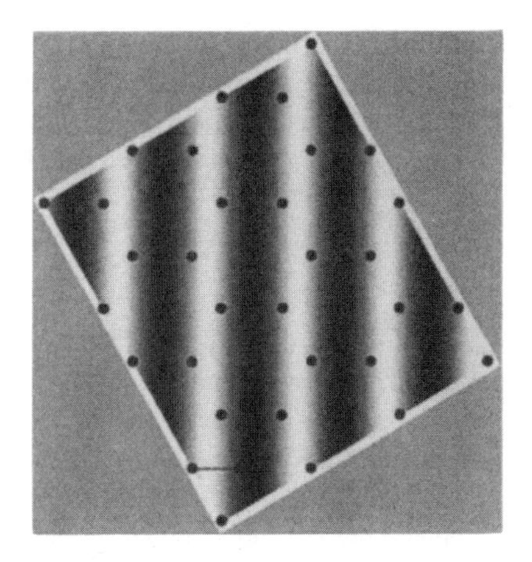

Figure 5: Complementary tunneling images obtained at the valence band edge (left) and conduction band edge (right) of a zigzag [17,0] nanotube. A superposition of these two images also gives a pattern which recovers the sixfold symmetry of the parent graphene sheet.

band state. For the [12,7] tube each the two band edge states produce complementary spiral stripes which wind in opposite directions around the tube surface. For tubes wrapped near the "zig-zag" configuration the complementary patters are dimer bonds oriented along the tube direction and "rings" of charge density which follow the zigzag bonds in the wrapping direction. In either case superposition of the positive and negative energy solutions recovers the full sixfold translational symmetry. In the notation of Equation (12) only the coefficient n_{10} is nonzero after the symmetrizing over positive and negative energy solutions. Note also that the symmetry breaking terms in equation (12) depend on the product of the energy E and the chiral index s. Thus the symmetries of the valence band and conduction band states are also exchanged if one exchanges the wrapping indices [M,N]. The the valence band edge states of a [12,7] and [7,12] tube are complementary in the sense described above.

SUMMARY

We have discussed the effects of electronic backscattering on atomic resolution images obtained with scanning tunneling microscopy of carbon nanotubes. The results show that the symmetries of the tunneling images can be quite strongly affected by elastic scattering from defects. We have identified three general effects that one can associate with elastic backscattering, and have classified them by the the local graphene symmetries which are broken or preserved in each case.

We should note that all the effects described here have been observed in STM images of carbon nanotubes obtained in various labs over the last two years[1-3,5]. In fact several of these phenomena have closely related counerparts in STM images of graphite surfaces, which also show very interesting symmetry breaking motifs near adsorbates and point defects[8,12]. For carbon nanotubes packed into ropes, essentially all images of which we are aware are very strongly influenced by these scattering effects. In fact images of carbon nanotubes in ropes are typically extremely complex, and indicate in our opinion that the low energy states are strongly scattered by various defects on the tube surface. In special cases one can observe a well defined $\sqrt{3} \times \sqrt{3}$ modulation of the electronic density in the defective tube segment which provides the putative scattering center. The nanotube rope one seems to be in a regime where Li > L >> Le where Li is the inelastic mean free path, L is the tube length (or at least the scanning range in a typical measurement) and Le is the elastic mean free path. This is a regime in which one expects to see localization effects in the transport at low temperature, as indeed has been observed experimentally[13]. Electrical transport is also likely to be strongly influenced by the nature of inter-tube contacts within a rope.

Many (though not all) published STM images of isolated carbon nanotubes show the "striped" motif as a generic effect. Furthermore the "tripled period" structure identified by Venema et al.[5] on STM studies of rope segments is easy to interpret using the large momentum backscattering model presented here. Remarkably, the measurements of Odom et al.[2], though they focused on the STM spectroscopy show, for their measurements on semiconducting tubes, images which are strikingly similar to those presented in this paper. Occasionally, the bias switching phenomenon for a semiconducting tube can be clearly identified in experimental data. We have recently reported such an observation in data collected by the Penn scanning microscopy group[7]. Although these measurements quite clearly show the switching effect, it has not yet been possible to simultaneously measure the tunneling conductance spectrum to correlate this effect with the insulating character of the tube. It would be very useful to extend these measurements in this direction. Likewise there has not yet been a systematic study of the variation of the local tunneling density of states at low bias as a function of the Fermi energy. This could be modified by an electrostatic gate or by chemical doping for example, and seems particularly relevant for correlating the transport properties of undoped and doped specimens. Finally, it would be most interesting to induce some of the density waves described here by controllably creating a defective site, for example by adsorption of an atomic or molecular species on the tube surface.

ACKNOWLEDGMENTS

This work was supported under DOE Grant DE-FG02-84ER45118 (EJM) and NSF Grant DMR 95-05425 (CLK) and by the NSF FRG program under Grant DMR 98-02560. We are pleased to thank our colleagues A.T. Johnson and W. Clauss with whom we have collaborated on the experimental aspects of this project.

REFERENCES

1. J.W.G. Wildoer, L.C. Venema, A.G. Rinzler, R.E. Snalley and C. Dekker, Nature 391, 59-62 (1998)

2. T.W. Odom, J.L. Huang, P. Kim and C.M. Lieber, Nature 391, 62 (1998)

3. W. Clauss, D. Bergeron and A.T. Johnson, Physical Review BN 58, R4266 (1998)

4. P. Kim, T.W. Odom, J.L. Hunag and C.M. Lieber, Physical Review Letters 82, 1225 (1999)

5. L. C. Venema, J.W.G. Wildoer, J. W. Janssen, S.J. Tans, H.L.J. Temminck Tuinstra, L.P. Kouwenhoven and V. Dekker, Science 283, 52 (1999)

6. C.L. Kane and E.J. Mele, Physical Review B 59, R12759 (1999)

7. W. Clauss, D. Bergeron, M. Freitag, C.L. Kane, E.J. Mele and A.T. Johnson Europhysics Letters (in press, 1999)

8. H. Mizes, Science 244, 599 (1989)

9. A. Rubio, D. Sanchez-Portal, E. Artacho, P. Ordejon and J.M. Soler, Physical Review Letters 82, 3520 (1999)

10. V. Meunier and P. Lambin, Physical Review Letters 81, 5588 (1998)

11. C.L. Kane and E.J. Mele, Physical Review Letters, 78, 1932 (1997)

12. J. Kushwerick, K.H. Kelly, H.P. Rust, N.J. Halas and P.J. Weiss, Science 273, 1371 (1996)

13. M.S. Fuhrer, U. Varadarajan, W. Holmes, P.L. Richards, P. Delaney, S.G. Louie and A. Zettl, in Electronic Properties of Novel Materials, AIP Conference Proc. 442 (H. Kuzmany, J. Fink, M. Mehring and S. Roth. eds.) 69 (1998)

QUANTUM TRANSPORT IN INHOMOGENEOUS MULTI-WALL NANOTUBES

S. Sanvito[1,2,3], Y.-K. Kwon[3], D. Tománek[3], and C.J. Lambert[1]

[1]*School of Physics and Chemistry, Lancaster University, Lancaster, LA1 4YB, UK*
[2]*DERA, Electronics Sector, Malvern, Worcs. WR14 3PS, UK*
[3]*Department of Physics and Astronomy, and Center for Fundamental Materials Research, Michigan State University, East Lansing, Michigan 48824-1116*

INTRODUCTION

Carbon nanotubes[1,2] are narrow seamless graphitic cylinders, which show an unusual combination of a nanometer-size diameter and millimeter-size length. This topology, combined with the absence of defects on a macroscopic scale, gives rise to uncommon electronic properties of individual single-wall nanotubes[3,4], which depending on their diameter and chirality, can be either metallic, semiconducting or insulating[5–7].

In this paper we focus attention only on metallic nanotubes and in particular on the so-called "armchair" nanotubes. An armchair nanotube is a graphite tube in which the hexagon rows are parallel to the tube axis. If n is the number of carbon dimers along the nanotube circumference the tube will be labeled as (n, n) nanotube. One of the most important properties of the armchair nanotubes is that they behave like a mono-dimensional metal and this is directly connected with their structure. The electronic wave-length in fact is quantized around the circumference of the tube because of the periodic boundary conditions. This gives rise to mini-bands along the tube axis and the tube is metallic or insulating whether or not one or more mini-bands cross the Fermi energy. In the case of armchair nanotubes two mini-bands along the tube axis cross the Fermi energy[8], therefore, according to scattering theory[9], the conductance is expected to be $2G_0$, where $G_0 = 2e^2/h \approx (12.9\ \text{k}\Omega)^{-1}$ is the quantum conductance. Direct evidence of the de-localization of the wave function along the tube axis has been already shown[10,11], while a direct measurement of the conductance quantization for single-wall nanotubes is still missed (for an introduction to electronic transport in carbon nanotubes see reference 12).

The situation for multi-wall nanotubes is rather different. A multi-wall nanotube consists of several single-wall nanotubes inside one another, forming a structure reminiscent of a "Russian doll". A section of a double-wall (5,5)@(10,10) armchair nanotube is presented in figure 1.

Recent measurements[13] of the conductance in multi-wall nanotubes have raised a significant controversy due to the observation of unexpected conductance values and of ballistic transport at temperatures far above room temperature. In these experiments

Science and Application of Nanotubes, edited by Tománek and Enbody
Kluwer Academic / Plenum Publishers, New York, 2000

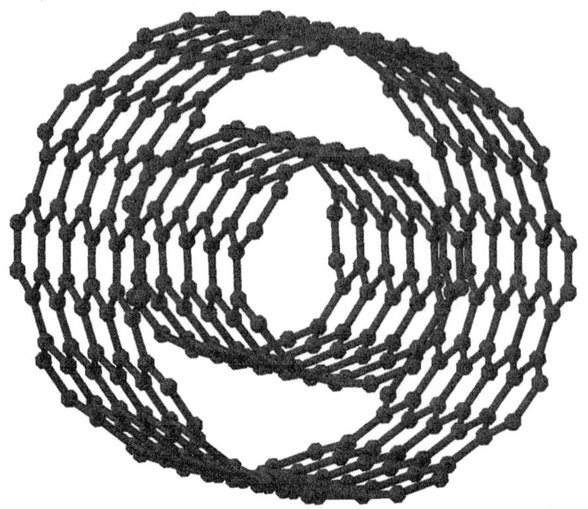

Figure 1: Section of a (5,5)@(10,10) "armchair" nanotube.

several multi-wall nanotubes are glued to a gold tip, which acts as the first electrode, with a colloidal silver paint. The second electrode is made by a copper bowl containing mercury, which provides a gentle contact with the nanotube. The tip is lowered into the mercury and the two-probe conductance is measured as a function of the immersion depth of the tubes into the mercury. The main feature of the experiments is that at room temperature the conductance shows a step-like dependence on the immersion depth, with a value of $0.5\,G_0$ for low immersion and $1\,G_0$ when the tip is further lowered. The value of $0.5\,G_0$ usually persists for small immersion depths ($\leq$ 40nm) and is completely absent in some samples, while the value $1\,G_0$ is found for very long immersion depths, up to 0.5μm. Nevertheless some anomalies have been found with conductances of $0.5\,G_0$ lasting for more than 500nm[13].

While the ballistic behavior up to high temperature can be explained by the almost complete absence of backward scattering[14], the presence of such conductance values is still not completely understood. In the absence of inter-tube interactions, if one assumes that m of the nanotubes forming the multi-wall nanotube are metallic and in contact with both the electrodes, then a conductance of $2mG_0$ is expected for the multi-wall nanotube. This means that even in the extreme case in which only one tube is metallic and in contact with the electrodes a conductance of $2G_0$ must be measured. Therefore the values $0.5\,G_0$ and $1\,G_0$ are largely unexpected. One possible explanation, provided by the authors of the experiments, is that only the outermost tube is responsible for the transport and that the anomalous conductance is the result of scattering to impurities. Nevertheless both these hypothesis may be challenged. The first is based on the assumption that, since mercury does not wet the innermost tubes, it does not provide an efficient electrical contact with the innermost part of the multi-wall nanotube. This may not be the case because the interaction between the different walls may be large and the motion of electrons across the structure efficient. As far as the second hypothesis concerns, it has been shown recently[15] that disorder averages over the tube's circumference, leading to an electron mean free path that increases with the nanotube diameter. Therefore single impurities affect transport only weakly, particularly in the nanotube forming the outermost shell, which has the largest diameter.

In this paper we address these puzzling measurements and show that the structural properties of multi-wall nanotubes can explain their peculiar transport. The electronic band structure of multi-wall carbon nanotubes[16–18], as well as single-wall ropes[19,20] is now well documented. More recently, it has been shown that pseudo-gaps form near the Fermi level in multi-wall nanotubes[18] due to inter-wall coupling, similar to the pseudo-gap formation in single-wall nanotube ropes[19,20]. Here we demonstrate that the unexpected transport properties of multi-wall nanotubes arise from the inter-wall interaction. This interaction may not only block some of the quantum conductance channels, but also redistribute the current non-uniformly over the individual tubes. When only the outermost tube is in contact with one of the voltage/current electrodes, then this forms a preferred current path and, because of inter-tube interaction, the conductance of the whole system will typically be smaller than $2G_0$.

The paper is organized as follows. In the next section we will briefly describe a general scattering technique to compute the transport properties of finite systems attached to semi-infinite contacts, both described by a tight-binding Hamiltonian. In the following section we will discuss the transport in infinite multi-wall nanotubes and understand which are the effects of the inter-tube interaction both on the dispersion and on the wave-function of the tube. Then we present the results for transport properties of inhomogeneous multi-wall nanotubes, giving an explanation of the experiments of reference 13. In this part we will consider different scenarios about the structure of the electrical contacts. At the end we will make some final remarks.

GENERAL SCATTERING TECHNIQUE

To determine transport properties of finite multi-wall nanotubes, we combine for the first time, a tight-binding parameterization determined by *ab-initio* calculations for simpler structures[21], with a scattering technique developed recently for magnetic multi-layers[22,23]. The use of a tight-binding model is justified by the necessity to deal with a system comprising a large number of degrees of freedom. This parameterization has been used to describe detailed electronic structure and total energy differences of systems with unit cells which are too large to handle accurately by *ab-initio* techniques. The electronic structure and superconducting properties of the doped C_{60} solid[24], the opening of a pseudo-gap near the Fermi level in a rope consisting of (10,10) nanotubes[20] and in (5,5)@(10,10) double-wall nanotubes[18] are some of the problems successfully tackled by this technique. The band structure energy functional is augmented by pairwise interactions describing both the closed-shell interatomic repulsion and the long-range attractive van der Waals interaction. This reproduces correctly the interlayer distance and the C_{33} modulus of graphite. Independent checks of this approach can be carried out by realizing that the translation and rotation of individual tubes are closely related to the shear motion of graphite. We expect that the energy barriers in tubes lie close to the graphite value which, due to the smaller unit cell, is easily accessible to *ab-initio* calculations[25].

The scattering technique that we used have been recently employed in studies of giant magnetoresistance[22,23] and ferromagnetic/superconductor structures[26]. It yields the quantum-mechanical scattering matrix S for a phase-coherent system attached to external reservoirs. The rôle of the reservoirs is to inject and collect incoherent electrons into the scattering region. The energy-dependent conductance $G(E)$ in the zero-temperature limit is computed by evaluating the Landauer-Büttiker formula[27]

$$G(E) = \frac{2e^2}{h} T(E) , \tag{1}$$

where $T(E)$ is the total transmission coefficient evaluated at the energy E (E_F in the case of zero-bias). The formula of equation (1) provides an exact relation between the conductance of a system and its scattering properties.

The transmission coefficient is evaluated using a scattering technique that combines a real space Green function calculation for the incoherent leads and a Gaussian elimination (ie "decimation") algorithm for the scattering region. A general scheme of the technique is presented in figure 2, where we indicate how a transport problem can be mapped onto a quantum mechanical scattering problem.

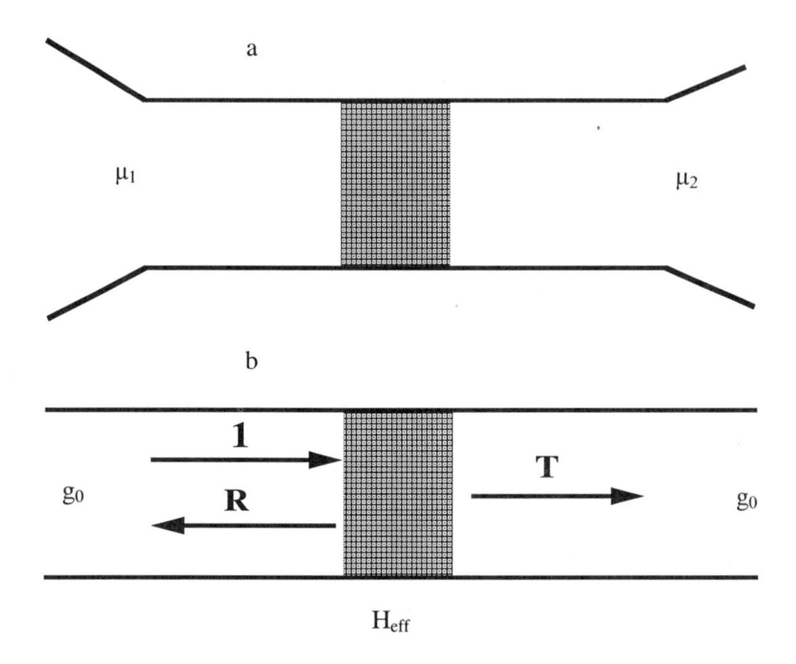

Figure 2: Scheme of the scattering calculation. The system (a) consists in two reservoirs with chemical potentials μ_1 and μ_2 separated by a scattering region. The problem is mapped by using the Landauer-Büttiker formalism onto a quantum mechanical scattering problem (b). The incoming scattering channels in the leads are calculated through the surface Green function g_0. The effective coupling matrix H_{eff} is computed by "decimating" the internal degrees of freedom of the scattering region. The total transmission T and reflection R coefficients are then calculated by solving exactly the Dyson's equation and by using a generalization of the Fisher-Lee relations.

Suppose the total Hamiltonian H for the whole system (nanotubes plus external leads) can be written

$$H = H_{\text{L}} + H_{\text{L-NT}} + H_{\text{NT}} + H_{\text{NT-R}} + H_{\text{R}} , \tag{2}$$

where H_{L} and H_{R} describe respectively the semi-infinite left-hand side and right-hand side lead, $H_{\text{L-NT}}$ and $H_{\text{NT-R}}$ are the coupling matrices between the leads and the nanotube and H_{NT} is the Hamiltonian of the nanotube. In what follows we will consider the leads themselves to be carbon nanotubes, whose number of walls depends on the position of the electrical contacts. This is justified when the transport bottleneck is formed by the nanotubes and not to the metal-nanotube contacts. As far as we know detailed *ab-initio* analysis of metal/nanotube interaction is still not available.

The surface Green function g_0^S of the leads are calculated by numerically evaluating the general semi-analytic formula given in reference 22. One of the key-points of such a calculation is to compute the scattering channels in the leads. Suppose z to be the direction of the transport and the Hamiltonian of the leads to be an infinite matrix of trigonal form with respect to such a direction, with the matrices H_0 and H_1 respectively in the diagonal and off-diagonal positions. Therefore the dispersion relation for electrons in a Bloch state

$$\psi_z = \frac{1}{v_k^{1/2}} e^{ikz} \phi_k \tag{3}$$

and moving along z with unit flux can be written as

$$(H_0 + H_1 e^{ik} + H_{-1} e^{-ik} - E)\phi_k = 0 \,, \tag{4}$$

where v_k is the group velocity corresponding to the state (3) and $H_{-1} = H_1^\dagger$ $(H_0 = H_0^\dagger)$. Note that the matrices H_0 and H_1 describe respectively the interaction within a unit cell and the interaction between adjacent cells. If a unit cell possesses M degrees of freedom, these matrices will be $M \times M$ matrices. Moreover ϕ_k is a M dimensional column vector which describes the transverse degrees of freedom of the Bloch-function. The Green function in the leads is constructed by adding up states of the form of equation (3) with k both real and imaginary, which means that the dispersion relation (4) must be solved for real energies in the form $k = k(E)$. This is the opposite to what is usually computed by ordinary band structure theory where one is interested in finding all the real energies $E = E(k)$ for a chosen real k-vector. Moreover in the calculation of $k = k(E)$ instead of solving the equation

$$\det\,(H_0 + H_1 e^{ik} + H_{-1} e^{-ik} - E) = 0 \,, \tag{5}$$

which involves the use of a root tracking algorithm in the complex plane, we map the problem onto an eigenvalue problem by defining the matrix $\mathcal{H}$

$$\mathcal{H} = \begin{pmatrix} -H_1^{-1}(H_0 - E) & -H_1^{-1}H_{-1} \\ \mathcal{I} & 0 \end{pmatrix} , \tag{6}$$

where $\mathcal{I}$ is the $M \times M$ identity matrix. The eigenvalues of $\mathcal{H}$ are the roots e^{ik} and the upper half of the eigenvectors of $\mathcal{H}$ are the corresponding eigenvectors ϕ_k.

The second part of the calculation involves computing an effective coupling matrix between the surfaces of the scattering region. Note that the purpose of a scattering technique is to calculate the S matrix between electrons in the leads. Therefore one is not interested in information regarding the internal degrees of freedom of the scattering region, but only in the resulting coupling between the external interfaces. This can be achieved by reducing the matrix $H_{\mathrm{L-NT}} + H_{\mathrm{NT}} + H_{\mathrm{NT-R}}$ to an effective coupling matrix H_{eff}. Suppose the total number of degrees of freedom of the Hamiltonian $H_{\mathrm{L-NT}} + H_{\mathrm{NT}} + H_{\mathrm{NT-R}}$ is N, and the number of degrees of freedom of the lead surfaces M. One can eliminate the $i = 1$ degree of freedom (not belonging to the external surfaces) by reducing the $N \times N$ total Hamiltonian to an $(N-1) \times (N-1)$ matrix with elements

$$H_{ij}^{(1)} = H_{ij} + \frac{H_{i1} H_{1j}}{E - H_{11}} \,. \tag{7}$$

Repeating this procedure l times we obtain the "decimated" Hamiltonian at l-th order

$$H_{ij}^{(l)} = H_{ij}^{(l-1)} + \frac{H_{il}^{(l-1)} H_{lj}^{(l-1)}}{E - H_{ll}^{(l-1)}} \,, \tag{8}$$

and finally after $N - M$ times, the effective Hamiltonian

$$H_{\text{eff}}(E) = \begin{pmatrix} H_{\text{L}}^*(E) & H_{\text{LR}}^*(E) \\ H_{\text{RL}}^*(E) & H_{\text{R}}^*(E) \end{pmatrix} . \tag{9}$$

In the equation (9) the matrices $H_{\text{L}}^*(E)$ and $H_{\text{R}}^*(E)$ describe the intra-surface couplings respectively in the left-hand side and right-hand side surfaces, and $H_{\text{LR}}^*(E)$ and $H_{\text{RL}}^*(E)$ describe the effective coupling between these surfaces. From the above equations it is clear that only matrix elements coupled to the eliminated degree of freedom are redefined. This exact recursive technique therefore turns out to be very efficient in the case of short-range interaction like the nearest neighbors tight-binding model considered here. Two important considerations must be made. Firstly we note that both the Green function calculation and the "decimation" require a fixed energy. Once this has been set the calculation is exact and does not use any approximation. Secondly the calculation of the Green function is completely decoupled by the calculation of the effective Hamiltonian for the scatterer. This can allow very efficient numerical optimizations, particularly in the study of disordered systems[28].

Once both the surface Green function of the leads g_0^{S} and the effective coupling Hamiltonian $H_{\text{eff}}(E)$ are computed then the total Green function G^{S} for the whole system (leads plus scattering region) are easily calculated by solving the Dyson's equation

$$G^{\text{S}}(E) = [(g_0^{\text{S}}(E))^{-1} - H_{\text{eff}}]^{-1} . \tag{10}$$

Finally the scattering matrix elements are extracted from G^{S} by using a generalization of the Fisher-Lee relations[29].

For the case of leads made by carbon nanotubes a final observation must be considered. The unit cell along the axis of the nanotube comprises two atomic planes, and since the hopping matrix between sequential unit cells H_1 is therefore singular, the dispersion relation cannot be calculated by using the equation (6). We avoid this complication by projecting out the non-coupled degrees of freedom between sequential cells before calculating the scattering channels. This has been done by using the "decimation" technique described above.

CONDUCTANCE IN MULTI-WALL NANOTUBES

For an homogeneous system $T(E)$ assumes integer values corresponding to the total number of open scattering channels at energy E. For individual (n, n) "armchair" tubes, this integer is further predicted to be even[8], with a conductance of $2G_0$ near the Fermi level. As an example, our results for the conductance $G(E)$ and the density of states of the $(10, 10)$ nanotube are shown in Fig. 3.

The main feature of an "armchair" nanotube is its true mono-dimensional metallic behavior. Note that the density of state shows mono-dimensional van Hove singularities which are due to the presence of dispersion-less mini-bands. This is reflected in the energy-dependent conductance which shows a typical step-like behavior. Such steps appear whenever the energy crosses a new mini-band, and therefore correspond to the van Hove singularities in fig 3a. It is crucial to note that in an infinite system every scattering channel gives the same contribution G_0 to the conductance independently from its dispersion and group velocity. The situation is rather different in an inhomogeneous system, where the scattering of electrons from low dispersion to high dispersion bands of different materials, can give rise to strong backward scattering and therefore

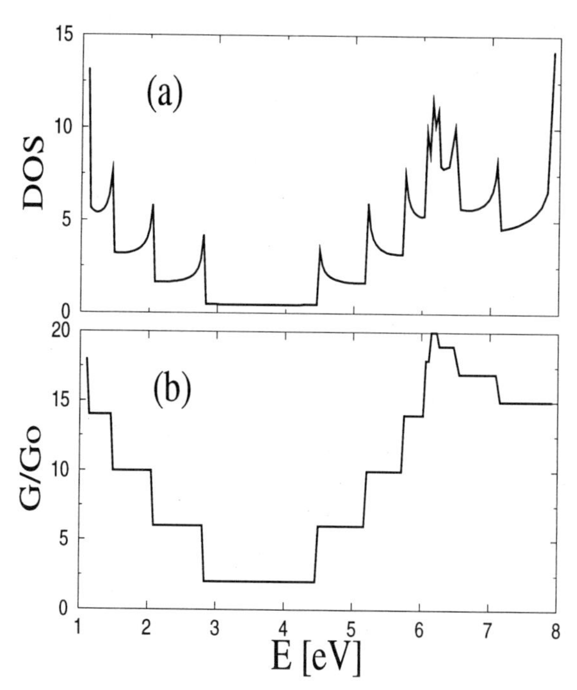

Figure 3: Single-wall (10,10) nanotube. (a) Local density of states. (b) Conductance as a function of energy. The Fermi level lies at 3.65 eV.

to a reduction of the conductance[22,23,28]. At the Fermi energy of an "armchair" nanotube (in this case $E_F = 3.65$ eV) only two scattering channels are present resulting in a conductance $2G_0$, which remains constant in an energy interval of approximately 1.5eV.

Consider now multi-wall nanotubes. As observed in the introduction, in the absence of inter-tube interactions, different tubes behave as conductors in parallel and the conductances are simply additive. Therefore, since the position of the Fermi energy does not change with the tube diameters we expect a conductance $2mG_0$ for a multi-wall nanotubes comprising m walls. Note also that the width of the energy region around the Fermi energy where the conductance is $2G_0$, depends only weakly on the tube diameters. The situation changes drastically when inter-tube interaction is switched on. In figures 4 and 5 we present the density of states and the conductance respectively for a (10,10)@(15,15) and for a (5,5)@(10,10)@(15,15) multi-wall nanotube.

In the figures we restricted the energy window to the region where the single-wall armchair nanotubes present conductances of $2G_0$. The main feature of both the figures is the presence of pseudo-gaps[18] which lower the conductance from the expected value $2mG_0$. In the case of a double-wall nanotube, this results in two regions where the conductance passes from $4G_0$ to $2G_0$, while in triple-wall nanotube the values $6G_0$, $4G_0$ and $2G_0$ are possible. Nevertheless both these results are still not consistent with the experimental observations of $1G_0$ and $0.5G_0$[13].

It is important to note that the presence of energy pseudo-gaps does not only lowers the conductance but also gives rise to two important effects. First it changes drastically the dispersion of the mini-bands close to the gaps. At the edge of the gaps in fact the dispersion passes from a linear to an almost dispersion-less parabolic-like structure. This is shown in figure 6 where we present the band structure along the

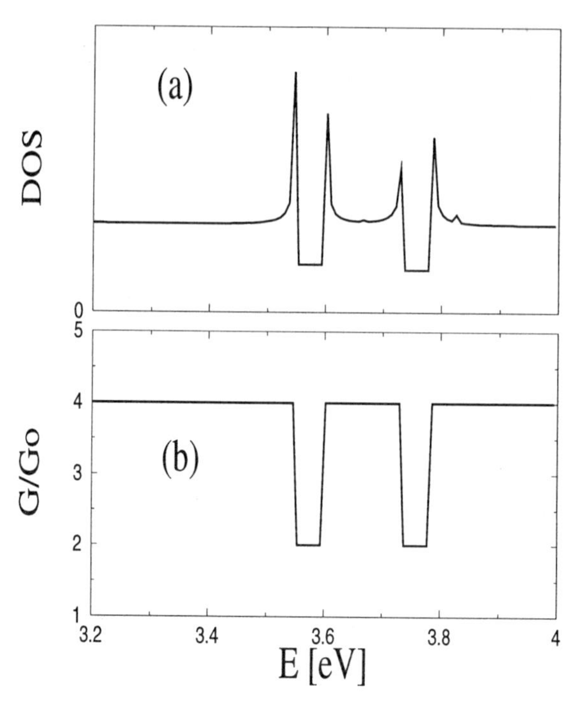

Figure 4: (a) Local density of states for a double-wall (10,10)@(15,15) nanotube. (b) Conductance as a function of energy.

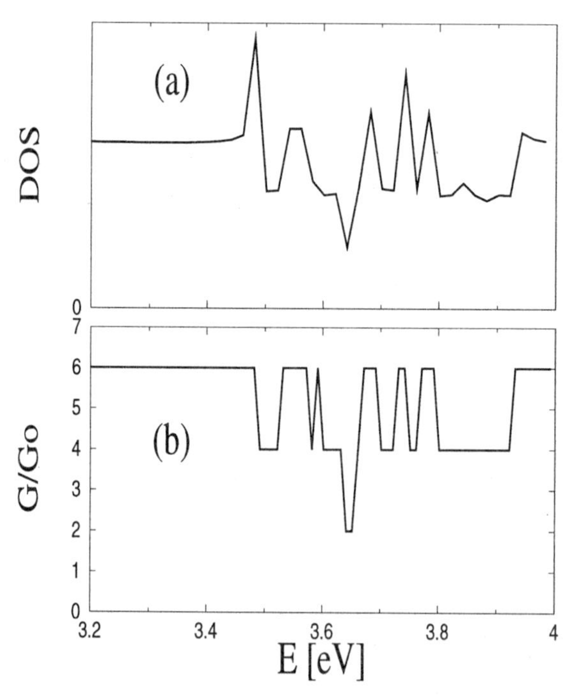

Figure 5: (a) Local density of states for a triple-wall (5,5)@(10,10)@(15,15) nanotube. (b) Conductance as a function of energy.

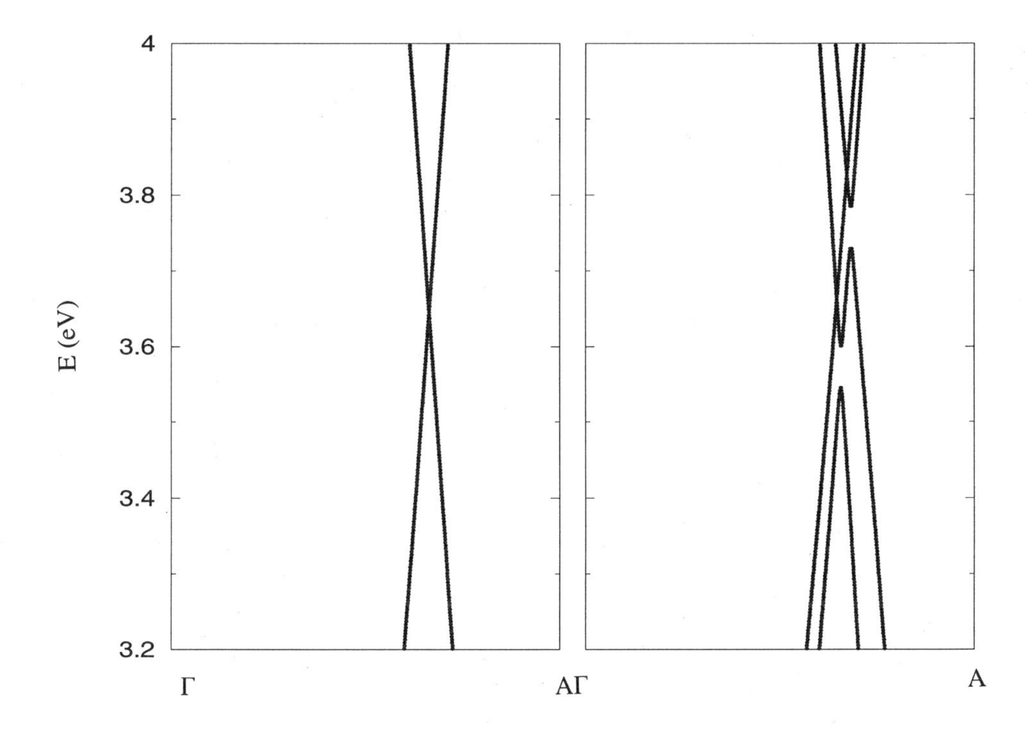

Figure 6: (a) Band structure along the tube axis for a (15,15) nanotube, with $E_F = 3.65$ eV. (b) Band structure along the tube axis for a (10,10)@(15,15) nanotube.

direction of the tube axis for a double-wall (10,10)@(15,15) nanotube (b) together with the band structure of a single-wall (15,15) nanotube.

Secondly the amplitude of the wave-function across the nanotubes changes. Far from the gaps, where the effects of the inter-tube interaction are weak, the wave-function is expected to have a uniform distribution across the different walls composing the nanotube. This is what is found in the case of non-interacting walls, whereas in the vicinity of a pseudo-gap, the distribution changes dramatically and the amplitude may be enhanced along some walls and reduced along some others. To demonstrate this effect in figure 7 we present the partial conductance across the two walls composing a (10,10)@(15,15) nanotube and across the three walls composing a (5,5)@(10,10)@(15,15) nanotube. The partial conductance is defined as the projection of the total conductance for an infinite multi-wall tube onto the degrees of freedom describing the individual walls. From the figure it is very clear that the amplitude of the wave-function (which is proportional to the partial conductance) is not uniform across the structure and depends critically on the energy.

Both the change in the dispersion and the non-uniform distribution of the amplitude of the wave-function across the tubes have drastic effects on the transport of heterogeneous systems, because it creates strong inhomogeneities along the structure, and therefore strong backward scattering. This aspect, which occurs in a multi-wall nanotube when one of the innermost walls closes, will be discussed in the next section.

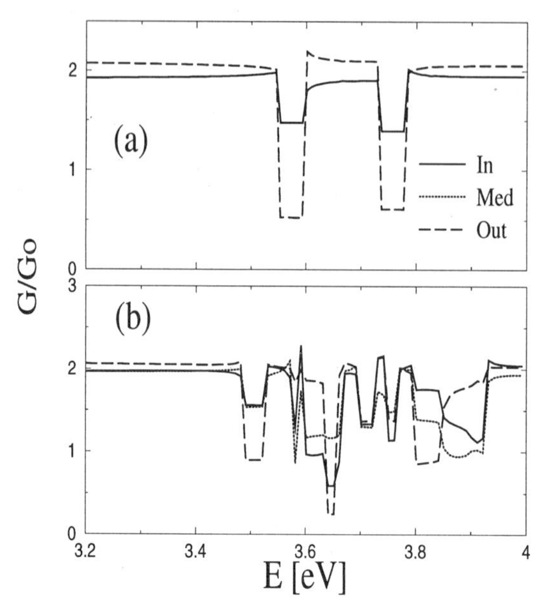

Figure 7: Partial conductance of (a) (10,10)@(15,15) and (b) (5,5)@(10,10)@(15,15) nanotube. The solid line, dotted and dashed lines represent the partial conductance respectively onto the innermost, the medium (only in the case of (5,5)@(10,10)@(15,15)) and the outermost tube. Note that within the pseudo-bandgaps the conductance does not distribute uniformly onto the different tubes.

TRANSPORT IN INHOMOGENEOUS MULTI-WALL NANOTUBES

In this section we will use the ideas developed above to describe the experiments of reference 13. Note that for inhomogeneous systems, where multi-wall nanotubes are contacted to the voltage/current probes, the conductance quantization in unit of $2G_0$ which we found also for multi-wall nanotube in presence of inter-wall interaction is evidently violated and fractional values of the conductance are allowed. One of the difficulties of the experiments, which use gold as one electrode and mercury as the other, is that not all tubes make contact with the electrodes. We have considered two different scenarios and have found that agreement with the experiments is obtained when we assume that only the outermost tube is in contact with the gold electrode, whereas the number of walls in contact with the mercury depends on the depth at which the tube is immersed into the liquid. This latter assumption may seem surprising, because the mercury does not wet the inner tubes. Nevertheless we believe that at equilibrium, the inter-tube interaction allows a uniform distribution of the chemical potential across the cross-section of the whole structure and therefore in the linear-response regime, the scattering problem reduces to a semi-infinite single-wall nanotube (the one in direct contact with gold) attached to a scattering region in which a variable number of walls are present (see fig.8a). Moreover a close analysis of the inter-tube matrix elements shows that these are of the same order of magnitude as the intra-wall ones. This means that electron transport between different walls may be efficient, as well as the electron feeding of the innermost walls from the electrons reservoirs.

Consider first the case in which only the outermost tube makes contact with the gold electrode. We argue that the step-like dependence of the conductance on the immersion depth is due to the fact that the scattering region makes contact with the

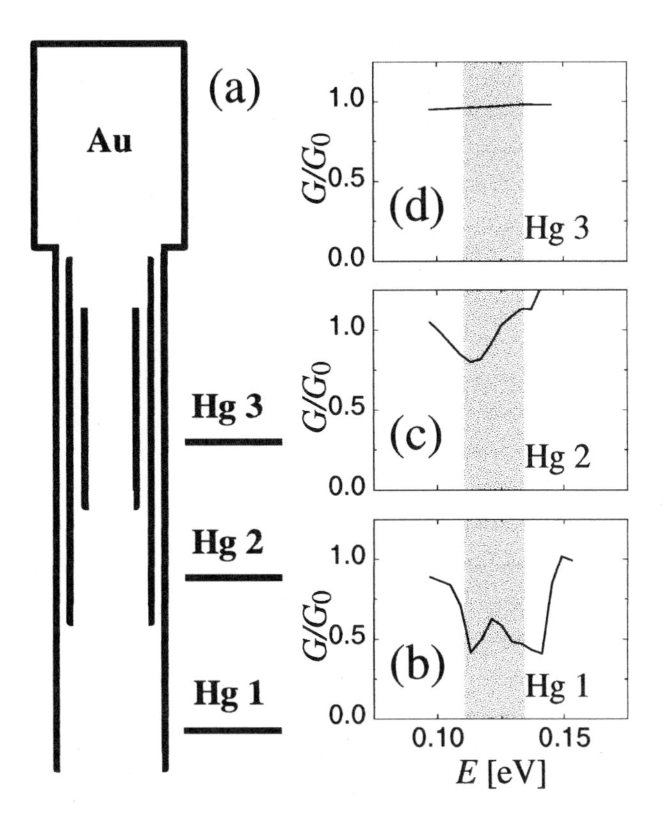

Figure 8: (a) Schematic geometry of the system in which only the outermost tube is contacted with the gold electrode for different immersion depths. (b)-(d) Conductance as a function of energy for the system of (a) at the immersion depths Hg1, Hg2, and Hg3. E is given with respect to E_F of the pristine (undoped) nanotube.

mercury reservoir via a multi-wall semi-infinite nanotube whose number of walls varies and depends on the immersion of the structure. For small immersion depths (such as Hg1 in fig.8a), only the outermost tube is in contact with mercury, because it is the only one with an end below the mercury level. A further lowering of the gold tip (to depths such as Hg2 and Hg3 in fig.8a) will sequentially place more inner walls into electrical contact with the mercury, thereby changing the conductance. We notice that the conductance of such a structure cannot be larger than that of the single-wall nanotube, which is the only tube in contact with the gold electrode.

In figure 8b we present the conductance as a function of energy for the inhomogeneous structure described in figure 8a. In all three cases, the simulated structure makes contact with the upper Au reservoir via a (15,15) nanotube, which forms the upper external lead, whereas the lower external lead contacting the Hg comprises either a single, double or triple-wall nanotube. The solid curve corresponds to a structure formed from a 200 atomic plane (AP) (5,5)@(10,10)@(15,15) triple-wall region, below which is attached to a 200 AP (10,10)@(15,15) double-wall region. The ends of the outer (15,15) nanotube are connected to semi-infinite (15,15) nanotubes, which form the external leads. The dashed curve corresponds to a structure formed from a 200 AP (5,5)@(10,10)@(15,15) triple-wall region. The upper end of the outer tube attached to a semi-infinite (15,15) nanotube, which forms the external lead contacting the Au reservoir. The lower end of the (10,10) and (15,15) nanotubes continue to infinity, and form a (10,10)@(15,15) external contact to the Hg reservoir. Finally the dot-dashed line shows the conductance of a (5,5)@(10,10)@(15,15) nanotube, which at the lower end makes direct contact with the Hg and at the upper end, the outer tube continues to infinity, thereby forming a (15,15) external contact to the Au reservoir. These situations correspond to immersion of the tube into the mercury at positions Hg1, Hg2 and Hg3 respectively, where either one wall and two walls are in electrical contact with the mercury.

In all the simulations, the ends of the finite-length tubes are left open and we do not include capping layers. We believe that the capping layers are not crucial to the description of the transport properties of inhomogeneous multi-wall nanotubes, since these are mainly determined by the mis-match of wave-vectors between different regions. Figure 8b shows clearly that in an energy window of about 0.05eV (indicated by vertical dashed lines), the conductance for the first structure is approximately $0.5G_0$, while for the latter two is of order $1G_0$. Note that such energy window is two times larger than the bias used in the experiments and also much larger than room temperature. This suggests that these results are quite robust and will survives both at room temperature and moderate biases. This remarkable result is in excellent agreement with the recent experiments of reference 13.

The scattering in such an inhomogeneous structure arises from the reasons pointed out in the previous section. In the energy window considered in fact the infinite (5,5)@(10,10)@(15,15) presents a large pseudo-gap with conductance $4G_0$. We therefore expect that at both the interfaces of the (5,5)@(10,10)@(15,15) region with respectively the (10,10)@(15,15) region and the (15,15) tube, the mismatch of either the transverse components of the wave-function ϕ_k and the longitudinal k-vectors will be large. This gives rise to the strong suppression of the conduction observed in the experiments. In figure 9 we present the conductance as a function of immersion depth in mercury for the structure described above. The conductance is calculated at zero-temperature in the zero bias limit and the energy has been set in the middle of the marked region of figure 8a (3.825eV). Note again that the agreement with the curve of experiments of reference 13 is very good.

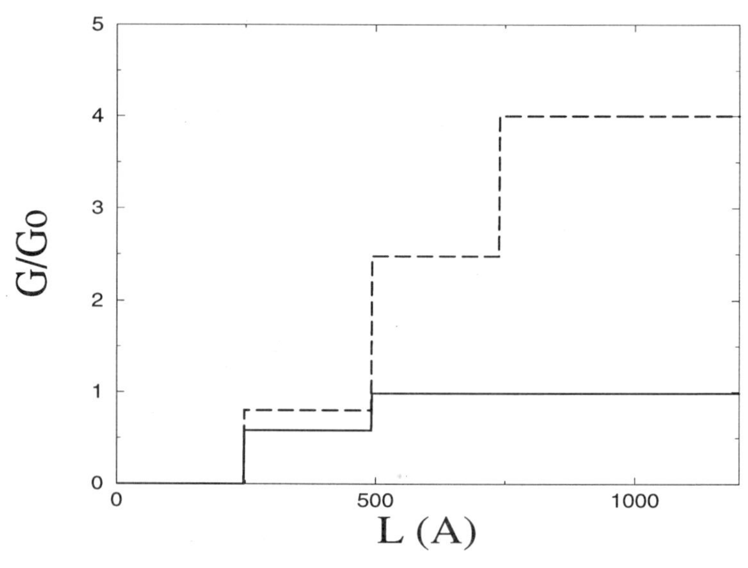

Figure 9: Conductance as a function of immersion depth. The solid curve corresponds to the structure of figure 8a and the dashed curve to that of figure 10a.

We now consider a second possible scenario, in which three tubes are in direct contact with the gold electrode. In this case the electrons are fed from gold into the structure directly along all the tubes. This contact can be simulated by a semi-infinite (5,5)@(10,10)@(15,15) nanotube with uniform chemical potential across the tubes. The structure considered is presented in figure 10a. In this case the upper bound of the conductance is no longer fixed by the single-wall tube to be $2G_0$ but can be as large as $6G_0$ and depends on the number of walls contacting the mercury. In figure 10b we show the conductance as a function of energy respectively for a 200 AP (10,10)@(15,15) nanotube sandwiched between a (15,15) and a (5,5)@(10,10)@(15,15) nanotube leads, for (10,10)@(15,15) nanotube lead in contact with a (5,5)@(10,10)@(15,15) nanotube lead, and for an infinite (5,5)@(10,10)@(15,15) nanotube. This again corresponds to the different levels of immersion Hg1, Hg2 and Hg3 in (Fig. 10a). Note that in the case in which the (5,5)@(10,10)@(15,15) nanotube is in direct contact with both the gold and the mercury electrodes its conductance corresponds to the number of opening scattering channels for the infinite triple-wall system.

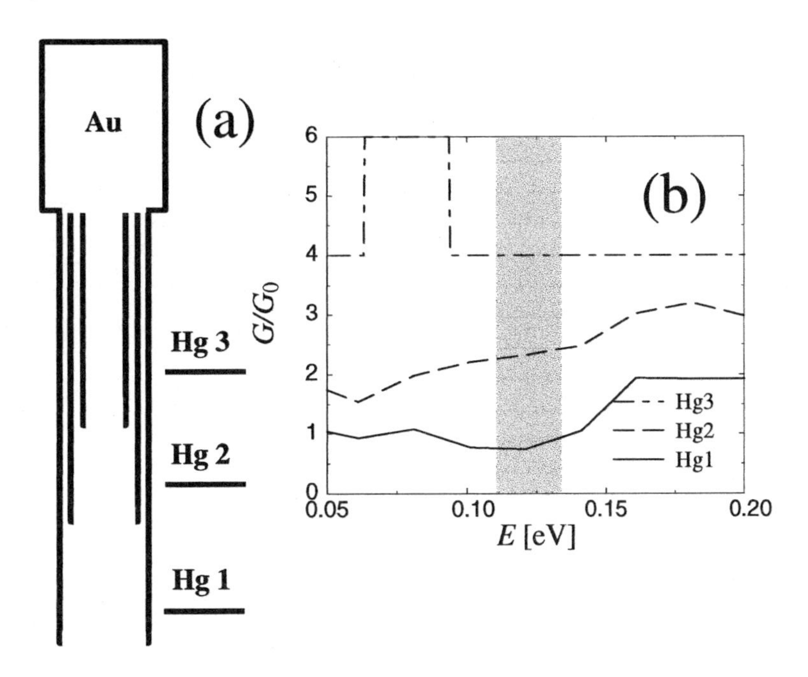

Figure 10: (a) Schematic geometry of the system in which three tubes tubes are contacted with the gold electrode for different immersion depths. (b) Conductance as a function of energy for the system of (a). E is given with respect to E_F of the pristine (undoped) nanotube.

Figure 10 shows that when all three tubes are electrically connected to the gold electrode, a much larger increase in the conductance occurs when a new wall is lowered below the mercury level, although this is still smaller than the value of $2G_0$, obtained for completely isolated tubes. In this case, corresponding to the different value of the immersion depth, we expect the conductance to be respectively $1G_0$, $2G_0$ and $4G_0$.

The large difference between the transport of the structures in figures 8a and 10a is therefore crucially dependent on the number of tubes which make a direct contact with the gold electrode. At the moment a complete description of the nanotube/metal interface is not available, although it will deserve further investigation both experimentally and theoretically.

CONCLUSIONS

To conclude we have presented a fully quantum scattering technique which yields the S matrix of inhomogeneous multi-wall nanotubes. We have shown that the inter-tube interaction drastically modifies transport, not only by opening pseudo-gaps close to the Fermi energy, but also by redistributing the amplitude of the transverse component of the wave-function across the multi-wall structure. These effects, when combined together, form a convincing explanation of puzzling experiments in which non-integer values of conductance have been found in multi-wall nanotubes[13]. To arrive at this quantitative description of the experiments, we have explored several possibilities regarding on the nature of the nanotube/metal interfaces. Only those calculations in which the outermost tube is in direct contact with the gold electrode showed good agreement with the experiments.

ACKNOWLEDGEMENTS

This work has been done in collaboration with the group of Prof. J.H. Jefferson at DERA Malvern, who is kindly acknowledged. SS acknowledges also the financial support by the DERA and the MSU-CMPT visitor fund. YKK and DT acknowledge financial support by the Office of Naval Research under Grant Number N00014-99-1-0252.

REFERENCES

1. S. Iijima, Nature **354**, 56 (1991).
2. For a general review, see M.S. Dresselhaus, G. Dresselhaus, and P.C. Eklund, *Science of Fullerenes and Carbon Nanotubes* (Academic Press Inc., 1996 San Diego), and references therein.
3. S. Iijima and T. Ichihashi, Nature **363**, 603 (1993).
4. D.S. Bethune, C.H. Kiang, M.S. de Vries, G. Gorman, R. Savoy, J. Vazquez, R. Beyers, Nature **363**, 605 (1993).
5. J.W. Mintmire, B.I. Dunlap, and C.T. White, Phys. Rev. Lett. **68**, 631 (1992).
6. R. Saito, M. Fujita, G. Dresselhaus, and M.S. Dresselhaus, Appl. Phys. Lett. **60**, 2204 (1992).
7. N. Hamada, S. Sawada, and A. Oshiyama, Phys. Rev. Lett. **68**, 1579 (1992).
8. L. Chico, L.X. Benedict, S.G. Louie and M.L. Cohen, Phys. Rev. B **54**, 2600 (1996), W. Tian and S. Datta, *ibid.* **49**, 5097 (1994), M.F. Lin and K.W.-K. Shung, *ibid.* **51**, 7592 (1995).
9. R. Landauer, Phil. Mag. **21**, 863 (1970).
10. S.J. Tans, M.H. Devoret, H. Dai, A. Thess, R.E. Smalley, L.J. Geerligs, C. Dekker, Nature **386**, 474 (1998).
11. M. Bockrath, D. H. Cobden, P. L. McEuen, N. G. Chopra, A. Zettl, A. Thess, R. E. Smalley Science **275**, 1922 (1997).
12. C. Dekker, Physics Today May 1999, 22 (1999).
13. S. Frank, P. Poncharal, Z.L. Wang, and Walt A. de Heer, Science **280**, 1744 (1998).
14. T. Ando, T. Nakanishi, J. Phys. Soc. Jpn. **67**, 1704 (1998).
15. C.T. White, T.N. Todorov, Nature **393**, 240 (1998).

16. Riichiro Saito, G. Dresselhaus, and M.S. Dresselhaus, J. Appl. Phys. **73**, 494 (1993).
17. Ph. Lambin, L. Philippe, J.C. Charlier, and J.P. Michenaud, Comput. Mater. Sci. **2**, 350 (1994).
18. Young-Kyun Kwon and David Tománek, Phys. Rev. B **58**, R16001 (1998).
19. P. Delaney, H.J. Choi, J. Ihm, S.G. Louie, and M.L. Cohen, Nature **391**, 466 (1998).
20. Young-Kyun Kwon, Susumu Saito, and David Tománek, Phys. Rev. B **58**, R13314 (1998).
21. D. Tománek and Michael A. Schluter, Phys. Rev. Lett. **67**, 2331 (1991).
22. S. Sanvito, C.J. Lambert, J.H. Jefferson, A.M. Bratkovsky, Phys. Rev. B **59**, 11936 (1999).
23. S. Sanvito, C.J. Lambert, J.H. Jefferson, A.M. Bratkovsky, J. Phys. C: Condens. Matter. **10**, L691 (1998).
24. M. Schluter, M. Lannoo, M. Needels, G.A. Baraff, and D. Tománek, Phys. Rev. Lett. **68**, 526 (1992).
25. J.-C. Charlier, X. Gonze, and J.-P. Michenaud, Europhys. Lett. **28**, 403 (1994); M.C. Schabel and J.L. Martins, Phys. Rev. B **46**, 7185 (1992).
26. F. Taddei, S. Sanvito, C.J. Lambert, J.H. Jefferson, Phys. Rev. Lett. **82**, 4938 (1999).
27. M. Büttiker, Y. Imry, R. Landauer, S. Pinhas, Phys. Rev. B **31**, 6207 (1985).
28. S. Sanvito, C.J. Lambert, J.H. Jefferson, to appear in Phys. Rev. B, and also cond-mat/9903381
29. C.J. Lambert, V.C. Hui, S.J. Robinson, J. Phys. C: Condens. Matter. **5**, 4187 (1993).

CONDUCTIVITY MEASUREMENTS OF CATALYTICALLY SYNTHESIZED CARBON NANOTUBES

M. Ahlskog[1], R.J.M. Vullers[1], E. Seynaeve[1], C. Van Haesendonck[1], A. Fonseca[2] and J. B.Nagy[2]

[1]Laboratorium voor Vaste-Stoffysica en Magnetisme, Katholieke Universiteit Leuven, Celestijnenlaan 200 D, B-3001 Leuven, Belgium
[2]Laboratoire de Résonance Magnétique Nucléaire, Facultés Universitaires Notre-Dame de la Paix (FUNDP), Rue de Bruxelles 61, B-5000 Namur, Belgium

ABSTRACT

Electrical transport measurements have been undertaken on individual multiwalled catalytically synthesized carbon nanotubes. Two different techniques have been employed: In the first one nanotubes have been deposited on top of goldelectrodes. In the second one, atomic force microscope nanolithography has been used to locally oxidize a Ti thin film structure deposited above a single nanotube, thereby defining a two probe electrode configuration. It is shown that the latter method results in a two orders of magnitude lower resistance than the former method, due to the reduced contact resistance.

INTRODUCTION

Very recently several innovative techniques have been developed to measure the transport properties of single carbon nanotubes [1], following and including a few pioneering works [2 and refs. therein] which showed that the carbon nanotube offers an ideal system on which to study mesoscopic transport phenomena in general and 1D-effects in particular. In addition the carbon nanotube in itself is a complex system with different structures being either semiconducting or metallic.

For electrical transport measurements it is imperative to understand and control the nature of the contact resistances between the electrode and the nanotube. The most generally, and most straightforwardly implementable, method is to deposit the nanotube on top of nobel metal (usually gold) electrodes. This, however, tends to result in a tunneling barrier between the metal electrode and the nanotube. The simplest alternative

Science and Application of Nanotubes, edited by Tománek and Enbody
Kluwer Academic / Plenum Publishers, New York, 2000

is to have the electrode cover the nanotube. Such a configuration has shown that, in a freestanding multiwalled carbon nanotube (MWNT), ballistic conduction can exist. On supported nanotubes, measurements with electrodes covering the nanotube have in some cases shown very low resistivities in carbon nanotubes. There is therefore reason to implement such fabrication methods where metal electrodes completely cover the nanotube.

In recent years the atomic force microscope (AFM) has been used for nanolithographic purposes with the demonstrated potential to define features on semiconductors and metal thin films with smaller dimensions than can be achieved with standard e-beam lithography [3]. We have developed the technique of field induced local oxidation with the AFM to fabricate mesoscale structures from metal (Ti) thin films [4]. In particular, we have used the technique to oxidize more complicated composite structures consisting of a Ti thin films covering gold islands, 5-20 nm in diameter, whereby the formed oxide covers complete gold islands. In the following, we show how this technique is applied to fabricate contacts on top of individual carbon nanotubes

NANOTUBE SYNTHESIS

The MWNTs have been produced by catalytic decomposition of acetylene, carried out at 700 °C in a flow reactor at atmospheric pressure [5]. The catalyst, "2.5 wt% Co/zeolite NaY" prepared by impregnation, was removed from the carbon material (nanotubes + amorphous carbon) by HF treatment.

RESULTS AND DISCUSSION

The purified carbon nanotube material is sonicated in isopropanol and is subsequently deposited onto a piece of Si/SiO_2 wafer onto which goldelectrodes with 0.5 μm spacing between the electrodes were fabricated with electron beam lithography.

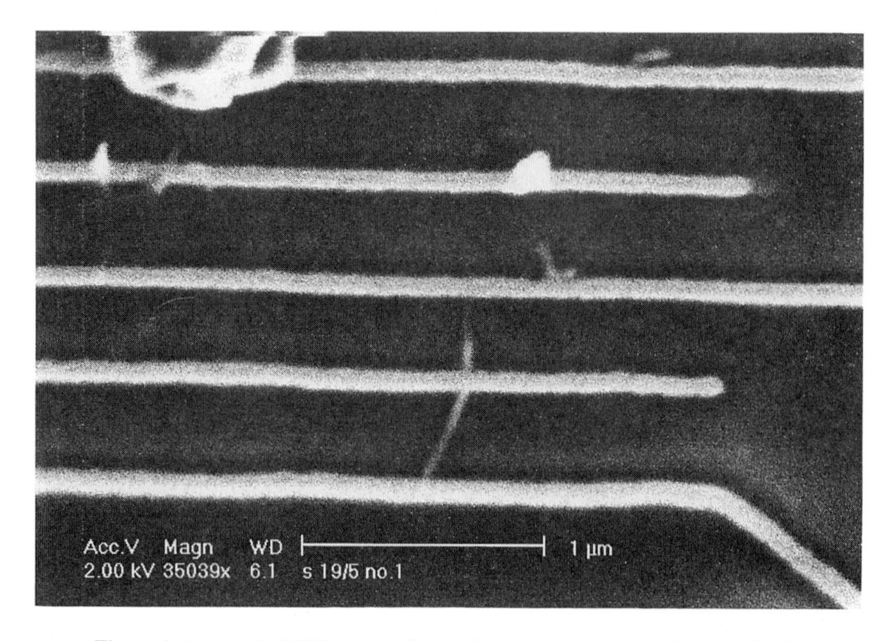

Figure 1. A single MWNT on top of two adjacent electrodes (and almost a third).

Nanotubes were deposited on top of these according to the procedure described in ref. [6]. A droplet of the nanotube solution was deposited on the substrate. After the droplet was removed and the sample dried a few among the adjacent electrode pairs were typically found to be connected either by one or more nanotubes. Single nanotubes were thus ready for electrical measurements.

The current-voltage (I-V) characteristics of these samples were measured between 4.2 K and room temperature. In fig. 2 are shown those of fig. 1. We obtained typically a room temperature resistance of the order of 100 kΩ with Ohmic current-voltage behaviour. At low temperatures the resistance increases and the I-V characteristics become increasingly non-linear. The zero-bias gap which appears at low temperatures is the stronger, the larger the room temperature resistance is (We have measured values in the range 50 kΩ - 1 MΩ). From these kind of measurements alone it is difficult to infer whether we are probing intrinsic behaviour or if contact resistance dominates the transport behaviour. We note however, that the two-terminal resistance is one order of magnitude smaller than what typically is observed for single wall nanotubes measured in a similar way [7].

To fabricate electrodes on top of an individual nanotube we use the technique of local oxidation with the AFM. This process is explained in fig. 3. Nanotubes are deposited from a solution onto a substrate with goldmarkers. An individual nanotube is searched with the AFM and its position recorded with respect to the nearest marker. Next a 2 μm × 4 μm film of 10 nm thick Ti, connected to larger contact pads, is deposited on top of the nanotube with conventional electron beam lithography.

In fig. 4 an AFM image is shown of a nanotube with Ti contacts fabricated with the method described in Fig. 3. A 500 nm wide oxide line is created perpendicular to the nanotube, thus separating the two sides of the Ti pattern which then form electrodes covering each end of the nanotube. The two-terminal resistance between the separate sides of the of the Ti structure was followed during oxidation.

In fig. 5 the resistance (R) is shown as a function of time which is roughly equivalent to the width of the oxide since the oxidation speed is constant. The relationship between the

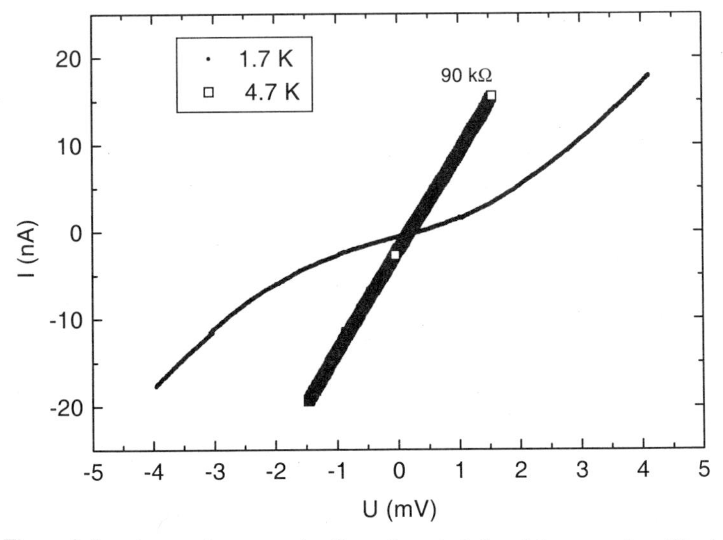

Figure 2. Low temperature current-voltage characteristics of the nanotube of fig. 1.

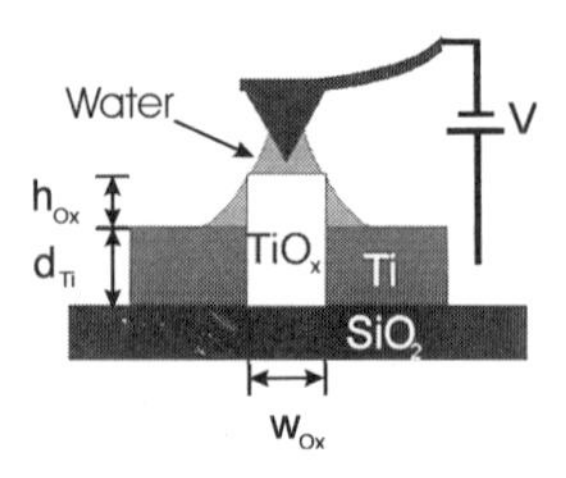

Figure 3. Schematic drawing of the AFM local oxidation technique. With AFM local oxidation it is possible, with a conducting AFM tip, to oxidize through a thin enough Ti film (a non-noble metal), by applying a voltage (V) between the scanning tip and the metal [3]. The oxide (TiO_x) protrudes with a height (h) above the film thickness (d). A water layer due to the ambient humidity is always present and is necessary for the oxidation process.

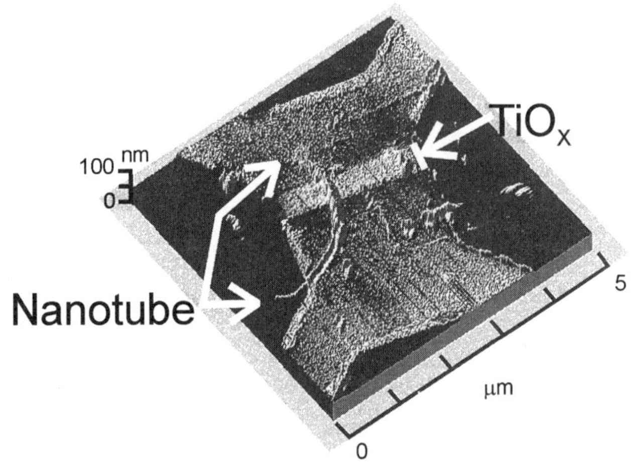

Figure 4. AFM image of a MWNT with Ti electrodes on top of it. The size of the central rectangular area is 2 µm × 4 µm. The lighter section going over the nanotube is the 500 nm wide oxide (TiO_x) which separates the two Ti electrodes from each other.

increase of R, ΔR, and time is linear except for the beginning. At a larger width than ~ 100 nm the oxide is perfectly insulating and ΔR therefore corresponds to the resistance of the nanotube covered by the oxide. After the AFM oxidation has been stopped (after 800 seconds) $\Delta R \cong 500\ \Omega$. This value is two orders of magnitude smaller than the two-terminal resistance measured for nanotubes on top of gold electrodes.

We proceeded to measure R down to a temperature of 4.2 K in a four point configuration, which however included some series resistance from the 2×4 µm Ti film (the contribution from this was estimated from separate measurements and subsequently subtracted). We thus measured a ≅ 20 % increase in R at 4.2 K, which is similar to the temperature dependence measured in a four point configuration on arc-discharge synthesized MWNTs by Bachtold et al. [8]. Some recent prominent work emphasizes the need to embed the nanotube inside the electrodes to probe intrinsic transport behaviour. These observations include ballistic conduction [9] and superconducting supercurrents [10]. In the latter case some very small two-terminal resistances were reported for SWNTs. Our method can also be realized in a genuine four probe configuration, which is essential for accurate low temperature transport measurements.

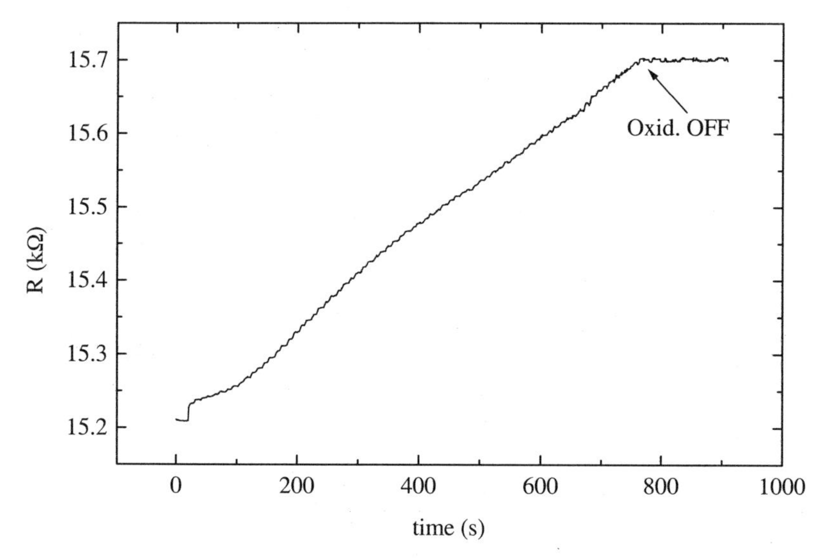

Figure 5. In-situ two terminal resistance measurement of the Ti structure of fig. 4 during oxidation. The moment when the oxidation of the Ti was stopped is indicated.

CONCLUSIONS

The method of contact fabricaton is crucial for obtaining correct information on the intrinsic transport properties of carbon nanotubes. This was shown by the drastic difference between the two-terminal resistance of MWNTs when the nanotube lies on top of electrodes or when it is covered by them. From these measurements it is further concluded that catalytically synthesized MWNTs can have a very low resistance ($\leq$ 1 kΩ/μm), comparable to those synthesized by the arc-discharge method.

ACKNOWLEDGEMENTS

The work at the K.U. Leuven was supported by the Fund for Scientific Research - Flanders (FWO) and by the Belgian Inter-University Attraction Poles (IUAP) research program. M.A also acknowledges the financial support from the Research Council of the K.U. Leuven. FUNDP is grateful to the Inter University Poles of Attraction on Reduced Dimensionality Systems (PAI-IUAP No. 4/10) for financial support.

REFERENCES

[1] Bulletin of the American Physical Society, **44**, No. 1, (1999).
[2] C. Dekker, Physics Today, May 1999, p.22.
[3] K. Matsumoto, Proceedings of the IEEE **85**, 612 (1997).
[4] R.J.M. Vullers, M. Ahlskog, M. Cannaerts, and C. Van Haesendonck, submitted.
[5] K. Hernadi, A. Fonseca, J. B.Nagy, D. Bernaerts, A. Fudala, and A. A. Lucas, Zeolites **17** (1996) 416.
[6] M. Ahlskog, E. Seynaeve, R.J.M. Vullers, and C. Van Haesendonck, J. Appl. Physics, **85**, 8432 (1999).
[7] A. Bezryadin, A.R.M. Verschueren, S.J. Tans, and C. Dekker, Phys. Rev. Lett., **80**, 4036 (1998).
[8] A. Bachtold, C. Strunk, J-P. Salvetat, J-M. Bonard, L. Forró, T. Nussbaumer, and C. Schönenberger.

[9] S. Frank, P. Poncharal, Z.L. Wang, and W.A. de Heer, Science, **280**, 1744 (1998).
[10] A. Yu. Kasumov, R. Deblock, M. Kociak, B. Reulet, H. Bouchiat, I.I. Khodos, Yu. B. Gorbatov, V.T. Volkov, C. Journet, and M. Burghard, Science, **284**, 1508 (1999).

Fabrication of full-color carbon-nanotubes field-emission displays: Large area, high brightness, and high stability

W.B. Choi , Y.H. Lee+, D.S. Chung, N.S.Lee and J.M. Kim

The National Creative Research Initiatives Center for Electron Emission Source, Samsung Advanced Institute of Technology, P.O. Box 111, Suwon 440-600, Korea
+Department of Physics, Jeonbuk National University, Jeonju 561-756, Korea
wbchoi@sait.samsung.co.kr, Jongkim@sait.samsung.co.kr

I. Introduction

Carbon nanotubes (CNTs), originally produced as by-products of fullerene synthesis, have remarkable mechanical, electronic, and magnetic properties that can be tailored in principle by varying diameters and chirality of CNTs and the number of concentric shells [1]. CNTs with extremely small diameters, hollowness, and chemical and mechanical strengths have provided a vast range of applications of nanotubes such as electron field emitters [2], room-temperature transistors [3], and vehicles for hydrogen storage [4]. In particular, there have been tremendous efforts in developing field emission displays (FEDs) using CNTs prepared by a suspension-filtering method [2], a CNTs/epoxy mixture [5, 6], chemical vapor deposition for vertical alignment [7, 8], and soot from arc-discharge [9]. CNT-FEDs have a strong potential to be applied to emissive devices including flat panel displays, cathode-ray tubes, back-lights for liquid crystal displays, outdoor displays, and traffic signals. A FED in a diode type was recently tested in a vacuum chamber for the size of 10 mm x 10 mm, where CNTs were non-aligned [5]. Yet, high brightness, uniformity, and high stability on a large area at low operating voltage have never been achieved in a fully integrated device level. Main difficulty in realizing FEDs using CNTs arises from the vertical alignment of a large number of CNTs on a large area at low processing temperature. Here we report integration processes of large-area, fully sealed full-color CNT-FEDs at low temperature.

II.Field Emission Display

Field emission is defined as "the emission of electrons from the surface of a condensed phase into another phase, usually a vacuum, under the action of a high electrostatic field" [10]. Since no thermal energy is necessary, field emission is called a "cold emission" process. Conventional FED depends on field emission from an array of small micro tip.(Fig 1) High electric field is present between gate electrode and tip emitter, field emission of electrons is occurred. The ejected electrons strike the phosphor-coated screen

Science and Application of Nanotubes, edited by Tománek and Enbody
Kluwer Academic / Plenum Publishers, New York, 2000

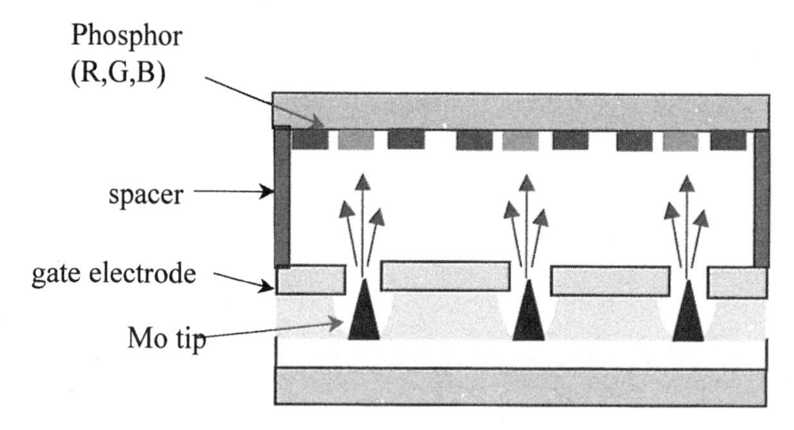

Fig. 1 Schematic diagram of field emission display, in which electrons from an array of millions of micro field emitters are propelled to the phosphor screen.

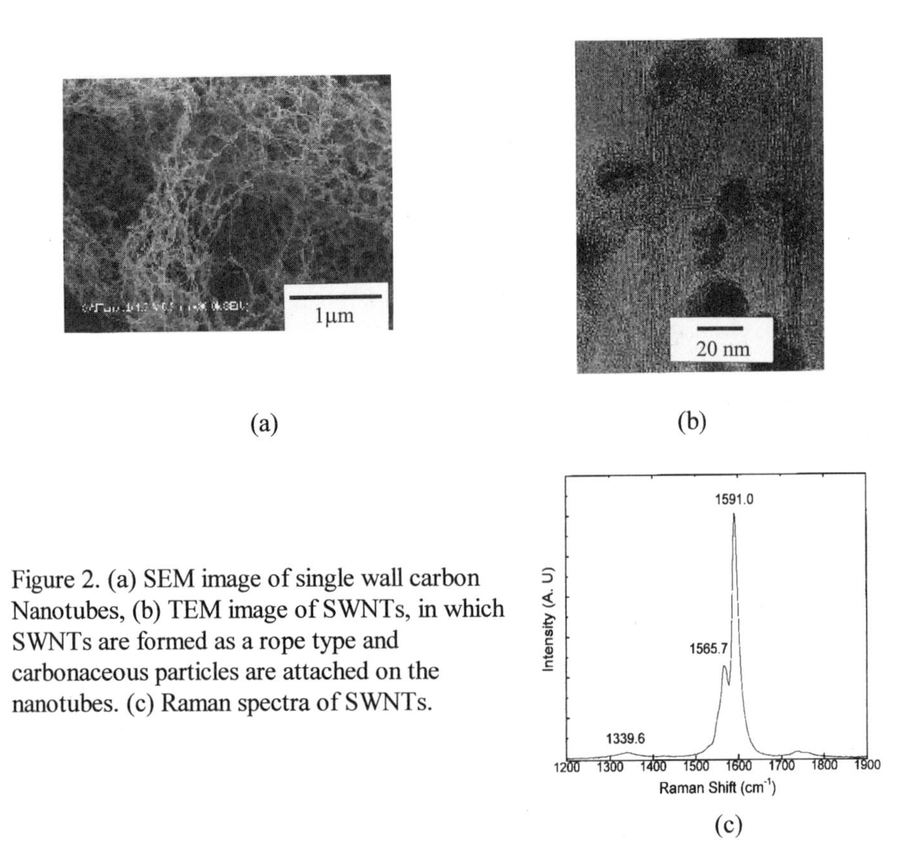

Figure 2. (a) SEM image of single wall carbon Nanotubes, (b) TEM image of SWNTs, in which SWNTs are formed as a rope type and carbonaceous particles are attached on the nanotubes. (c) Raman spectra of SWNTs.

and light is emitted. Most conventional FEDs operate at 300 to 5000 V, as opposed to the 30000 V used by CRTs. Recently, FED with high brightness as bright as CRT has been achieved by many researchers. The main difficulties are developing high efficient low-voltage phosphor and making cost-effective FEDs [11].

III. Fabrication process of CNT-FED

CNTs were synthesized by a conventional arc-discharge as described elsewhere [12]. As-grown single-wall CNTs with a diameter of 1.4 nm were purified and cut into short pieces in a mixture of sulfuric and nitric acids (1:1) at 100 °C [13]. The purified CNTs were rinsed with distilled water, followed by drying, and then further dispersed in isopropyl alcohol by sonication. Scanning electron microscope (SEM) images showed that most carbonaceous particles were removed by purification and typical lengths of CNTs were about 0.5-2 μm [14]. After drying at 150 °C, CNTs were mixed with a slurry of metal powders and organic vehicles.

Figure 2(a) show a the scanning electron microscope (SEM) image of SWNTs. Figure 2(b) shows transmission electron microscopy (TEM) image of as-fabricated SWNTs. Bundles of SWNTs with diameters of about 1.6 nm are clearly seen. Metal particles were attached at the edge of the SWNT bundles. Figure 2(c) shows the Raman spectrum of SWNTs by using Ar excitation (λ=514.5 nm). The Raman spectrum clearly shows the G-line at 1591 cm^{-1} which originates from the graphitic sheets, with extra peak near 1566 cm^{-1}, indicating the general trends of SWNTs [15, 16]. The broad peaks near 1750 cm^{-1} result from the second order Raman scatterings, coupled between the breathing modes near 180 cm^{-1} and G-lines [16]. The broad peaks near 1340 cm^{-1} indicate the existence of defective graphitic layers and/or some carbonaceous particles, which remain even after the purification procedure.

The CNT-FED in a diode type is shown in Fig. 3. The panel structure consists of two glass plates: stripes of CNTs on the patterned cathode glass and phosphor-coated indium tin oxide (ITO) stripes on the anode glass. The cathode and anode stripes are arranged perpendicular to each other, resulting in the formation of pixels at their intersections. The spacing between two glass plates is kept by 200μm.

The paste of well-dispersed CNTs was squeezed onto the metal-patterned sodalime glass through the metal mesh of 20 μm in size, and subsequently heat-treated at 300 °C for 20 min in order to burn out the organic binders. Finally, the mixture of CNTs and metal powders was strongly adhered onto the metal-patterned stripes. A subsequent surface

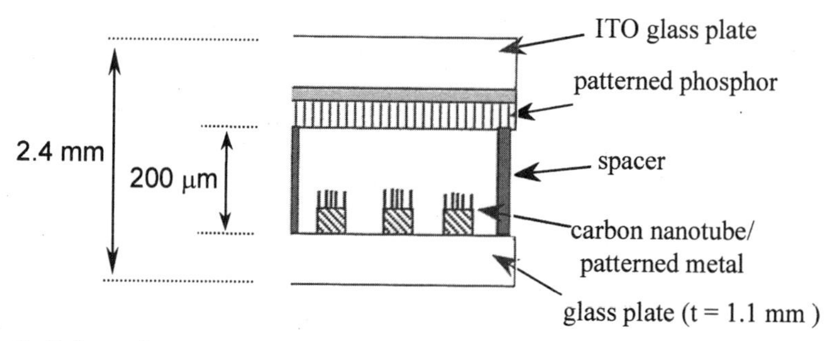

Figure 3. Schematic structure of the fully sealed 128 lines matrix-addressable carbon nanotube flat panel display.

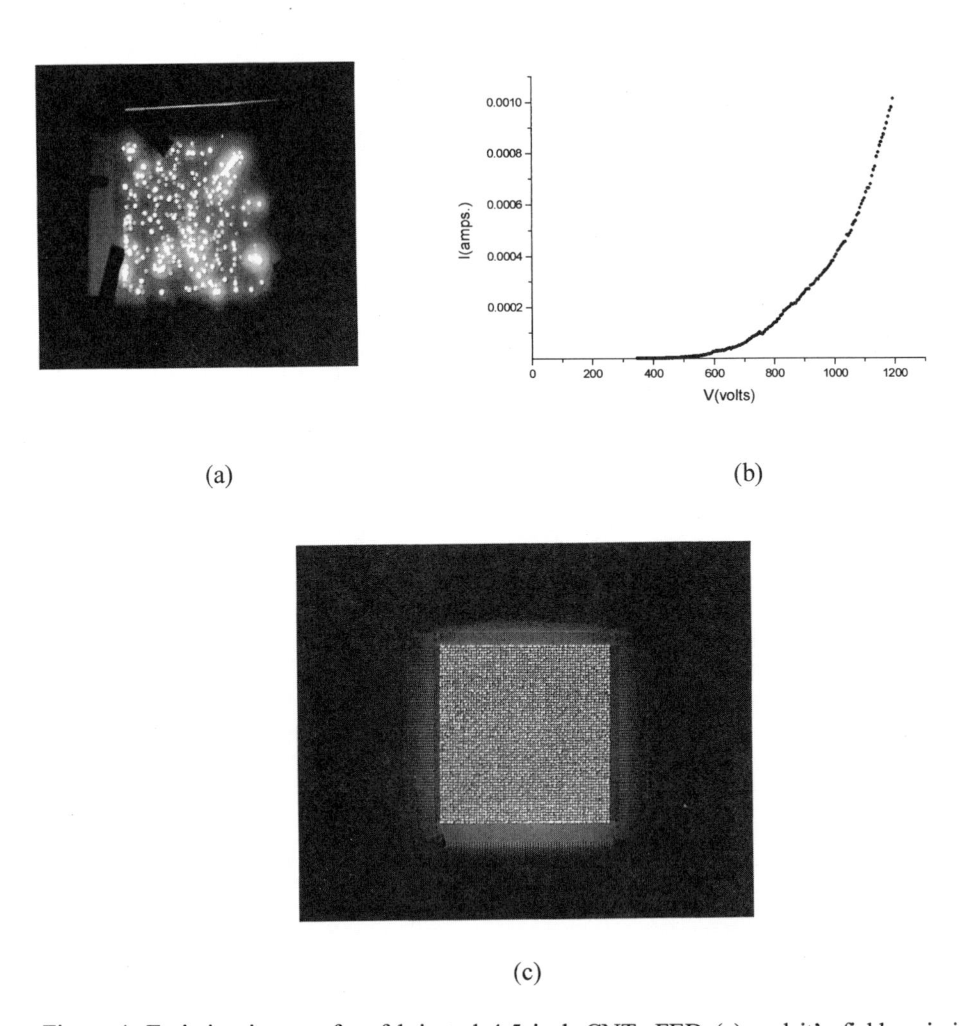

Figure 4. Emission image of as-fabricated 4.5 inch CNTs-FED (a) and it's field emission characteristic. Uniform emission image of surface treated 4.5 inch CNT-FEDs (c).

rubbing process removed metal powders on the uppermost surface, making CNTs protrude from the top surface.

For the anode plate, the Y_2O_2S:Eu, ZnS:Cu,Al, and ZnS:Ag,Cl phosphors were screened with a thickness of 6-10 μm on the ITO-coated sodalime glass for red, green, and blue colors, respectively. Following assembling of cathode and anode plates with spacers in-between, the panel was sealed at 415 °C using frit glass in an ambient of highly purified Ar gas [17]. The panel was evacuated down to the pressure level of $1x10^{-7}$ Torr. Non-evaporable getters of a Ti-Zr-V-Fe alloy were activated during the final heat-exhausting procedure at 330 °C for 6 hours, finally leading to the complete fabrication of 4.5-inch fully sealed CNT-FEDs. Emission currents were measured in dc and pulse modes at voltages up to 800 V. The resultant brightness was measured by a luminance calorimeter.

IV. Emission properties of CNT-FED

A. Imaging of emission site

Many researchers have been reported on the field emission properties of CNTs. However, their emission uniformity did not confirmed by direct imaging experiment. The CNT cathode with high emissivity could not promise uniform emission sites (Fig. 4). For display applications, emission uniformity is so important that one has to check the emission image by phosphor screen. The emission uniformity depends on the distribution of CNTs, number of active tips, geometry of CNTs, and bias mode. Well-aligned CNTs cathode could be obtained by slurry squeezing technique and special surface treatment. The uniform emission image was successfully achieved on the 4.5 inch CNTs cathode as shown in Fig. 4(c).

B. Emission characteristics of a 4.5 inch CNT-FED

Figure 5(a) shows a red, green, and blue color bar image of the CNT-FED with 128 cathode lines that is matrix-addressable in a diode mode. A very uniform and stable emission image over the entire 4.5-inch panel was obtained. The brightness of 1800 cd/m^2

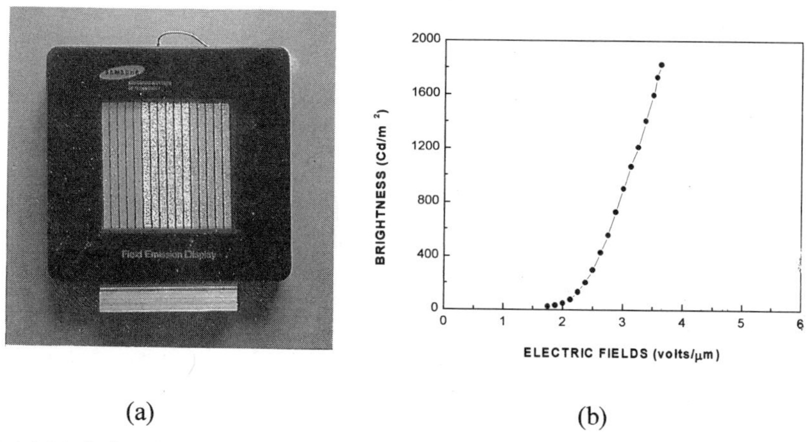

(a) (b)

Fig. 5 (a) Emission image of the fully sealed 4.5-inch CNT-FED. A red, green and blue color bar image is shown with high brightness at the applied field of 3.5 V/μm ac. A half of the cathode lines are biased for this display. (b) Brightness of the 4.5 CNT-FEDs as a function of electric fields.

at 800 V or 4 V/μm (duty: 1/4, frequency: 15.7 kHz) was achieved on the green phosphor (Fig.5 (b)). The fabricated CNT-FED showed unusually high brightness at the low operating voltage, compared to that (300 cd/m^2 at 6 kV) of Spindt-type FEDs [11]. Such high and uniform brightness over a large area implies that the CNTs are well aligned, uniformly distributed with a high number density, and highly efficient in emitting electrons. The whole fabrication processes were fully scalable and reproducible.

In order to analyze the emission uniformity, the 4.5-inch screen was divided into nine domains. Brightness and current density were measured in the nine different domains at the anode voltage of 700 V using green phosphor. The brightness and current density in Fig. 6(a) and (b) vary from domain to domain by about 50 %. Such variation may be attributed to i) different lengths of CNTs protruding from the surface, ii) different degree of alignment on the surface, iii) various chirality, iv) non-uniform distribution of CNTs, and v) non-uniform deposition of phosphor layers from domain to domain. The fluctuation of the brightness does not necessarily coincide with that of the current density of the corresponding domains, due to the non-uniform thickness of phosphor layers. Nevertheless the resultant fluctuation of brightness even at the current stage is acceptable for low-end display applications.

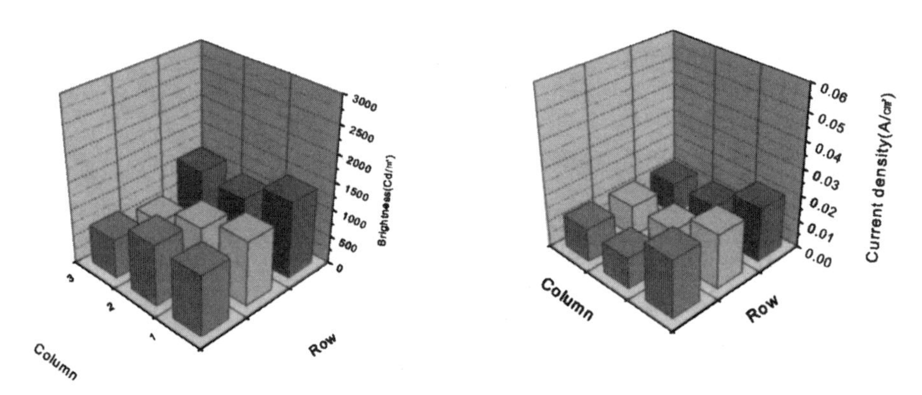

Figure 6. Brightness and current density distribution of a 4.5 inch CNT-FED at nine different domains. The 4.5 inch screen is divided by nine sectors on the green phosphor anode for a purpose of area-by-area analysis.

C. Emission current-voltage characteristics

Emission current-voltage (I-V) curves from the nine different domains of the CNT-FED are shown in Fig. 7(a). Currents start increasing at 1.5 – 3 V/μm and saturate near 3 –5 V/μm, which varies from domain to domain. Despite the fact that currents were measured with phosphor layers, the onset electric fields (for emission current of 10 nA) were quite low (1.5-3 V/μm) and the currents were saturated to about 1 mA [18]. Emission characteristics were analyzed by applying the Fowler-Nordheim (F-N) equation, $I = aV^2 \exp(-b/V)$, where a and b are constants [19]. Figure. 7(b) shows the F-N plots, $\log (I/V^2)$ versus $1/V$, for the nine domains. The local electric field (E_l) can be related to the field enhancement factor (β) and macroscopic field (E_m) by $E_l = \beta E_m$. The field enhancement factor can be calculated either from the slope of the F-N plot if the work function of the emitter is known or from measuring the geometry of a CNT tip.

In order to analyze the effects of field distribution, the emission currents were measured by changing cathode-to-anode distances (Fig.8). The emission currents as a

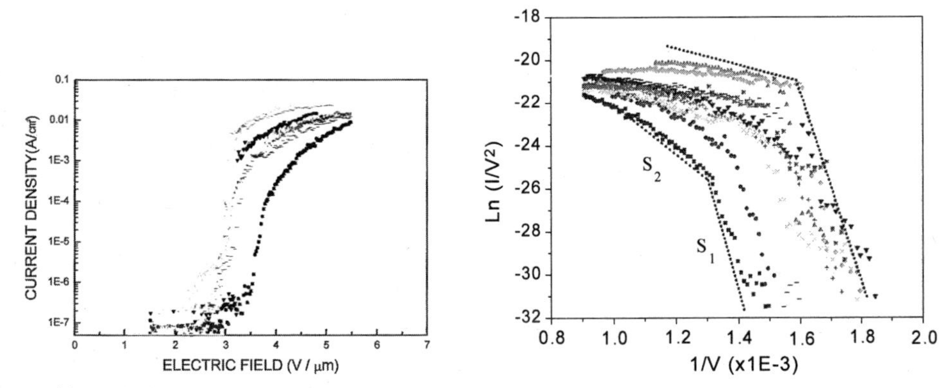

Fig. 7 (a) Emission I-V curves and (b) Fowler-Nordheim plots of the nine different sectors of the 4.5 inch CNT-FED.

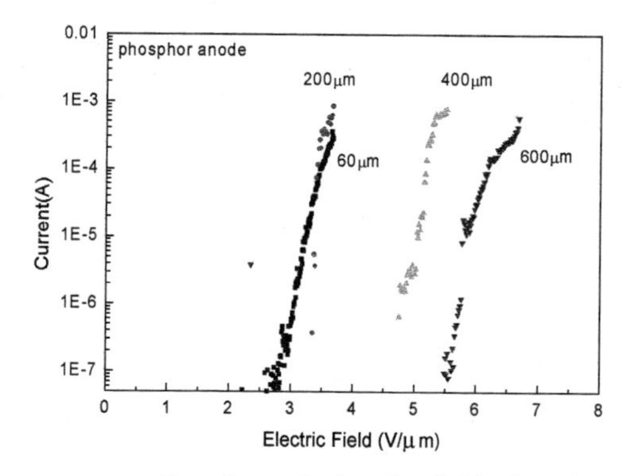

Fig. 8 Emission currents as a function of electric fields by changing cathode-to-anode distances.

function of voltages were insensitive to distance variation below 200 μm, while the emission currents increased linearly with the gap greater than 200μm. Therefore, a point-to-sphere electric-field model is more suitable for the gap of 200μm, instead of a plane-to-plane model. The field enhancement factor is $\beta = 1/kr$ (r: tip radius, k ~ 10 [20]) for the point-to-sphere model, where the unit of β is cm^{-1}. The tip radii r were obtained to be 30 nm ~60 nm by the relation of $r=1/(k\beta)$. The calculated tip radii were in the same order as the measured ones, which were shown in the SEM observations as a form of bundle instead of a single nanotube, as given in Fig. 9. Assuming the work function of CNTs to be 5 eV, same as that of graphite or C_{60} [21], the field enhancement factors in the low field region S_1 were determined to be in the range of 17000 - 33000 for the different domains.

At the high field region of S_2 in Fig. 7(b), the slopes of the F-N plots decrease, exhibiting a different emission behavior from that of the low field region S_1. This deviation from the F-N law at the high field region is usually attributed to the space charge effect [22]. It has been also argued that localized states at CNT tips play a role in suppressing the electron emission, which was supported by the light emission from CNTs [23]. The light emission from CNTs on the transparent ITO glass anode was observed with the brightness of 2 cd/m^2 only at a much higher field region (5 V/μm), which seems to be ascribed to burning-out of CNT edges, as observed in other experiments [24]. Therefore, the suppression of electron emission by the localized states appears to be negligible and then be mainly contributed by the space charge effect and/or resistive heating even in the case of CNTs.

Figure 9 shows a cross-sectional SEM image of a CNT cathode. It clearly shows that CNT bundles are firmly adhered onto the metal electrode and aligned mostly perpendicular to the substrate. The density of CNT bundles from the SEM measurements was 5-10 /μm^2, about 100 times larger than the typical density of microtips in conventional Spindt-type FEDs [11]. It was estimated that the average emission current from a single nanotube bundle was an order of pico-amperes at 3 V/μm. The CNTs were well aligned over the entire patterned area of the 4.5-inch panel. Vertical alignment of CNTs was solely achieved by i) paste squeezing through the metal mesh, ii) surface rubbing, and /or iii) conditioning by electric field. The former aligned CNTs have been grown by chemical vapor deposition at high temperatures over 700 ^{0}C [25]. The display applications using sodalime glass, however, require low-temperature processing below 500 °C. During our processes of CNT-FED fabrication, the temperatures were kept below 415 °C.

Fig. 9 Cross-sectional SEM image of CNT cathode. CNTs are aligned perpendicular to the substrate and firmly embedded into the metal electrode.

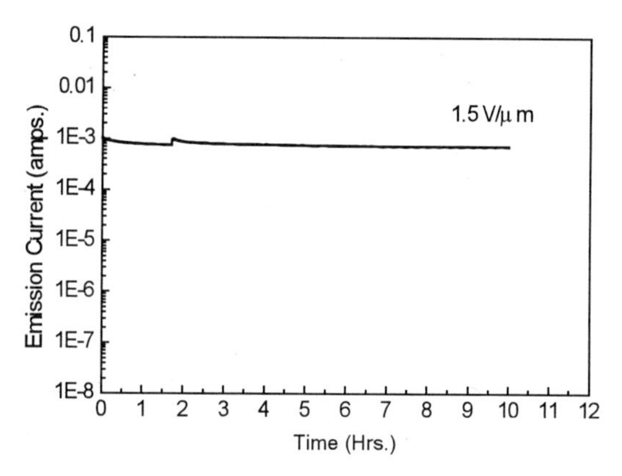

Fig. 10 Current stability of the 4.5 inch CNT cathode at DC-mode. A bare ITO glass anode was used for this experiment in order to avoid voltage drop and/or outgassing.

Emission stability of CNTs over the 4.5–inch cathode was tested by measuring the current fluctuation with time at a dc mode using the bare ITO glass in order to remove the voltage drop and/or degassing by phosphor layers. In the Fig. 10, the data were acquired by averaging 10 data points per second. The current fluctuation at 0.1mA was found to be less than 10 % over the 4.5-in. panel. Currents fluctuate occasionally at the initial stage for higher voltages that is probably due to the field stress effect. In the extended aging, however, the CNTs show negligible current deviations, indicating high stability in the electron emission.

To our knowledge, the fully-sealed CNT-FEDs in a 4.5-inch size were fabricated for the first time on glass at low temperature, where CNTs were vertically aligned using the paste squeeze and surface rubbing techniques [26, 27]. The resultant high luminance of 1800 cd/m^2 at 4 V/μm with stable electron emission was achieved. The whole processes are scalable and cost-effective in fabricating CNT-FEDs.

Acknowledgment This work was supported by the National Creative Research Initiatives Fund

References

[1] T.W.Ebbesen, *Carbon Nanotubes* (CRC Press,Inc., New York, 1997), chap. 9.
[2] W.A. de Heer et al., *Science* **270,** 1179 (1995).
[3] S. J. Tans et al., *Nature* **393**, 49 (1998).
[4] D.S. Bethune et al., *Nature* **363**, 605 (1993).
[5] Q.H. Wang et al., *Appl. Phys. Lett.* **72,** 2912 (1998).
[6] P. G. Collins and A. Zettl, *Appl. Phys. Lett.* **69,** 1969 (1996).
[7] S. Fan et al., *Science* **283**, 512 (1999).
[8] C.J. Lee et al., *Phy. Rev. Lett.*, May 1999 (in submission).
[9]Y. Saito et al., *Jpn. J. Appl. Phys.* **36**, L1340 (1997).
[10] R. Gormer "Field emission and field ionization', American Institute of Physics, New York , (1993).
[11] B.R. Chalamala et al., IEEE Spectrum, April 42 (1998).

[12] S. Iijima, *Nature* **354**, 56 (1991).
[13] Jie Liu et al., *Science* **280**, 1253 (1998).
[14]. Our SEM results showed that the length of CNTs was shortened and impurities of carbonaceous and catalytic particles were removed during the purification process. Raman analyses confirmed the characteristic peak of SWNTs (1580 cm^{-1}) is unchanged during purification process.
[15] S.L. Fang, A.M. Rao, P.C. Eklund, P. Nikolaev, A.G. Rinzler, and R.E. Smalley, J. Mater. Res. **13**, 2405 (1998).
[16] C.J. Lee, D.W. Kim, T.J.Lee, Y.C. Choi, Y.S. Park, W.S. Kim, W.B. Choi, N.S. Lee, J.M. Kim, Y.G. Choi, S.C. Yu, and Y.H. Lee, submitted to Appl. Phys. Lett. April (1999).
[17] The result of thermal analyses of our CNTs revealed that the dissociation of CNTs started at 500^0C in an inert gas, and the weight loss from dissociation was over 40% at 800^0C. In order to avoid thermal reaction of CNTs, the processing temperature should be kept below 500°C.
[18] The anode currents strongly depend on the type of anode and preparation condition. A bare ITO glass and a phosphor-deposited ITO glass were used in this experiment. We found that the turn-on field for the phosphor-deposited ITO glass was around 1.5-3 V/μm, whereas that for the bare ITO glass was around 1 V/μm.
[19] I. Brodie and C.A. Spindt, *Advances in electronics and electron physics, Vacuum Microelectronics*, (Academic Press, Inc., 1992) , vol.83, pp. 91-95.
[20] P.G. Collins and A. Zettl, *Physical Review B* **55,** 9391 (1997).
[21] B.W. Gadzuk et al., *Acad. Sci.* **B 278**, 659 (1974).
[22] G.N.A. van Veen, *J. Vac. Sci. Technol.* **B12,** 655 (1994).
[23] J. Bonard et al., *Phys. Rev. Lett.* **81,** 1441 (1998).
[24] A. G. Rinzler et al., *Science* **269,** 1550 (1995).
[25] Z.F. Ren et al., *Science* **282,** 1105 (1998).
[26] W.B. Choi, D.S. Chung, S.H. Park, and J.M. Kim, L2.1 Digest of Technical Papers, Society for Information Display International Symposium, San Jose California, May 18-20 1999.
[27] W.B. Choi, D.S. Chung, H.Y. Kim, J.H. Kang, S.H. Park, and J.M. Kim, in Proceedings of MRS Spring Meeting, San Francisco (1999).

FREE SPACE CONSTRUCTION WITH CARBON NANOTUBES

George D. Skidmore, Matthew Ellis, and Jim Von Ehr

Zyvex LLC
1321 North Plano Road Suite 200
Richardson, Texas 75081

ABSTRACT

Construction of carbon nanotube assemblies within scanning (SEM) and transmission electron microscopes (TEM) is being pursued. Inside the electron microscopes, the imaging and construction operations can be accomplished with the electron beam. Multiple degree of freedom actuators with sub-nanometer resolution built specifically for these microscopes allow for manipulation of nanotubes in three dimensions. Electron beam induced deposition (EBID) of both the "contaminating" background gases and controllably introduced gas precursors facilitates the attachment of nanotubes to each other and to manipulator probes. Voltage pulses and mechanical strain cut nanotubes to length. Simple structures have been built demonstrating these capabilities.

INTRODUCTION

The field of nanotechnology currently encompasses many areas of research in different scientific disciplines. The ultimate goal of molecular nanotechnology is the controlled manufacturing of structures at the molecular scale with a corresponding level of precision. How this goal will be achieved is yet unknown. The techniques of chemical synthesis and increasingly complex methods of self-assembly routinely achieve this goal in a limited manner.[1] Progress is occurring with the utilization of bio-molecules such as DNA.[2] Current efforts in scaling down microfabrication are also leading towards nanotechnology.[3] Ultimately, success will result from a combination of many different approaches. In any case, advances in micro and nanomanipulation of small components and molecules drives the pursuit of nanomanipulation of carbon nanotubes within scanning and transmission electron microscopes.

A feasible path to nanotechnology appears to be the directed assembly of molecules and atoms via a nanomanipulator, such as a scanning probe microscope (SPM). The SPM combines two necessary components of nanomanipulation into a single instrument. The first component is high-resolution imaging. The second component is sub-nanometer

Science and Application of Nanotubes, edited by Tománek and Enbody
Kluwer Academic / Plenum Publishers, New York, 2000

positioning. In fact, individual atoms and molecules have been moved via the STM.[4,5] But, for truly three-dimensional manipulation, an SPM has not yet proven very useful. Electron microscopes provide high-resolution imaging. When combined with multi-degree of freedom manipulators with sub-nanometer positioning resolution, an electron microscope meets the same two requirements and may thus be a good tool for three-dimensional manipulation of certain small parts and molecules. Because carbon nanotubes are capable of withstanding the electron beam they are excellent molecules for testing some nanomanipulation protocols.

EXPERIMENTAL APPARATUS

Electron microscopes can be made capable of performing the procedures needed to manipulate and image carbon nanotubes in three dimensions. Scanning probe microscopes have successfully imaged and manipulated nanotubes but this work is confined to a surface due to the nature of the instruments.[6] Within an electron microscope, nanotubes may be manipulated freely in space using as many degrees of freedom as allowed by the actuators. At Zyvex, two SEMs, a JEOL T-300 and a Hitachi S-900, and a TEM, a JEOL 2000FX, now have the capability to manipulate and image carbon nanotubes. Each microscope has actuators built specifically for performing multiple degree of freedom manipulation with sub-nanometer resolution. Since the different microscopes have complimentary imaging and manipulating capabilities it is useful to mount the same samples into each of the microscopes. Special sample holders have been made for the manipulating probes; these fit into all of the microscopes simplifying transfer of samples from one instrument to another. Thus, the assemblies built in the SEMs can be examined in the TEM and vice versa.

JEOL T-300

The T-300 is a medium resolution SEM with a large sample chamber. This instrument has been previously used for nanotube manipulation and the results have been described elsewhere.[7] A quick review of the instrument features are: a hot tungsten filament electron source, a minimum spot size of 6 nm as specified by the manufacturer, a diffusion pumped column and chamber with no cold trap, and a large sample chamber with no load lock system. The primary disadvantage results from the tungsten filament electron source and the 6 nm spot size. The low brightness of the electron source and the low resolution of the instrument preclude the imaging of the smallest individual single wall nanotubes. Without a cold trap or chamber load lock, maintaining samples free from contamination is not possible. This residual contamination in the chamber enhances the growth of small structures and the attachment of nanotubes to the structures by electron beam deposition. When introducing gaseous metal precursors for electron beam induced deposition, incorporation of carbon in the metal growths is virtually assured. The lack of a load lock mechanism also contributes to contamination of the chamber but having open access to the sample chamber simplifies the installation of manipulators and samples.

Modifications have been made to the microscope since the previous report of nanotube manipulation.[7] Another three-degree of freedom manipulator complements the previous two-probe setup allowing for three-probe work. Each probe may be biased independently of the other probes for performing electrical measurements. An external gas manifold and inlet provide for the controlled introduction of gases into the sample chamber for electron beam induced deposition. Figure 1 shows the microscope manipulation platform in its current configuration. The new manipulator provides coarse positioning in x, y, and z with a resolution of less than 50 nm. This manipulator also has a piezoelectric tube actuator for probe positioning with sub-nanometer resolution in three dimensions.

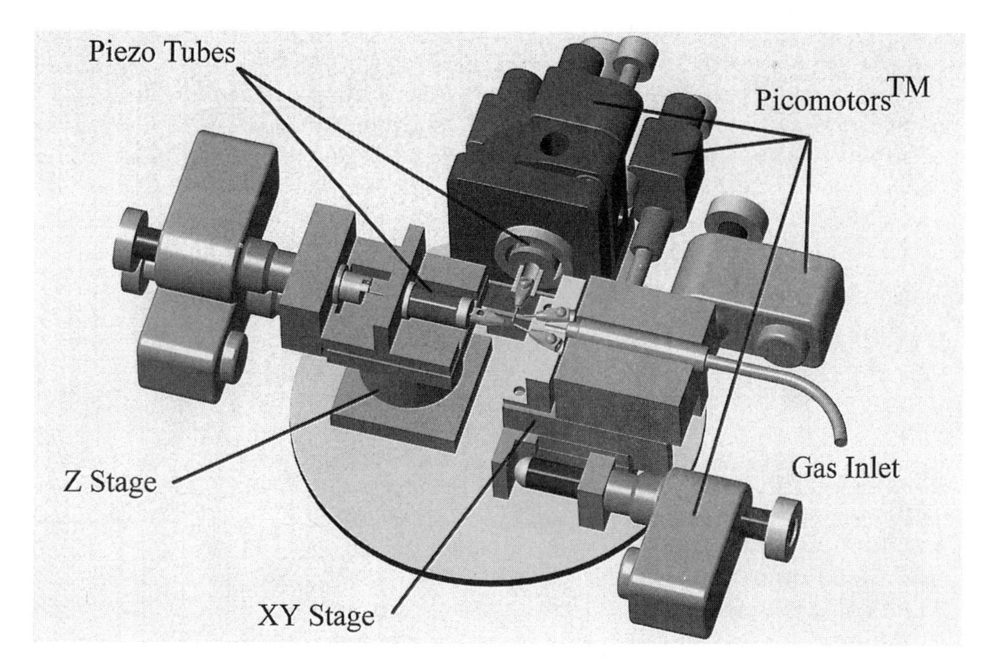

Figure 1. T-300 manipulator with multiple probes and gas precursor inlet.

Figure 2 shows a closer view of the T-300 manipulator probes. In this drawing three of the probes are tungsten needles and one probe is an AFM cantilever. The probe holders are interchangeable so cantilevers or tungsten needles may reside in any of the four locations shown. Between the two probes on the right is the gas precursor inlet. This gas inlet and these two probes are mounted on a x-y stage driven by two Picomotors.[8] The other two probes, shown in the left and upper center of the drawing have coarse positioning via Picomotors and fine positioning via piezoelectric tubes. The probe on the left has one rotational degree of freedom and one linear degree of freedom. The probe in the upper section of the drawing has three degrees of freedom.

A small gas manifold external to the sample chamber and a small gas inlet allow controlled amounts of metal carrying gas precursors into the sample chamber for electron beam induced deposition (EBID). This gas admission enhances the growth rate to more than twice the rate from the background contamination alone. Energy dispersive x-ray spectrometer (EDS) data taken in the Hitachi S-900 described below indicates that posts grown with WF_6 gas flowing contain some amount of tungsten. Also, similar work done with $TiCl_4$ shows posts with a titanium composition. Unfortunately, the lack of a low element detector window in the EDS inhibits quantitative determination of the materials content grown by EBID.

Hitachi S-900

The S-900 is a sample-in-lens, cold field emission SEM. The pole pieces of the objective lens limit the available sample volume such that only one articulating, electrically isolated probe can be installed along with another stationary probe that is electrically connected to the microscope ground. Although the microscope has limits regarding sample volume, the sample-in-lens setup allows for unparalleled resolution, 0.7 nm as stated by the manufacturer. Furthermore, the S-900 has an oil-free pumping system with a turbo-molecular pump and a liquid nitrogen cold trap shrouding the sample area. When using the

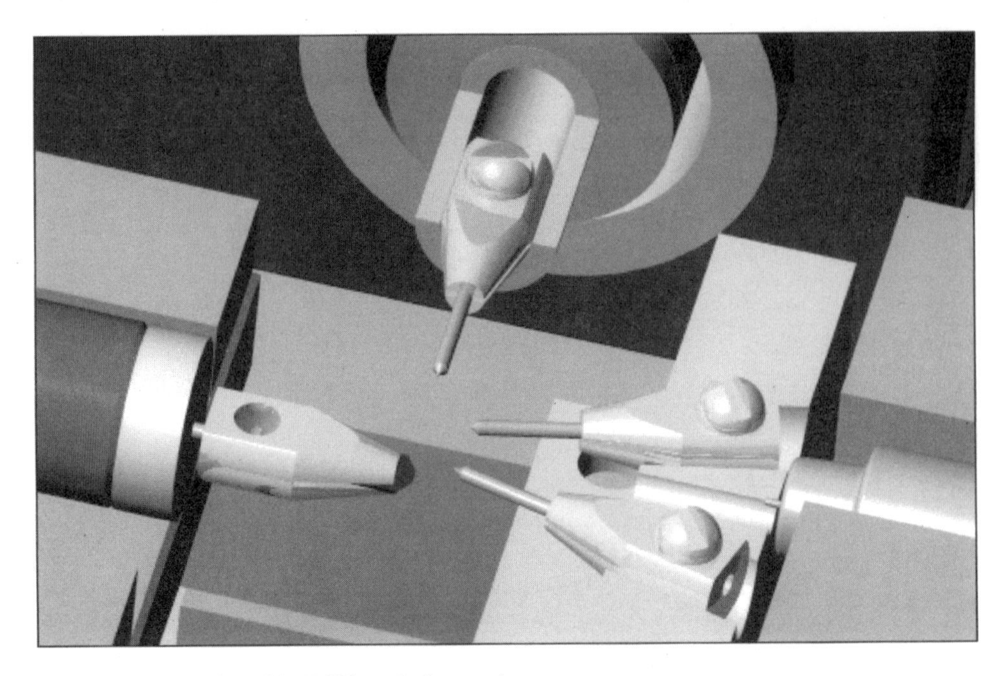

Figure 2. Close up view of the T-300 manipulator probes.

cold trap the, contamination in this scope can be brought to imperceptible levels; when not using the cold trap the contamination can be used for tube attachments and EBID growths. An attached energy dispersive x-ray spectrometer provides compositional analysis of sample materials. This EDS system, a Tracor Northern detector with Noran Voyager controller, contains a Be detector window that does not allow for compositional analysis of elements with atomic numbers below that of Sodium.

Figure 3 presents a drawing of the manipulator built for this machine. Shown in a cutaway is the main sample rod, which is hollowed to contain a concentric sliding rod that is the manipulator. Mounted on the sliding rod are two piezoelectric tubes, one with solid electrodes for z-motion, where z is defined along the rod axis, and another with quadrant electrodes for x and y motion. The fine motion from the piezotubes provides 1.75 microns of travel in the z direction and 60 microns of travel in x and y with 150 volts applied to the electrodes. For coarse positioning the piezotube and sliding rod assembly, a micrometer screw mounted at the end of the sample rod provides approximately 12 mm of travel in the z direction. This assembly also pivots about the O-ring seal allowing coarse positioning in the x and y directions of about 1 mm. The rod can rotate about the z-axis providing a rotational degree of freedom. With the current manipulator design, sample vibration is a problem because the sample is cantilevered from the O-ring seal by 10 cm. Having the samples cantilevered at this length results in vibrations of about 2 nm that are detectable in the SEM.

JEOL 2000FX

The 2000FX is a 200 kV TEM with ASID20 scanning attachment and transmitted electron bright field and secondary electron detectors. With a LaB_6 electron source, the 0.34 nm graphite lattice is easily resolved demonstrating that the instrument is capable of determining the type of nanotubes being manipulated, either ropes of tubes, multi-wall nanotubes (MWNTs), or single-wall nanotubes (SWNTs). Pumping the sample chamber is

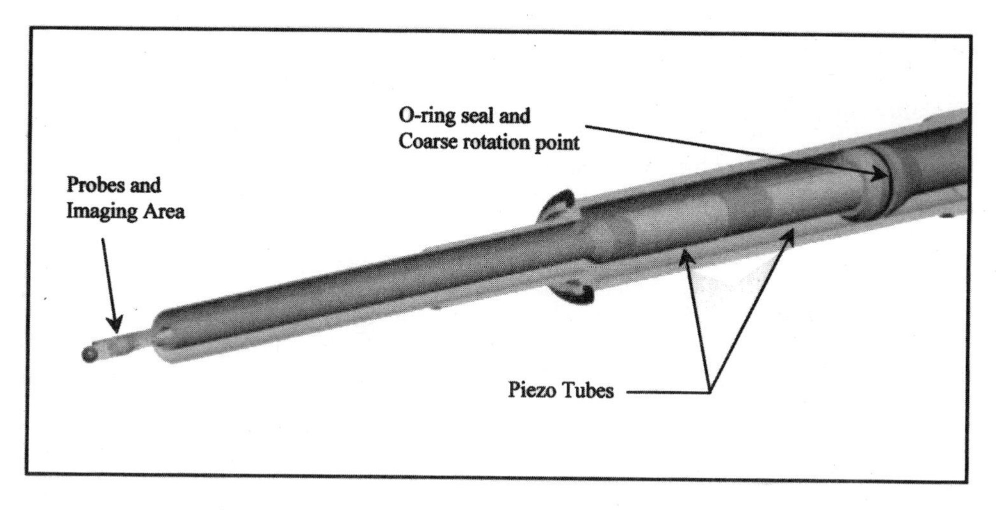

Figure 3. Section view of the Hitachi S-900 sample rod with the piezoelectric tube manipulator inside.

an ion pump that maintains the sample vacuum at $5x10^{-5}$ Pa during imaging. There is also a liquid nitrogen cooled shroud around the sample area to limit or allow contamination growth as desired. Similar to other TEMs and the S-900 SEM, the instrument has a limited sample volume. Thus, the manipulator for this instrument shares much in common with the S-900 manipulator.

Figure 4 shows the manipulator built for the 2000FX. It outward appearance is similar to the manipulator for the S-900. But, the manipulator differs in that the movable probe is decoupled from its externally controlled coarse positioning mechanisms. This decoupling reduces the vibration of the manipulated probe to as yet undetectable levels. The range of motion in x and y (perpendicular to the rod axis) is 1 mm of coarse travel and 30 microns of fine travel. In the z direction (along the rod axis) the coarse travel is 6 mm and the fine range of travel is 1.5 microns.

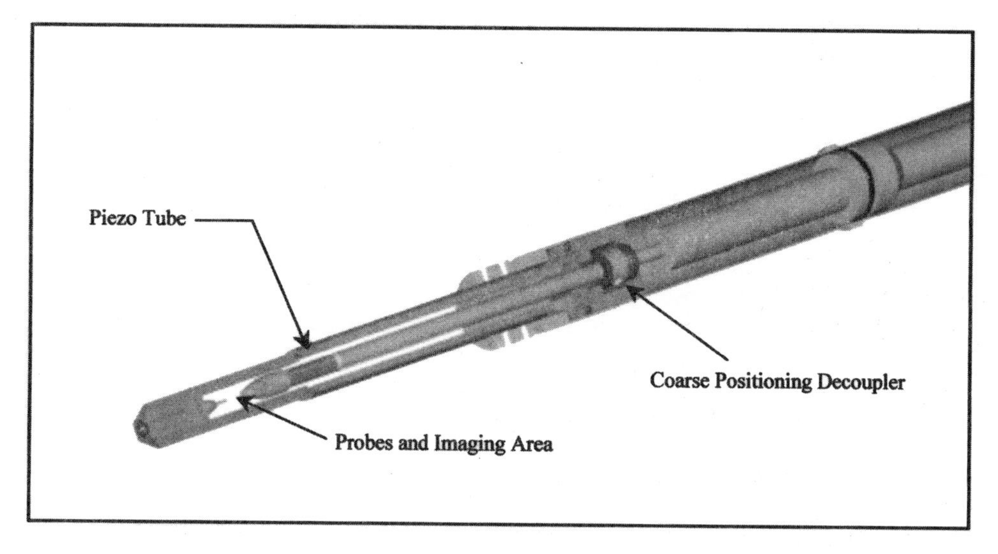

Figure 4. Section view of the sample rod for the JEOL 2000FX TEM. This view shows the piezoelectric tube and the decoupling mechanism.

EXPERIMENT

Nanotube Sources

Three different nanotube sources, single wall tubes and multi-wall tubes in dry form and nanotube "paper", have been used for manipulation thus far. The nanotube "paper" consists of ropes of single wall nanotube material matted together and pressed into a sheet. The result is a sheet of nanotubes that can be torn and frayed. Examining a frayed edge reveals the paper to be made of many ropes of nanotubes. The rope diameters range from 30 nm to a few hundred nanometers with lengths of several microns. These ropes can be pulled from the frayed edge and manipulated but completely removing a rope from the paper by purely mechanical pulling is difficult. Cutting the ropes from the paper with a third probe is the easiest method of removal. Single wall nanotubes produced via the arc method with a CoS catalyst are the second source material.[9] These individual single wall nanotubes are too small to see in the T-300 but are easily used in the S-900 and the TEM. They are generally coated with amorphous material from the production process when used unpurified. Multi-wall nanotube material available from Aldrich is the third source.[10] This material is provided in the as-formed, unpurified state. These multi-wall tubes are loosely bound to each other making them more easily removable by mechanical pulling alone.

Manipulation Probes

The manipulation probes are either etched tungsten needles, silicon AFM cantilevers, or silicon nitride AFM cantilevers with conductive coating. Tungsten probes may be etched very long and sharp with tip radii of less than 50 nm, making them easy to position to the desired tube with no interference. Silicon nitride cantilevers have low spring constants (0.01 N/m to 0.1 N/m) so they are easily deflected during manipulations. This allows for measure of the forces being exerted. These cantilevers are electrically insulating and must be coated with gold or chromium for electrical conductivity. Silicon cantilevers with spring constants from 0.1 N/m to 100 N/m are also used, the 100 N/m type are not easily deflected, have sharp protrusions, and have large enough conductivity to be used as received making them the most popular manipulation probe.

EBID, Attachment, and Cutting

Electron microscopists became aware of electron beam induced deposition soon after the invention of the electron microscope. Scientists working with these early instruments used EBID to produce small deposits for determining the astigmatism of the microscopes. The chemicals they were depositing were basically residual contamination from their samples and pump oil from the instrument vacuum systems. More recently, several experimental advancements have been made in the controlled growth of nanostructures with this method. A variety of gas precursors have been purposefully introduced into the chambers. Some potential applications of EBID are high-resolution lithography and the growth of three-dimensional structures with sub-micron feature sizes. Koops et al. provide an excellent review of EBID.[11]

EBID can produce three-dimensional shapes at any location on the probe where the beam can be highly focused. Sweeping the electron beam determines the shape of the deposit in the image plane. Controlling the dwell of the electron beam determines how much growth occurs normal to the image plane and the width of the resulting structures. Dwell times on the order of minutes can produce EBID "posts" several microns long. With these techniques shapes can be fabricated in the T-300 with feature sizes less than 100 nm. In the S-900, the feature size of these deposits can be less than 10 nm. A thin layer of evaporated chromium is used to ensure that these structures are conductive. Figure 5A shows S-900 images of EBID posts made in the S-900, these posts are 10-15 nm at the base, 250-300 nm long, and culminate in tips of radius 4 nm. Figure 5B shows a TEM image of another set of posts after coating with 20 nm of thermally sublimed chromium.

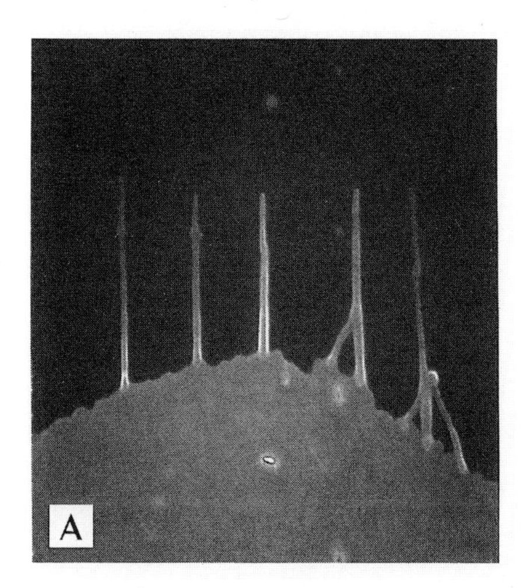

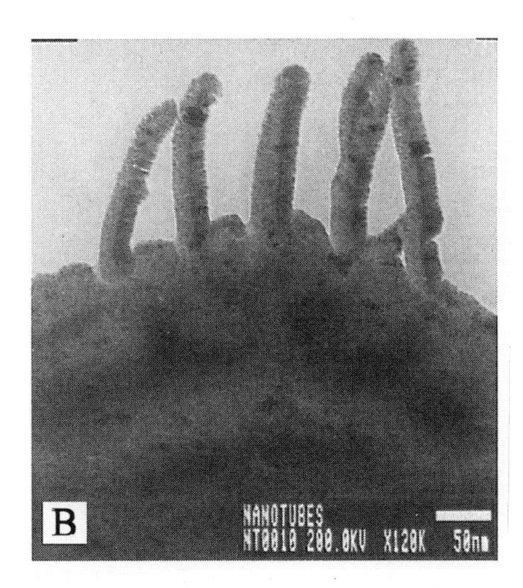

Figure 5. EBID produced posts for nanotube attachments shown in the as grown state by SEM in photo A. Photo B shows a TEM image of EBID posts after chromium deposition

After any two nano-objects have been positioned, attachment is made using EBID. Positioning an EBID growth in contact with a nanotube, or contacting two nanotubes together, generally forms a weak attachment between the two via van der Waals forces. Figure 6 shows images in a tube attachment series where the first image shows the two tubes near each other, the second image shows, at higher magnification, the two tubes after they have been brought into contact. The stationary probe is on the left with five EBID posts visible; the movable probe has several nanotubes protruding. After contact, slight movement of the probe can verify contact. This is shown in the third and fourth images of the sequence where the movable probe has been displaced. The fourth image is at even higher magnification where the single-wall nanotubes are visible within the amorphous carbon coating. To strengthen the attachment, material must now be deposited using EBID. In the SEMs, positioning the focused beam at the junction deposits material and makes a stronger bond. In the TEM, the entire area has been imaged for several minutes depositing material around the entire field of view. Moving the base verifies a stronger attachment and the tube can usually be released from its former attachment point in the boule by mechanical pulling. Failure to form the union can happen in two ways. If a tube is too well adhered to the boule it will slip from the EBID growth and remain where it was. It is usually best to leave such a tube and attempt to pick another. In other failed attempts, more than one tube will be removed from the nest. This procedure of positioning the tubes, depositing material, and pulling the tube free is the basic construction method for nanotube attachment.

The nanotube sources used in these experiments provide tubes and ropes of various lengths. Building determinate structures requires tubes of specific lengths. Thus, a cutting operation is necessary for shortening nanotubes and separating nanotube ropes from boules. Other groups have reported cutting nanotubes on a surface by applying voltage pulses to the imaging probe of an STM.[12] In the work presented here the nanotubes are cut in free

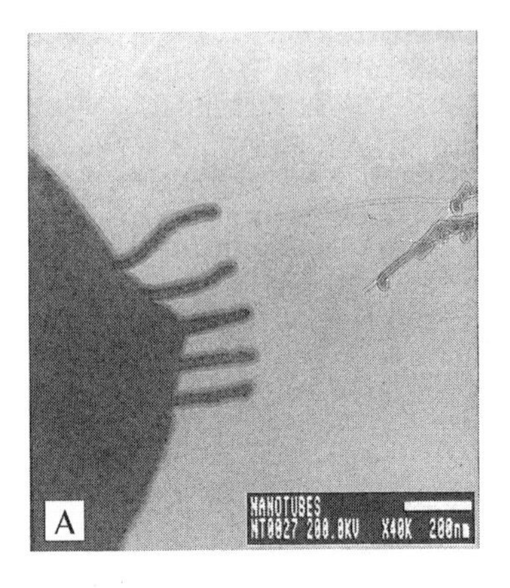

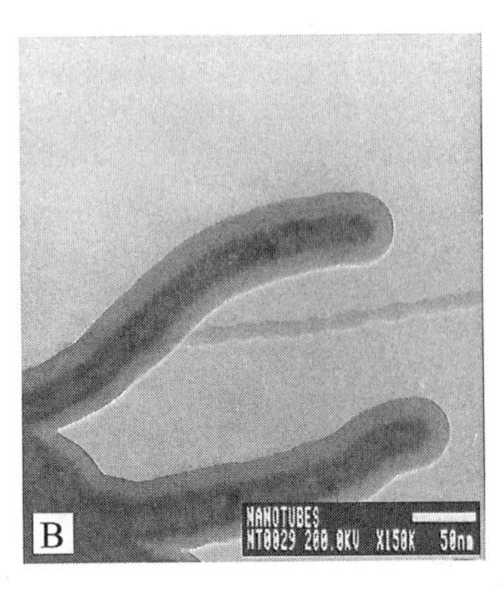

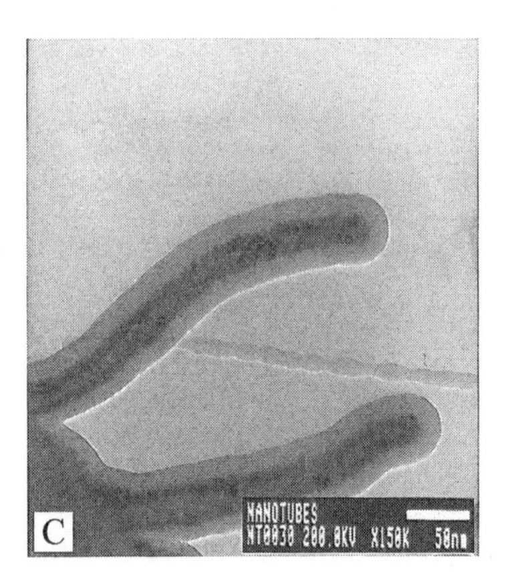

Figure 6. Attaching a nanotube to an EBID post.

space by applying voltage pulses to the articulating probes. Two different procedures have successfully cut nanotubes. The first procedure uses two probes and the second procedure utilizes three. Two-probe cutting is done with the nanotube attached to the probe with the applied voltage; a three-probe cut is done with the nanotube attached to two points while a third voltage-pulsing probe is brought into close proximity. A trimming procedure is done by bringing two nanotube ends into proximity and applying a voltage pulse.

Images outlining a two-probe cut in the TEM are shown in Figure 7. On the left is the stationary probe with an attached nanotube having a blob of carbon near its end; on the right is the nanotube nest, where a tube running vertical will be used for the other attachment point. Figure 7A shows the nanotubes before they are positioned, Figure 7B shows the nanotubes in contact with each other. After a voltage pulse of 5 volts lasting 50 microseconds, the image shown as Figure 7C was taken showing that the tube has been cut and the blob of carbon has been transferred to the vertical tube. Figure 7D shows the blob at higher magnification. Two-probe cutting is not a well-controlled operation. The cut might occur anywhere along the length of either nanotube or rope of nanotubes. The mechanism that causes the cutting is unknown. It is likely due to resistive heating at a junction where one tube ends and current must pass from one tube to another. Further exploration is warranted.

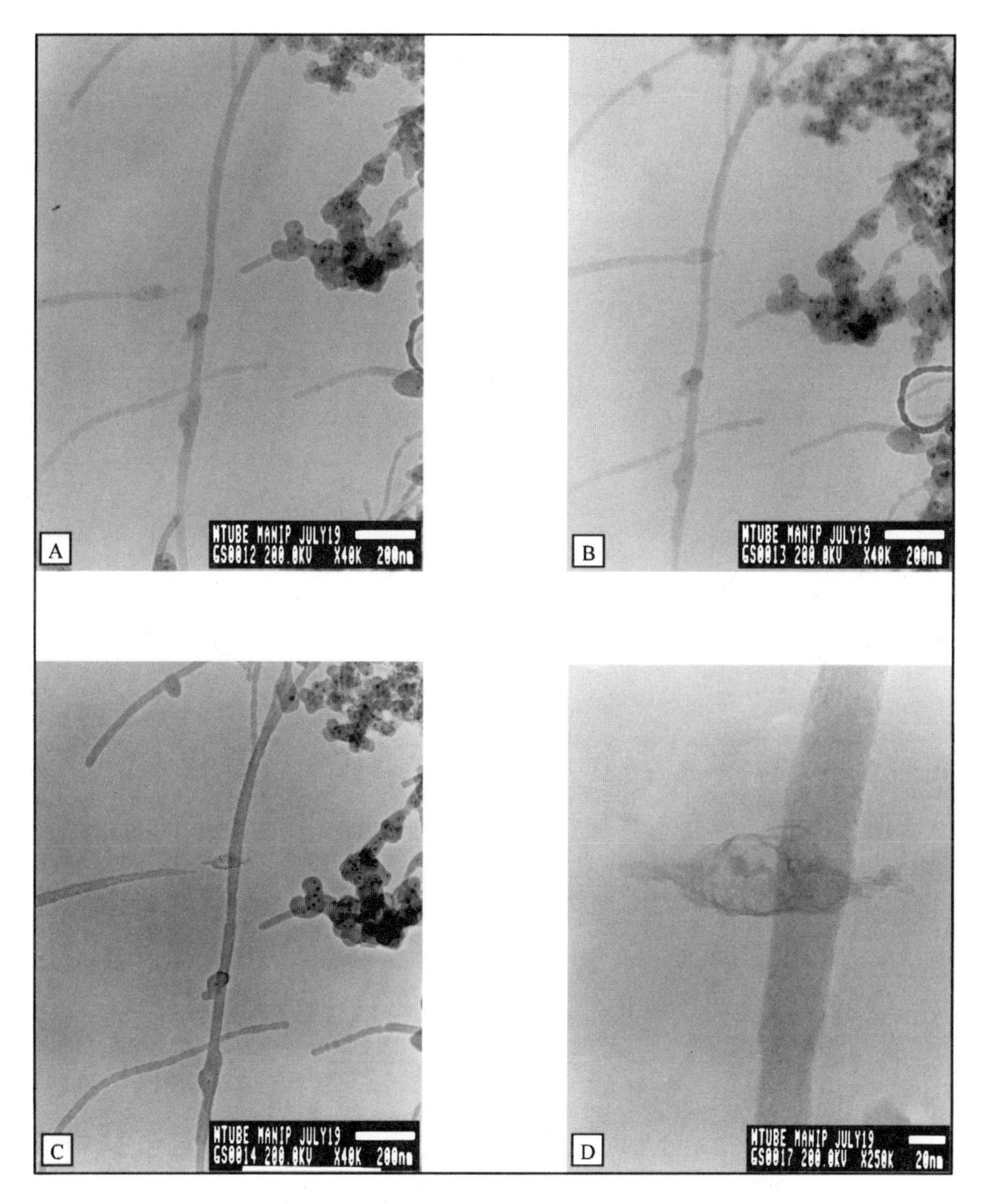

Figure 7. Sequence of TEM images showing a two-probe voltage probe cut of a nanotube.

Using a third probe to cut nanotubes suspended between two other probes provides more control for cutting nanotubes. Figure 8 shows before and after images of a nanotube rope that was cut with a third probe. The rope was suspended between an EBID growth on an AFM cantilever and the nanotube paper attached to a tungsten needle. In Figure 8A the nanotube rope, coming straight down in the upper left of the picture, is connected at one end to the nanotube paper attached to a tungsten probe (not shown in the photo). The other end of the rope is attached to an EBID structure, shown in the lower right of Figure 8A, grown on the end of an AFM cantilever. After this rope was connected to the EBID structure, a third tungsten probe, shown in the lower left of the Figure 8A, was brought in close proximity to the nanotube rope. A square voltage pulse of 15 volts and 12 microseconds applied to this third probe cut this nanotube rope. The result is shown in Figure 8B. The third probe is visible in the lower left of the image. The nanotube rope has now been cut and the section remaining attached to the EBID is visible in the center of the photo.

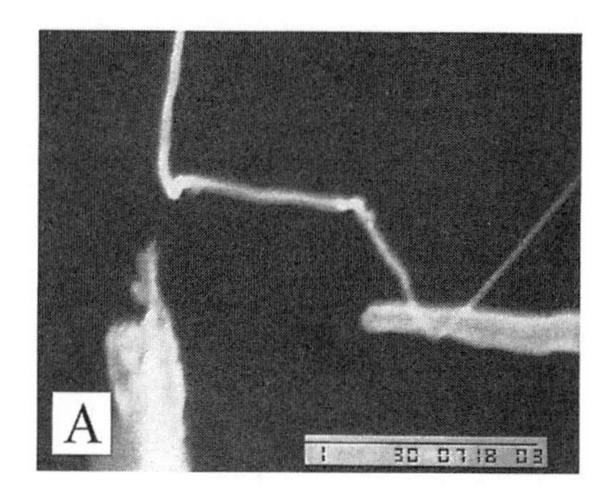

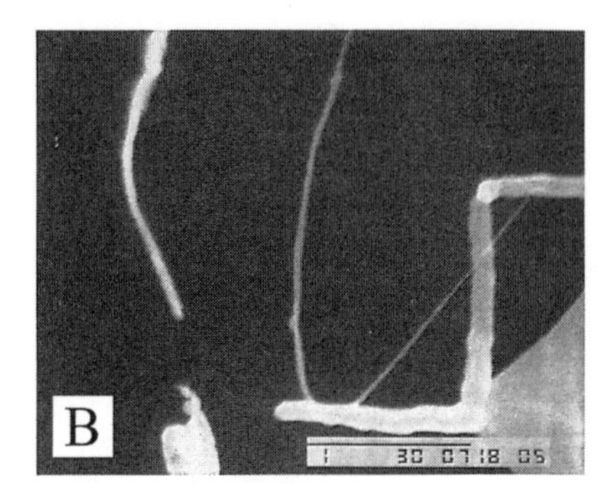

Figure 8. Before and after SEM images of a three-probe cut of a nanotube rope.

Trimming or shortening a nanotube or rope of nanotubes is done in a manner similar to the two-probe cutting of nanotubes. When two nanotubes are brought into close proximity while continuously voltage pulsing, both of the nanotubes are usually shortened. Currently this procedure can trim one or both tubes in an uncontrolled manner. Most of the nanotubes were trimmed or cut with voltage pulses of amplitude exceeding 5 volts. The amount of trimming probably depends on the pulse width and amplitude but more work is necessary before any quantitative data can be presented. Threshold voltages for cutting and trimming nanotubes have yet to be determined. Furthermore, without TEM images confirming the structure (single-wall, multi-wall, or rope) of the cut nanotubes, quantitative data for threshold voltages and pulse widths will be of questionable scientific value.

RESULTANT GEOMETRIES

Multiple Tube Assembly

Assemblies with multiple tubes have been built using the attachment procedure outlined above. Figure 9 shows SEM images of an assembly of ten MWNTs and two EBID support growths. The assembly started with the EBID posts shown in Figure 5A. After chromium coating the EBID posts, the nanotubes were attached from a boule of Aldrich MWNT material. Figure 9A shows the assembly after seven tubes have been attached. As the tubes are not supplied in a regular manner it is difficult to build an aesthetically pleasing structure. The structure shown here only demonstrates that multiple tube assemblies can be built but the quality of the structures needs improvement.

Weaving a SWNT Rope

Figure 10 is a sequence of SEM images showing a rope of nanotubes being woven between two EBID posts. The posts were grown on an AFM cantilever using WF_6 as a precursor gas. Figure 10A shows the posts just after being grown. The scale bar in the bottom right is one micron thus each post just over a micron tall and they are separated by

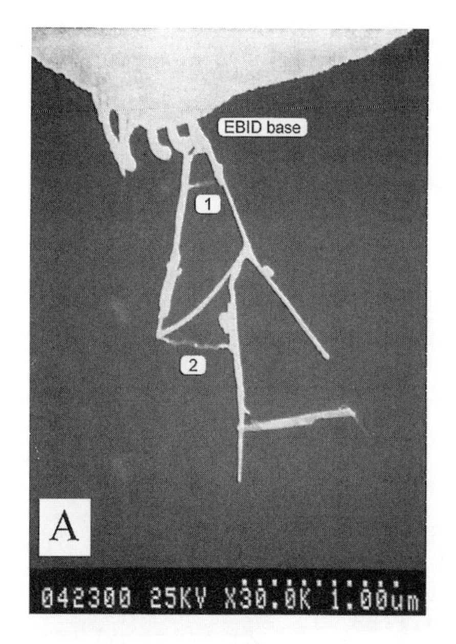

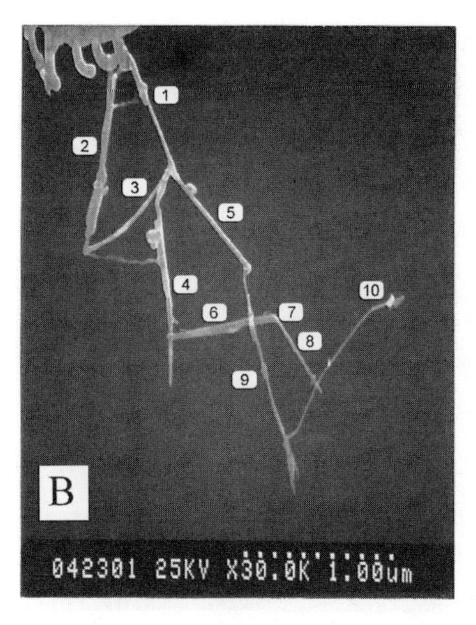

Figure 9. SEM photos of a multiple nanotube assembly constructed within the S-900. The two EBID supports are identified along with the EBID base in A. The ten nanotubes are identified in B.

about 1.25 microns. In Figure 10B the rope of tubes has been attached to the base of the left post and woven between the posts and around the right post. Figure 10C shows the rope of tubes woven between both posts several times. Figure 10D shows the woven structure after the rope of tubes has been slightly strained. The right EBID post has been deformed and is bent to the left from the tension placed on the nanotube rope. A lower magnification view of the structure is in Figure 10E. This scale bar in this image represents 10 microns. The nanotube and EBID structure is directly in the center of the photo. The large structure in the bottom half of the picture is the AFM cantilever. The brighter structure in the upper left of the photo is the nanotube paper. Finally, Figure 10F shows the nanotube weave after the rope has been broken free from the nanotube paper.

CONCLUSIONS

Manipulating nanotubes to build simple assemblies under electron microscopes has been demonstrated. Positioning tubes in contact with EBID posts and with other tubes, combined with EBID, has enabled multiple tube assemblies to be made. Weaving nanotube ropes around EBID posts shows that manipulation alone may be used for assembling structures. Voltage pulses applied to the probes have enabled nanotube cutting and trimming. The construction fundamentals demonstrated here gave a glimpse into what might be possible in free-space nanotube construction.

Further improvements in this work are foreseeable. Controlled nanotube starting material, such as those grown from arrays of catalyst, would allow for easier and possibly automated access to source material. Higher degree of freedom manipulators and more probes in a smaller area would open the possibilities even further. The highest resolution electron microscopes do not have large sample volumes making it difficult to install more than the described two-probe arrangement, but the addition of MEMS-type actuators may allow for greater probe density. Higher degree of freedom manipulators would allow tubes

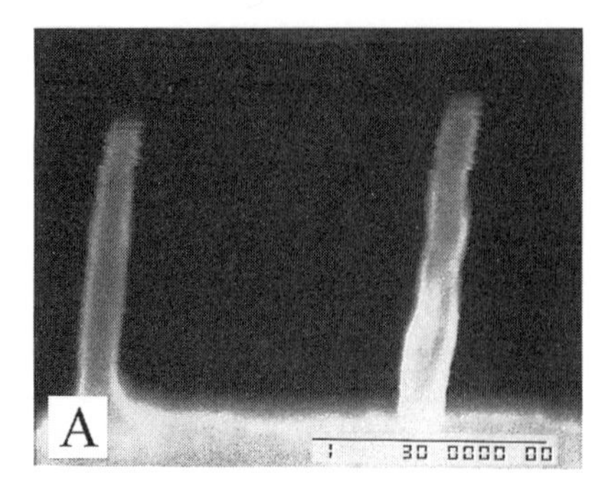

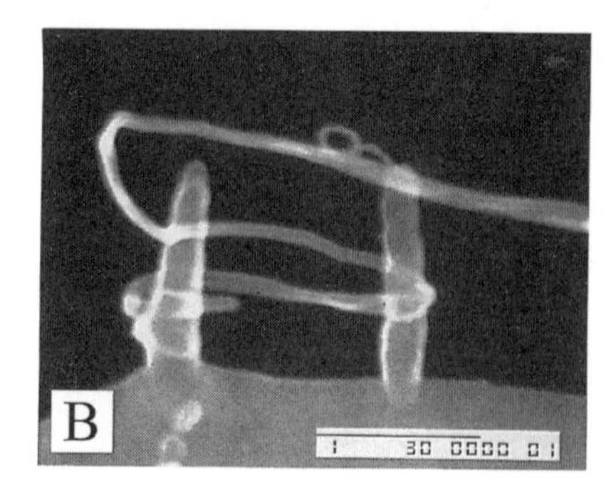

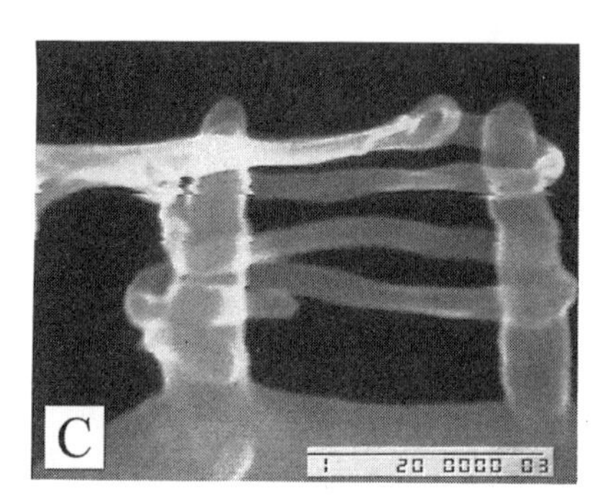

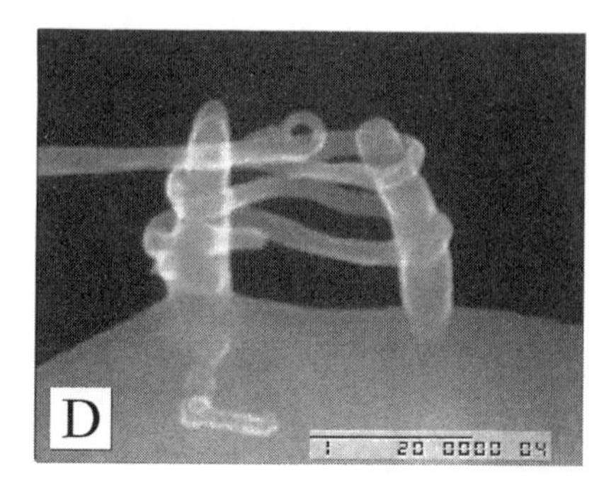

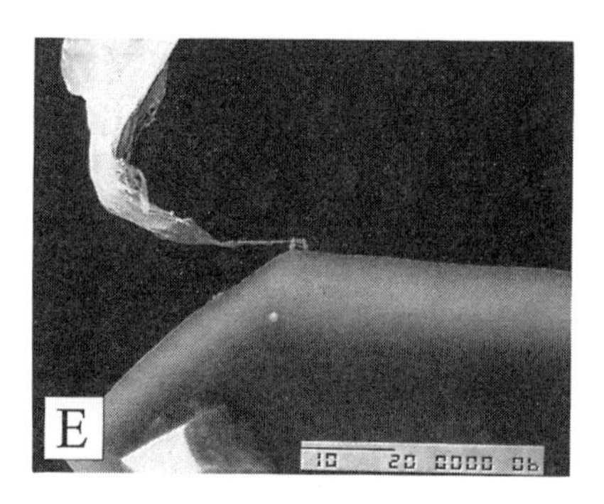

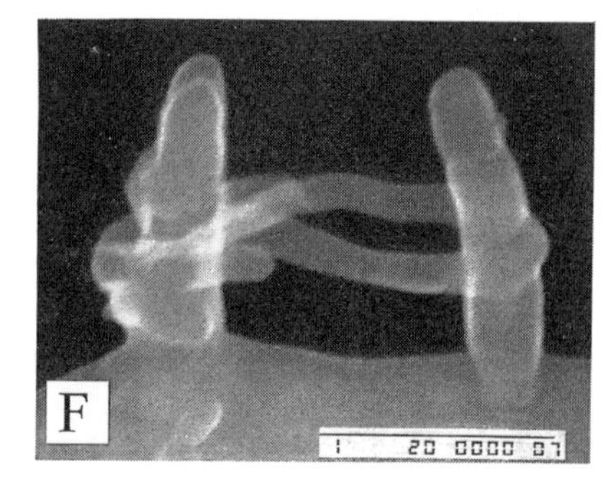

Figure 10. Sequence of SEM photos showing a rope of nanotubes being woven between two posts grown using electron beam induced deposition.

to be twisted during imaging and construction assembly; all of the rotation mechanisms described here are only useful for coarse positioning.

Several experiments on suspended single nanotubes might now be pursued. Tensile strength measurements could now be taken on tubes of known structure. Two-point electrical contact is easily formed making it possible to measure electrical properties of stressed or bent tubes at room temperature. Technologically interesting structures may also be made. The multiple tube assembly shown here would be considered a passive mechanical structure; any arrangement of tubes into a passive structure could now be built limited only by time. Active structures, providing nanoscale actuation, are more difficult to construct but may also be pursued.

ACKNOWLEDGEMENTS

The authors wish to thank Mark Dyer and Christof Baur for discussions and assistance in this work.

REFERENCES

1. M. Gomez-Lopez, J. A. Preece, J. F. Stoddart, "The Art and Science of Self-Assembling Molecular Machines," *Nanotechnology*. 7:183-192 (1996).

2. C. Mao, W. Sun, Z. Shen, N. C. Seeman, "A nanomechanical device based on the B-Z transition of DNA," *Nature*. 397:144-146 (1999).

3. D. W. Carr, L. Sekaric, H. G. Craighead, "Measurement of nanomechanical resonant structures in single-crystal silicon", *J. Vac. Sci. Technol. B*. 16:3821 (1998).

4. P. Avouris, "Manipulation of Matter at the Atomic and Molecular Levels," *Acc. Chem. Res.* 28:95-102 (1995).

5. J. A. Stroscio, D. M. Eigler, "Atomic and Molecular Manipulation with the Scanning Tunneling Microscope," *Science*. 254:1319-1326 (1991).

6. T. Hertel, R. Martel, P. Avouris, "Manipulation of Individual Carbon Nanotubes and Their Interaction with Surfaces," *J. Phys. Chem. B*. 102(6):910-915 (1998).

7. M. Yu, M. J. Dyer, G. D. Skidmore, H. W. Rohrs, X. Lu, K. D. Ausman, J. Von Ehr, R. S. Ruoff, "Three-Dimensional Manipulation of Carbon Nanotubes under a Scanning Electron Microscope," *Nanotechnology*. 10:244-252 (1999)

8. Picomotor™ is a registered trademark of New Focus Inc. 2630 Walsh Avenue, Santa Clara, CA 95051.

9. D. S. Bethune, C.-H. Kiang, M.S. de Vries, G. Gorman, R. Savoy, J. Vazquez, R. Beyers, "Cobalt-Catalyzed Growth of Carbon Nanotubes with single-atomic-layer Walls," *Nature*. 363:605 (1993).

10. Aldrich product number 41,300-3. Aldrich Chemical Company P.O. Box 2060 Milwaukee, WI 53201.

11. H. W. P. Koops, J. Kretz, M. Rudolph, M. Weber, G. Dahm, K. L. Lee, "Characterization and Application of Materials Grown by Electron-Beam-Induced Deposition," *Jpn. J. Appl. Phys.* 33:7099-7107 (1994).

12. L. C. Venema, J. W. G. Wildoer, H. L. J. Temminck Tuinstra, C. Dekker, A. G. Rinzler, R. E. Smalley, "Length control of individual carbon nanotubes by nanostructuring with the scanning tunneling microscope," *Appl. Phys. Lett.* 71:2629 (1997).

PARTICIPANTS

Dr. Markus Erik Ahlskog
Low Temperature Lab.
Helsinki Univ. of Technology
Otakaari 3A
P.O. Box 2200
Espoo, FIN-02015
Finland
358-9-4512964 (phone)
358-9-4512969 (fax)
markus@boojum.hut.fi (email)
http://boojum.hut.fi/nano/ (web)

Dr. Bruce William Alphenaar
Hitachi Cambridge Laboratory
Madingley Road
Cambridge, CB3 0HE
United Kingdom
+44-1223-442905 (phone)
+44-1223-467942 (fax)
alphenaar@phy.cam.ac.uk (email)

Dr. Walter E. Alvarez
CEMS - University of Oklahoma
100 E. Boyd
Energy Center, Room T335
Norman, OK 73019
USA
405-325-4940 (phone)
405-325-5813 (fax)
catalwa@mailhost.ecn.ou.edu (email)

Dr. M P Anantram
NASA Ames Research Center
Mail Stop: T27A-1
Moffett Field, CA 94086
USA
(650)-604-1852 (phone)
anant@nas.nasa.gov (email)
http://www.nas.nasa.gov/~anant (web)

Dr. Phaedon Avouris
IBM Research Division
T. J. Watson Research Center
P. O. Box 218
Yorktown Heights, New York 10598
USA
914-945-2722 (phone)
914-945-4531 (fax)
avouris@us.ibm.com (email)
http://www.research.ibm.com/
nanoscience (web)

Prof. Virginia Marian Ayres
Michigan State University
Department of Electrical & Computer
Engineering
East Lansing, MI 48824
USA
517-355-5236 (phone)
517-353-1980 (fax)
ayresv@egr.msu.edu (email)
http://www.egr.msu.edu/ece/Information/
People/FacultyStaff/Ayres.htm (web)

Adrian Bachtold
Institute of Physics
University of Basel
Klingelbergstrasse 82
CH-4056 Basel
Switzerland
++41-61-267-3679 (phone)
++41-61-267-3784 (fax)
bachtold@ubaclu.unibas.ch (email)

Dr. Miroslaw Bartkowiak
The University of Tennessee
Dept. of Physics and Astronomy
200 South College
Knoxville, TN 37996-1501

USA
423/974-0768 (phone)
423/974-6378 (fax)
ORNL, Solid State Division
Bldg 3025, MS-6030, PO Box 2008
Oak Ridge, TN 37831-6030
USA
+1 (423) 576 5977 (phone)
+1 (423) 574 4143 (fax)
bartkowiakm@ornl.gov (email)

Prof. Inder P. Batra
University of Illinois at Chicago
UIC Physics (MC 273)
845 W. Taylor, Rm. 2236
Chicago, IL 60607
USA
312-413-2798 (phone)
312-996-9016 (fax)
ipbatra@uic.edu (email)
http://www.uic.edu/ (web)

Dr. Ray H. Baughman
Allied Signal, Inc.
Research & Technology
101 Columbia Road
Morristown, NY 07962-1021
U.S.A.
973-455-2375 (phone)
973-455-5991 (fax)
ray.baughman@alliedsignal.com (email)

Mr. Rick C. Becker
IBADEX
57 N. Putnam St
Danvens, MA 01923
USA
978-777-5500 (phone)
978-777-6575 (fax)
rbecker@ibadex.com (email)
http://www.Refractor.com (web)

Mr. Savas Berber
Michigan State University
Department of Physics and Astronomy
East Lansing, MI 48824-1116
USA
+1-517-355-0230 (phone)
+1-517-353-4500 (fax)
berber@pa.msu.edu (email)
http://www.pa.msu.edu/~berber/ (web)

Dr. Peter Berghuis
Argonne National Laboratory
MSD, 223
9700 South Cass Avenue
Argonne, IL 60439
USA
1 630 252 3996 (phone)
1 630 252 9595 (fax)
berghuis@anl.gov (email)
http://www.anl.gov (web)

Prof. Jerry Bernholc
NC State University
Department of Physics
Box 8202
Raleigh, NC 27695-8202
USA
+1 (919) 515-3126 (phone)
+1 (919) 515-7331 (fax)
bernholc@ncsu.edu (email)

Dr. Richard Hermann Joseph Blunk
General Motors Corporation
R&D Center
30500 Mound Rd.
Warren, Michigan 48090-9050
U.S.A.
810-986-1967 (phone)
810-986-2244 (fax)
rblunk@gmr.com (email)

Mr. Mark Brehob
Michigan State University
Department of Computer Science and Engineering
2319-15 E. Jolly Road
Lansing, MI 48910
USA
517-393-9057 (phone)
brehob@cse.msu.edu (email)
http://www.cse.msu.edu/~brehob (web)

Mr. Peter Robert Butzloff
University of North Texas
Materials Science Dept.
P.O. Box 305310
Denton, TX 76302-5310
USA
817-280-3197 (phone)
817-280-2324 (fax)
Pbutzloff@bellhelicopter.textron.com (email)

Mr. Attila Caglar
University of Bonn
Dept. of Applied Mathematics
Wegelerstr. 6
Bonn, D-53115
Germany
+ 49 228 73 2748 (phone)
+ 49 228 73 7527 (fax)
caglar@iam.uni-bonn.de (email)
http://wwwwissrech.iam.uni-bonn.de/people/caglar.html (web)

Dr. Mei Cai
General Motors Corp.
R&D Center
30500 Mound Road, Bldg. 1-3
Warren, MI 48090-9055
U.S.A.
810-986-2870 (phone)
810-986-0294 (fax)
mcai@gmr.com (email)

Dr. Jean-Christophe Charlier
University of Louvain
Physico-Chemistry and Physics of Materials (PCPM)
Place Croix du Sud 1 (Boltzmann)
Louvain-la-Neuve, 1348
BELGIUM
(32) 10 47 33 59 (phone)
(32) 10 47 34 52 (fax)
charlier@pcpm.ucl.ac.be (email)
http://bohr.pcpm.ucl.ac.be/~charlier/ (web)

Dr. Ying Chen
The Australian National University
Department of EME
RSPHYSSE
Canberra, ACT 0200
Australia
61 02 6249 0511 (fax)
ying.chen@anu.edu.au (email)

Dr. Hui-Ming Cheng
Chinese Academy of Sciences
Institute of Metal Research
72 Wenhua Road
Shenyang, Liaoning 110015
China
086-24-23991831 (phone)
086-24-23891320 (fax)
cheng@imr.ac.cn (email)
http://www.imr.ac.cn (web)

Dr. WonBong Choi
Samsung Advanced Institute of Technology
Display Lab. P.O.Box 111
Suwon, 440-600
Korea
82-331-280-9316 (phone)
82-331-280-9349 (fax)
wbchoi@sait.samsung.co.kr (email)

Mr. Deukseok Chung
Samsung Advanced Institute of Technology
PO Box 111
Suwon, 440-600
Korea
82-231-280-9316 (phone)
82-331-280-9349 (fax)
dschung@sait.samsung.co.kr (email)

Dr. David Henry Cobden
Niels Bohr Institute
Oersted Laboratory
Universitetsparken 5
Copenhagen N., 2100
Denmark
+45 3532 0420 (phone)
+45 3532 0460 (fax)
cobden@fys.ku.dk (email)
http://ntserv.fys.ku.dk/nanolab/people/David.htm (web)

Dr. Alan C. Cooper
Air Products and Chemicals, Inc.
7201 Hamilton Blvd.
Mail Stop R3102
Allentown, PA 18195-1501
USA
(610) 481-2607 (phone)
(610) 481-7719 (fax)
cooperac@apci.com (email)

Ms. Carole Ann Cooper
UMIST, Manchester
Manchester Materials Science Centre
Grosvenor Street
Manchester, Lancashire M1 7HS
United Kingdom
+44 0161 236 3311 ex (phone)
+44 0161 200 3586 (fax)
mblx3cac@mh1.mcc.ac.uk (email)

carole@periscope84.freeserve.co.uk
(email)

Prof. Pavel Nikolaevich D'yachkov
Institute of General and Inorganic
Chemistry
Russian Academy of Sciences
Leninskii pr. 31
Moscow, 117907
Russia
(7-095) 955 4820 (phone)
(7-095) 954 1279 (fax)
dyachkov@mxo2.nifhi.ac.ru (email)

Prof. Cees Dekker
Delft University of Technology
Department of Applied Sciences and
DIMES
Lorentzweg 1, 2628 CJ Delft
The Netherlands
+31 15 2786094 (phone)
+31 15 2617868 (fax)
dekker@qt.tn.tudelft.nl (email)
http://vortex.tn.tudelft.nl/~dekker (web)

Prof. Mildred Dresselhaus
Massachusetts Institute of Technology
Room 13-3005
Cambridge, MA 02139
USA
617-253-6864 (phone)
617-253-6827 (fax)
millie@mgm.mit.edu (email)

Prof. Gene Dresselhaus
Massachusetts Institute of Technology
Room 13-3005
Cambridge, MA 02139
USA
617-253-6864 (phone)
617-253-6827 (fax)
gene@mgm.mit.edu (email)

Prof. Lawrence T. Drzal
Michigan State University
Composite Materials & Structures Center
2100 EB
East Lansing, MI 48824-1226
USA
517-353-5466 (phone)
517-432-1634 (fax)
drzal@egr.msu.edu (email)
http://cmscsun.egr.msu.edu/profiles/
person.info/drzal.html (web)

Dr. Georg Stefan Duesberg
Trinity College, Dublin and
Max-Planck-Institut, Stuttgart
Heisenbergstr. 1
D-70569 Stuttgart
Germany
++49-711-689-1343 (phone)
++49-711-689-1010 (fax)
duesberg@klizix.mpi-stuttgart.mpg.de
(email)

James R. Von Ehr
Zyvex LLC
1321 N. Plano Road
Suite 200
Richardson, TX 75081
USA
972/235-7881 x211 (phone)
972/235-7882 (fax)
jvonehr@zyvex.com (email)
http://www.zyvex.com (web)

Prof. Richard J. Enbody
Michigan State University
Department of Computer Science
3115 Engineering Building
East Lansing, Michigan 48824-1226
USA
+1-517-353-3389 (phone)
+1-517-432-1061 (fax)
enbody@cse.msu.edu (email)
http://www.cse.msu.edu/~enbody/ (web)

Mr. Leo S. Fifield
Allied Signal, Inc.
101 Columbia Rd.
Morristown, NJ 07962
USA
973-455-5409 (phone)
lfifield@u.washington.edu (email)

Prof. John E. Fischer
University of Pennsylvania
Materials Science Dept
3231 Walnut St.
Philadelphia, PA 19104-6272
USA
215 898 6924 (phone)
215-573-2128 (fax)
fischer@sol1.lrsm.upenn.edu (email)

Dr. Laszlo Forro
Swiss Federal Institute for Technology
Physics Dept/IGA
Lausanne, CH-1015
Switzerland
(41 21) 693 4306 (phone-office)
(41 21) 693 4360 (phone-lab)
(41 21) 693 4470 (fax)
Laszlo.forro@epfl.ch (email)

Dr. Ricardo Fuentes
The Dow Chemical Company
2030 Dow Center
USA
Midland, MI 48674
USA
517-636-0201 (phone)
517-636-7085 (fax)
rfuentes@dow.com (email)

Mr. Hiroyuki Fukushima
Michigan State University
Composite Materials and Structures
Center
2100 Engineering Building
East Lansing, MI 48824-1226
U.S.A.
517-353-4708 (phone)
517-432-1634 (fax)
fukushi3@pilot.msu.edu (email)

Mr. Elton Daniel Graugnard
Purdue University
1396 Physics Building
West Lafayette, IN 47907-1396
USA
1-765-494-5386 (phone)
1-765-494-3032 (fax)
elton@physics.purdue.edu (email)
www.physics.purdue.edu/nanophys/
(web)

Ms. Barr Halevi
University Of Colorado, Boulder
P.O.Box 0424
Boulder, Colorado 80309-0424
USA
001-303-492-5463 (phone)
001-303-492-4341 (fax)
barr.halevi@colorado.edu (email)

Mr. In Taek Han
Samsung Advanced Institute of
Technology
san #14, Nong Seo-Ri, kiheung Eup
Yong-In, Kyoung Ki-Do 440-600
Korea
82-331-280-9347 (phone)
82-331-280-9349 (fax)
ithan@sait.samsung.co.kr (email)

Prof. Walter de Heer
Georgia Institute of Technology
School of Physics
Atlanta, GA 30332
USA
404-894-7879 (phone)
404-894-7452 (fax)
deheer@electra.physics.gatech.edu
(email)

Eng. Wim J. Hellemans
European Patent Office
Willem de Merode Str. 3
Delft, 2624 LC
Netherlands
31-70-340-2445 (phone)
31-70-340-3016 (fax)

Dr. Mihai Horoi
Central Michigan University
Dow 203
Mt. Pleasant, MI 48859
USA
517-774-2185 (phone)
517-774-2697 (fax)
Horoi@phy.cmich.edu (email)

Prof. Jisoon Ihm
Seoul National University
Dept. of Physics
Seoul, 151-742
Korea
822-880-6614 (phone)
822-884-3002 (fax)
jihm@snu.ac.kr (email)

Prof. Sumio Iijima
Meijo University and NEC
34 Miyukigaoka
Tsukuba, Ibaraki 305-8501
Japan
81-298-501117 (phone)
81-298-56-6136 (fax)

s-iijima@frl.cl.nec.co.jp (email)

Prof. Chakram Sampath Jayanthi
University of Louisville
Department of Physics
Louisville, KY 40292
USA
502-852-3335 (phone)
502-852-0742 (fax)
csjaya01@fellini.physics.louisville.edu (email)
http://www.physics.louisville.edu (web)

Dr. Sungho Jin
Bell Labs, Lucent Technologies
Rm. 1A-123
600 Mountain Ave
Murray Hill
New Jersey 07974
USA
908-582-4076 (phone)
908-582-3609 (fax)
jin@lucent.com (email)

Dr. Karl Johnson
University of Pittsburgh
Department of Chemical Engineering
1249 Benedum Engineering Hall
Pittsburgh, PA 15261
USA
412 624 5644 (phone)
412 624 9639 (fax)
karlj+@pitt.edu (email)
http://www.engrng.pitt.edu/~chewww/johnson.html (web)

Prof. Thomas A. Kaplan
Michigan State University
Physics-Astronomy Dept.
East Lansing, MI 48824-1116
USA
517-353-8644 (phone)
517-353-0690 (fax)
kaplan@pa.msu.edu (email)

Dr. Tapas Kar
Southern Illinois University
Department of Chemistry
Forestry Bldg F118
Carbondale, IL 62901
USA
1-618-453-6485 (phone)
1-618-453-6408 (fax)
tapaskar@siu.edu (email)

Ms. Elliot B. Kennel
Applied Sciences, Inc.
141 W. Xenia Ave.
PO Box 579
Cedarville, OH 45314-0579
USA
1-937-766-2020 x 134 (phone)
1-937-766-9260 (fax)
EKennel@Apsci.com (email)

Prof. Kathleen V. Kilway
University of Missouri-Kansas City
205 Spencer Chemistry Building
5100 Rockhill Road
Kansas City, MO 64110-2499
USA
(816)235-2289 (phone)
(816)235-5502 (fax)
kilwayk@umkc.edu (email)

Mr. Boonyarach Kitiyanan
CEMS, University of Oklahoma
100 E.Boyd, Rm T-335
Norman, OK 73019
USA
405-3253738 (phone)
405-3255813 (fax)
bkitiyan@ou.edu (email)

Dr. Tomasz Kostyrko
University of Tennessee & ORNL
ORNL, Solid State Division
Bldg 3025, MS-6030, PO Box 2008
Oak Ridge, TN 37831-6030
USA
+1 (423) 574 5232 (phone)
+1 (423) 574 4143 (fax)
tkos@dlin.ssd.ornl.gov (email)

Dr. Petr Kral
University of Toronto
Department of Physics
60 St. George Street
Toronto, Ontario M5S 1A7
Canada
(416) 978-1546 (phone)
(416) 978-2537 (fax)
kral@cheetah.physics.utoronto.ca (email)

Mr. Sriramanan Krishnamurthy
Univ. of Notre Dame

Electrical Engineering Dept
275 Fitzpatrick Hall
Notre Dame, IN 46556
USA
(219) 631-9011 (phone)
(219) 631-4393 (fax)
Krishnamurthy.1@nd.edu (email)

Ms. Anya Kuznetsova
University of Pittsburgh
Surface Science Center
135 Eberly Hall
Pittsburgh, PA 15260
USA
(412)6385413 (phone)
(412)6246003 (fax)
anyak+@pitt.edu (email)

Mr. Young-Kyun Kwon
Michigan State University
Department of Physics and Astronomy
East Lansing, MI 48824-1116
USA
+1-517-355-0230 (phone)
+1-517-353-4500 (fax)
ykkwon@pa.msu.edu (email)
http://www.pa.msu.edu/~ykkwon/ (web)

Prof. Philippe Lambin
Laboratoire de physique du solide
Facultes Universitaires N-D Paix
61 Rue de Bruxelles
Namur, B-5000,
Belgium
+32(0)81-724710 (phone)
+32(0)81-724707 (fax)
philippe.lambin@scf.fundp.ac.be (email)

Mr. Istvan Laszlo
Technical University of Budapest
Dep. of Theor. Physics
Budapest, H-1521
Hungary
36-1-463-4110 (phone)
36-1-463-3567 (fax)
laszlo@phy.bme.hu (email)

Dr. Christophe Laurent
LCMI/ESA CNRS 5070
Bat. 2R1, Universite Paul-Sabatier
Toulouse, F-31062
France
(33) 561 55 61 22 (phone)
(33) 561 55 61 63 (fax)
laurent@iris.ups-tlse.fr (email)

Dr. Francois Leonard
IBM T.J. Watson Research Center
Rt. 134
Yorktown Heights, NY 10598
USA
914-945-2260 (phone)
914-945-4506 (fax)
fleonard@us.ibm.com (email)

Dr. Jun Li
Institute of Materials Research &
Engineering
Blk S7, Level 3
National University of Singapore
Singapore, 119260
Singapore
(65) 874-8109 (phone)
(65) 872-0785 (fax)
j-li@imre.org.sg (email)

Dr. Lei Liu
University of Louisville
Department of Physics
Louisville, KY 40292
USA
502-852-7754 (phone)
502-852-0742 (fax)
l0liu001@fellini.physics.louisville.edu
(email)
http://www.physics.louisville.edu (web)

Dr. Annick Loiseau
Laboratoire d'Etudes des Microstructures
Unite Mixte Onera-Cnrs
O.N.E.R.A., B.P. 72,
29 av. de la Division Leclerc
Chatillon, F-92322,
France
33 (0) 1 46 73 44 53 (phone)
33 (0) 1 46 73 41 55 (fax)
33 (0) 1 46 73 41 42 (fax)
loiseau@onera.fr (email)

Prof. Steven G. Louie
University of California at Berkeley
Department of Physics
366 LeConte Hall #7300
Berkeley, CA 94720-7300
USA
510-642-1709 (phone)

510-643-9473 (fax)
sglouie@uclink.berkeley.edu (email)
http://tiger.berkeley.edu/louie (web)

Dr. Douglas H. Lowndes
Oak Ridge National Laboratory
Solid State Division
P.O. Box 2008
Oak Ridge, TN 37831-6056
USA
423-574-6306 (phone)
423-576-3676 (fax)
vdh@ornl.gov (email)

Dr. David E. Luzzi
University of Pennsylvania
Dept. of Materials Science and Engineering
3231 Walnut St
Philadelphia, PA 19104
USA
1-215-898-8366 (phone)
1-215-573-2128 (fax)
luzzi@lrsm.upenn.edu (email)

Prof. Subhendra D. Mahanti
Michigan State University
Department of Physics and Astronomy
East Lansung, Michigan 48824
USA
517-355-9701 (phone)
517-353-4500 (fax)
mahanti@pa.msu.edu (email)

Mr. Vladimir Mancevski
Xidex Corporation
8906 Wall Street
Ste. 105
Austin, TX 78754
USA
512-339-0608 (phone)
512-339-9497 (fax)
vam@xidex.com (email)
http://www.xidex.com (web)

Prof. Thomas J. Manning
Valdosta State University
Chemistry Department
1500 Patterson
Valdosta, GA 31698
USA
912-333-7178 (phone)
912-333-7389 (fax)
tmanning@valdosta.edu (email)
http://www.valdosta.edu/~tmanning/research/ (web)

Mr. Hajime Matsumura
Institute for Solid State Physics,
University of Tokyo
7-22-1 Roppongi Minato-ku
Tokyo, 106-8666
Japan
+81-3-3478-6811 (phone)
+81-3-3402-7326 (fax)
hmatsu@kodama.issp.u-tokyo.ac.jp (email)

Prof. Eugene J. Mele
University of Pennsylvania
David Rittenhouse Labs
209 S. 33rd Street
Philadelphia, PA 19104
USA
(215)-898-3135 (phone)
(215)-898-2010 (fax)
mele@mele.physics.upenn.edu (email)
http://dept.physics.upenn.edu/facultyinfo/mele.html (web)

Dr. Vladimir I. Merkulov
Oak Ridge National Laboratory
Solid State Division
P.O. Box 2008
Oak Ridge, TN 37831-6056
USA
423-574-6306 (phone)
423-576-3676 (fax)
merkulov@solid.ssd.ornl.gov (email)

Mr. Vincent Meunier
University of Namur
Rue de Bruxelles, 61
Namur, B5000
BELGIUM
3281724702 (phone)
3281724707 (fax)
Vincent.Meunier@fundp.ac.be (email)
http://www.scf.fundp.ac.be/~vmeunier/ (web)

Prof. Aldo Dante Migone
Southern Illinois University Carbondale
Department of Physics
Carbondale, IL 62901
USA

(618)-453-2044 (phone)
(618)-453-1056 (fax)
aldo@physics.siu.edu (email)

Prof. Gary E. Miracle
Colby College
Waterville, ME 04901
USA
(207) 872-3429 (phone)
gemiracl@colby.edu (email)

Dr. Katerina Moloni
Univ. of Wisconsin
ERB 1107
1500 Engineering Dr.
Madison, WI 53706
USA
608-265-4119 (phone)
608-265-4118 (fax)
moloni@mrgcvd.engr.wisc.edu (email)
http://mrgcvd.engr.wisc.edu/moloni (web)

Mr. Gunnar Moos
Fritz-Haber-Institut der MPG
Abteilung PC
Faradayweg 4-6
Berlin, D-14195
Germany
+49-30-8413-5504 (phone)
+49-30-8413-5377 (fax)
moos@fhi-berlin.mpg.de (email)

Mr. Rentaro Mori
Toyota Technical Center, USA
1555 Woodridge, RR#7
Ann Arbor, MI 48105
USA
734-995-7133 (phone)
734-995-3684 (fax)
rmori@ttc-usa.com (email)

Mr. Joerg Muster
Max-Planck-Institut fuer Festkoerperforschung
Heisenbergstr. 1
Stuttgart, D-70569
Germany
++49 711 689 1432 (phone)
++49 711 689 1010 (fax)
muster@klizix.mpi-stuttgart.mpg.de (email)

Dr. Takeshi Nakanishi
Institute of Physical and Chemical Research (RIKEN)
2-1 Hirosawa
Wako, Saitama 351-0198
Japan
+81-48-467-9599 (phone)
+81-48-467-5087 (fax)
nkns@postman.riken.go.jp (email)

Dr. Gerard Newman
University of Oklahoma
School of Chemcial Engineering & Material Science
100 E. Boyd, Rm T-335
Norman, OK 73019
USA
(405) 325-2833 (phone)
(405) 325-5813 (fax)
newman@mailhost.ecn.ou.edu (email)

Mr. Pavel Nikolaev
NASA/Johnson Space Center
P.O. Box 58561
Mail Stop C-61
Houston, TX 77258
USA
281-483-5946 (phone)
281-483-1605 (fax)
pnikolae@ems.jsc.nasa.gov (email)
http://www.jsc.nasa.gov/ea/em/nano (web)

Mr. Christoph Nuetzenadel
University of Fribourg
Physics Institute
Fribourg, CH-1700
Switzerland
++41 26 300 9090 (phone)
++41 26 300 9747 (fax)
christoph.nuetzenadel@unifr.ch (email)

Dr. Elena Dmitrievna Obraztsova
General Physics Institute, RAS
Natural Sciences Center
38 Vavilov Street
Moscow, 117942
Russia
7(095)132 8206 (phone)
7(095)135 0270 (fax)
elobr@kapella.gpi.ru (email)

Dr. Alexander Vladimirovich Okotrub
Institute of Inorganic Chemistry SB RAS
pr. Ak. Lavrenteva, 3
Novosibirsk, 630090
Russia
(3832)341366 (phone)
(3832)344489 (fax)
spectrum@che.nsk.su (email)

Dr. Charles Olk
GM Research & Planning Center
Materials Processes Lab
MC 480-106-224
Warren, MI 48090-9055
USA
810 986-0611 (phone)
810 986-3091 (fax)
colk@gmr.com (email)

Mr. Pedro Jose de Pablo
Universidad Autonoma Madrid
C-III
Canto Blanco
Madrid, 28043
Spain
34-91-397-4754 (phone)
34-91-397-3961 (fax)
p.j.depablo@uam.es (email)

Dr. Jayasree Pattanayak
Southern Illinois University at Carbondale
Department of Chemistry and Biochemistry
Mail Code 4409
Carbondale, IL 62901-4409
USA
618-453-5721 (phone)
618-453-6408 (fax)
jaypad@siu.edu (email)
jaya@risky3.thchem.siu.edu (email)

Dr. Ton Peijs
Queen Mary and Westfield College
University of London, Department of Materials
Mile End Road
London, E1 4NS
UK
+44-1719755281 (phone)
+44-1819818904 (fax)
t.peijs@qmw.ac.uk (email)
http://www.materials.qmw.ac.uk (web)

Dr. Frank Channa Peiris
University of Notre Dame
Department of Electrical Engineering
Notre Dame, IN 46556
USA
219-631-5972 (phone)
219-631-5952 (fax)
fpeiris@boron.helios.nd.edu (email)

Prof. Annick Percheron-Guigan
C.N.R.S.
UPR 209 LCMTR CNRS
2-8 rue Henri Dunant
THAIS, 94320
FRANCE
33 1 49781204 (phone)
33 1 49781203 (fax)
apg@glvt-cnrs.fr (email)

Mr. Per Rugaard Poulsen
Oersted Laboratory
Universitetsparken 5
Copenhagen, DK-2100
Denmark
45 3555 3833 (phone)
45 3532 0460 (fax)
prpbn@post.tele.dk (email)

Dr. Apparao M. Rao
University of Kentucky
Center for Applied Energy Research
2540 Research Park Drive
Lexington, KY 40511
USA
1-606-257 0216 (phone)
1-606-257-0220 (fax)
rao@pop.uky.edu (email)

Mr. Ionel Adrian Rata
Central Michigan University
Department of Physics
1522 S. Mission, 214N
Mt. Pleasant, MI 48858
USA
(517)774-2788 (phone)
rata@phy.cmich.edu (email)

Dr. Maja Remskar
J.Stefan Institute
Jamova 39
Ljubljana, 1000,
Slovenia
386 61 1773 728 (phone)

386 61 219 385 (fax)
maja.remskar@ijs.si (email)

Prof. Zhifeng Ren
Boston College
Department of Physics
140 Commonwealth Ave
Chestnut Hill, MA 02467
USA
617-552-2832 (phone)
617-552-8478 (fax)
renzh@bc.edu (email)
http://ph99.bc.edu/faculty/Ren.html (web)

Dr. Jane Repko
Michigan State University
101 Physics and Astronomy
East Lansing, Michigan 48824-1116
USA
517-483-1107 (phone)
repko_j@pa.msu.edu (email)

Prof. Susumu Saito
Tokyo Institute of Technology
Department of Physics
2-12-1 Oh-okayama
Meguro-ku, Tokyo 152
Japan
+81-3-5734-2070 (phone)
+81-3-5734-2739 (fax)
saito@stat.phys.titech.ac.jp (email)

Prof. Riichiro Saito
Univ. of Electro-Communications
Department of Electronic Eng.
1-5-1 Chofugaoka
Chofu, Tokyo 182-8585
Japan
+81-424-43-5148 (phone)
+81-424-43-5210 (fax)
rsaito@ee.uec.ac.jp (email)
http://flex.ee.uec.ac.jp (web)

Ms. Darlene Salman
Michigan State University
113 Physics and Astronomy
East Lansing, Michigan 48824-1116
USA
5173534503 (phone)
5173534500 (fax)
salman@pa.msu.edu (email)

Mr. Jan Sandler
Technical University Hamburg-Harburg
Polymer Composites 5-09/1
Denickestrasse 15
Hamburg, 21073
Germany
0049-40-42878-2400 (phone)
0049-40-42878-2002 (fax)
sandler@tu-harburg.de (email)
http://www.tu-harburg.de/kvweb/ (web)

Mr. Stefano Sanvito
Lancaster University
School of Physics and Chemistry
Lancaster, Lancs. LA1 4YB
U.K.
+44-1684-894211 (phone)
sanvito@dera.gov.uk (email)

Prof. Louis Schlapbach
University of Fribourg
Physics Institute
Perolles
Fribourg, CH-1700
Switzerland
+41 26 300 9066 (phone)
+41 26 300 9747 (fax)
Louis.Schlapbach@unifr.ch (email)
http://www.unifr.ch/physics/fk/ (web)

Dr. Peter Schmidt
Office of Naval Research
800 N. Quincy St
Arlington, VA 22217-5660
USA
703-696-4362 (phone)
703-696-6887 (fax)
schmidp@onr.navy.mil (email)

Ms. Anthony Schultz
CCNY
6063 Broadway Apt. #4C
Bronx, NY 10471
USA
(718) 543-6647 (phone)
trismegis@hotmail.com (email)

Mr. Pedro Serra
Centro de Fisica Computacional
Department of Physics
University of Coimbra
3000 Coimbra
Portugal

+351-39-410622 (phone)
+351-39-829158 (fax)
serra@pa.msu.edu> (email)
serra@teor.fis.uc.pt (email)
http://www.pa.msu.edu/~serra/ (web)

Mrs. Priscilla Simonis
Facultis Universitaires Notre-Dame de la Paix
rue de Bruxelles, 61
Namur, 5000
Belgium
3281724712 (phone)
3281724707 (fax)
priscilla.simonis@fundp.ac.be (email)

Dr. Vahan Vladimiri Simonyan
University of Pittsburgh
1242 Benedum Hall
Pittsburgh, PA 15261
USA
(412) 624-5644 (phone)
(412) 624-9639 (fax)
vsim+@pitt.edu (email)

Mr. Navjot Singh
GE Research & Development
One Research Circle
Building K-1, Room 5A71
Niskayuna, NY 12309
USA
518-387-7240 (phone)
518-387-7403 (fax)
singhna@crd.ge.com (email)

Dr. George D. Skidmore
Zyvex LLC
1321 N. Plano Road
Suite 200
Richardson, TX 75081
USA
972/235-7881 x216 (phone)
972/235-7882 (fax)
gskidmore@zyvex.com (email)
http://www.zyvex.com (web)

Mr. Oliver Smiljanic
INRS
1650 ld Linel Barlet
Vareumer, Quebec 53x152
Canada
(650) 929-8902 (phone)
(650) 929-8237 (fax)
smiljani@inrs-ener.uquebec.ca (email)

Dr. Gregory Snider
University of Notre Dame
275 Fitzpatrick Hall
Electrical Engineering Dept.
Notre Dame, IN 46556
USA
(219) 631-4148 (phone)
(219) 631-4393 (fax)
snider.7@nd.edu (email)

Dr. Geoffrey M Spinks
University of Wollongong
Intelligent Polymer Research Institute
Northfields Ave
Wollongong, NSW 2522
Australia
61 2 4221 3010 (phone)
61 2 4221 3114 (fax)
geoff_spinks@uow.edu.au (email)
http://www.uow.edu.au/science/research/ipri (web)

Dr. Deepak Srivastava
NASA Ames Research Center
MS T27A-1
Moffett Field, CA 94035-1000
USA
(650) 604-3486 (phone)
(650) 604-3957 (fax)
deepak@nas.nasa.gov (email)
http://science.nas.nasa.gov/~deepak/home.html (web)

Dr. Shekhar Subramoney
DuPont Company
Central Research and Development
P. O. Box 80228
Wilmington, DE 19880-0228
USA
(302) 695-2992 (phone)
(302) 695-1351 (fax)
shekhar.subramoney@usa.dupont.com (email)

Mr. Saikat Talapatra
Southern Illinois University Carbondale
Department of Physics
Carbondale, IL 62901
USA
(618)-453-2044 (phone)
(618)-453-1056 (fax)

saikatt@hotmail.com (email)

Dr. Cher-Dip Tan
Wright Materials Research Co.
318 Amherst Bend
Dayton, Ohio 45440-4441
USA
(937) 643-0007 (phone)
(937) 293-4108 (fax)
cherdip-tan@wrightmat.com (email)

Dr. Cha-Mei Tang
Creatv MicroTech, Inc.
11609 Lake Potomac Drive
Potomac, MD 20854
USA
(301)983-1650 (phone)
(301)983-6264 (fax)
cmtinc@yahoo.com (email)

Mr. Daniel G. Tekleab
Clemson University
Department of Physics and Astronomy
Laboratory for Nanotechnology
Clemson, SC 29634
USA
864-656-4447 (phone)
864-656-0805 (fax)
dteklea@clemson.edu (email)

Prof. Michael F. Thorpe
Michigan State University
Department of Physics and Astronomy
East Lansing, MI 48824-1116
U.S.A.
517-355-9279 (phone)
517-353-0690 (fax)
thorpe@pa.msu.edu (email)
http://www.pa.msu.edu/~thorpe/ (web)

Dr. Gary G. Tibbetts
General Motors R & D Center
Materials and Processes Lab
30500 Mound Rd.
Warren, MI 48009
USA
810-986-0655 (phone)
810-986-3091 (fax)
gtibbett@gmr.com (email)

Prof. David Tomanek
Michigan State University
Department of Physics and Astronomy
East Lansing, MI 48824-1116
USA
+1-517-355-9702 (phone)
+1-517-353-4500 (fax)
tomanek@pa.msu.edu (email)
http://www.pa.msu.edu/~tomanek/ (web)

Prof. Mojtaba Vaziri
University of Michigan-Flint
Department of Physics
Flint, MI 48502
USA
810-762-3409 (phone)
810-766-6780 (fax)
mvaz@umich.edu (email)

Mr. Brian Lawrence Walsh
Purdue University
School of Electrical and Computer Engineering
1285 Electrical Engineering Bldg.
P.O.Box 484
West Lafayette, IN 47907
USA
(765) 494-5386 (phone)
(765) 494-0706 (fax)
blwalsh@ecn.purdue.edu (email)
http://www.physics.purdue.edu/~bwalsh (web)

Ms. Sarah E Weber
Southern Illinois University Carbondale
Department of Physics
Carbondale, IL 62901
USA
(618)-453-2044 (phone)
(618)-453-1056 (fax)
sarah777@siu.edu (email)

Mr. Keith Andrew Williams
University of Kentucky
ASTECC A361
Lexington, KY 40506
USA
606-257-3019 (phone)
606-257-9485 (fax)
keith@monica.gws.uky.edu (email)
http://jovette.gws.uky.edu/~keith (web)

Prof. Shi-Yu Wu
University of Louisville
Department of Physics
Louisville, KY 40292

USA
502-852-3335 (phone)
502-852-0742 (fax)
sywu0001@fellini.physics.louisville.edu (email)
http://www.physics.louisville.edu (web)

Mr. Yongqiang Xue
School of Electrical and Computer Engineering
Purdue University
1285 EE Building
Purdue University
West Lafayette, IN 47906
USA
(765)494-3365 (phone)
yxue@ecn.purdue.edu (email)

Prof. Tokio Yamabe
Dept.Molecular Engineering
Kyoto University
Yoshida-honmachi, Sakyo-ku,
Kyoto, 606-8501
Japan
+81-75-753-5684 (phone)
+81-75-751-7279 (fax)
yamabe@mee3.moleng.kyoto-u.ac.jp (email)

Ms. Mina Yoon
Seoul National University
Dept. of Physics
Seoul, 151-742
Korea
082-0344-9133527 (phone)
082-0344-918-3527 (fax)
mnyoon@phya.snu.ac.kr (email)

Dr. Masako Yudasaka
Japan Science Technology Corporation, ICORP
c/o NEC Corporation
34 Miyukigaoka
Tsukuba, Ibaraki 305-8501
Japan
81-298-50-1109 (phone)
81-298-50-1366 (fax)
yudasaka@frl.cl.nec.co.jp (email)

Dr. Anvar A. Zakhidov
Allied Signal, Inc.
Research & Technology
101 Columbia Road
Morristown, NJ 07962-1021
U.S.A.
(973)-455-3935 (phone)
(973)-455-5991 (fax)
Anvar.Zakhidov@alliedsignal.com (email)

Prof. Alex Zettl
University of California at Berkeley
Department of Physics
366 Leconte Hall #7300
Berkeley, CA 94720-7300
USA
510-642-4939 (phone)
510-643-8793 (fax)
azettl@physics.berkeley.edu (email)

Dr. Chongwu Zhou
Stanford Univ.
Dept. of Chemistry
Stanford Univ.
Stanford, CA 94305
USA
650-725-9156 (phone)
650-725-0259 (fax)
chongwu@leland.stanford.edu (email)

Ms. Marina A Zhuravleva
MSU
745 Burcham Dr #92
East Lansing, MI 48823
USA
517-333-2119 (phone)
zhuravle@pilot.msu.edu (email)

Dr. Ronald Ziolo
Xerox Corp.
Univ Barcelona Xerox Laboratory
Fisica Fonamental, Av. Diagonal 647
08028 Barcelona, 008028
Spain
34-639 90 91 64 (phone)
34-934 02 11 49 (fax)
rziolo@compuserve.com (email)
rziolo@ubxlab.com (email)

GLOSSARY OF COMMON ABBREVIATIONS

A@(n,m)	system A enclosed in an (n,m) nanotube
AFM	atomic force microscopy
CNT	carbon nanotube
CVD	chemical vapor deposition
DOS	density of states
DWNT	double-wall nanotube
EG	exfoliated graphite
ESR	electron spin resonance
FEM	field emission microscopy
GIC	graphite intercalation compounds
HREM	high-resolution electron microscopy
HRTEM	high-resolution transmission electron microscopy
ICP	inductively coupled plasma
IR	infrared spectroscopy
LDOS	local density of states
MR	magnetoresistance
MWNT	multi-wall nanotube
(n,m)	topological characterization of a nanotube by the chiral vector $n\vec{a}_1 + m\vec{a}_2$, where $\vec{a}_1$ and $\vec{a}_2$ span the graphite lattice
SEM	scanning electron microscopy
SPM	scanning probe microscopy (includes STM, AFM, etc.)
STM	scanning tunneling microscopy
STS	scanning tunneling spectroscopy
SWNT	single-wall nanotube
TEM	transmission electron microscopy
TGA	thermogravimetric analysis
XRD	x-ray diffraction

INDEX